Instructor's Resou

to accompany

DIGITAL FUNDAMENTALS

Seventh Edition

Thomas L. Floyd

Prentice Hall
Upper Saddle River, New Jersey   Columbus, Ohio

Pearson Education
Upper Saddle River, New Jersey 07458

Printed in the United States of America

10 9 8 7 6 5 4 3

ISBN: 0-13-084667-8

PREFACE

This seventh edition of the *Instructor's Resource Manual* has been updated to reflect changes in *Digital Fundamentals*, 7th edition and to provide additional instructor support and resources. This IRM is divided into six parts: Part 1- Problem Solutions, Part 2- System Applications Solutions, Part 3-System Applications Worksheet Masters, Part 4-EWB Circuit Simulation Results, Part 5- Overview of IEEE Std. 91-1984, and Part 6-Introduction to CUPL.

The solutions to all of the end-of-chapter problems with the exception of the EWB troubleshooting problems are found in Part 1. The solutions given reflect the primary approach to working the problems, but not necessarily the only approach. These pages may be copied for student handouts, if you wish to do this.

The solutions to all of the Sytem Applications, with the exception of the EWB circuit board simulations, are found in Part 2. These solutions typically show one approach to implementing the required logic and are not necessarily the only approach. As you know, the purpose of the System Applications is to show "real world" applications of some of the topics covered in the text and to acquaint the student with actual devices used in a circuit and their relationship to schematic representations. They are not intended to be laboratory projects.

Worksheet masters for the circuit boards and other diagrams in the System Applications are found in Part 3. These can be copied for the students to use in "tracing out" the printed circuit boards to develop a schematic.

The results of CD-ROM EWB circuit simulations of the end-of-chapter EWB troubleshooting problems and the System Application circuit board simulations are found in Part 4. The instrument connections and the measurement results are shown by EWB screen-capture diagrams.

The IEEE std. 91-1984 is given in Part 5. This material presents and explains the logic symbols and labels specified in the standard. As you know, many of these standard symbols are presented throughout the textbook.

A brief introduction to the CUPL programming language for PLDs is found in Part 6. CUPL and ABEL are two major PLD programming languages. This material provides preliminary information for those wishing to compare CUPL to ABEL or to present CUPL as an alternative. The website address from which you can download a fully functional version of CUPL is http://www.logicaldevices.com

You may wish to download the ABEL software. The latest software for programming the GAL16V8 PLD discussed in this text and used in Buchla's lab manual can be obtained free of charge from Lattice Semiconductor Corporation at http://www.latticesemi.com.

1 Before you can download the software, you will need to set up an account and register.
2. Once registered, click on Downloadable Software and User Manuals.
3. Choose ispEXPERTSystem7.1 Starter Software. This will bring up a description of the files and complete installation instructions. Download time varies with connection speed.
4. After the software has been downloaded, you will need a license to run the software. The license can also be downloaded from the web by clicking on Request for a six month license.
5. The license is locked to your computer hard drive and you will need to provide your hard drive number. All necessary information to do this is clearly mentioned on the web page. If you have any problems during the download and installation of the software, call the Lattice technical support line at 1-800-FASTGAL,

NOTE: The circuit restrictions password for the EWB troubleshooting circuits on the DF7 CD-ROM is **book** (all lowercase). If this does not work, try entering **BOOK** (all uppercase).

TABLE OF CONTENTS

PART 1: PROBLEM SOLUTIONS

PART 2: SYSTEM APPLICATIONS SOLUTIONS

PART 1

Problem Solutions

CHAPTER 1
INTRODUCTORY DIGITAL CONCEPTS

1. Digital data can be transmitted and stored more efficiently and reliably than analog data. Also, digital circuits are simpler to implement and there is a greater immunity to noisy environments.

2. Pressure is an analog quantity.

3. HIGH = 1; LOW = 0. See Figure 1-1.

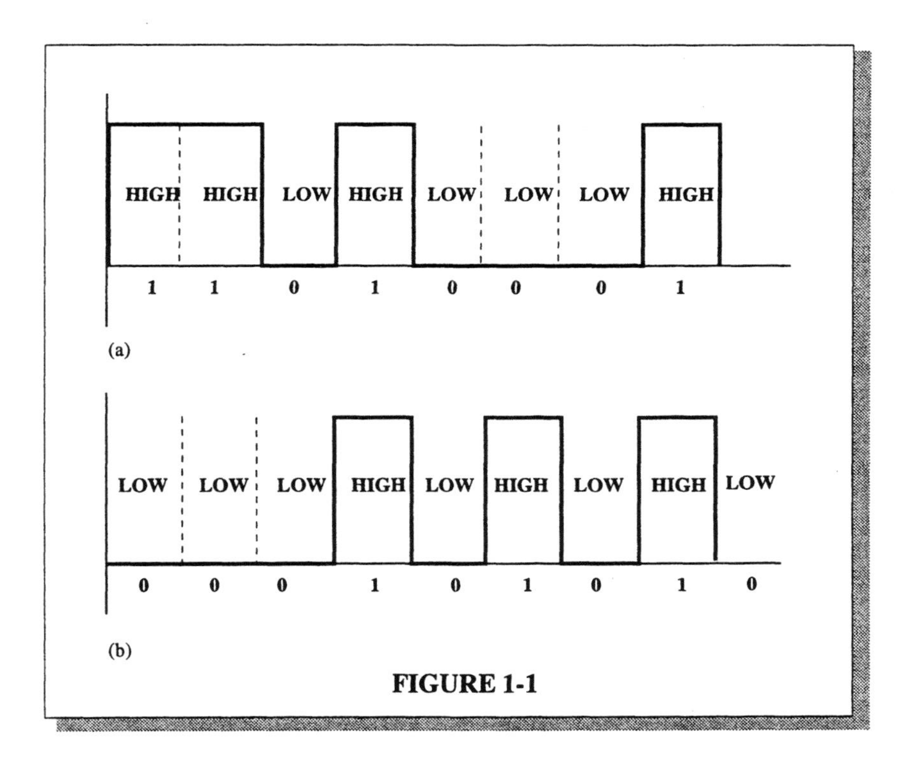

FIGURE 1-1

4. A 1 is a HIGH and a 0 is a LOW :
(a) HIGH, LOW, HIGH, HIGH, HIGH, LOW, HIGH
(b) HIGH, HIGH, HIGH, LOW, HIGH, LOW, LOW, HIGH

5. See Figure 1-2.

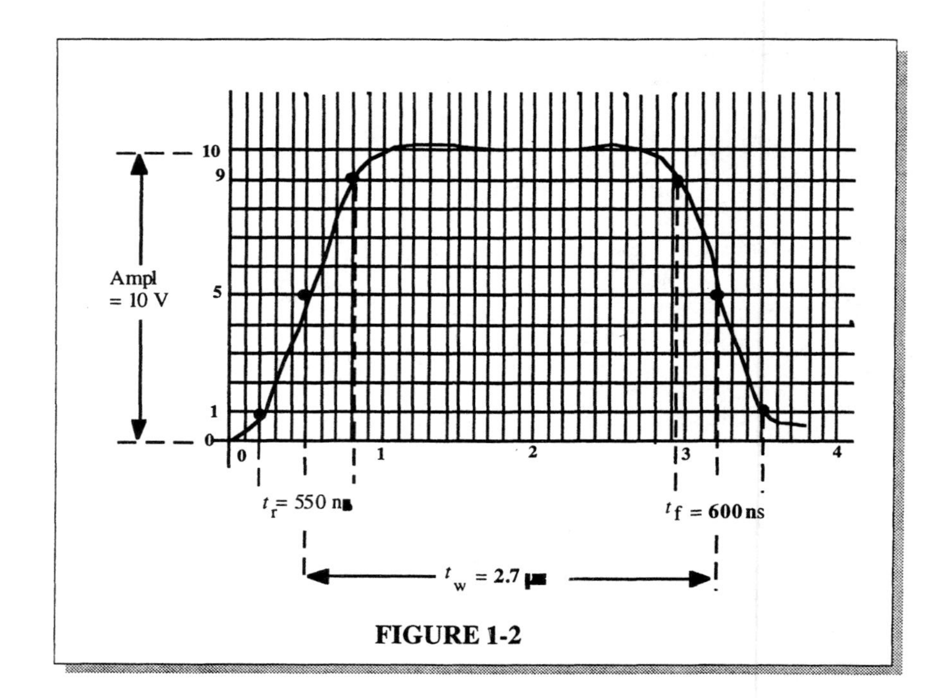

FIGURE 1-2

6. $T = 4$ ms. See Figure 1-3.

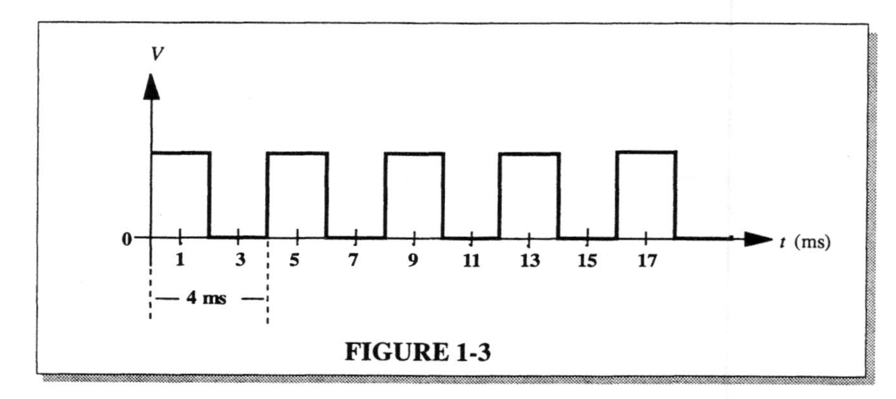

FIGURE 1-3

7.

$$f = \frac{1}{T} = \frac{1}{4 \text{ ms}} = 0.25 \text{ kHz} = 250 \text{ Hz}$$

8. The waveform in Figure 1-45 is **periodic** because it repeats at a fixed interval.

9. $t_w = 2$ ms; $T = 4$ ms

$$\% \text{ duty cycle} = \left(\frac{t_w}{T}\right) 100 = \left(\frac{2 \text{ ms}}{4 \text{ ms}}\right) 100 = 50 \%$$

10. See Figure 1-4.

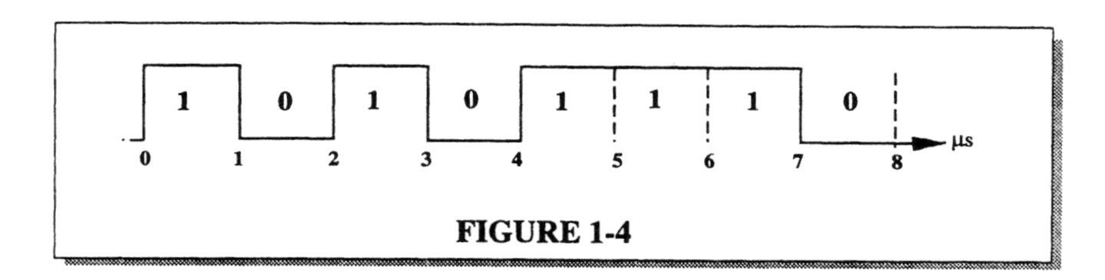

FIGURE 1-4

11. Each bit time = 1 μs
Serial transfer time = (8 bits)(1 μs/bit) = 8 μs

Parallel transfer time = 1 bit time = 1 μs

12. An AND gate produces a HIGH output only when *all* of its inputs are HIGH.

13. AND gate. See Figure 1-5.

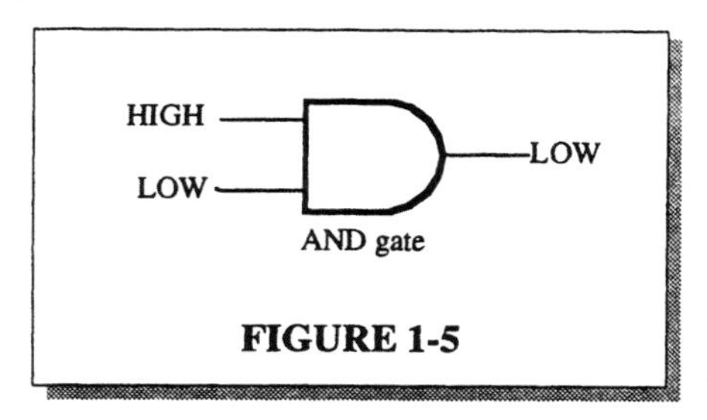

FIGURE 1-5

14. An OR gate produces a HIGH output when *either or both* inputs are HIGH. An exclusive-OR gate produces a HIGH if one input is HIGH and the other LOW.

15. See Figure 1-6.

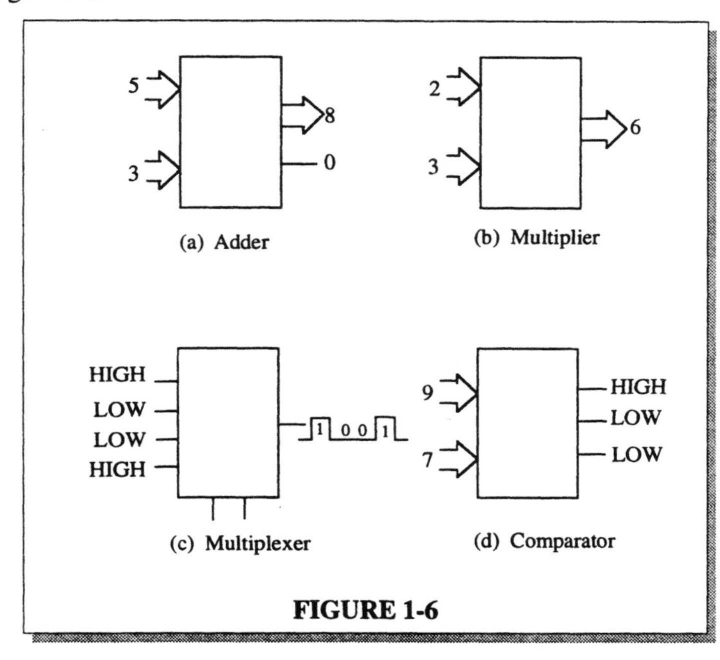

FIGURE 1-6

16. $T = \frac{1}{10\text{ kHz}} = 100\ \mu\text{s}$

$\text{Pulses counted} = \frac{100\text{ ms}}{100\ \mu\text{s}} = 1000$

17. See Figure 1-7.

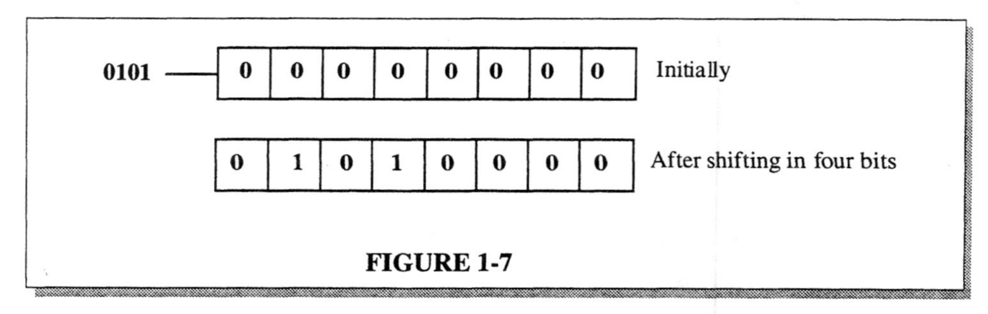

FIGURE 1-7

18. Circuits with complexities of from 100 to 1000 equivalent gates are classified as large scale integration (LSI).

19. The pins of an SMT are soldered to the pads on the surface of a pc board, whereas the pins of a DIP feed through and are soldered to the opposite side. Pin spacing on SMTs is less than on DIPs and therefore SMT packages are physically smaller and require less surface area on a pc board.

20. See Figure 1-8.

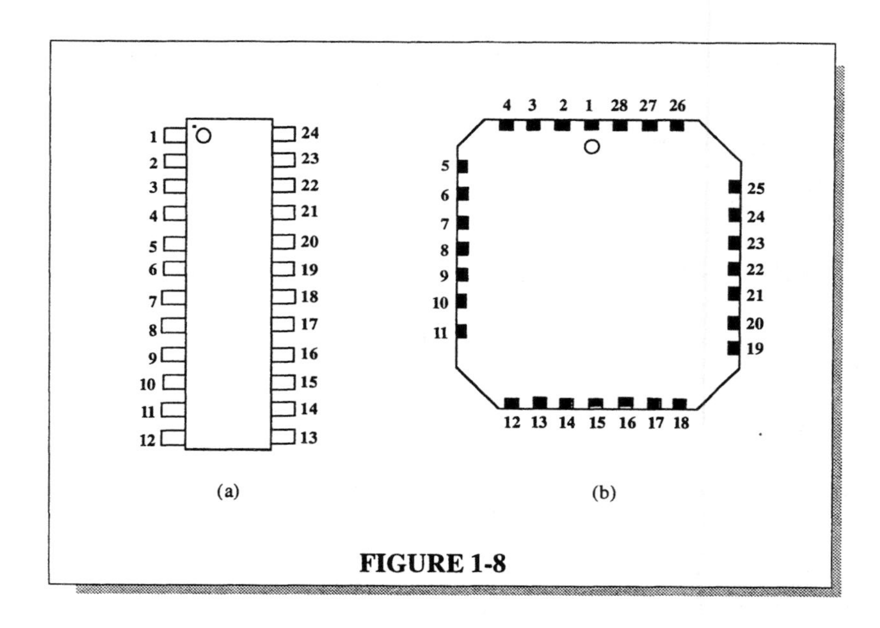

FIGURE 1-8

21. Amplitude = top of pulse minus base line
$V = 8\text{ V} - 1\text{ V} = 7\text{ V}$

22. A flashing probe lamp indicates a continuous sequence of pulses (pulse train).

23. A digital system is a combination of logic elements and functions arranged and interconnected to perform specified tasks.

24. The binary number representing the total number of tablets is converted from parallel to serial form by the multiplexer and sent, one bit at a time,to the remote location where the demultiplexer converts the serial number back to parallel form for decoding and display.

25. A new number of tablets per bottle can be entered with the key pad.

CHAPTER 2
NUMBER SYSTEMS, OPERATIONS, AND CODES

1.

(a) $1386 = 1X10^3 + 3X10^2 + 8X10^1 + 6X10^0$
$= 1X1000 + 3X100 + 8X10 + 6X1$
The digit 6 has a weight of $10^0 = 1$

(b) $54,692 = 5X10^4 + 4X10^3 + 6X10^2 + 9X10^1 + 2X10^0$
$= 5X10,000 + 4X1000 + 6X100 + 9X10 + 2X1$
The digit 6 has a weight of $10^2 = 100$

(c) $671,920 = 6X10^5 + 7X10^4 + 1X10^3 + 9X10^2 + 2X10^1 + 0X10^0$
$= 6X100,000 + 7X10,000 + 1X1000 + 9X100 + 2X10 + 0X1$
The digit 6 has a weight of $10^5 = 100,000$

2.

(a) $10 = 10^1$
(b) $100 = 10^2$
(c) $10,000 = 10^4$
(d) $1,000,000 = 10^6$

3.

(a) $471 = 4X10^2 + 7X10^1 + 1X10^0$
$= 4X100 + 7X10 + 1X1$
$= 400 + 70 + 1$

(b) $9,356 = 9X10^3 + 3X10^2 + 5X10^1 + 6X10^0$
$= 9X1000 + 3X100 + 5X10 + 6X1$
$= 9,000 + 300 + 50 + 6$

(c) $125,000 = 1X10^5 + 2X10^4 + 5X10^3$
$= 1X100,000 + 2X10,000 + 5X1000$
$= 100,000 + 20,000 + 5,000$

4. The highest four-digit decimal number is 9999.

5.

(a) $11 = 1X2^1 + 1X2^0 = 2 + 1 = 3$
(b) $100 = 1X2^2 + 0X2^1 + 0X2^0 = 4$
(c) $111 = 1X2^2 + 1X2^1 + 1X2^0 = 4 + 2 + 1 = 7$
(d) $1000 = 1X2^3 + 0X2^2 + 0X2^1 + 0X2^0 = 8$
(e) $1001 = 1X2^3 + 0X2^2 + 0X2^1 + 1X2^0 = 8 + 1 = 9$
(f) $1100 = 1X2^3 + 1X2^2 + 0X2^1 + 0X2^0 = 8 + 4 = 12$
(g) $1011 = 1X2^3 + 0X2^2 + 1X2^1 + 1X2^0 = 8 + 2 + 1 = 11$
(h) $1111 = 1X2^3 + 1X2^2 + 1X2^1 + 1X2^0 = 8 + 4 + 2 + 1 = 15$

6.

(a) $1110 = 1X2^3 + 1X2^2 + 1X2^1 = 8 + 4 + 2 = 14$
(b) $1010 = 1X2^3 + 1X2^1 = 8 + 2 = 10$
(c) $11100 = 1X2^4 + 1X2^3 + 1X2^2 = 16 + 8 + 4 = 28$
(d) $10000 = 1X2^4 = 16$
(e) $10101 = 1X2^4 + 1X2^2 + 1X2^0 = 16 + 4 + 1 = 21$
(f) $11101 = 1X2^4 + 1X2^3 + 1X2^2 + 1X2^0 = 16 + 8 + 4 + 1 = 29$
(g) $10111 = 1X2^4 + 1X2^2 + 1X2^1 + 1X2^0 = 16 + 4 + 2 + 1 = 23$
(h) $11111 = 1X2^4 + 1X2^3 + 1X2^2 + 1X2^1 + 1X2^0 = 16 + 8 + 4 + 2 + 1 = 31$

7.

(a) $110011.11 = 1X2^5 + 1X2^4 + 1X2^1 + 1X2^0 + 1X2^{-1} + 1X2^{-2}$
$= 32 + 16 + 2 + 1 + 0.5 + 0.25 = 51.75$

(b) $101010.01 = 1X2^5 + 1X2^3 + 1X2^1 + 1X2^{-2} = 32 + 8 + 2 + 0.25$
$= 42.25$

(c) $1000001.111 = 1X2^6 + 1X2^0 + 1X2^{-1} + 1X2^{-2} + 1X2^{-3}$
$= 64 + 1 + 0.5 + 0.25 + 0.125 = 65.875$

(d) $1111000.101 = 1X2^6 + 1X2^5 + 1X2^4 + 1X2^3 + 1X2^{-1} + 1X2^{-3}$
$= 64 + 32 + 16 + 8 + 0.5 + 0.125 = 120.625$

(e) $1011100.10101 = 1X2^6 + 1X2^4 + 1X2^3 + 1X2^2 + 1X2^{-1} + 1X2^{-3}$
$+ 1X2^{-5} = 64 + 16 + 8 + 4 + 0.5 + 0.125$
$+ 0.03125 = 92.65625$

(f) $1110001.0001 = 1X2^6 + 1X2^5 + 1X2^4 + 1X2^0 + 1X2^{-4}$
$= 64 + 32 + 16 + 1 + 0.0625 = 113.0625$

(g) $1011010.1010 = 1X2^6 + 1X2^4 + 1X2^3 + 1X2^1 + 1X2^{-1} + 1X2^{-3}$
$= 64 + 16 + 8 + 2 + 0.5 + 0.125 = 90.625$

(h) $1111111.11111 = 1X2^6 + 1X2^5 + 1X2^4 + 1X2^3 + 1X2^2 + 1X2^1$
$+ 1X2^0 + 1X2^{-1} + 1X2^{-2} + 1X2^{-3}$
$+ 1X2^{-4} + 1X2^{-5} = 64 + 32 + 16 + 8 + 4 + 2$
$+ 1 + 0.5 + 0.25 + 0.125 + 0.0625 + 0.03125$
$= 127.96875$

8.

(a) $2^2 - 1 = 3$
(b) $2^3 - 1 = 7$
(c) $2^4 - 1 = 15$
(d) $2^5 - 1 = 31$
(e) $2^6 - 1 = 63$
(f) $2^7 - 1 = 127$
(g) $2^8 - 1 = 255$
(h) $2^9 - 1 = 511$
(i) $2^{10} - 1 = 1023$
(j) $2^{11} - 1 = 2047$

9.

(a) $(2^4 - 1) < 17 < (2^5 - 1)$; 5 bits
(b) $(2^5 - 1) < 35 < (2^6 - 1)$; 6 bits
(c) $(2^5 - 1) < 49 < (2^6 - 1)$; 6 bits
(d) $(2^6 - 1) < 68 < (2^7 - 1)$; 7 bits
(e) $(2^6 - 1) < 81 < (2^7 - 1)$; 7 bits
(f) $(2^6 - 1) < 114 < (2^7 - 1)$; 7 bits
(g) $(2^7 - 1) < 132 < (2^8 - 1)$; 8 bits
(h) $(2^7 - 1) < 205 < (2^8 - 1)$; 8 bits

10.

(a) 0 through 7:
000, 001, 010, 011, 100, 101, 110, 111

(b) 8 through 15:
1000, 1001, 1010, 1011, 1100, 1101, 1110, 1111

(c) 16 through 31:
10000. 10001, 10010, 10011, 10100, 10101, 10110, 10111, 11000, 11001, 11010,11011, 11100, 11101, 11110, 11111

(d) 32 through 63:
100000, 100001, 100010, 100011, 100100, 100101, 100110, 100111, 101000, 101001, 101010, 101011, 101100, 101101, 101110, 101111, 110000, 110001, 110010, 110011, 110100, 110101, 110110, 110111, 111000, 111001, 111010, 111011,111100, 111101, 111110, 111111

(e) 64 through 75:
1000000, 1000001, 1000010, 1000011, 1000100, 1000101, 1000110, 1000111, 1001000, 1001001, 1001010, 1001011

11.

(a) $10 = 8 + 2 = 2^3 + 2^1 = 1010$

(b) $17 = 16 + 1 = 2^4 + 2^0 = 10001$

(c) $24 = 16 + 8 = 2^4 + 2^3 = 11000$

(d) $48 = 32 + 16 = 2^5 + 2^4 = 110000$

(e) $61 = 32 + 16 + 8 + 4 + 1 = 2^5 + 2^4 + 2^3 + 2^2 + 2^0 = 111101$

(f) $93 = 64 + 16 + 8 + 4 + 1 = 2^6 + 2^4 + 2^3 + 2^2 + 2^0 = 1011101$

(g) $125 = 64 + 32 + 16 + 8 + 4 + 1 = 2^6 + 2^5 + 2^4 + 2^3 + 2^2 + 2^0$
$= 1111101$

(h) $186 = 128 + 32 + 16 + 8 + 2 = 2^7 + 2^5 + 2^4 + 2^3 + 2^1$
$= 10111010$

12.

(a) $0.32 \cong 0.00 + 0.25 + 0.0625 + 0.0 + 0.0 + 0.0078125 = 0.0101001$

(b) $0.246 \cong 0.0 + 0.0 + 0.125 + 0.0625 + 0.03125 + 0.015625 = 0.001111$

(c) $0.0981 \cong 0.0 + 0.0 + 0.0 + 0.0625 + 0.03125 + 0.0 + 0.0 + 0.00390625 = 0.00011001$

13.

(a)
$\frac{15}{2} = 7, \quad R = 1$ (LSB)
$\frac{7}{2} = 3, \quad R = 1$
$\frac{3}{2} = 1, \quad R = 1$
$\frac{1}{2} = 0, \quad R = 1$ (MSB)

(b)
$\frac{21}{2} = 10, \quad R = 1$ (LSB)
$\frac{10}{2} = 5, \quad R = 0$
$\frac{5}{2} = 2, \quad R = 1$
$\frac{2}{2} = 1, \quad R = 0$
$\frac{1}{2} = 0, \quad R = 1$ (MSB)

(c)
$\frac{28}{2} = 14, \quad R = 0$ (LSB)
$\frac{14}{2} = 7, \quad R = 0$
$\frac{7}{2} = 3, \quad R = 1$
$\frac{3}{2} = 1, \quad R = 1$
$\frac{1}{2} = 0, \quad R = 1$ (MSB)

(d)
$\frac{34}{2} = 17, \quad R = 0$ (LSB)
$\frac{17}{2} = 8, \quad R = 1$
$\frac{8}{2} = 4, \quad R = 0$
$\frac{4}{2} = 2, \quad R = 0$
$\frac{2}{2} = 1, \quad R = 0$
$\frac{1}{2} = 0, \quad R = 1$ (MSB)

(e)
$\frac{40}{2} = 20, \quad R = 0$ (LSB)
$\frac{20}{2} = 10, \quad R = 0$
$\frac{10}{2} = 5, \quad R = 0$
$\frac{5}{2} = 2, \quad R = 1$
$\frac{2}{2} = 1, \quad R = 0$
$\frac{1}{2} = 0, \quad R = 1$ (MSB)

(f)
$\frac{59}{2} = 29, \quad R = 1$ (LSB)
$\frac{29}{2} = 14, \quad R = 1$
$\frac{14}{2} = 7, \quad R = 0$
$\frac{7}{2} = 3, \quad R = 1$
$\frac{3}{2} = 1, \quad R = 1$
$\frac{1}{2} = 0, \quad R = 1$ (MSB)

(g)
$\frac{65}{2} = 32, \quad R = 1$ (LSB)
$\frac{32}{2} = 16, \quad R = 0$
$\frac{16}{2} = 8, \quad R = 0$
$\frac{8}{2} = 4, \quad R = 0$
$\frac{4}{2} = 2, \quad R = 0$
$\frac{2}{2} = 1, \quad R = 0$
$\frac{1}{2} = 0, \quad R = 1$ (MSB)

(h)
$\frac{73}{2} = 36, \quad R = 1$ (LSB)
$\frac{36}{2} = 18, \quad R = 0$
$\frac{18}{2} = 9, \quad R = 0$
$\frac{9}{2} = 4, \quad R = 1$
$\frac{4}{2} = 2, \quad R = 0$
$\frac{2}{2} = 1, \quad R = 0$
$\frac{1}{2} = 0, \quad R = 1$ (MSB)

14.

(a)

$0.98 \times 2 = 1.96$ 1 (MSB)
$0.96 \times 2 = 1.92$ 1
$0.92 \times 2 = 1.84$ 1
$0.84 \times 2 = 1.68$ 1
$0.68 \times 2 = 1.36$ 1
$0.36 \times 2 = 0.72$ 0
continue if more accuracy is desired
0.111110

(b)

$0.347 \times 2 = 0.694$ 0 (MSB)
$0.694 \times 2 = 1.388$ 1
$0.388 \times 2 = 0.776$ 0
$0.776 \times 2 = 1.552$ 1
$0.552 \times 2 = 1.104$ 1
$0.104 \times 2 = 0.208$ 0
$0.208 \times 2 = 0.416$ 0
continue if more accuracy is desired
0.0101100

(c)

$0.9028 \times 2 = 1.8056$ 1 (MSB)
$0.8056 \times 2 = 1.6112$ 1
$0.6112 \times 2 = 1.2224$ 1
$0.2224 \times 2 = 0.4448$ 0
$0.4448 \times 2 = 0.8896$ 0
$0.8896 \times 2 = 1.7792$ 1
$0.7792 \times 2 = 1.5584$ 1
continue if more accuracy is desired
0.1110011

15.

(a) $\begin{array}{r} 11 \\ +\underline{01} \\ 100 \end{array}$ (b) $\begin{array}{r} 10 \\ +\underline{10} \\ 100 \end{array}$ (c) $\begin{array}{r} 101 \\ +\underline{011} \\ 1000 \end{array}$

(d) $\begin{array}{r} 111 \\ +\underline{110} \\ 1101 \end{array}$ (e) $\begin{array}{r} 1001 \\ +\underline{0101} \\ 1110 \end{array}$ (f) $\begin{array}{r} 1101 \\ +\underline{1011} \\ 11000 \end{array}$

16.

(a) $\begin{array}{r} 11 \\ -\underline{01} \\ 10 \end{array}$ (b) $\begin{array}{r} 101 \\ -\underline{100} \\ 001 \end{array}$ (c) $\begin{array}{r} 110 \\ -\underline{101} \\ 001 \end{array}$

(d) $\begin{array}{r} 1110 \\ -\underline{0011} \\ 1011 \end{array}$ (e) $\begin{array}{r} 1100 \\ -\underline{1001} \\ 0011 \end{array}$ (f) $\begin{array}{r} 11010 \\ -\underline{10111} \\ 00011 \end{array}$

17.

(a) $\begin{array}{r} 11 \\ \underline{\text{X}\ \ 11} \\ 11 \\ \underline{11\ } \\ 1001 \end{array}$ (b) $\begin{array}{r} 100 \\ \underline{\text{X}\ \ 10} \\ 000 \\ \underline{100\ } \\ 1000 \end{array}$ (c) $\begin{array}{r} 111 \\ \underline{\text{X}\ 101} \\ 111 \\ 000\ \\ \underline{111\ \ } \\ 100011 \end{array}$ (d) $\begin{array}{r} 1001 \\ \underline{\text{X}\ 110} \\ 0000 \\ 1001\ \\ \underline{1001\ \ } \\ 110110 \end{array}$

(e) $\begin{array}{r} 1101 \\ \underline{\text{X}\ \ 1101} \\ 1101 \\ 0000\ \\ 1101\ \ \\ \underline{1101\ \ \ } \\ 10101001 \end{array}$ (f) $\begin{array}{r} 1110 \\ \underline{\text{X}\ \ 1101} \\ 1110 \\ 0000\ \\ 1110\ \ \\ \underline{1110\ \ \ } \\ 10110110 \end{array}$

18. (a) $\frac{100}{10} = 010$ (b) $\frac{1001}{0011} = 0011$ (c) $\frac{1100}{0100} = 0011$

19.
(a) The 1s complement of 101 is 010
(b) The 1s complement of 110 is 001
(c) The 1s complement of 1010 is 0101
(d) The 1s complement of 11010111 is 00101000
(e) The 1s complement of 1110101 is 0001010
(f) The 1s complement of 00001 is 11110

20. Take the 1s complement and add 1:

(a) 01 + 1 = 10
(b) 000 + 1 = 001
(c) 0110 + 1 = 0111
(d) 0010 + 1 = 0011
(e) 00011 + 1 = 00100
(f) 01100 + 1 = 01101
(g) 01001111 + 1 = 01010000
(h) 11000010 + 1 = 11000011

21.
(a) Magnitude of 29 = 0011101
+29 = 00011101

(b) Magnitude of 85 = 1010101
-85 = 11010101

(c) Magnitude of 100_{10} = 1100100
+100 = 01100100

(d) Magnitude of 123 = 1111011
-123 = 11111011

22.
(a) Magnitude of 34 = 0100010
-34 = 11011101

(b) Magnitude of 57 = 0111001
+57 = 00111001

(c) Magnitude of 99 = 1100011
-99 = 10011100

(d) Magnitude of 115 = 1110011
+115 = 01110011

23.
(a) Magnitude of 12 = 1100
+12 = 00001100

(b) Magnitude of 68 = 1000100
-68 = 10111100

(c) Magnitude of 101_{10} = 1100101
$+101_{10}$ = 01100101

(d) Magnitude of 125 = 1111101
-125 = 10000011

24. (a) 10011001 = -25 (b) 01110100 = +116 (b) 10111111 = -63

25.
(a) 10011001 = -(01100110) = -102
(b) 01110100 = +(1110100) = +116
(c) 10111111 = -(1000000) = -64

26.
(a) 10011001 = -(1100111) = -103
(b) 01110100 = +(1110100) = +116
(c) 10111111 = -(1000001) = -65

27.

(a) 0111110000101011 → sign = 0

1.111110000101011 $\times 2^{14}$ → exponent = 127 + 14 = 141 = 10001101

Mantissa = 11110000101011000000000

01000110111111000010101100000000

(b) 100110000011000 → sign = 1

1.10000011000 $\times 2^{11}$ → exponent = 127 + 11 = 138 = 10001010

Mantissa = 10000001100000000000000

11000101010000001100000000000000

28.

(a) 11000000101001001110001000000000

Sign = 1

Exponent = 10000001 = 129 − 127 = 2

Mantissa = 1.01001001110001 $\times 2^2$ = 101.001001110001

− 101.001001110001 = **− 5.15258789**

(b) 01100110010000111110100100000000

Sign = 0

Exponent = 11001100 = 204 − 127 = 77

Mantissa = 1.100001111101001

1.100001111101001 $\times$ **2^{77}**

29.

(a) 33 = 00100001
15 = 00001111

```
  00100001
+ 00001111
----------
  00110000
```

(b) 56 = 00111000
27 = 00011011
− 27 = 11100101

```
  00111000
+ 11100101
----------
  00011101
```

(c) 46 = 00101110
− 46 = 11010010
25 = 00011001

```
  11010010
+ 00011001
----------
  11101011
```

(d) 110_{10} = 01101110
− 110_{10} = 10010010
84 = 01010100
− 84 = 10101100

```
  10010010
+ 10101100
----------
 100111110
```

30.

(a)

```
  00010110
+ 00110011
----------
  01001001
```

(b)

```
  01110000
+ 10101111
----------
 100011111
```

31.

(a)

```
  10001100
+ 00111001
----------
  11000101
```

(b)

```
  11011001
+ 11100111
----------
  11000000
```

32.

(a)

```
  00110011
- 00010000
----------
```

```
  00110011
+ 11110000
----------
 100100011   (leading 1 crossed out)
```

(b)

```
  01100101
- 11101000
----------
```

```
  01100101
+ 00011000
----------
  01111101
```

33.

```
   01101010            01101010
 × 11110001          × 00001111
 ----------          ----------
                       01101010
                      01101010
                     ----------
                      100111110
                     01101010
                     ----------
                     1011100110
                    01101010
                    -----------
                    11000110110
```

Changing to 2s complement with sign : 100111001010

34.

$$\frac{01000100}{00011001} = 00000010$$

$$\frac{68}{25} = 2 \text{ remainder of } 18$$

35.

(a) $38_{16} = 0011\ 1000$
(b) $59_{16} = 0101\ 1001$
(c) $A14_{16} = 1010\ 0001\ 0100$
(d) $5C8_{16} = 0101\ 1100\ 1000$
(e) $4100_{16} = 0100\ 0001\ 0000\ 0000$
(f) $\text{FB17}_{16} = 1111\ 1011\ 0001\ 0111$
(g) $8A9\text{D}_{16} = 1000\ 1010\ 1001\ 1101$

36.

(a) $1110 = \text{E}_{16}$
(b) $10 = 2_{16}$
(c) $0001\ 0111 = 17_{16}$
(d) $1010\ 0110 = \text{A6}_{16}$
(e) $0011\ 1111\ 0000 = \text{3F0}_{16}$
(f) $1001\ 1000\ 0010 = 982_{16}$

37.

(a) $23_{16} = 2 \times 16^1 + 3 \times 16^0 = 32 + 3 = 35$
(b) $92_{16} = 9 \times 16^1 + 2 \times 16^0 = 144 + 2 = 146$
(c) $1\text{A}_{16} = 1 \times 16^1 + 10 \times 16^0 = 16 + 10 = 26$
(d) $8\text{D}_{16} = 8 \times 16^1 + 13 \times 16^0 = 128 + 13 = 141$
(e) $\text{F3}_{16} = 15 \times 16^1 + 3 \times 16^0 = 240 + 3 = 243$
(f) $\text{EB}_{16} = 14 \times 16^1 + 11 \times 16^0 = 224 + 11 = 235$
(g) $5\text{C}2_{16} = 5 \times 16^2 + 12 \times 16^1 + 2 \times 16^0 = 1280 + 192 + 2 = 1474$
(h) $700_{16} = 7 \times 16^2 = 1792$

38.

(a) $\frac{8}{16} = 0$, remainder $= 8$
hexadecimal number $= 8_{16}$

(b) $\frac{14}{16} = 0$, remainder $= 14 = E_{16}$
hexadecimal number $= E_{16}$

(c) $\frac{33}{16} = 2$, remainder $= 1$ (LSD)
$\frac{2}{16} = 0$, remainder $= 2$
hexadecimal number $= 21_{16}$

(d) $\frac{52}{16} = 3$, remainder $= 4$ (LSD)
$\frac{3}{16} = 0$, remainder $= 3$
hexadecimal number $= 34_{16}$

(e) $\frac{284}{16} = 17$, remainder $= 12 = C_{16}$ (LSD)
$\frac{17}{16} = 1$, remainder $= 1$
$\frac{1}{16} = 0$, remainder $= 1$
hexadecimal number $= 11C_{16}$

(f) $\frac{2890}{16} = 180$, remainder $= 10 = A_{16}$ (LSD)
$\frac{180}{16} = 11$, remainder $= 4$
$\frac{11}{16} = 0$, remainder $= 11 = B_{16}$
hexadecimal number $= B4A_{16}$

(g) $\frac{4019}{16} = 251$, remainder $= 3$ (LSD)
$\frac{251}{16} = 15$, remainder $= 11 = B_{16}$
$\frac{15}{16} = 0$, remainder $= 15 = F_{16}$
hexadecimal number $= FB3_{16}$

(h) $\frac{6500}{16} = 406$, remainder $= 4$ (LSD)
$\frac{406}{16} = 25$, remainder $= 6$
$\frac{25}{16} = 1$, remainder $= 9$
$\frac{1}{16} = 0$, remainder $= 1$
hexadecimal number $= 1964_{16}$

39.

(a) $37_{16} + 29_{16} = 60_{16}$
(b) $A0_{16} + 6B_{16} = 10B_{16}$
(c) $FF_{16} + BB_{16} = 1BA_{16}$

40.

(a) $51_{16} - 40_{16} = 11_{16}$
(b) $C8_{16} - 3A_{16} = 8E_{16}$
(c) $FD_{16} - 88_{16} = 75_{16}$

41.

(a) $12_8 = 1 \times 8^1 + 2 \times 8^0 = 8 + 2 = 10$
(b) $27_8 = 2 \times 8^1 + 7 \times 8^0 = 16 + 7 = 23$
(c) $56_8 = 5 \times 8^1 + 6 \times 8^0 = 40 + 6 = 46$
(d) $64_8 = 6 \times 8^1 + 4 \times 8^0 = 48 + 4 = 52$
(e) $103_8 = 1 \times 8^2 + 3 \times 8^0 = 64 + 3 = 67$
(f) $557_8 = 5 \times 8^2 + 5 \times 8^1 + 7 \times 8^0 = 320 + 40 + 7 = 367$
(g) $163_8 = 1 \times 8^2 + 6 \times 8^1 + 3 \times 8^0 = 64 + 48 + 3 = 115$
(h) $1024_8 = 1 \times 8^3 + 2 \times 8^1 + 4 \times 8^0 = 512 + 16 + 4 = 532$
(i) $7765_8 = 7 \times 8^3 + 7 \times 8^2 + 6 \times 8^1 + 5 \times 8^0 = 3584 + 448 + 48 + 5 = 4085$

42.

(a) $\frac{15}{8} = 1$, remainder = 7 (LSD)

$\frac{1}{8} = 0$, remainder = 1

octal number = 17_8

(b) $\frac{27}{8} = 3$, remainder = 3 (LSD)

$\frac{3}{8} = 0$, remainder = 3

octal number = 33_8

(c) $\frac{46}{8} = 5$, remainder = 6 (LSD)

$\frac{5}{8} = 0$, remainder = 5

octal number = 56_8

(d) $\frac{70}{8} = 8$, remainder = 6 (LSD)

$\frac{8}{8} = 1$, remainder = 0

$\frac{1}{8} = 0$, remainder = 1

octal number = 106_8

(e) $\frac{100}{8} = 12$, remainder = 4 (LSD)

$\frac{12}{8} = 1$, remainder = 4

$\frac{1}{8} = 0$, remainder = 1

octal number = 144_8

(f) $\frac{142}{8} = 17$, remainder = 6 (LSD)

$\frac{17}{8} = 2$, remainder = 1

$\frac{2}{8} = 0$, remainder = 2

octal number = 216_8

(g) $\frac{219}{8} = 27$, remainder = 3 (LSD)

$\frac{27}{8} = 3$, remainder = 3

$\frac{3}{8} = 0$, remainder = 3

octal number = 333_8

(h) $\frac{435}{8} = 54$, remainder = 3 (LSD)

$\frac{54}{8} = 6$, remainder = 6

$\frac{6}{8} = 0$, remainder = 6

octal number = 663_8

43.

(a) 13_8 = 001 011
(b) 57_8 = 101 111
(c) 101_8 = 001 000 001
(d) 321_8 = 011 010 001
(e) 540_8 = 101 100 000
(f) 4653_8 = 100 110 101 011
(g) 13271_8 = 001 011 010 111 001
(h) 45600_8 = 100 101 110 000 000
(i) 100213_8 = 001 000 000 010 001 011

44.

(a) 111 = 7_8
(b) 010 = 2_8
(c) 110 111 = 67_8
(d) 101 010 = 52_8
(e) 001 100 = 14_8
(f) 001 011 110 = 136_8
(g) 101 100 011 001 = 5431_8
(h) 010 110 000 011 = 2603_8
(i) 111 111 101 111 000 = 77570_8

45.

(a) 10 = 0001 0000
(b) 13 = 0001 0011
(c) 18 = 0001 1000
(d) 21 = 0010 0001
(e) 25 = 0010 0101
(f) 36 = 0011 0110
(g) 44 = 0100 0100
(h) 57 = 0101 0111
(i) 69 = 0110 1001
(j) 98 = 1001 1000
(k) 125 = 0001 0010 0101
(l) 156 = 0001 0101 0110

46.

(a) 10 = 1010_2 4 bits binary, 8 bits BCD
(b) 13 = 1101_2 4 bits binary, 8 bits BCD
(c) 18 = 10010_2 5 bits binary, 8 bits BCD
(d) 21 = 10101_2 5 bits binary, 8 bits BCD
(e) 25 = 11001_2 5 bits binary, 8 bits BCD
(f) 36 = 100100_2 6 bits binary, 8 bits BCD
(g) 44 = 101100_2 6 bits binary, 8 bits BCD
(h) 57 = 111001_2 6 bits binary, 8 bits BCD
(i) 69 = 1000101_2 7 bits binary, 8 bits BCD
(j) 98 = 1100010_2 7 bits binary, 8 bits BCD
(k) 125 = 1111101_2 7 bits binary, 12 bits BCD
(l) 156 = 10011100_2 8 bits binary, 12 bits BCD

47.

(a) 104 = 0001 0000 0100
(b) 128 = 0001 0010 1000
(c) 132 = 0001 0011 0010
(d) 150 = 0001 0101 0000
(e) 186 = 0001 1000 0110
(f) 210 = 0010 0001 0000
(g) 359 = 0011 0101 1001
(h) 547 = 0101 0100 0111
(i) 1051 = 0001 0000 0101 0001

48.

(a) 0001 = 1
(b) 0110 = 6
(c) 1001 = 9
(d) 0001 1000 = 18
(e) 0001 1001 = 19
(f) 0011 0010 = 32
(g) 0100 0101 = 45
(h) 1001 1000 = 98
(i) 1000 0111 0000 = 870

49.

(a) 1000 0000 = 80
(b) 0010 0011 0111 = 237
(c) 0011 0100 0110 = 346
(d) 0100 0010 0001 = 421
(e) 0111 0101 0100 = 754
(f) 1000 0000 0000 = 800
(g) 1001 0111 1000 = 978
(h) 0001 0110 1000 0011 = 1683
(i) 1001 0000 0001 1000 = 9018
(j) 0110 0110 0110 0111 = 6667

50.

(a)
$$\begin{array}{r} 0010 \\ +\,0001 \\ \hline 0011 \end{array}$$

(b)
$$\begin{array}{r} 0101 \\ +\,0011 \\ \hline 1000 \end{array}$$

(c)
$$\begin{array}{r} 0111 \\ +\,0010 \\ \hline 1001 \end{array}$$

(d)
$$\begin{array}{r} 1000 \\ +\,0001 \\ \hline 1001 \end{array}$$

(e)
$$\begin{array}{r} 00011000 \\ +\,00010001 \\ \hline 00101001 \end{array}$$

(f)
$$\begin{array}{r} 01100100 \\ +\,00110011 \\ \hline 10010111 \end{array}$$

(g)
$$\begin{array}{r} 01000000 \\ +\,01000111 \\ \hline 10000111 \end{array}$$

(h)
$$\begin{array}{r} 10000101 \\ +\,01000111 \\ \hline 10000111 \end{array}$$

51.

(a)
$$\begin{array}{rl} 1000 & \\ +\,0110 & \\ \hline 1110 & \textit{invalid} \\ +\,0110 & \\ \hline 00010100 & \end{array}$$

(b)
$$\begin{array}{rl} 0111 & \\ +\,0101 & \\ \hline 1100 & \textit{invalid} \\ +\,0110 & \\ \hline 00010010 & \end{array}$$

(c)
$$\begin{array}{rl} 1001 & \\ +\,1000 & \\ \hline 10001 & \textit{invalid} \\ +\,0110 & \\ \hline 00010111 & \end{array}$$

(d)
$$\begin{array}{rl} 1001 & \\ +\,0111 & \\ \hline 10000 & \textit{invalid} \\ +\,0110 & \\ \hline 00010110 & \end{array}$$

(e)
$$\begin{array}{rl} 00100101 & \\ +\,00100111 & \\ \hline 01001100 & \textit{invalid} \\ +\,0110 & \\ \hline 01010010 & \end{array}$$

(f)
$$\begin{array}{rl} 01010001 & \\ +\,01011000 & \\ \hline 10101001 & \textit{invalid} \\ +\,0110 & \\ \hline 000100001001 & \end{array}$$

(g)
$$\begin{array}{rl} 10011000 & \\ +\,10010111 & \\ \hline 100101111 & \textit{invalid} \\ +\,01100110 & \\ \hline 000110010101 & \end{array}$$

(h)
$$\begin{array}{rl} 010101100001 & \\ +\,011100001000 & \\ \hline 110001101001 & \textit{invalid} \\ +\,0110 & \\ \hline 0001001001101001 & \end{array}$$

52.

(a) 4 + 3

```
  0100
+ 0011
  0111
```

(b) 5 + 2

```
  0101
+ 0010
  0111
```

(c) 6 + 4

```
    0110
  + 0100
    1010
  + 0110
00010000
```

(d) 17 + 12

```
  00010111
+ 00100010
  00101001
```

(e) 28 + 23

```
  00101000
+ 00100011
  01001011
    + 0110
  01010001
```

(f) 65 + 58

```
   01100101
+  01011000
   10111101
+  01100110
000100100011
```

(g) 113 + 101

```
  000100010011
+ 000100000001
  001000010100
```

(h) 295 + 157

```
  001010010101
+ 000101010111
  001111101100
  + 01100110
  010001010010
```

53. The gray code makes only one bit change at a time when going from one num ber in the sequence to the next number .

Gray for 1111_2 = 1000

Gray for 0000_2 = 0000

54.

(a)

1 + 1 + 0 + 1 + 1 *Binary*

1 0 1 1 0 *Gray*

(b)

1 + 0 + 0 + 1 + 0 + 1 + 0 *Binary*

1 1 0 1 1 1 1 *Gray*

(c)

1 + 1 + 1 + 1 + 0 + 1 + 1 + 1 + 0 + 1 + 1 + 1 + 0 *Binary*

1 0 0 0 1 1 0 0 1 1 0 0 1 *Gray*

55.

(a)

1 0 1 0 *Gray*

1 1 0 0 *Binary*

(b)

0 0 0 1 0 *Gray*

0 0 0 1 1 *Binary*

(c)

1 1 0 0 0 0 1 0 0 0 1 *Gray*

1 0 0 0 0 0 1 1 1 1 0 *Binary*

56.

(a) 1 → 00110001

(b) 3 → 00110011

(c) 6 → 00110110

(d) 10 → 0011000100110000

(e) 18 → 0011000100111000

(e) 29 → 0011001000111001

(g) 56 → 0011010100110110

(g) 75 → 0011011100110101

(i) 107 → 001100010011000000110111

57.

(a) 0011000 $\rightarrow$ **0** (b) 1001010 $\rightarrow$ **J**

(c) 0111101 $\rightarrow$ = (d) 0100011 $\rightarrow$ #

(e) 0111110 $\rightarrow$ > (f) 1000010 $\rightarrow$ **B**

58.

1001000 1100101 1101100 1101100 1101111 0101110 0100000
H e l l o .
1001000 1101111 1110111 0100000 1100001 1110010 1100101
H o w a r e
0100000 1111001 1101111 1110101 0111111
y 0 u ?

59.

1001000 1100101 1101100 1101100 1101111 0101110 0100000
48 65 6C 6C 6F 2E 20
1001000 1101111 1110111 0100000 1100001 1110010 1100101
48 6F 77 20 61 72 65
0100000 1111001 1101111 1110101 0111111
20 79 6F 75 3F

60.

30 INPUT A, B

3	0110011	33_{16}
0	0110000	30_{16}
SP	0100000	20_{16}
I	1001001	49_{16}
N	1001110	$4E_{16}$
P	1010000	50_{16}
U	1010101	55_{16}
T	1010100	54_{16}
SP	0100000	20_{16}
A	1000001	41_{16}
,	0101100	$2C_{16}$
B	1000010	42_{16}

61. Code (b) 011101010 has five 1s, so it is in error.

62. Codes (a) 11110110 and (c) 01010101010101010 are in error because they have an even number of 1s.

63. (a) 1 10100100 (b) 0 00001001 (c) 1 11111110

64. The system is limited to handling 8 bits or 2 BCD digits so 99 is the maximum number of tablets per bottle.

65.

(a) Register A stores the number of tablets per bottle: 00100101 (BCD 25).

(b) The A input of the comparator: 00011001 (binary 25)

(c) The counter holds the number of tablets in the current bottle: 00001111 (binary 15).

(d) Register B contains the number of tablets in 999 bottles: 0110000110010010 (binary 24,975)

(e) The output of the adder is the total number of tablets bottled: 0110000110100001 (binry 24,990)

66. Maximum total tablets $= 2^{16} - 1 = 65,535$
The capacity of Register B must be increased to more than 16 bits.
The other circuity would have to handle more than 16 bits

CHAPTER 3

LOGIC GATES

1. See Figure 3-1.

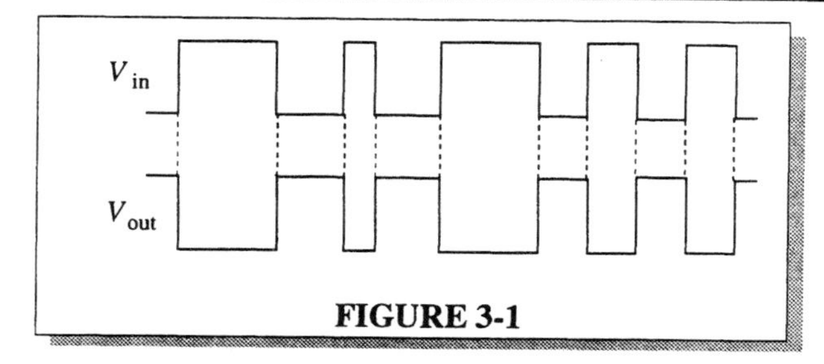

FIGURE 3-1

2. B- LOW, C- HIGH, D- LOW, E- HIGH, F- LOW

3. See Figure 3-2.

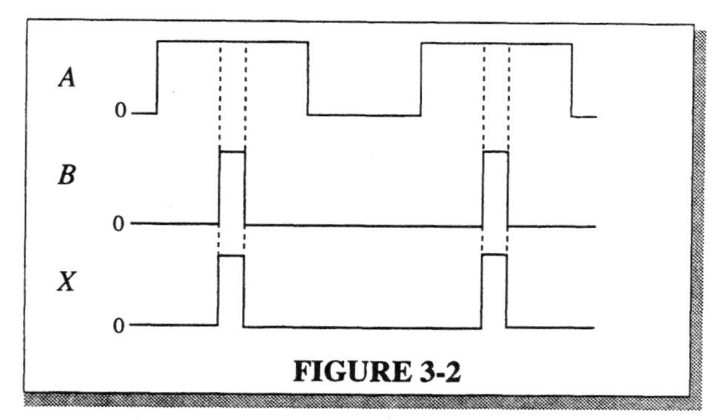

FIGURE 3-2

4. See Figure 3-3.

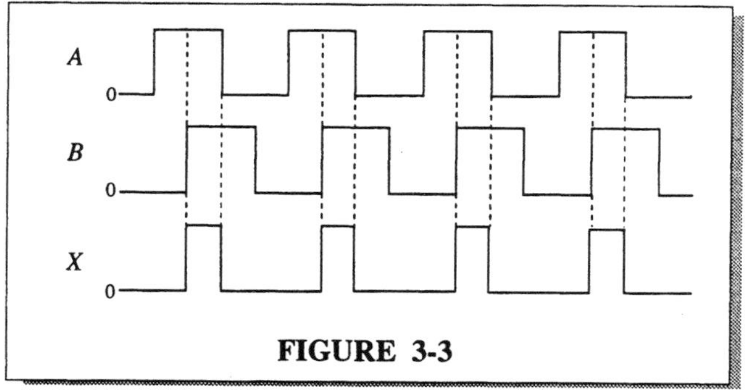

FIGURE 3-3

5. See Figure 3-4.

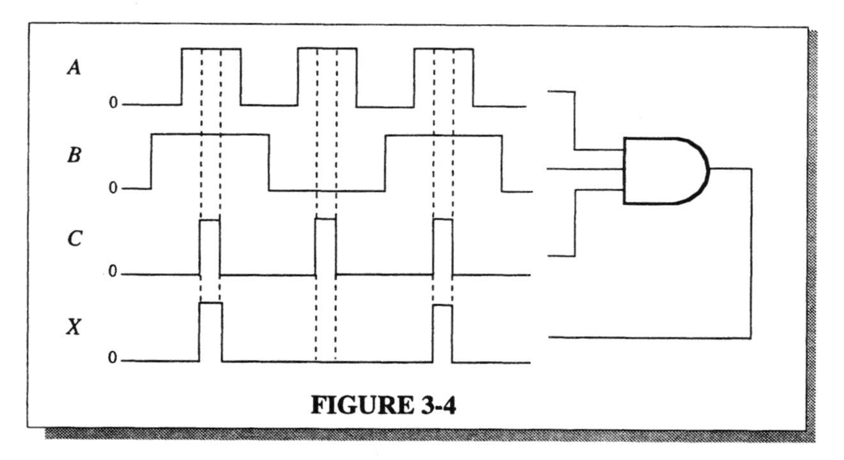

FIGURE 3-4

6. See Figure 3-5.

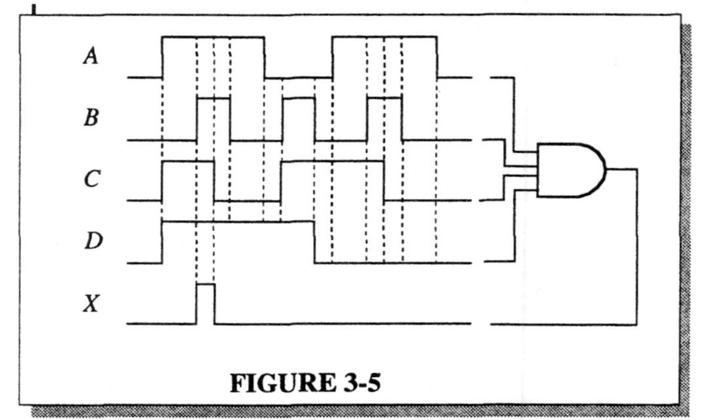

FIGURE 3-5

7. See Figure 3-6.

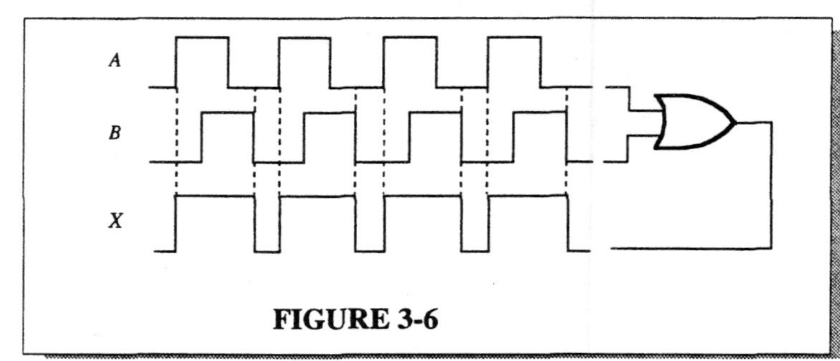

FIGURE 3-6

8. See Figure 3-7.

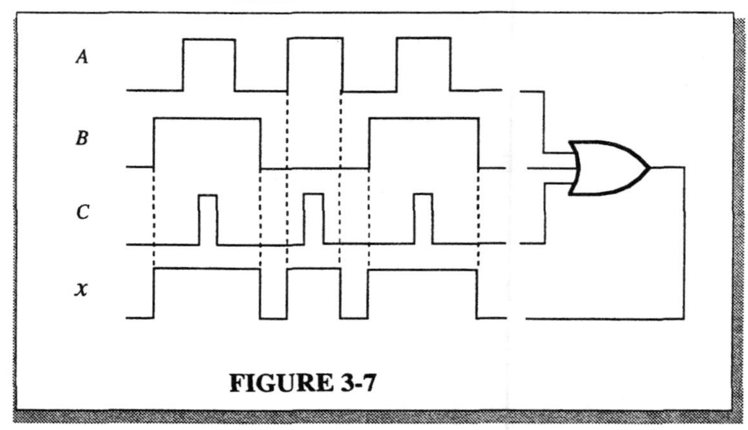

FIGURE 3-7

9. See Figure 3-8.

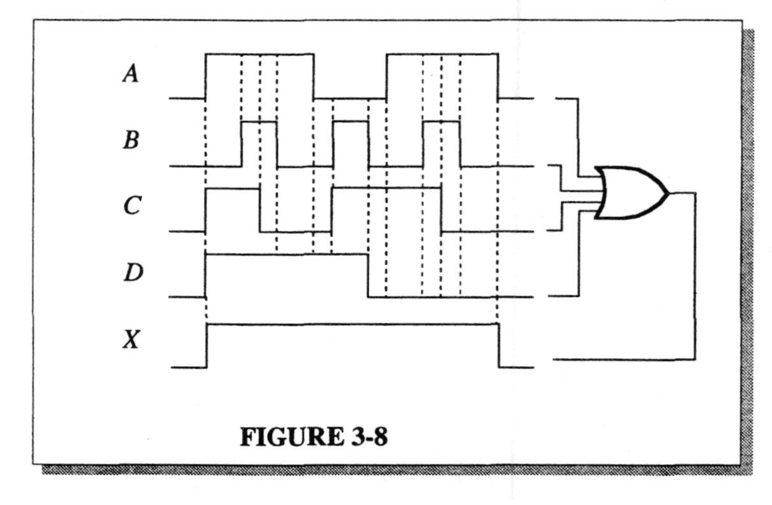

FIGURE 3-8

10. See Figure 3-9.

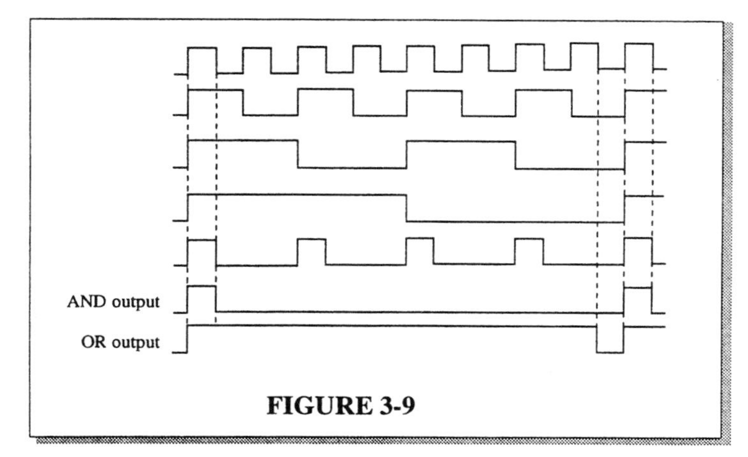

FIGURE 3-9

11. See Figure 3-10.

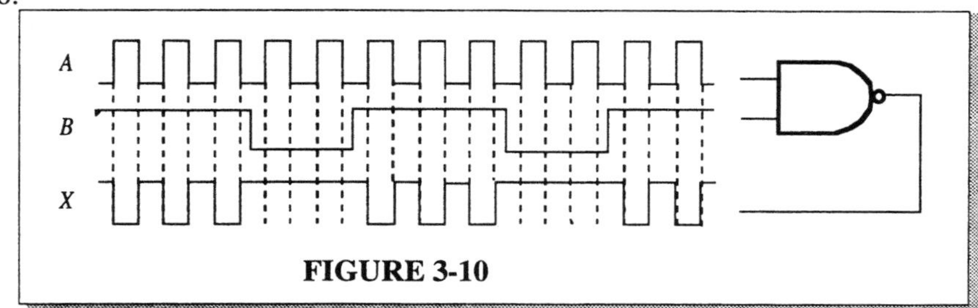

FIGURE 3-10

12. See Figure 3-11.

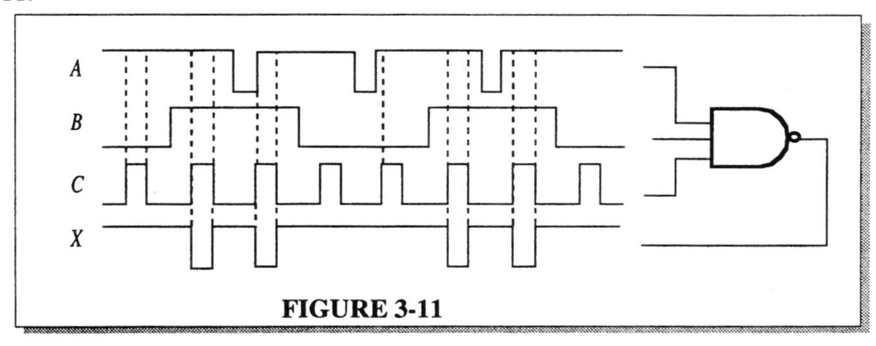

FIGURE 3-11

13. See Figure 3-12.

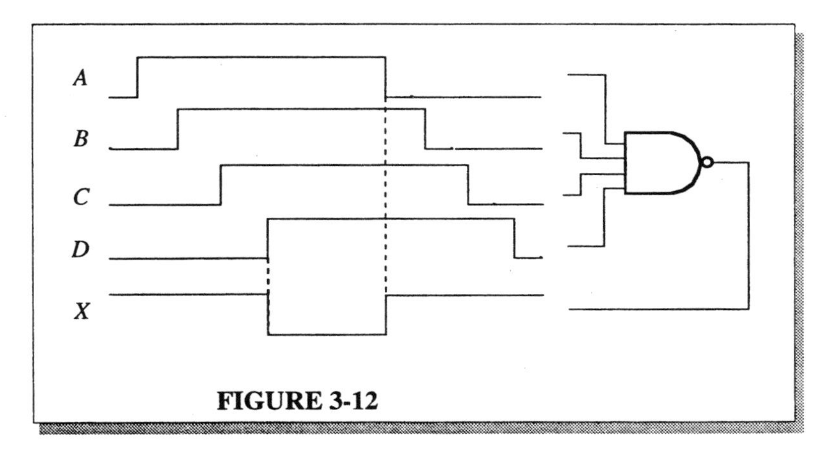

FIGURE 3-12

14. See Figure 3-13.

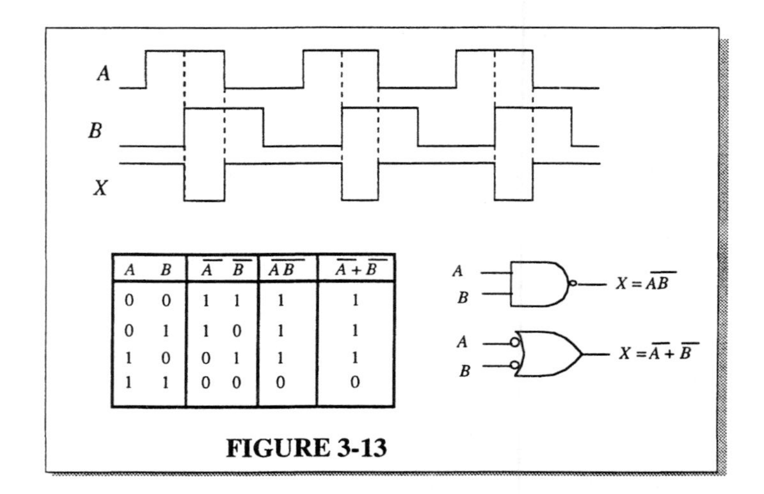

A	B	$\overline{A}$	$\overline{B}$	$\overline{AB}$	$\overline{A}+\overline{B}$
0	0	1	1	1	1
0	1	1	0	1	1
1	0	0	1	1	1
1	1	0	0	0	0

FIGURE 3-13

15. See Figure 3-14.

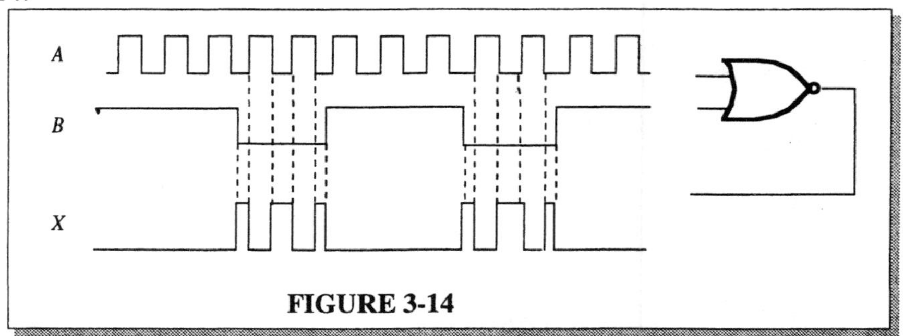

FIGURE 3-14

16. See Figure 3-15.

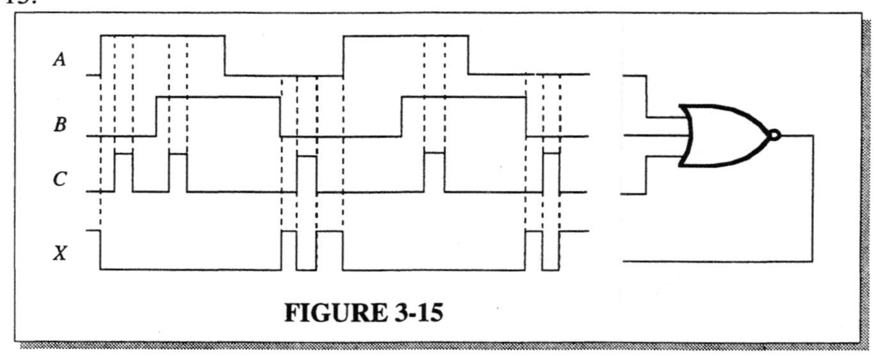

FIGURE 3-15

17. See Figure 3-16.

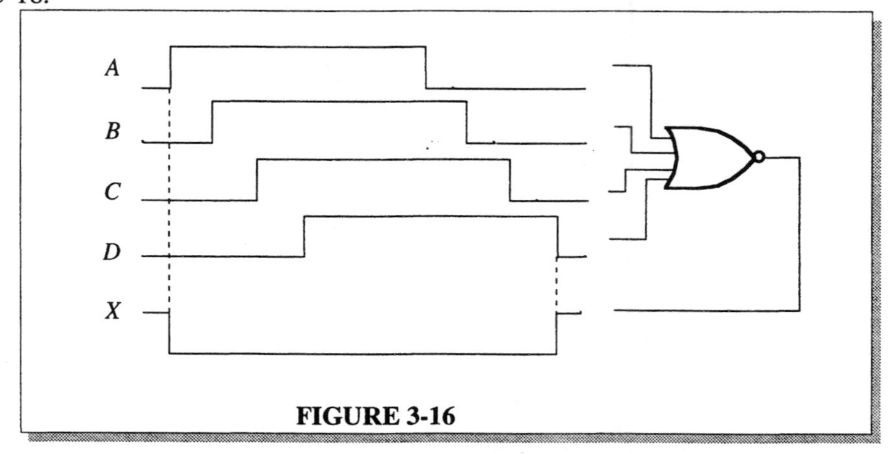

FIGURE 3-16

18. See Figure 3-17.

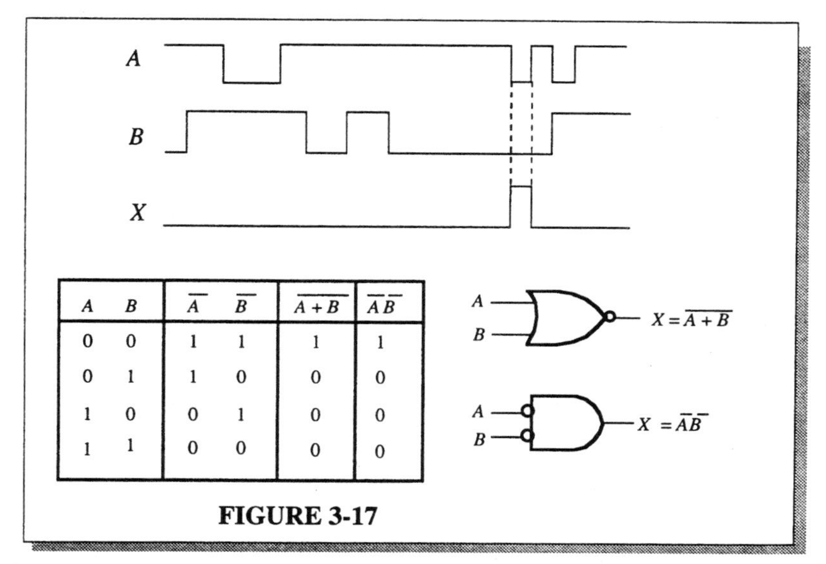

A	B	$\bar{A}$	$\bar{B}$	$\overline{A+B}$	$\bar{A}\bar{B}$
0	0	1	1	1	1
0	1	1	0	0	0
1	0	0	1	0	0
1	1	0	0	0	0

FIGURE 3-17

19. The output of the XOR gate is HIGH only when one input is HIGH. The output of the OR gate is HIGH any time one or more inputs are HIGH.

$$\text{XOR} = A\bar{B} + \bar{A}B$$

$$\text{OR} = A + B$$

20. See Figure 3-18.

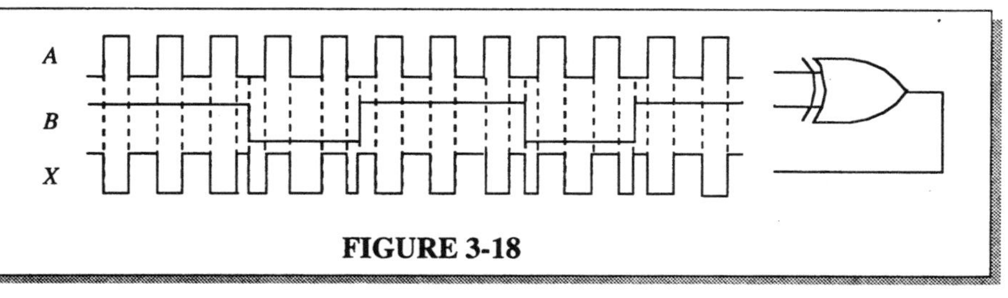

FIGURE 3-18

21. See Figure 3-19.

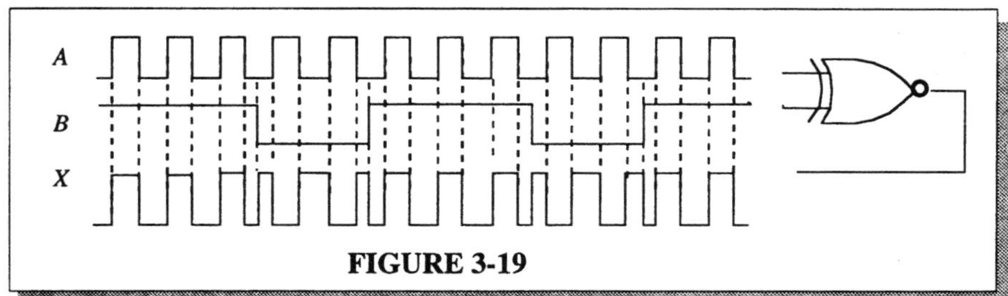

FIGURE 3-19

22. See Figure 3-20.

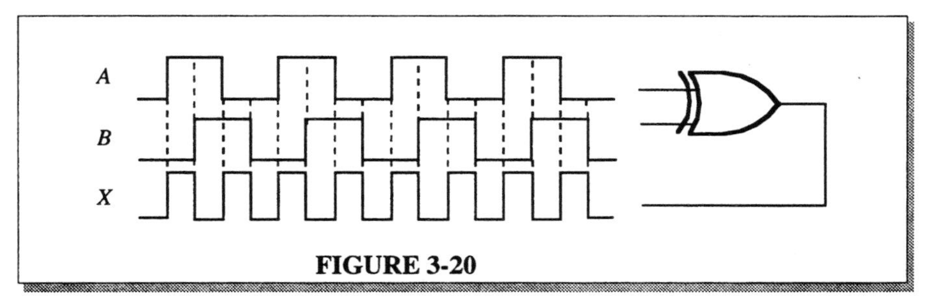

FIGURE 3-20

23. The power dissipation of CMOS **increases** with frequency.

24.

(*a*) $P = \left(\frac{I_{CCH} + I_{CCL}}{2}\right)V_{CC} = \left(\frac{1.6\text{ mA} + 4.4\text{ mA}}{2}\right)5.5\ V = 16.5\text{ mW}$

(*b*) $V_{OH(\min)} = 2.7\text{ V}$

(*c*) $t_{pLH} = t_{pHL} = 15\text{ ns}$

(*d*) $V_{OL} = 0.4\text{ V (max)}$

(*e*) @ $V_{CC} = 2\text{ V},\ t_{pHL} = t_{pLH} = 75\ ns$; @ $V_{CC} = 6\text{ V},\ t_{pHL} = t_{pLH} = 13\ ns$

25. See Figure 3-21.

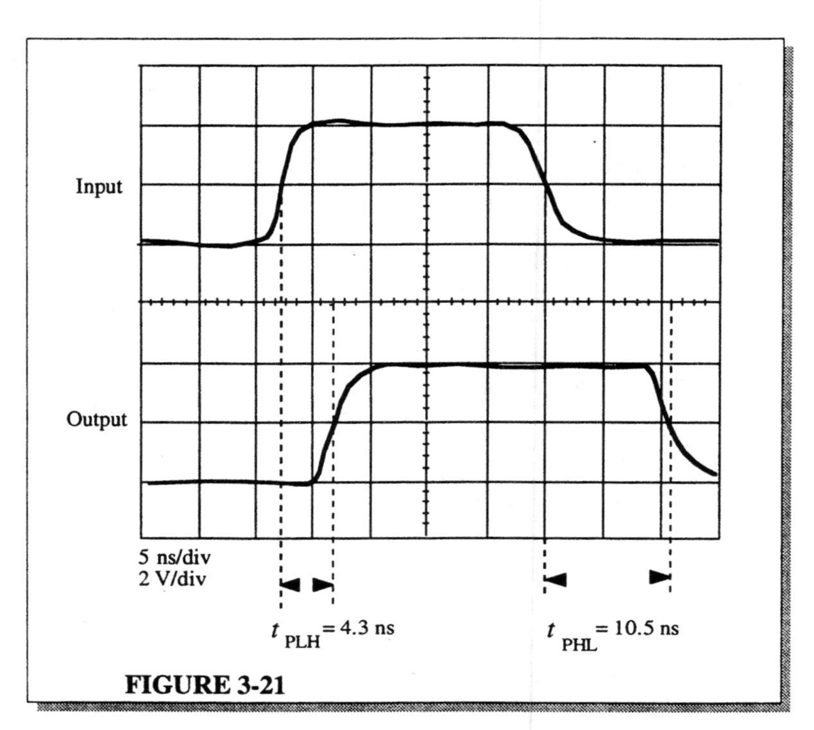

FIGURE 3-21

26. Gate A can be operated at the highest frequency because it has shorter propagation delay times than Gate B.

27. $P_D = V_{CC}I_C = (5\text{ V})(4\text{ mA}) = 20\text{ mW}$

28. $I_{CCH} = 4\text{ mA}$; $P_D = (5\text{ V})(4\text{ mA}) = 20\text{ mW}$

29.

(a) Nand gate OK
(b) AND gate faulty
(c) NAND gate faulty
(d) NOR gate OK
(e) XOR gate faulty
(f) XOR gate OK

30. (a) NAND gate faulty. Input A open.
(b) NOR gate faulty. Input B shorted to ground.
(c) NAND gate OK
(d) XOR gate faulty. Input A open.

31. (a) The gate does not respond to pulses on either input when the other input is HIGH. It is unlikely that both inputs are open. The most probable fault is that the output is stuck in the LOW state (shorted to ground, perhaps) although it could be open.

(b) The first check indicates proper operation. The second check results in no output activity indicating that input 2 is open.

32. The timer input to the AND gate is open. Check for 30-second HIGH level on this input when ignition is turned on.

33. An open seat-belt input to the AND gate will act like a constant HIGH just as if the seat-belt were unbuckled.

34. The remaining problem is to keep the extra tablet from dropping into the bottle rather than just masking it.

35. See Figure 3-22.

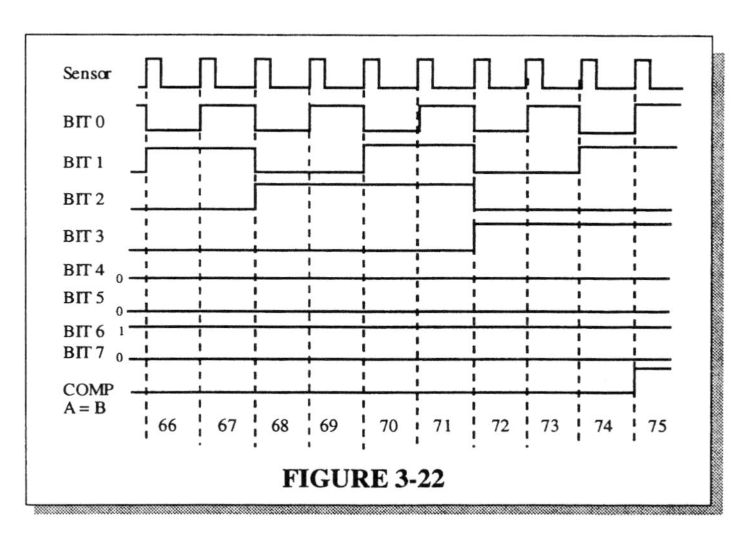

FIGURE 3-22

36. For each of the settings, inputs *a* through *h* are as follows:

(a) 30 tablets	(b) 75 tablets/bottle:	(c) 99 tablets/bottle:
d a b c j e f g	*d a b c j e f g*	*d a b c j e f g*
0 0 0 1 1 1 1 0	0 1 0 0 1 0 1 1	0 1 1 0 0 0 1 1

37. Eighteen tablets instead of fifty would be placed in each bottle if BIT 5 from the counter were stuck at a 1 or open which could be caused by an open input to the corresponding XOR gate. Also, if the *b* input from the BCD-to-binary code converter were shorted to ground and appeared as a 0.

38. One possible approach: Use two sets of 8 switches. One set simulates the BIT0 through BIT7 inputs from the system counter. The other set simulates the a through h inputs from the BCD-to-binary code converter. The switches are sequentially set for numbers 0 through 99 and the comparator out is checked for a HIGH for eachset of numbers. An easier way is to use binary counters to sequence through all possible numbers, but counters havenot been covered at this point.

39. See Figure 3-23.

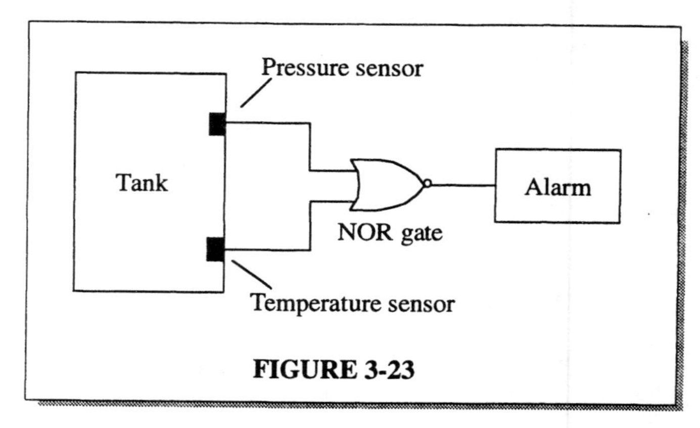

FIGURE 3-23

40. See Figure 3-24

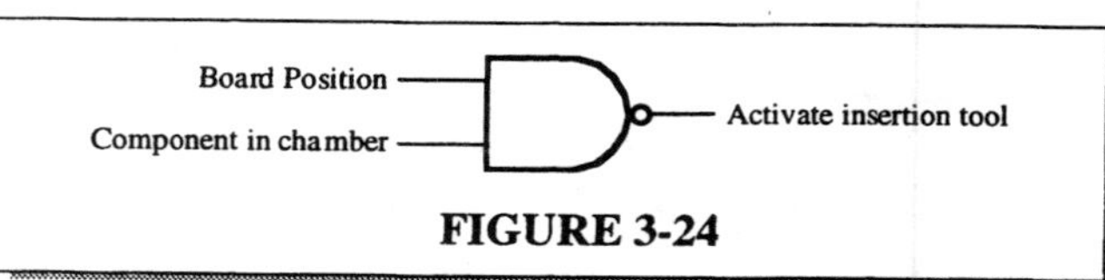

FIGURE 3-24

41. Add an inverter to the Enable input line of the AND gate as shown in Figure 3-25.

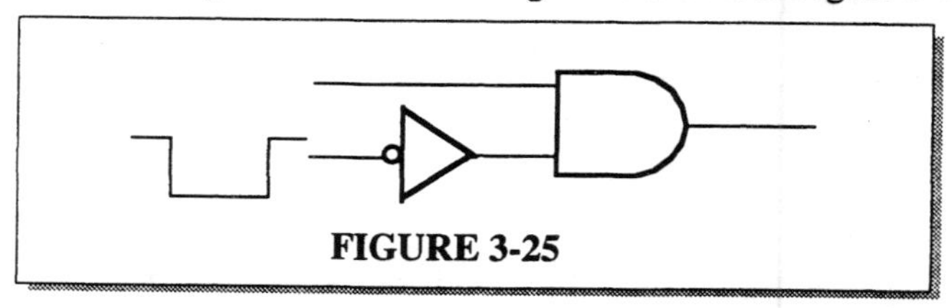

FIGURE 3-25

42. See Figure 3-26.

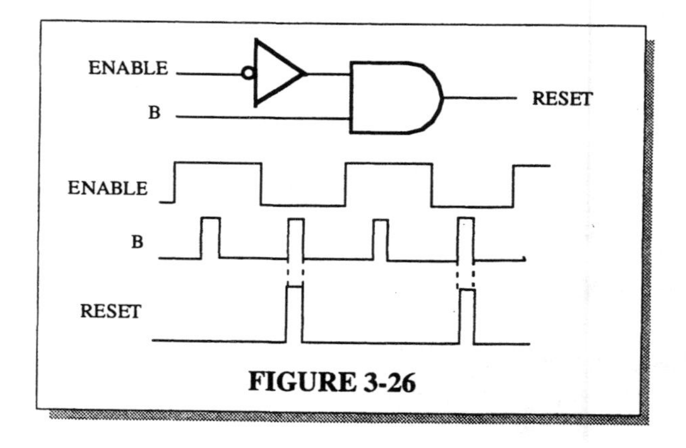

FIGURE 3-26

43. See Figure 3-27.

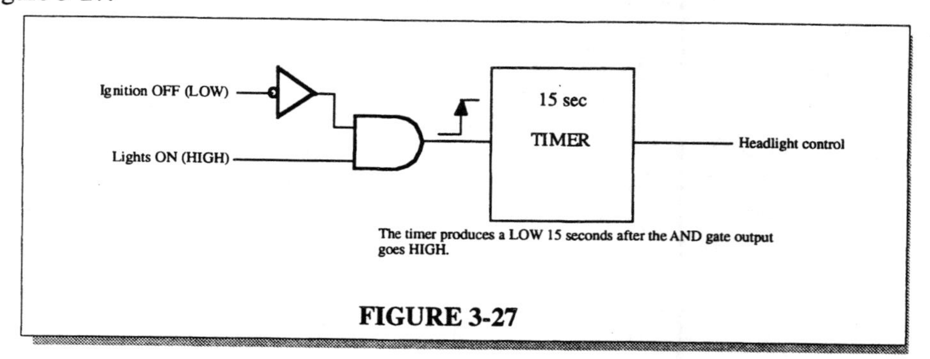

FIGURE 3-27

44. See Figure 3-28.

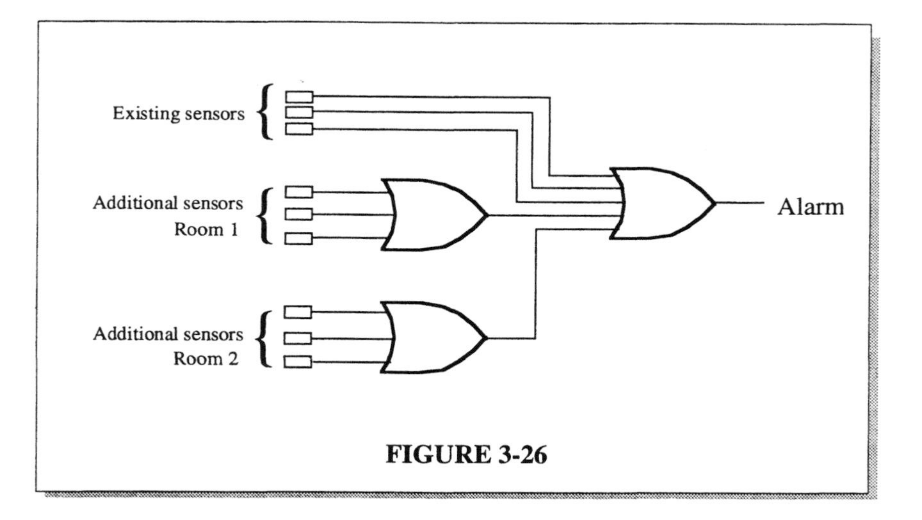

FIGURE 3-26

45. The valve and conveyor line must be taken from the NAND gate output. The register line still comes from the inverter.

CHAPTER 4
BOOLEAN ALGEBRA

1. $X = A + B + C + D$
This is an OR configuration.

2. $Y = ABCDE$

3. $Z = \overline{A} + \overline{B} + \overline{C}$

4.
(a) $0 + 0 + 1 = 1$
(b) $1 + 1 + 1 = 1$
(c) $1 \cdot 0 \cdot 0 = 0$
(d) $1 \cdot 1 \cdot 1 = 1$
(e) $1 \cdot 0 \cdot 1 = 0$
(f) $1 \cdot 1 + 0 \cdot 1 \cdot 1 = 1 + 0 = 1$

5.
(a) $AB = 1$ when $A = 1,\ B = 1$
(b) $A\overline{B}C = 1$ when $A = 1,\ B = 0,\ C = 1$
(c) $A + B = 0$ when $A = 0,\ B = 0$
(d) $\overline{A} + B + \overline{C} = 0$ when $A = 1,\ B = 0,\ C = 1$
(e) $\overline{A} + \overline{B} + C = 0$ when $A = 1,\ B = 1,\ C = 0$
(f) $\overline{A} + B = 0$ when $A = 1,\ B = 0$
(g) $A\overline{B}\overline{C} = 1$ when $A = 1,\ B = 0,\ C = 0$

6.
(a) $X = (A + B)C + B$

A	B	C	$A+B$	$(A+B)C$	X
0	0	0	0	0	0
0	0	1	0	0	0
0	1	0	1	0	1
0	1	1	1	1	1
1	0	0	1	0	0
1	0	1	1	1	1
1	1	0	1	0	1
1	1	1	1	1	1

(b) $X = (\overline{A + B})C$

A	B	C	$\overline{A+B}$	X
0	0	0	1	0
0	0	1	1	1
0	1	0	0	0
0	1	1	0	0
1	0	0	0	0
1	0	1	0	0
1	1	0	0	0
1	1	1	0	0

(c) $X = ABC + AB$

A	B	C	$A\overline{B}C$	AB	X
0	0	0	0	0	0
0	0	1	0	0	0
0	1	0	0	0	0
0	1	1	0	0	0
1	0	0	0	0	0
1	0	1	1	0	1
1	1	0	0	1	1
1	1	1	0	1	1

(d) $X = (A + B)(\overline{A} + B)$

A	B	$A + B$	$\overline{A} + B$	X
0	0	0	1	0
0	1	1	1	1
1	0	1	0	0
1	1	1	1	1

(e) $X = (A + BC)(\overline{B} + \overline{C})$

A	B	C	$A + BC$	$\overline{B} + \overline{C}$	X
0	0	0	0	1	0
0	0	1	0	1	0
0	1	0	0	1	0
0	1	1	1	0	0
1	0	0	1	1	1
1	0	1	1	1	1
1	1	0	1	1	1
1	1	1	1	0	0

7. (a) Commutative law of addition .
(b) Commutative law of multiplication.
(c) Distributive law.

8. Refer to Table 4-1 in the textbook.

(a) Rule 9: $\overline{\overline{A}} = A$

(b) Rule 8: $A\overline{A} = 0$ (applied to 1st and 3rd terms)

(c) Rule 5: $A + A = A$

(d) Rule 6: $A + \overline{A} = 1$

(e) Rule 10: $A + AB = A$

(f) Rule 11 $A + \overline{A}B = A + B$ (applied to 1st and 3rd terms)

9.

(a) $\overline{A+\overline{B}} = \overline{A}\,\overline{\overline{B}} = \mathbf{\overline{A}B}$

(b) $\overline{\overline{A}B} = \overline{\overline{A}} + \overline{B} = \mathbf{A + \overline{B}}$

(c) $\overline{A+B+C} = \mathbf{\overline{A}\,\overline{B}\,\overline{C}}$

(d) $\overline{ABC} = \mathbf{\overline{A} + \overline{B} + \overline{C}}$

(e) $\overline{A(B+C)} = \overline{A} + (\overline{B+C}) = \mathbf{\overline{A} + \overline{B}\,\overline{C}}$

(f) $\overline{AB} + \overline{CD} = \mathbf{\overline{A} + \overline{B} + \overline{C} + \overline{D}}$

(g) $\overline{AB+CD} = (\overline{AB})(\overline{CD}) = \mathbf{(\overline{A} + \overline{B})(\overline{C} + \overline{D})}$

(h) $\overline{(A+\overline{B})(\overline{C}+D)} = \overline{A+\overline{B}} + \overline{\overline{C}+D} = \mathbf{\overline{A}B + C\overline{D}}$

10.

(a) $\overline{A\overline{B}(C+\overline{D})} = \overline{A\overline{B}} + (\overline{C+\overline{D}}) = \mathbf{\overline{A} + B + \overline{C}D}$

(b) $\overline{AB(CD+EF)} = \overline{AB} + (\overline{CD+EF}) = \overline{A} + \overline{B} + (\overline{CD})(\overline{EF})$

$= \mathbf{\overline{A} + \overline{B} + (\overline{C} + \overline{D})(\overline{E} + \overline{F})}$

(c) $\overline{(A+\overline{B}+C+\overline{D})} + \overline{ABC\overline{D}} = \mathbf{\overline{A}B\overline{C}D + \overline{A} + \overline{B} + \overline{C} + D}$

(d) $\overline{\overline{(\overline{A}+B+C+D)}\;\overline{(A\overline{B}\overline{C}D)}} = \overline{(A\overline{B}\overline{C}\overline{D})(\overline{A}+B+C+\overline{D})}$

$= \overline{A\overline{B}\overline{C}\overline{D}} + \overline{\overline{A}+B+C+\overline{D}} = \mathbf{\overline{A} + B + C + D + A\overline{B}\overline{C}D}$

(e) $\overline{\overline{AB}(CD+\overline{E}F)(\overline{AB}+\overline{CD})} = \overline{\overline{AB}} + (\overline{CD+\overline{E}F}) + (\overline{\overline{AB}+\overline{CD}})$

$= AB + (\overline{CD})(\overline{\overline{E}F}) + (\overline{\overline{AB}})(\overline{\overline{CD}})$

$= \mathbf{AB + (\overline{C} + \overline{D})(E + \overline{F}) + ABCD}$

11.

(a) $\overline{\overline{(\overline{ABC})(\overline{EFG})} + \overline{(\overline{HIJ})(\overline{KLM})}} = \overline{\overline{\overline{ABC}} + \overline{\overline{EFG}} + \overline{\overline{HIJ}} + \overline{\overline{KLM}}}$

$= \overline{ABC + EFG + HIJ + KLM} = (\overline{ABC})(\overline{EFG})(\overline{HIJ})(\overline{KLM})$

$= \mathbf{(\overline{A} + \overline{B} + \overline{C})(\overline{E} + \overline{F} + \overline{G})(\overline{H} + \overline{I} + \overline{J})(\overline{K} + \overline{L} + \overline{M})}$

(b) $\overline{(A + \overline{B\overline{C}} + CD)} + \overline{\overline{BC}} = \overline{A}(\overline{\overline{B\overline{C}}})(\overline{CD}) + BC = \overline{A}(B\overline{C})(\overline{CD}) + BC$

$= \overline{A}B\overline{C}(\overline{C} + \overline{D}) + BC = \overline{A}B\overline{C} + \overline{A}B\overline{C}\,\overline{D} + BC = \overline{A}B\overline{C}(1 + \overline{D}) + BC$

$= \mathbf{\overline{A}B\overline{C} + BC}$

(c) $\overline{\overline{\overline{(A+B)}\,\overline{(C+D)}\,\overline{(E+F)}\,\overline{(G+H)}}}$

$= \overline{(A+B)}\,\overline{(C+D)}\,\overline{(E+F)}\,\overline{(G+H)} = \mathbf{\overline{A}\,\overline{B}\,\overline{C}\,\overline{D}\,\overline{E}\,\overline{F}\,\overline{G}\,\overline{H}}$

12. (a) $AB = X$
(b) $\overline{A} = X$
(c) $A + B = X$
(d) $A + B + C = X$

13. See Figure 4-1.

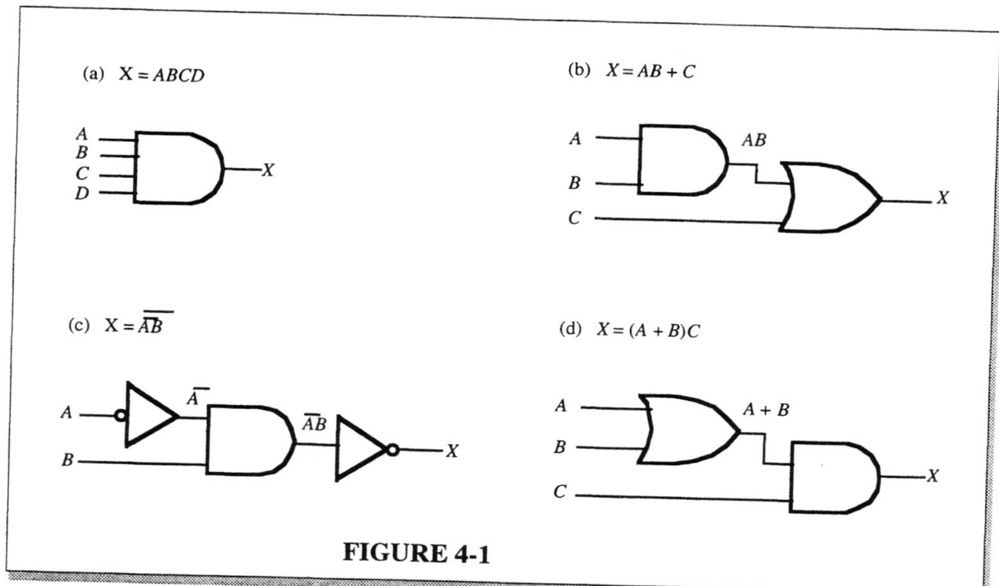

FIGURE 4-1

14. See Figure 4-2.

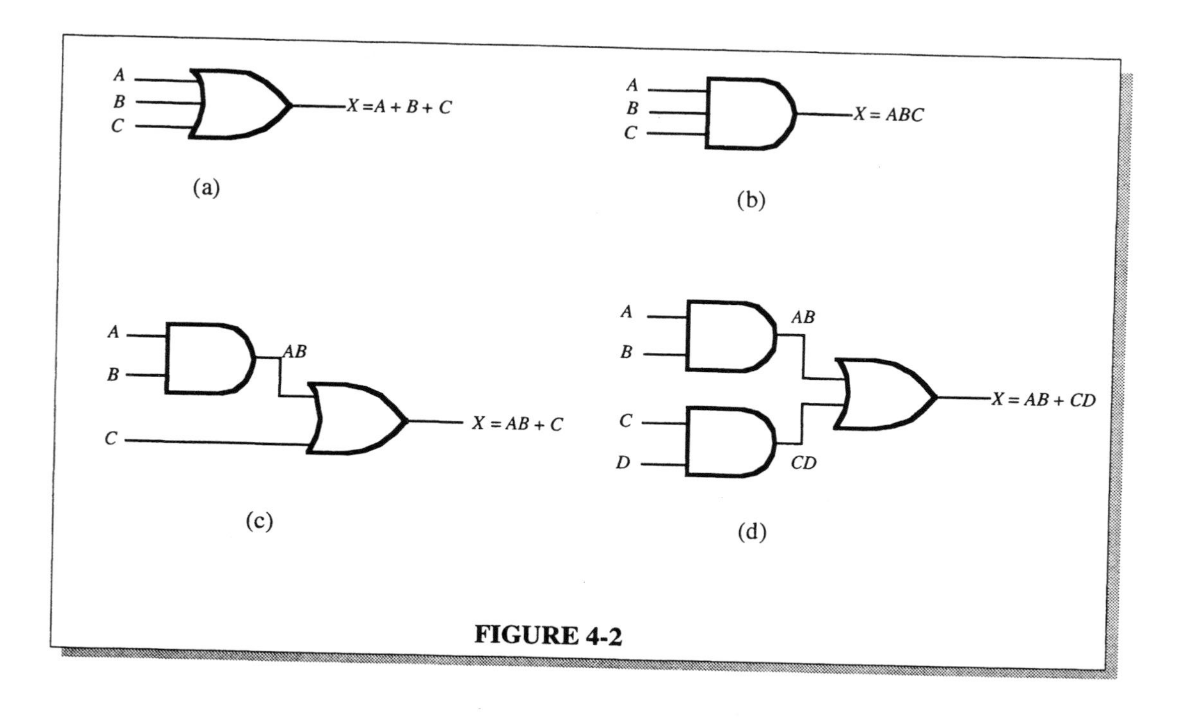

FIGURE 4-2

15. See Figure 4-3

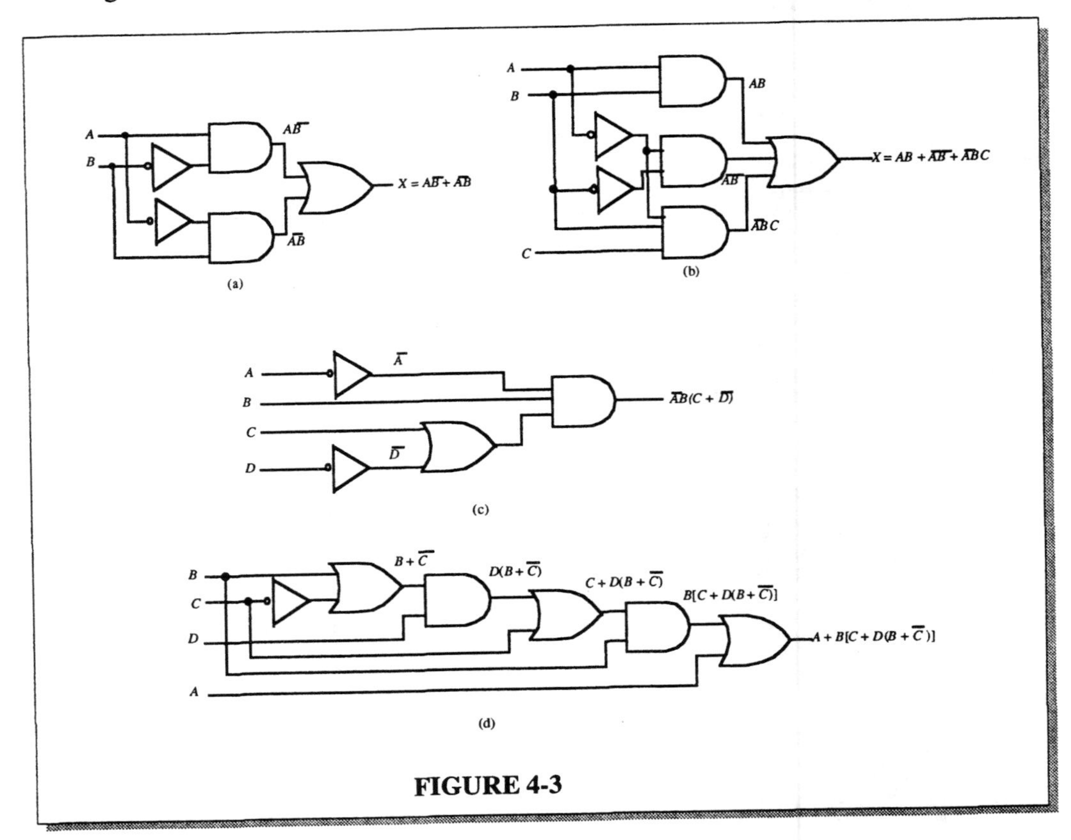

FIGURE 4-3

16.

(a) $X = A + B$

A	B	X
0	0	0
0	1	1
1	0	1
1	1	1

(b) $X = AB$

A	B	X
0	0	0
0	1	0
1	0	0
1	1	1

(c) $X = AB + BC$

A	B	C	X
0	0	0	0
0	0	1	0
0	1	0	0
0	1	1	1
1	0	0	0
1	0	1	0
1	1	0	1
1	1	1	1

(d) $X = (A + B)C$

A	B	C	X
0	0	0	0
0	0	1	0
0	1	0	0
0	1	1	1
1	0	0	0
1	0	1	1
1	1	0	0
1	1	1	1

(e) $X = (A + B)(\overline{B} + C)$

A	B	C	$A + B$	$\overline{B} + C$	X
0	0	0	0	1	0
0	0	1	0	1	0
0	1	0	1	0	0
0	1	1	1	1	1
1	0	0	1	1	1
1	0	1	1	1	1
1	1	0	1	0	0
1	1	1	1	1	1

17. (a) $A(A + B) = AA + BB = A + AB = A(1 + B) = \mathbf{A}$

(b) $A(\overline{A} + AB) = A\overline{A} + AAB = 0 + AB = \mathbf{AB}$

(c) $BC + \overline{B}C = C(B + \overline{B}) = C\,1 = \mathbf{C}$

(d) $A(A + \overline{A}B) = AA + A\overline{A}B = A + 0\,B = A + 0 = \mathbf{A}$

(e) $A\overline{B}C + \overline{A}BC + \overline{A}\,\overline{B}C = A\overline{B}C + \overline{A}C(B + \overline{B}) = A\overline{B}C + \overline{A}C\ \ 1$

$= A\overline{B}C + \overline{A}C = C(\overline{A} + A\overline{B}) = C(\overline{A} + \overline{B}) = \mathbf{\overline{A}C + \overline{B}C}$

18. (a) $(A + \overline{B})(A + C) = AA + AC + AB + BC = A + AC + A\overline{B} + \overline{B}C$

$= A(1 + C + \overline{B}) + \overline{B}C = A{\cdot}1 + \overline{B}C = \mathbf{A + \overline{B}C}$

(b) $\overline{A}B + \overline{A}B\overline{C} + \overline{A}BCD + \overline{A}B\overline{C}\,\overline{D}E = \overline{A}B(1 + \overline{C} + CD + \overline{C}\,\overline{D}E) = \overline{A}B\ \ 1$

$= \mathbf{\overline{A}B}$

(c) $AB + \overline{AB}C + A = AB + (\overline{A} + \overline{B})C + A = AB + \overline{A}C + \overline{B}C + A$

$= A(B + 1) + \overline{A}C + \overline{B}C = A + \overline{A}C + \overline{B}C = A + C + \overline{B}C = A + C(1 + \overline{B})$

$= \mathbf{A + C}$

(d) $(A + \overline{A})(AB + AB\overline{C}) = AAB + AAB\overline{C} + \overline{A}AB + \overline{A}AB\overline{C}$

$= AB + AB\overline{C} + 0 + 0 = AB(1 + \overline{C}) = \mathbf{AB}$

(e) $AB + (\overline{A} + \overline{B})C + AB = AB + \overline{A}C + \overline{B}C + AB = AB + (\overline{A}\ \ + \overline{B}\,)C$

$= AB + \overline{AB}C = AB + C$

19. (a) $BD + B(D + E) + \overline{D}(D + F) = BD + BD + BE + \overline{D}D + \overline{D}F$

$= BD + BE + 0 + \overline{D}F = \mathbf{BD + BE + \overline{D}F}$

(b) $\overline{A}\,\overline{B}C + (\overline{A + B + \overline{C}}) + \overline{A}\,\overline{B}\,\overline{C}D = \overline{A}\,\overline{B}C + \overline{A}\,\overline{B}C + \overline{A}\,\overline{B}\,\overline{C}D = \overline{A}\,\overline{B}C + \overline{A}\,\overline{B}\,\overline{C}D$

$= \overline{A}\,\overline{B}(C + \overline{C}D) = \overline{A}\,\overline{B}(C + D) = \mathbf{\overline{A}\,\overline{B}C + \overline{A}\,\overline{B}D}$

(c) $(B + BC)(B + \overline{B}C)(B + D) = B(1 + C)(B + C)(B + D)$

$= B(B + C)(B + D) = (BB + BC)(B + D) = (B + BC)(B + D)$

$= B(1 + C)(B + D) = B(B + D) = BB + BD = B + BD = B(1 + D) = \mathbf{B}$

(d) $ABCD + AB(\overline{CD}) + (\overline{AB})CD = ABCD + AB(\overline{C} + \overline{D}) + (\overline{A} + \overline{B})CD$

$= ABCD + AB\overline{C} + AB\overline{D} + \overline{A}CD + \overline{B}CD$

$= CD(AB + \overline{A} + \overline{B}) + AB\overline{C} + AB\overline{D} = CD(B + \overline{A} + \overline{B}) + AB\overline{C} + AB\overline{D}$

$= CD(1 + \overline{A}) + AB\overline{C} + AB\overline{D} = CD + AB\overline{C} + AB\overline{D} = CD + AB(\overline{CD}) = \mathbf{CD + AB}$

(e) $ABC[AB + \overline{C}(BC + AC)] = ABABC + ABC\overline{C}(BC + AC)$

$= ABC + 0(BC + AC) = \mathbf{ABC}$

20. First develop the Boolean expression for the output of eachgate network and simplify.

(a) See Figure 4-4

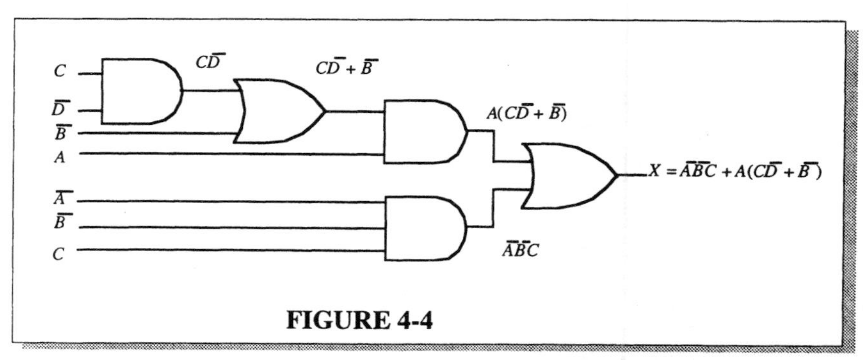

FIGURE 4-4

$X = \overline{A}\,\overline{B}C + A(C\overline{D} + \overline{B}) = \overline{A}\,\overline{B}C + AC\overline{D} + A\overline{B} = \overline{B}(A + \overline{A}C) + AC\overline{D}$

$= \overline{B}(A + C) + AC\overline{D} = \mathbf{A\overline{B} + \overline{B}C + AC\overline{D}}$

(b) See Figure 4-5.

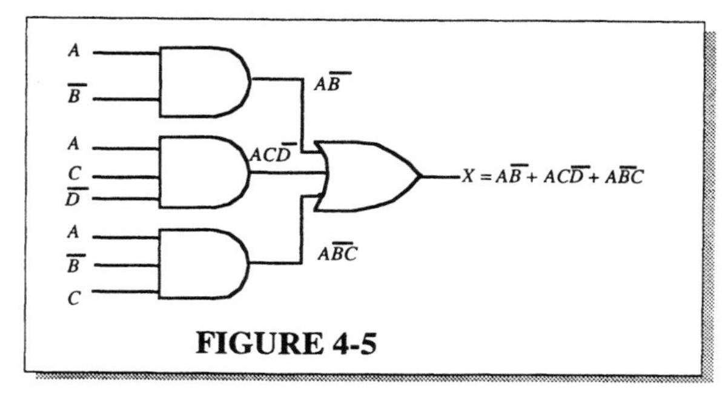

FIGURE 4-5

$$X = A\overline{B} + AC\overline{D} + A\overline{B}C = A\overline{B}(1 + C) + AC\overline{D} = \mathbf{A\overline{B} + AC\overline{D}}$$

(c) See Figure 4-6.

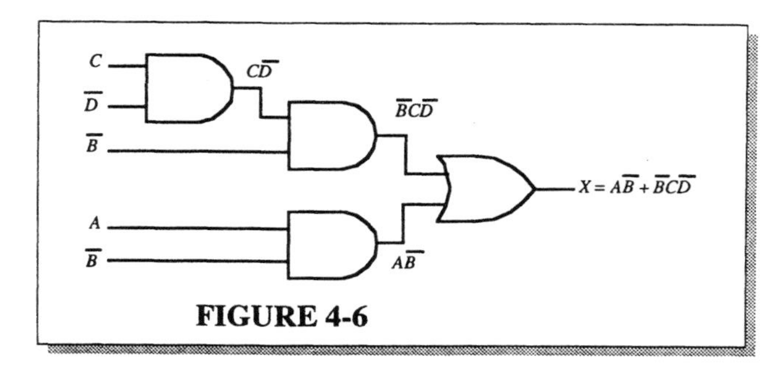

FIGURE 4-6

$X = \mathbf{A\overline{B} + \overline{B}C\overline{D}}$ No further simplification is possible

(d) See Figure 4-7.

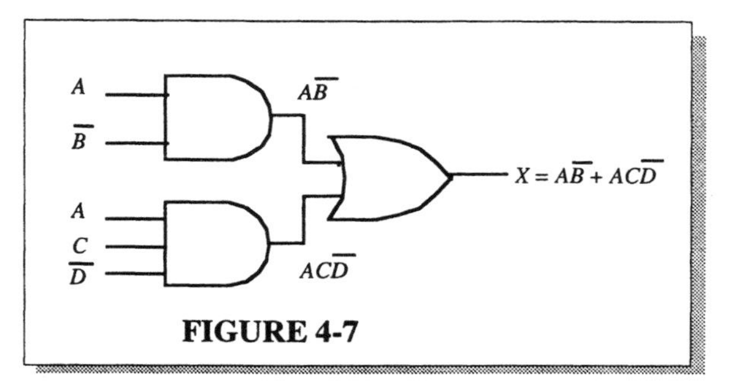

FIGURE 4-7

$X = \mathbf{A\overline{B} + AC\overline{D}}$ No further simplification is possible

21. (a) $(A + B)(C + \overline{B}) = \boldsymbol{AC + BC + B\overline{B} + A\overline{B} = AC + BC + A\overline{B}}$

(b) $(A + \overline{B}C)C = \boldsymbol{AC + \overline{B}CC = AC + \overline{B}C}$

(c) $(A + C)(AB + AC) = AAB + AAC + ABC + ACC = AB + AC + ABC + ACC$

$= (AB + AC)(1 + C) = AB + AC$

22. (a) $AB + CD(A\overline{B} + CD) = AB + A\overline{B}CD + CDCD = AB + A\overline{B}CD + CD$

$= AB\ (A\overline{B} + 1)CD = \boldsymbol{AB + CD}$

(b) $AB(\overline{B}\overline{C} + BD) = AB\overline{B}\overline{C} + ABBD = 0 + ABD = \boldsymbol{ABD}$

(c) $A + B[AC + (B + \overline{C})D] = A + ABC + (B + \overline{C})BD$

$= A + ABC + BD + B\overline{C}D = A(1 + BC) + BD + B\overline{C}D = A + BD(1 + \overline{C})$

$= \boldsymbol{A + BD}$

23. (a) The domain is A, B, C

The standard SOP is: $A\overline{B}C + A\overline{B}\,\overline{C} + ABC + \overline{A}BC$

(b) The domain is A, B, C

The standard SOP is: $ABC + A\overline{B}C + \overline{A}\,\overline{B}C$

(c) The domain is A, B, C

The standard SOP is: $ABC + AB\overline{C} + A\overline{B}C$

24. (a) $AB + CD = ABCD + ABC\overline{D} + AB\overline{C}D + AB\overline{C}\,\overline{D} + \overline{A}\,\overline{B}CD + \overline{A}BCD + A\overline{B}CD$

(b) $ABD = ABCD + AB\overline{C}D$

(c) $A + BD = A\overline{B}\,\overline{C}\,\overline{D} + A\overline{B}\,\overline{C}D + A\overline{B}C\overline{D} + A\overline{B}CD + AB\overline{C}\,\overline{D} + AB\overline{C}D$
$+ ABC\overline{D} + ABCD + \overline{A}B\overline{C}D + \overline{A}BCD$

25. (a) $A\overline{B}C + A\overline{B}\,\overline{C} + ABC + \overline{A}BC$: 101 + 100 + 111 + 011

(b) $ABC + A\overline{B}C + \overline{A}\,\overline{B}C$: 111 + 101 + 001

(c) $ABC + AB\overline{C} + A\overline{B}C$: 111 + 110 + 101

26. (a) $ABCD + ABC\overline{D} + AB\overline{C}D + AB\overline{C}\,\overline{D} + \overline{A}\,\overline{B}CD + \overline{A}BCD + A\overline{B}CD$:
1111 + 1110 + 1101 + 1100 + 0011 + 0111 + 1011

(b) $ABCD + AB\overline{C}D$: 1111 + 1101

(c) $A\overline{B}\,\overline{C}\,\overline{D} + A\overline{B}\,\overline{C}D + A\overline{B}C\overline{D} + A\overline{B}CD + AB\overline{C}\,\overline{D} + AB\overline{C}D$
$+ ABC\overline{D} + ABCD + \overline{A}B\overline{C}D + \overline{A}BCD$:
1000 + 1001 + 1010 + 1011 + 1100 + 1101 + 1110 + 1111 + 0101 + 0111

27.

(a) $(A+B+C)(A+B+\bar{C})(A+\bar{B}+C)(\bar{A}+\bar{B}+C)$

(b) $(A+B+C)(A+\bar{B}+C)(A+\bar{B}+\bar{C})(\bar{A}+B+C)(\bar{A}+\bar{B}+C)$

(c) $(A+B+C)(A+B+\bar{C})(A+\bar{B}+C)(A+\bar{B}+\bar{C})(\bar{A}+B+C)$

28.

(a) $(A+B+C+D)(A+B+C+\bar{D})(A+B+\bar{C}+D)(A+\bar{B}+C+D)(A+\bar{B}+C+\bar{D})$
$(A+\bar{B}+\bar{C}+D)(\bar{A}+B+C+D)(\bar{A}+B+C+\bar{D})(\bar{A}+B+\bar{C}+D)$

(b) $(A+B+C+D)(A+B+C+\bar{D})(A+B+\bar{C}+D)(A+B+\bar{C}+\bar{D})(A+\bar{B}+C+D)(A+\bar{B}+C+\bar{D})$
$(A+\bar{B}+\bar{C}+D)(A+\bar{B}+\bar{C}+\bar{D})(\bar{A}+B+C+D)(\bar{A}+B+C+\bar{D})(\bar{A}+B+\bar{C}+D)(\bar{A}+B+\bar{C}+\bar{D})$
$(\bar{A}+\bar{B}+C+D)(\bar{A}+\bar{B}+\bar{C}+D)$

(c) $(A+B+C+D)(A+B+C+\bar{D})(A+B+\bar{C}+D)(A+B+\bar{C}+\bar{D})(A+\bar{B}+C+D)(A+\bar{B}+\bar{C}+D)$

29.

(a)

A	B	C	X
0	0	0	0
0	0	1	0
0	1	0	1
0	1	1	0
1	0	0	0
1	0	1	1
1	1	0	0
1	1	1	1

(b)

X	Y	Z	W
0	0	0	1
0	0	1	1
0	1	0	0
0	1	1	1
1	0	0	0
1	0	1	1
1	1	0	1
1	1	1	0

30.

(a)

A	B	C	D	X
0	0	0	0	1
0	0	0	1	0
0	0	1	0	0
0	0	1	1	0
0	1	0	0	0
0	1	0	1	1
0	1	1	0	1
0	1	1	1	0
1	0	0	0	0
1	0	0	1	1
1	0	1	0	0
1	0	1	1	0
1	1	0	0	0
1	1	0	1	0
1	1	1	0	0
1	1	1	1	0

(b)

W	X	Y	Z	U
0	0	0	0	0
0	0	0	1	0
0	0	1	0	0
0	0	1	1	0
0	1	0	0	0
0	1	0	1	0
0	1	1	0	0
0	1	1	1	1
1	0	0	0	0
1	0	0	1	0
1	0	1	0	0
1	0	1	1	1
1	1	0	0	0
1	1	0	1	1
1	1	1	0	1
1	1	1	1	1

31. (a) $\overline{A}B + AB\overline{C} + \overline{A}\,\overline{C} + A\overline{B}C = \overline{A}BC + \overline{A}B\overline{C} + AB\overline{C} + \overline{A}\,\overline{B}\,\overline{C} + A\overline{B}C$

(b) $$\overline{X} + Y\overline{Z} + WZ + X\overline{Y}Z = W\overline{X}YZ + \overline{W}\,\overline{X}\,\overline{Y}\,\overline{Z} + \overline{W}\,\overline{X}\,\overline{Y}Z + \overline{W}\,\overline{X}Y\overline{Z} + \overline{W}\,\overline{X}YZ$$
$$+ \overline{W}X\overline{Y}Z + \overline{W}XY\overline{Z} + W\overline{X}\,\overline{Y}\,\overline{Z} + W\overline{X}\,\overline{Y}Z$$
$$+ W\overline{X}Y\overline{Z} + W\overline{X}YZ + WX\overline{Y}Z + WXY\overline{Z} + WXYZ$$

A	B	C	X
0	0	0	1
0	0	1	0
0	1	0	1
0	1	1	1
1	0	0	0
1	0	1	1
1	1	0	1
1	1	1	0

W	X	Y	Z	U
0	0	0	0	1
0	0	0	1	1
0	0	1	0	1
0	0	1	1	1
0	1	0	0	0
0	1	0	1	1
0	1	1	0	1
0	1	1	1	0
1	0	0	0	1
1	0	0	1	1
1	0	1	0	1
1	0	1	1	1
1	1	0	0	0
1	1	0	1	1
1	1	1	0	1
1	1	1	1	1

32. (a)

A	B	C	X
0	0	0	0
0	0	1	1
0	1	0	0
0	1	1	1
1	0	0	1
1	0	1	1
1	1	0	1
1	1	1	0

(b)

A	B	C	D	X
0	0	0	0	1
0	0	0	1	1
0	0	1	0	1
0	0	1	1	1
0	1	0	0	1
0	1	0	1	0
0	1	1	0	0
0	1	1	1	1
1	0	0	0	1
1	0	0	1	0
1	0	1	0	0
1	0	1	1	1
1	1	0	0	1
1	1	0	1	1
1	1	1	0	1
1	1	1	1	1

33. (a)

A	B	C	X
0	0	0	0
0	0	1	0
0	1	0	0
0	1	1	1
1	0	0	1
1	0	1	1
1	1	0	1
1	1	1	1

(b)

A	B	C	D	X
0	0	0	0	1
0	0	0	1	0
0	0	1	0	1
0	0	1	1	1
0	1	0	0	0
0	1	0	1	0
0	1	1	0	0
0	1	1	1	0
1	0	0	0	1
1	0	0	1	0
1	0	1	0	0
1	0	1	1	1
1	1	0	0	1
1	1	0	1	1
1	1	1	0	1
1	1	1	1	1

34. (a) $X = \overline{A}\,\overline{B}\,C + A\,\overline{B}\,\overline{C} + A\,\overline{B}\,C + A\,B\,C$

$X = (A + B + C)(A + \overline{B} + C)(A + \overline{B} + \overline{C})(\overline{A} + \overline{B} + C)$

(b) $X = A\,B\,\overline{C} + A\,\overline{B}C + A\,B\,C$

$X = (A + B + C)(A + B + \overline{C})(A + \overline{B} + C)(A + \overline{B} + \overline{C})\,(\overline{A} + B + C)$

(c) $X = \overline{A}\,\overline{B}\,\overline{C}\,\overline{D} + \overline{A}\,\overline{B}\,\overline{C}\,D + \overline{A}\,\overline{B}\,C\,D + \overline{A}\,B\,\overline{C}\,D + \overline{A}\,B\,C\,\overline{D} + A\,\overline{B}\,\overline{C}\,D + A\,B\,\overline{C}\,\overline{D}$

$X = (A + B + \overline{C} + D)(A + \overline{B} + C + D)(A + \overline{B} + \overline{C} + \overline{D})(\overline{A} + B + C + D)(\overline{A} + B + \overline{C} + D)$
$(\overline{A} + B + \overline{C} + \overline{D})(\overline{A} + \overline{B} + C + \overline{D})(\overline{A} + \overline{B} + \overline{C} + D)(\overline{A} + \overline{B} + \overline{C} + \overline{D})$

(d) $X = \overline{A}\,\overline{B}\,C\,\overline{D} + \overline{A}\,B\,\overline{C}\,\overline{D} + \overline{A}\,B\,\overline{C}\,D + \overline{A}\,B\,C\,D + A\,\overline{B}\,C\,D + A\,B\,\overline{C}\,\overline{D} + A\,B\,C\,D$

$X = (A + B + C + D)(A + B + C + \overline{D})(A + B + \overline{C} + \overline{D})(A + \overline{B} + \overline{C} + D)(\overline{A} + B + C + D)$
$(\overline{A} + B + C + \overline{D})(\overline{A} + B + \overline{C} + D)(\overline{A} + \overline{B} + C + \overline{D})(\overline{A} + \overline{B} + \overline{C} + D)$

35. See Figure 4-8.

36. See Figure 4-9.

37. See Figure 4-10.

AB \ C	0	1
00	000	001
01	010	011
11	110	111
10	100	101

FIGURE 4-8

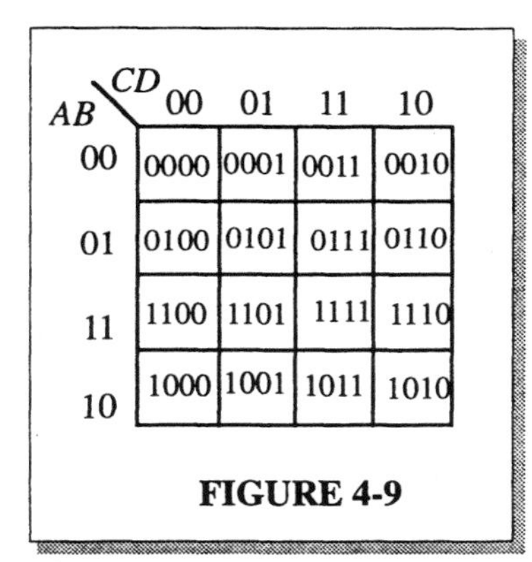

FIGURE 4-9

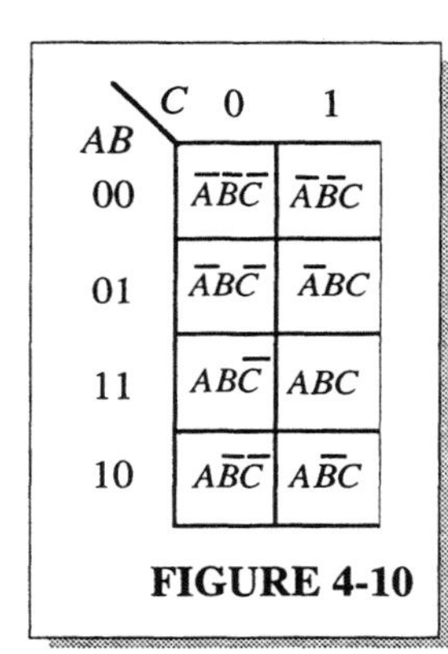

FIGURE 4-10

38. See Figure 4-11.

(a) $X = \overline{A}\,\overline{B}\,\overline{C} + \overline{A}\,\overline{B}\,C + A\,\overline{B}\,C$

AB \ C	0	1
00	1	1
01		
11		
10		1

$X = \overline{A}\overline{B} + \overline{B}C$

(b) $X = AC(B + C) = ABC + A\overline{B}C$

AB \ C	0	1
00		
01		
11		1
10		1

$X = AC$

(c) $X = \overline{A}(BC + B\overline{C}) + A(BC + B\overline{C})$
$= \overline{A}BC + \overline{A}B\overline{C} + ABC + AB\overline{C}$

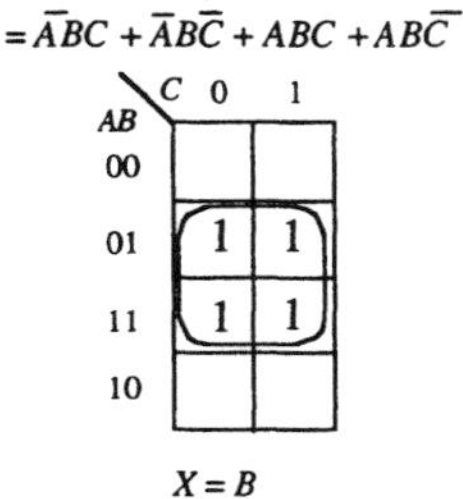

$X = B$

(d) $X = \overline{A}\,\overline{B}\,\overline{C} + A\,\overline{B}\,\overline{C} + \overline{A}\,B\,\overline{C} + A\,B\,\overline{C}$

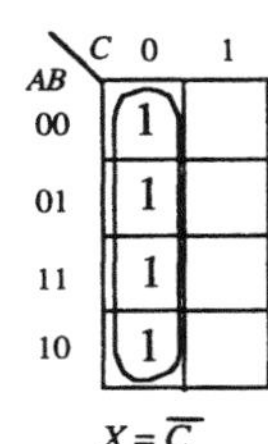

$X = \overline{C}$

FIGURE 4-11

39. See Figure 4-12.

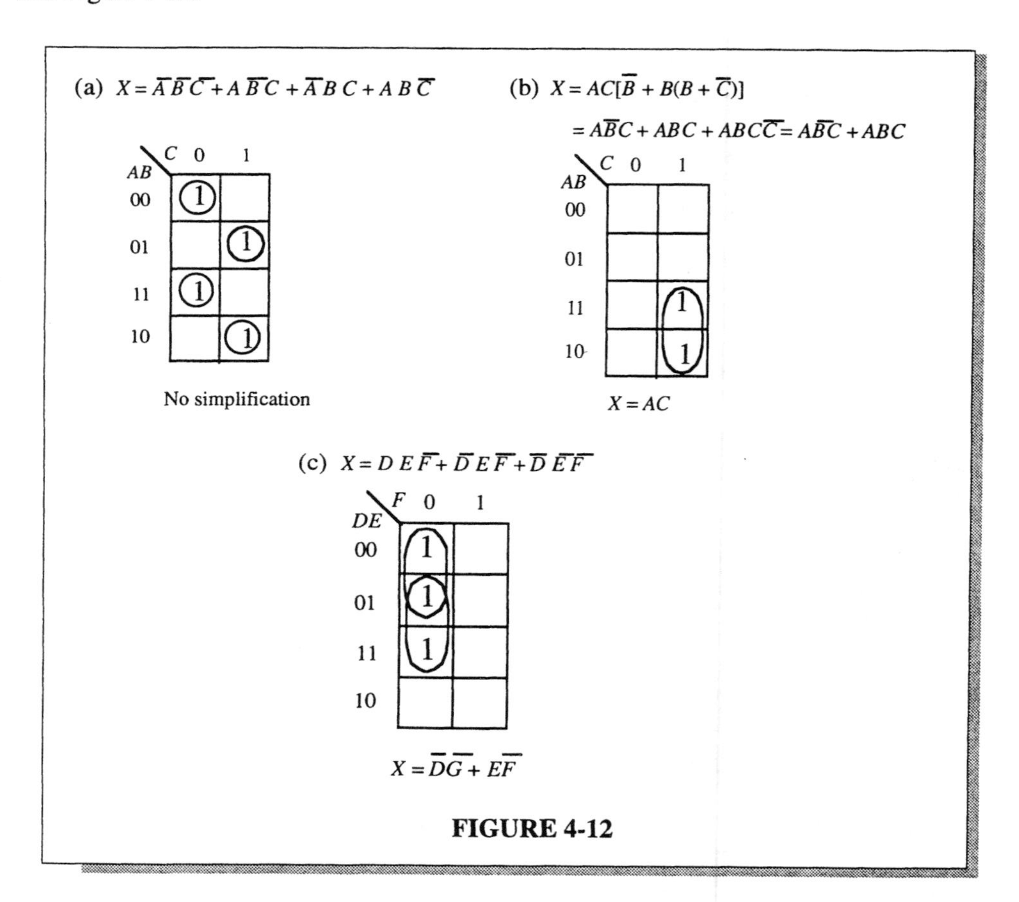

FIGURE 4-12

40. (a) $AB + A\,\bar{B}C + ABC = AB(C + \bar{C}) + A\bar{B}C + ABC$

$= ABC + AB\bar{C} + A\bar{B}C + ABC$

$= ABC + A\bar{B}C + AB\bar{C}$

(b) $A + BC = A(B + \bar{B})(C + \bar{C}) + (\bar{A} + A)BC = (AB + A\bar{B})(C + \bar{C}) + (\bar{A} + A)BC$

$= ABC + AB\bar{C} + A\bar{B}C + A\bar{B}\bar{C} + \bar{A}BC + ABC$

$= ABC + AB\bar{C} + A\bar{B}C + A\bar{B}\bar{C} + \bar{A}BC$

(c) $A\,\bar{B}\,\bar{C}\,D + AC\bar{D} + B\bar{C}D + \bar{A}BC\bar{D}$

$= A\,\bar{B}\,\bar{C}\,D + A(B + \bar{B})CD + (A + \bar{A})B\bar{C}D + \bar{A}BC\bar{D}$

$= A\,\bar{B}\,\bar{C}\,D + ABC\bar{D} + A\bar{B}C\bar{D} + AB\bar{C}D + \bar{A}B\bar{C}D + \bar{A}BC\bar{D}$

(d) $A\bar{B} + A\bar{B}\,\bar{C}D + CD + B\bar{C}D + ABCD$

$= A\bar{B}(C + \bar{C})(D + \bar{D}) + A\bar{B}\,\bar{C}D + (A + \bar{A})(B + \bar{B})CD + (A + \bar{A})B\bar{C}D + ABCD$

$= A\bar{B}\bar{C}\bar{D} + A\bar{B}\bar{C}D + AB\,C\bar{D} + ABCD + A\bar{B}\bar{C}D + ABCD + A\bar{B}CD + \bar{A}BCD + \bar{A}\bar{B}CD + AB\bar{C}D + \bar{A}B\bar{C}D + AB$

$= A\bar{B}\bar{C}\bar{D} + A\bar{B}\bar{C}D + ABC\bar{D} + AB\,CD + A\bar{B}CD + \bar{A}BCD + \bar{A}\,\bar{B}CD + AB\bar{C}D + \bar{A}B\bar{C}D$

$= \bar{A}\bar{B}CD + \bar{A}B\bar{C}D + \bar{A}BCD + A\bar{B}C\bar{D} + A\bar{B}\bar{C}D + A\bar{B}CD + AB\bar{C}D + ABC\overline{D + ABCD}$

41. See Figure 4-13.

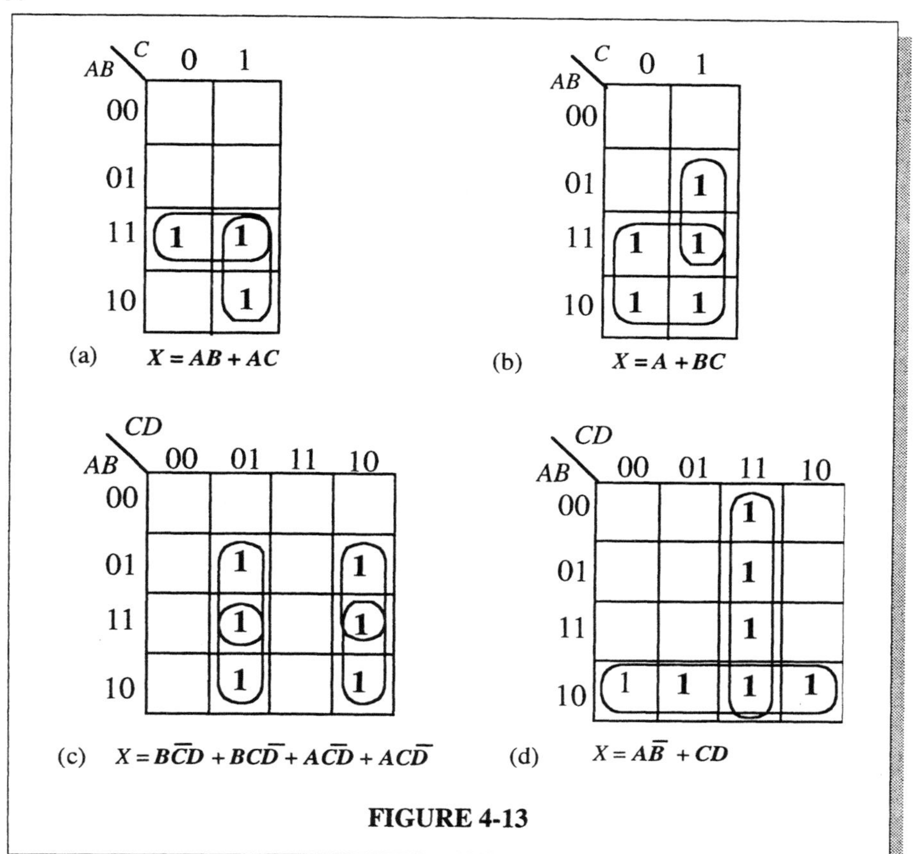

FIGURE 4-13

42. See Figure 4-14.

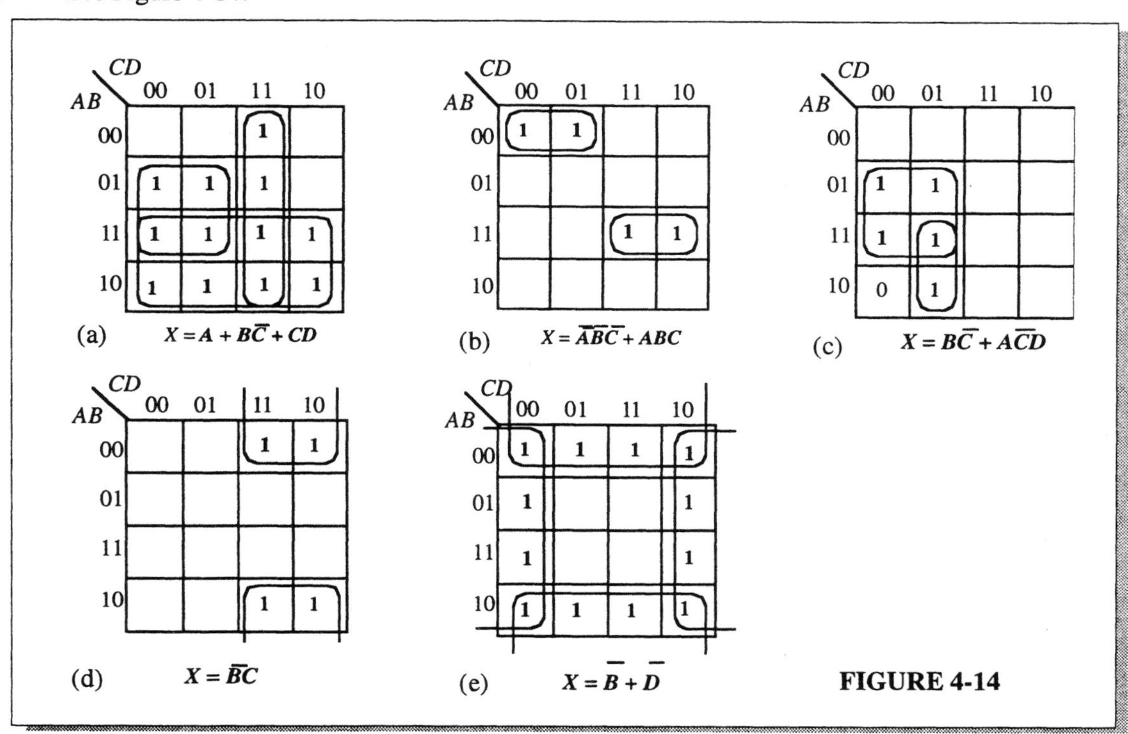

FIGURE 4-14

43. Plot the 1s from Table 4-54 in the text on the map as shown in Figure 4-15 and simplify.

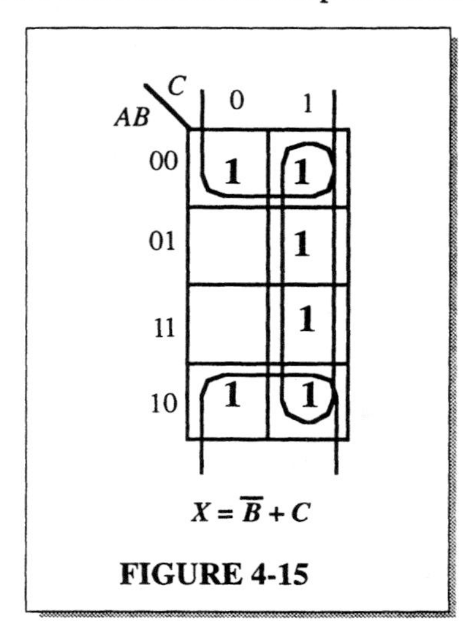

FIGURE 4-15

44. Plot the 1s from Table 4-55 in the text on the map as shown in Figure 4-16 and simplify.

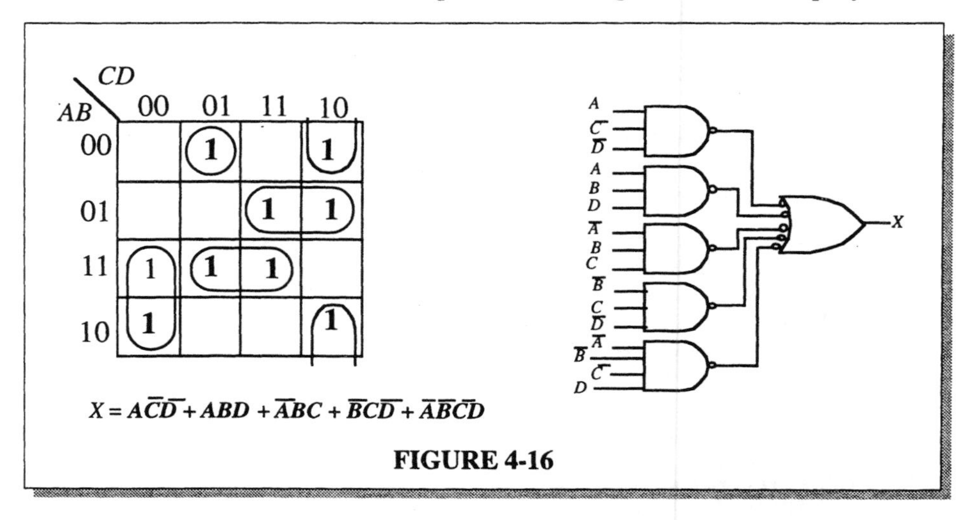

FIGURE 4-16

45. See Figure 4-17.

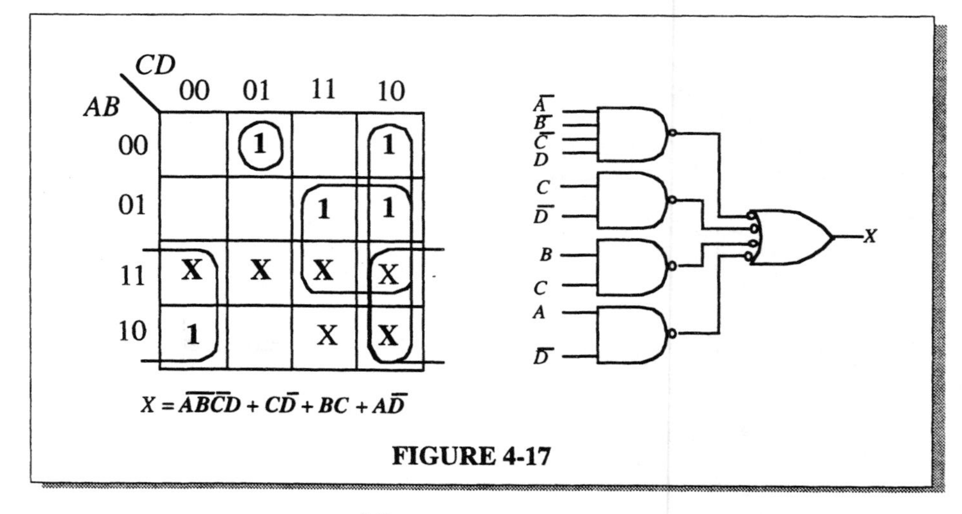

FIGURE 4-17

46. See Figure 4-18.

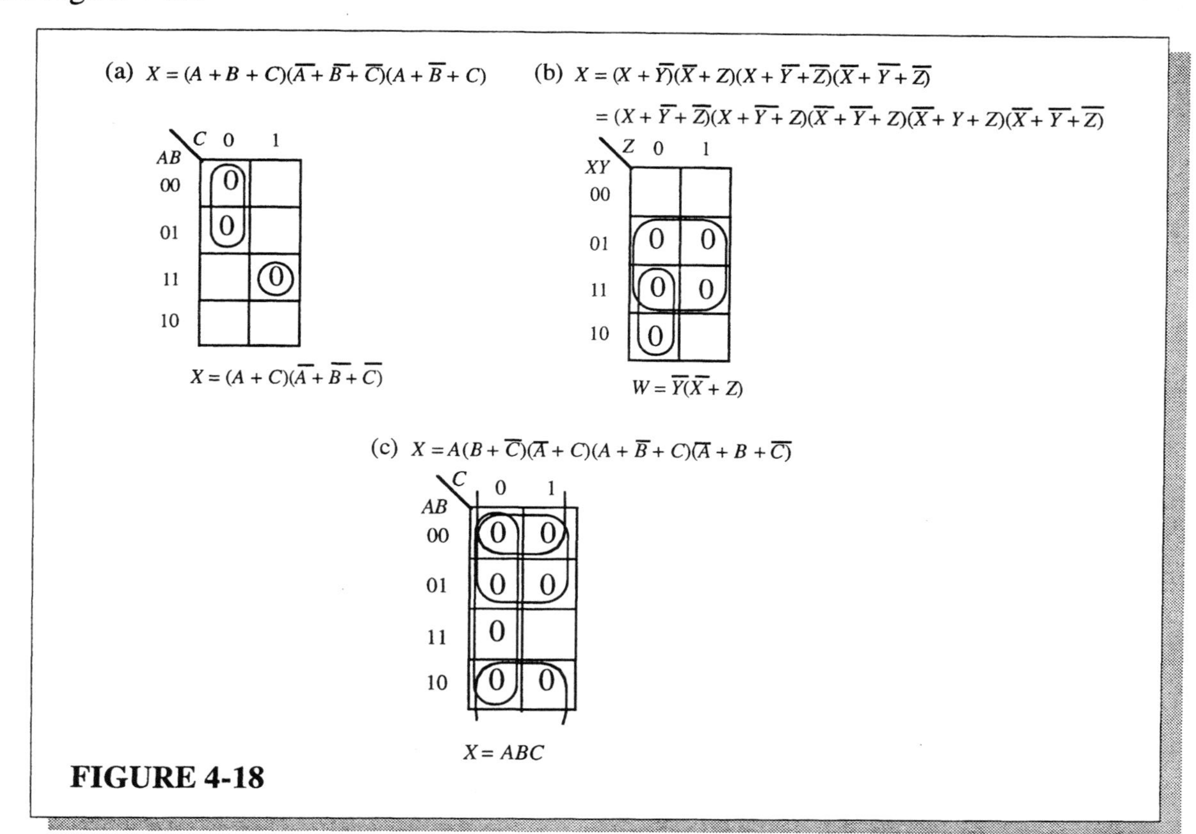

FIGURE 4-18

47. See Figure 4-19.

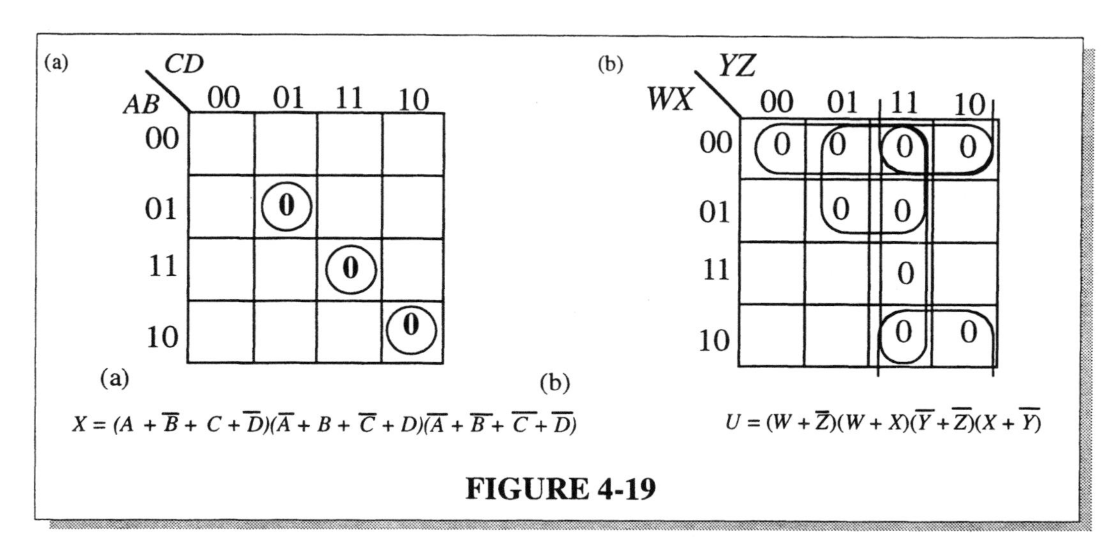

FIGURE 4-19

48. See Figure 4-20

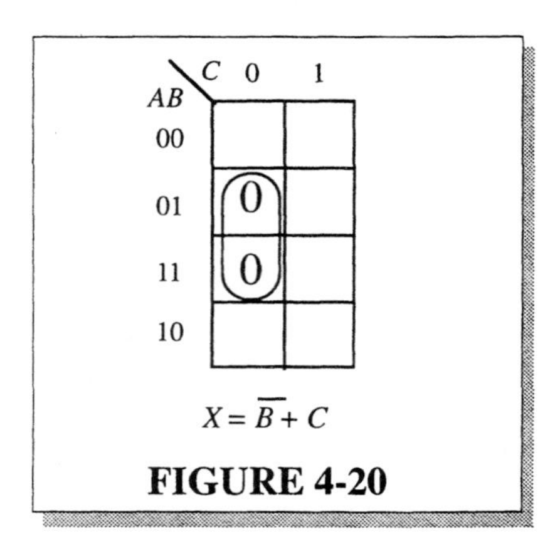

FIGURE 4-20

49. See Figure 4-21.

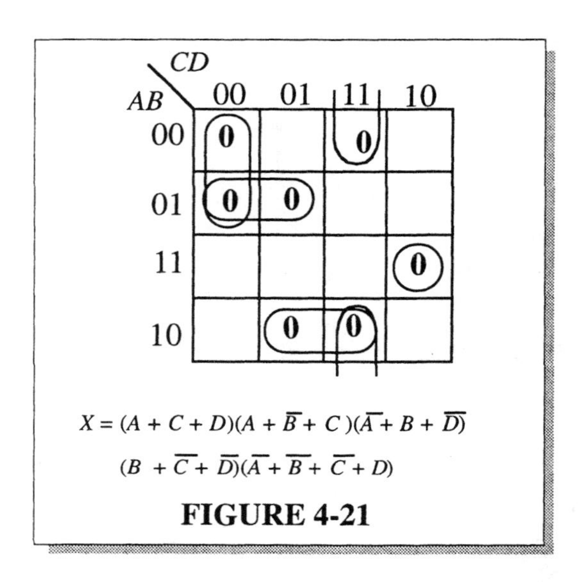

FIGURE 4-21

50. See Figure 4-22.

(a) $(A + \overline{B})(A + \overline{C})(\overline{A} + \overline{B} + C)$

AB \ C	0	1
00	1	0
01	0	0
11	0	1
10	1	1

$X = A\,C + \overline{B}\,\overline{C}$

(b) $(\overline{A} + B)(\overline{A} + \overline{B} + \overline{C})(B + \overline{C} + D)(A + \overline{B} + C + \overline{D})$

AB \ CD	00	01	11	10
00	1	1	1	0
01	1	0	1	1
11	1	1	0	0
10	0	0	0	0

$X = \overline{A}\,\overline{C}\,\overline{D} + AB\overline{C} + \overline{A}\,\overline{B}D + \overline{A}BC$

FIGURE 4-22

51. See Figure 4-23.

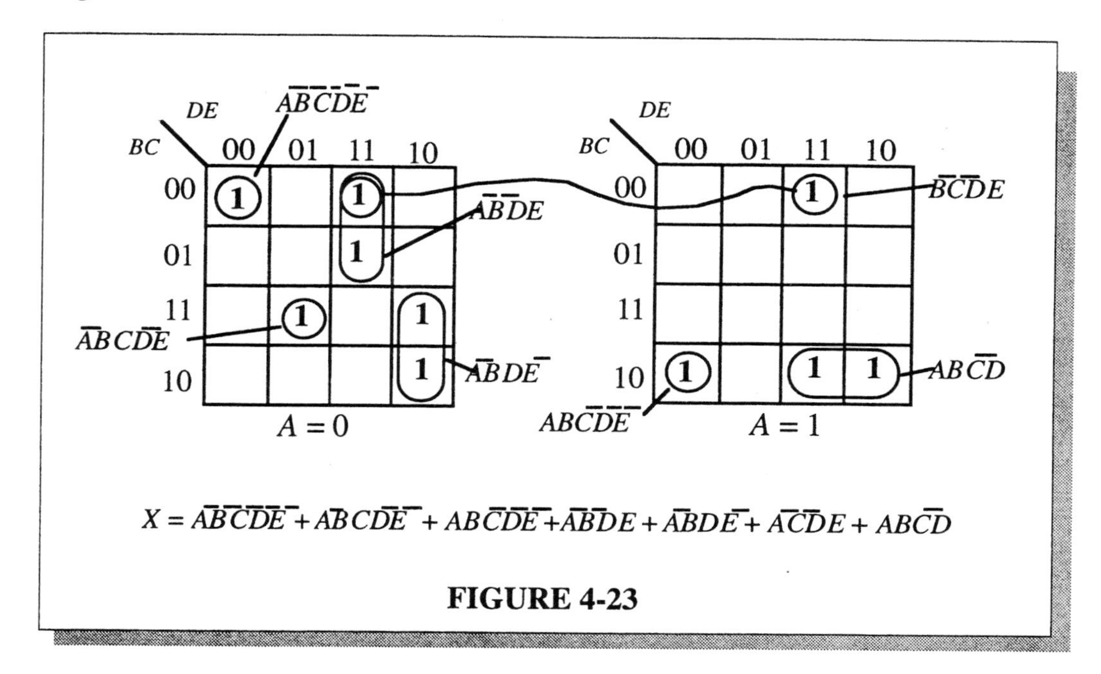

FIGURE 4-23

52. See Figure 4-24.

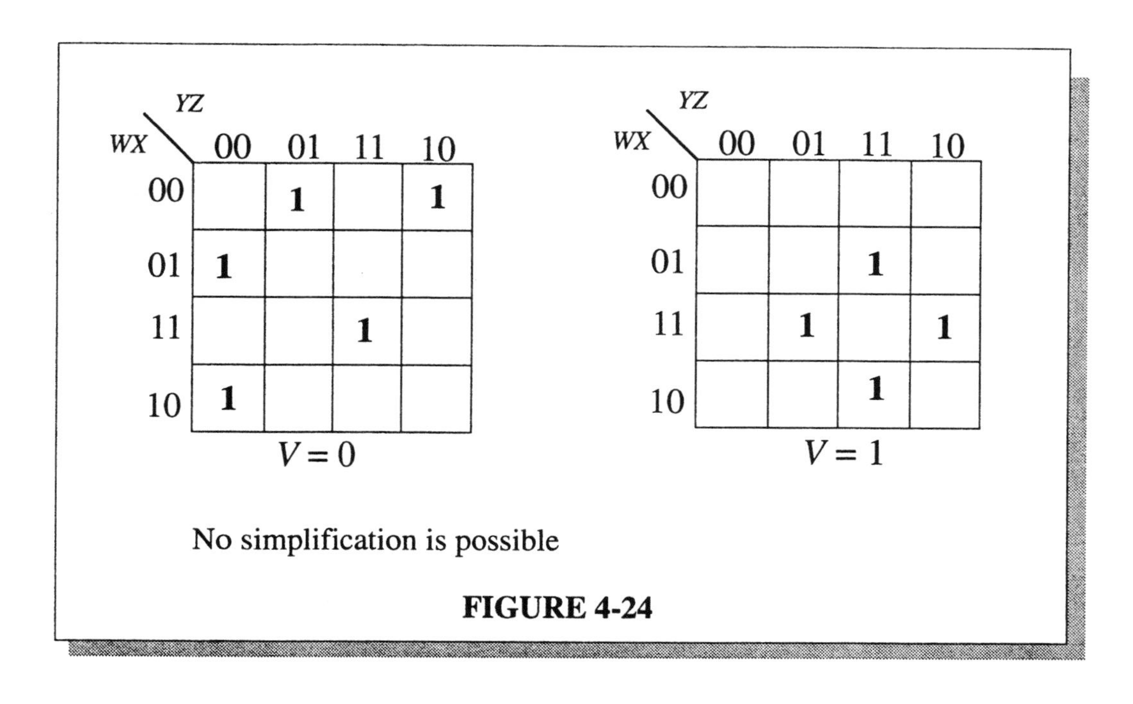

FIGURE 4-24

53. An LEDdisplay is more suitable for low-light conditions because LEDs emit light and LCDs do not.

54. The codes 1010, 1011, 1100, 1101, 1110, and 1111 correspond to non-decimal digit values and are not used in the BCD code.

55. The standard SOP expression for segment b is:

$b = \bar{D}\,\bar{C}\,\bar{B}\,\bar{A} + \bar{D}\,\bar{C}\,\bar{B}A + \bar{D}\,\bar{C}B\bar{A} + \bar{D}\,\bar{C}BA + \bar{D}C\bar{B}\,\bar{A} + \bar{D}CBA + D\bar{C}\,\bar{B}\,\bar{A} + D\bar{C}\,\bar{B}A$

This expression is minimized in Figure 4-25.

There are 4 fewer gates and the same number of inverters as a result of minimization

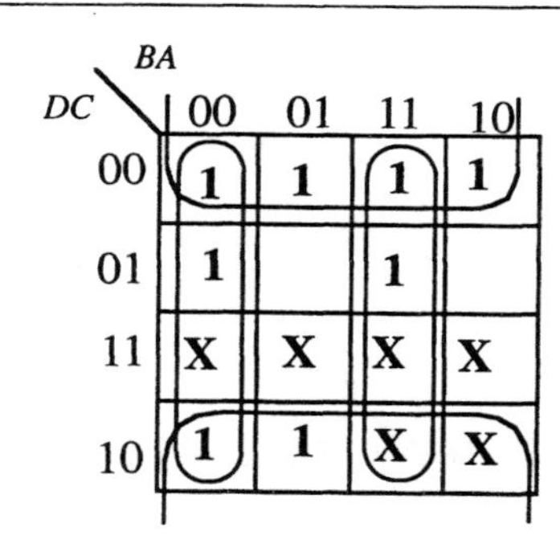

$b = \bar{C} + \bar{B}\bar{A} + BA$

The standard expression requires eight 4-input AND gates, one 8-input OR gate, and 4 inverters.

The minimum expression requires two 2-input AND gates, one 3-input OR gate, and 3 inverters.

FIGURE 4-25

56. The standard SOP expression for segment c is:

$c = \bar{D}\,\bar{C}\,\bar{B}\,\bar{A} + \bar{D}\,\bar{C}\,\bar{B}A + \bar{D}\,\bar{C}BA + \bar{D}C\,\bar{B}\,\bar{A} + D\bar{C}B\,\bar{A} + \bar{D}CB\overline{A} + \bar{D}CBA + D\bar{C}\,\bar{B}\,\bar{A} + D\bar{C}\,\bar{B}A$

This expression is minimized in Figure 4-26.

There are 9 fewer gates and 3 fewers inverters as a result of minimization

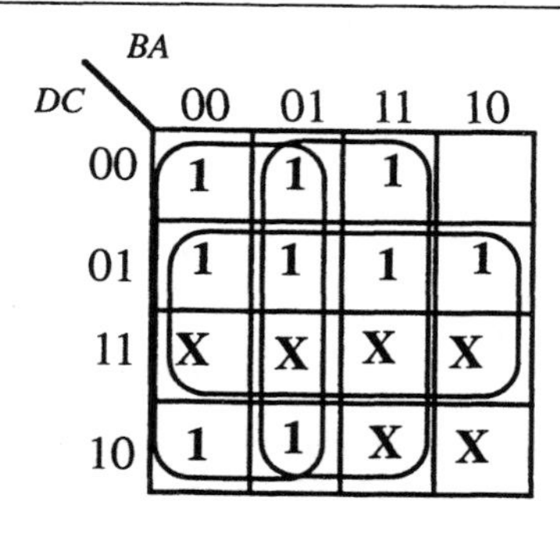

$c = A + \bar{B} + C$

The standard expression requires nine 4-input AND gates, one 9-input OR gate, and 4 inverters.

The minimum expression requires one 3-input OR gate, and 1 inverter.

FIGURE 4-26

The standard SOP expression for segment d is:

$d = \bar{D}\bar{C}\bar{B}\bar{A} + \bar{D}\bar{C}B\bar{A} + \bar{D}\bar{C}BA + \bar{D}C\bar{B}A + \bar{D}CB\bar{A} + D\bar{C}\bar{B}\bar{A} + D\bar{C}\bar{B}A$

This expression is minimized in Figure 4-27.

There are 3 fewer gates and 1 less inverter as a result of minimization

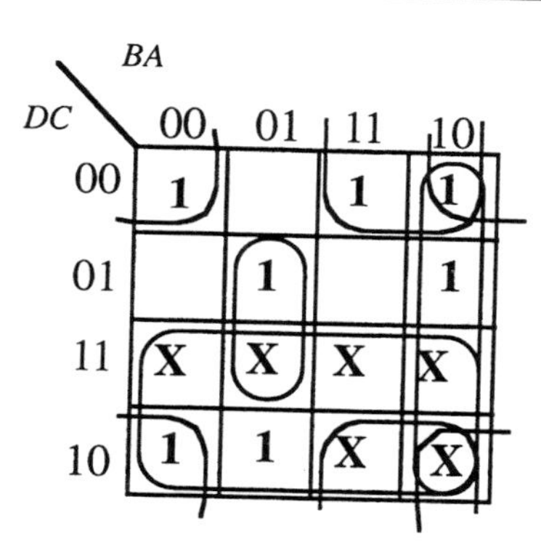

$d = D + \bar{C}\bar{A} + \bar{C}B + B\bar{A} + C\bar{B}A$

The standard expression requires seven 4-input AND gates, one 7-input OR gate, and 4 inverters.

The minimum expression requires three 2-input AND gates, one 3-input AND gate, one 5-input OR gate, and 3 inverters.

FIGURE 4-27

The standard SOP expression for segment e is:

$e = \bar{D}\bar{C}\bar{B}\bar{A} + \bar{D}\bar{C}B\bar{A} + \bar{D}CB\bar{A} + D\bar{C}\bar{B}\bar{A}$

This expression is minimized in Figure 4-28.

There are 2 fewer gates and 2 less inverters as a result of minimization

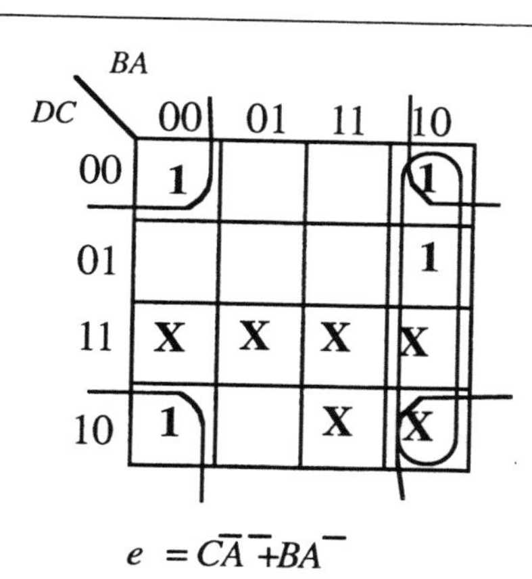

$e = \bar{C}\bar{A} + B\bar{A}$

The standard expression requires four 4-input AND gates, one 4-input OR gate, and 4 inverters.

The minimum expression requires two 2-input AND gates, one 2-input OR gate, and 2 inverters.

FIGURE 4-28

The standard SOP expression for segment f is:

$f = \bar{D}\,\bar{C}\,\bar{B}\,\bar{A} + \bar{D}C\bar{B}\,\bar{A} + \bar{D}C\bar{B}A + \bar{D}CB\overline{A} + D\bar{C}\,\bar{B}\,\bar{A} + D\bar{C}\,\bar{B}\,A$

This expression is minimized in Figure 4-29.

There are 3 fewer gates and 2 less inverters as a result of minimization.

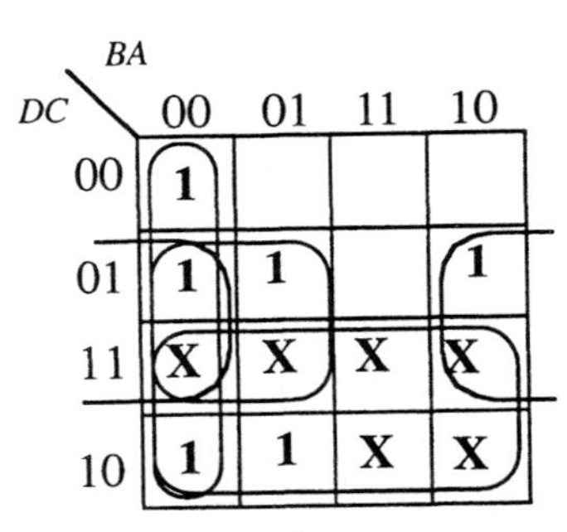

$f = C\bar{A} + B\bar{A} + C\bar{B} + D$

The standard expression requires six 4-input AND gates, one 6-input OR gate, and 4 inverters.

The minimum expression requires three 2-input AND gates, one 4-input OR gate, and 2 inverters.

FIGURE 4-29

The standard SOP expression for segment g is:

$g = \bar{D}\,\bar{C}B\bar{A} + \bar{D}\,\bar{C}BA + \bar{D}C\bar{B}\,\bar{A} + \bar{D}C\bar{B}A + \bar{D}CB\overline{A} + D\bar{C}\,\bar{B}\,\bar{A} + D\bar{C}\,\bar{B}\,A$

This expression is minimized in Figure 4-30.

There are 4 fewer gates and 1 less inverter as a result of minimization.

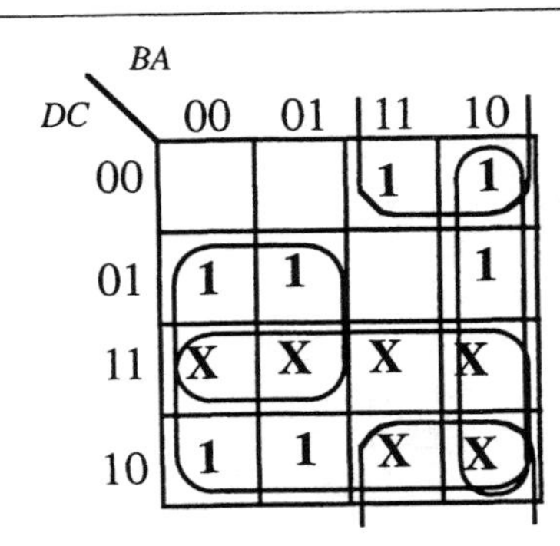

$g = \bar{C}B + B\bar{A} + C\bar{B} + D$

The standard expression requires seven 4-input AND gates, one 7-input OR gate, and 4 inverters.

The minimum expression requires three 2-input AND gates, one 4-input OR gate, and 3 inverters.

FIGURE 4-30

57. Connect the OR gate output for each segment to an inverter and then use the inverter output to drive the segment.

58. See Figure 4-31. The POS implementation requires one 3-input OR gate, one 4-input OR gate, one 2-input AND gate, and 2 inverters. The SOP implementation (see Figure 4-47 in text) requires two 2-input AND gates, one 4-input OR gate, and 2 inverters.

DC \ BA	00	01	11	10
00	1	0	1	1
01	0	1	1	1
11	X	X	X	X
10	1	1	X	X

$a = (\bar{C} + B + A)(D + C + B + \bar{A})$

FIGURE 4-31

59. See Figure 4-32. The POS implementation of segment b requires two 3-input OR gates, one 2-input AND gate, and 3 inverters.

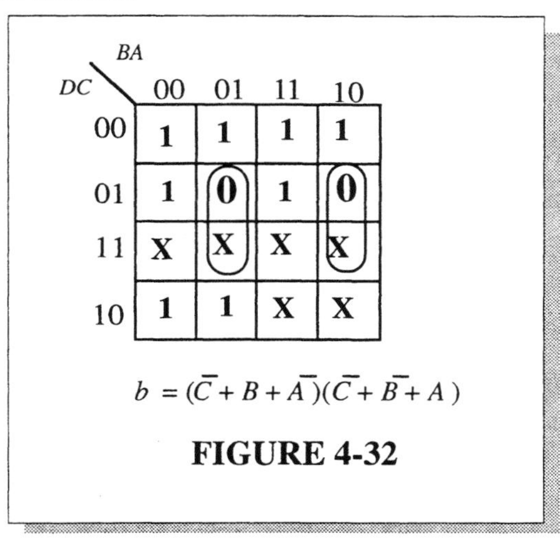

$b = (\bar{C} + B + \bar{A})(\bar{C} + \bar{B} + A)$

FIGURE 4-32

See Figure 4-33. The POS implementation of segment *c* requires one 3-input OR gate, and 1 inverters.

DC \ BA	00	01	11	10
00	1	1	1	0
01	1	1	1	1
11	XX		X	X
10	1	1	X	X

$c = C + \bar{B} + A$

FIGURE 4-33

See Figure 4-34. The POS implementation of segment d requires one 4-input OR gate, two 3-input OR gates, one 3-input AND gate, and 3 inverters.

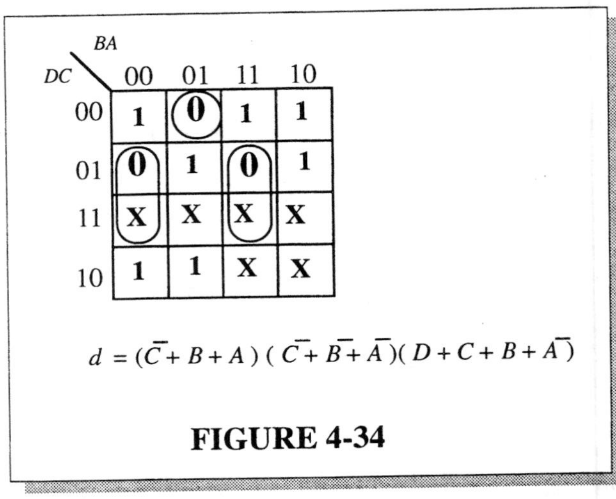

$$d = (\overline{C} + B + A)(\overline{C} + \overline{B} + \overline{A})(D + C + B + \overline{A})$$

FIGURE 4-34

See Figure 4-35. The POS implementation of segment e requires one 2-input OR gate, one 2-input AND gate, and 2 inverters.

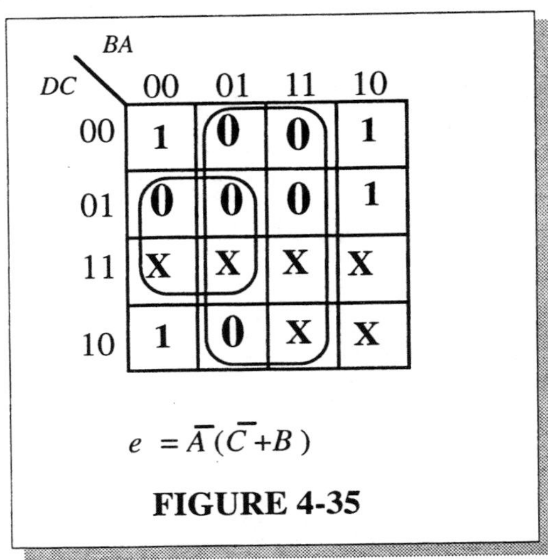

$$e = \overline{A}(\overline{C} + B)$$

FIGURE 4-35

See Figure 4-36. The POS implementation of segment f requires three 2-input OR gates, one 3-input AND gate, and 2 inverters.

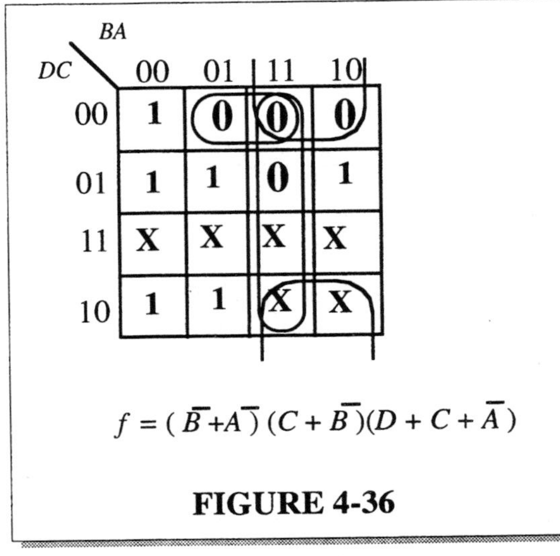

$$f = (\overline{B} + \overline{A})(C + \overline{B})(D + C + \overline{A})$$

FIGURE 4-36

See Figure 4-37. The POS implementation of segment g requires two 3-input OR gates, one 2-input AND gate, and 3 inverters.

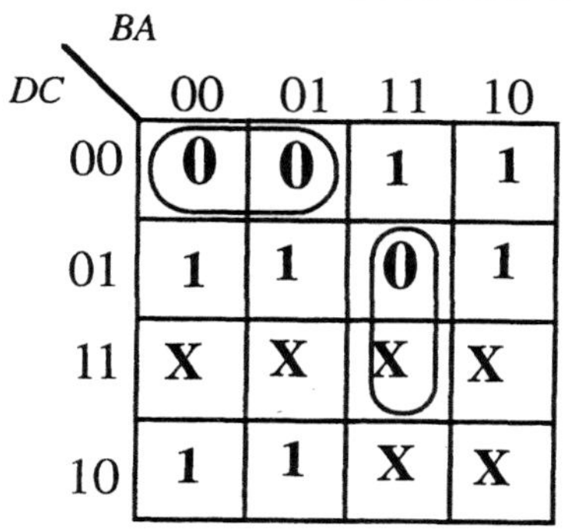

DC \ BA	00	01	11	10
00	0	0	1	1
01	1	1	0	1
11	X	X	X	X
10	1	1	X	X

$$g = (D + C + B)(\overline{C} + \overline{B} + \overline{A})$$

FIGURE 4-37

60 For the SOP implementation of the 7-segment decoding logic:

4 inverters: 1 7404
15 2-input AND gates: 4 7408s with 1 spare gate
1 3-input AND gate: 1 7411 with 2 spare gates
2 3-input OR gates: 4 7432 gates
3 4-input OR gates: 9 7432 gates
1 5 input OR gate: 4 7432 gates
1 2-input OR gate: 1 7432 gate
Total ICs for the SOP: one 7404, four 7408s, one 7411, and five 7432s

For the POS implementation of the 7-segment decoding logic:

4 inverters: 1 7404
3 2-input AND gates: 1 7408s with 1 spare gate
2 3-input AND gate: 1 7411 with 1 spare gates
10 3-input OR gates: 20 7432 gates
1 4-input OR gates: 3 7432 gates
3 2-input OR gate: 3 7432 gates
Total ICs for the POS: one 7404, one 7408s, one 7411, and seven 7432s

CHAPTER 5

COMBINATIONAL LOGIC

1. See Figure 5-1.

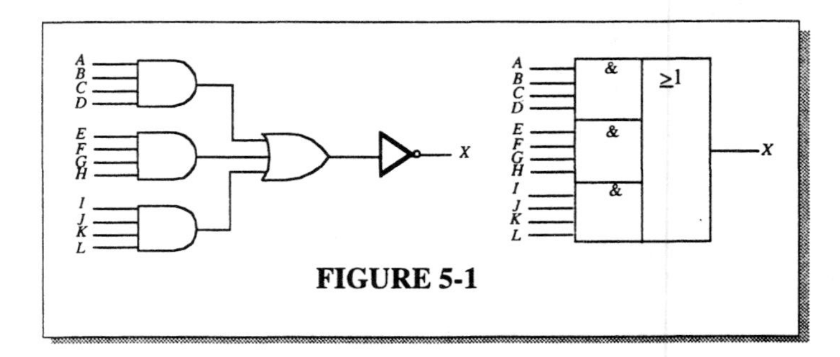

FIGURE 5-1

2.

(a) $X = \overline{A}B + \overline{A} + AC$

(b) $X = \overline{\overline{A}B + \overline{A}CD + DB\overline{D}}$

3.

(a) $X = ABB$

(b) $X = AB + B$

(c) $X = \overline{A} + B$

(d) $X = (A + B) + AB$

(e) $X = \overline{\overline{\overline{AB}}\,\overline{B}\,C}$

(f) $X = (A + B)(\overline{B} + C)$

4. See Figure 5-2 for the circuit corresponding to each expression.

(a) $X = (A + B)(C + D) = AC + AD + BC + BD$

(b) $X = \overline{\overline{\overline{AB}C} + \overline{CD}} = (\overline{AB}C)(CD) = (\bar{A} + \bar{B})CCD = \bar{A}CD + \bar{B}CD$

(c) $X = (AB + C)D + E = ABD + CD + E$

(d) $X = \overline{(\bar{A} + B)(\overline{BC})} + D = \overline{(\bar{A} + B)}\,\overline{(\overline{BC})} + D = \bar{A} + B + BC + D = \bar{A} + B + D$

(e) $X = (\overline{\overline{\overline{AB}} + \bar{C}})D + \bar{E} = (AB + \bar{C})D + \bar{E} = ABD + \bar{C}D + \bar{E}$

(f) $X = \overline{(\overline{\overline{AB}} + \overline{\overline{CD}})(\overline{\overline{EF}} + \overline{\overline{GH}})} = \overline{(AB + CD)(EF + GH)} = \overline{(AB + CD)} + \overline{(EF + GH)} = (\overline{AB})(\overline{CD}) + (\overline{EF})(\overline{GH})$
$(\bar{A} + \bar{B})(\bar{C} + \bar{D}) + (\bar{E} + \bar{F})(\bar{G} + \bar{H}) = \bar{A}\,\bar{C} + \bar{B}\,\bar{C} + \bar{A}\,\bar{D} + \bar{B}\,\bar{D} + \bar{E}\,\bar{G} + \bar{F}\,\bar{G} + \bar{E}\,\bar{H} + \bar{F}\,\bar{H}$

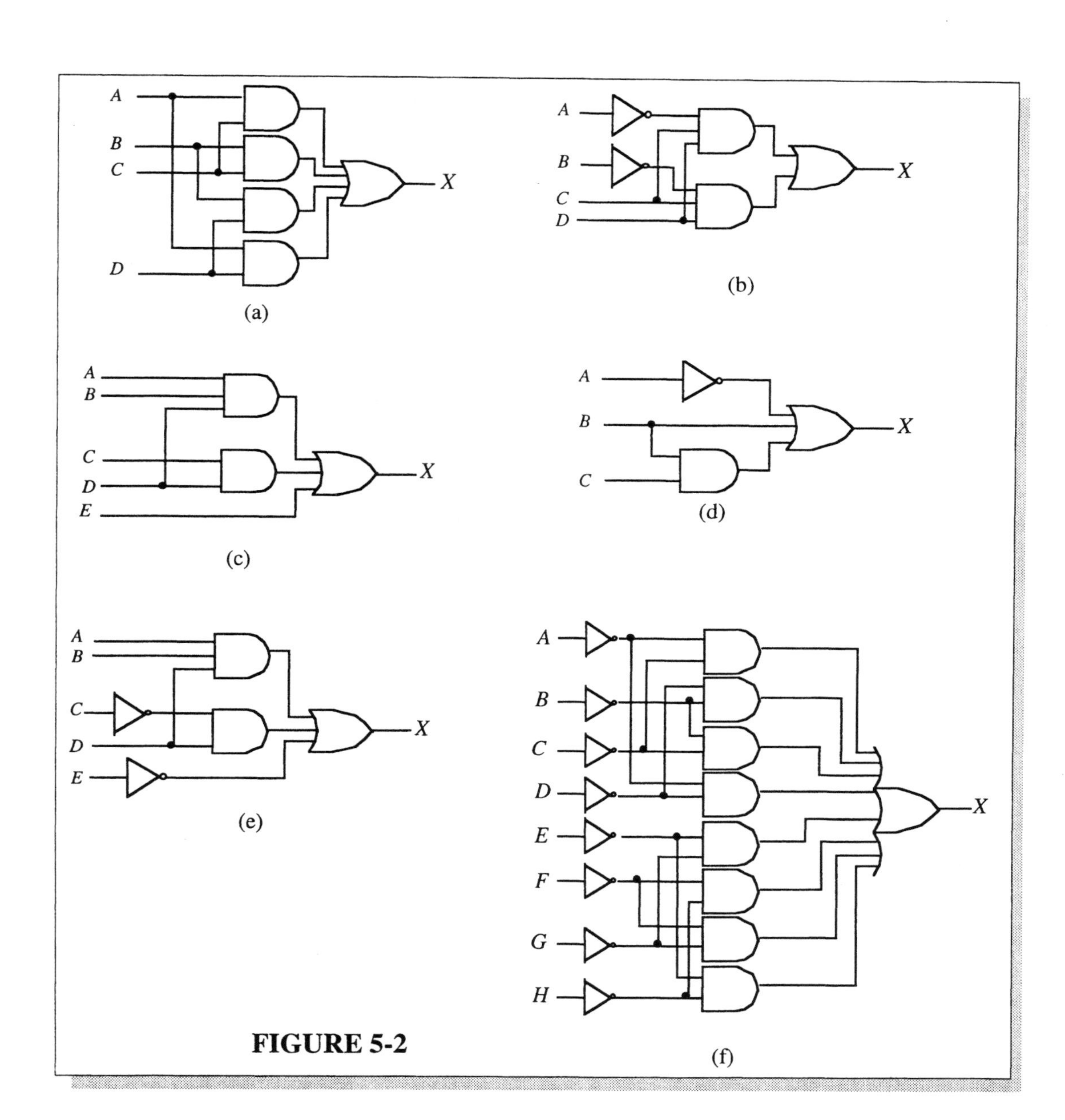

FIGURE 5-2

5. (a) $X = ABB$

A	B	X
0	0	0
0	1	0
1	0	0
1	1	1

(b) $X = AB + B$

A	B	X
0	0	0
0	1	1
1	0	0
1	1	1

(c) $X = \overline{A} + B$

A	B	X
0	0	1
0	1	1
1	0	0
1	1	1

(d) $X = (A + B) + AB$

A	B	X
0	0	0
0	1	1
1	0	1
1	1	1

(e) $X = \overline{\overline{\overline{AB}}\,\overline{B}C}$

A	B	C	X
0	0	0	0
0	0	1	0
0	1	0	0
0	1	1	0
1	0	0	0
1	0	1	0
1	1	0	0
1	1	1	0

(f) $X = (A + B)(\overline{B} + C)$

A	B	C	X
0	0	0	0
0	0	1	0
0	1	0	0
0	1	1	1
1	0	0	1
1	0	1	1
1	1	0	0
1	1	1	1

6.

(a) $X = (A + B)(C + D)$

A	B	C	D	X
0	0	0	0	0
0	0	0	1	0
0	0	1	0	0
0	0	1	1	0
0	1	0	0	0
0	1	0	1	1
0	1	1	0	1
0	1	1	1	1
1	0	0	0	0
1	0	0	1	1
1	0	1	0	1
1	0	1	1	1
1	1	0	0	0
1	1	0	1	1
1	1	1	0	1
1	1	1	1	1

(b) $X = \overline{\overline{\overline{AB}\,C} + \overline{CD}}$

A	B	C	D	X
0	0	0	0	0
0	0	0	1	0
0	0	1	0	0
0	0	1	1	1
0	1	0	0	0
0	1	0	1	0
0	1	1	0	0
0	1	1	1	1
1	0	0	0	0
1	0	0	1	0
1	0	1	0	0
1	0	1	1	1
1	1	0	0	0
1	1	0	1	0
1	1	1	0	0
1	1	1	1	0

(c) $X = (AB + C)D + E$

A	B	C	D	E	X	A	B	C	D	E	X
0	0	0	0	0	0	1	0	0	0	0	0
0	0	0	0	1	1	1	0	0	0	1	1
0	0	0	1	0	0	1	0	0	1	0	0
0	0	0	1	1	1	1	0	0	1	1	1
0	0	1	0	0	0	1	0	1	0	0	0
0	0	1	0	1	1	1	0	1	0	1	1
0	0	1	1	0	1	1	0	1	1	0	1
0	0	1	1	1	1	1	0	1	1	1	1
0	1	0	0	0	0	1	1	0	0	0	0
0	1	0	0	1	1	1	1	0	0	1	1
0	1	0	1	0	0	1	1	0	1	0	1
0	1	0	1	1	1	1	1	0	1	1	1
0	1	1	0	0	0	1	1	1	0	0	0
0	1	1	0	1	1	1	1	1	0	1	1
0	1	1	1	0	1	1	1	1	1	0	1
0	1	1	1	1	1	1	1	1	1	1	1

(d) $X = \overline{\overline{(\overline{A} + B)(\overline{BC})}} + D$

A	B	C	D	X
0	0	0	0	1
0	0	0	1	1
0	0	1	0	1
0	0	1	1	1
0	1	0	0	1
0	1	0	1	1
0	1	1	0	1
0	1	1	1	1
1	0	0	0	0
1	0	0	1	1
1	0	1	0	0
1	0	1	1	1
1	1	0	0	1
1	1	0	1	1
1	1	1	0	1
1	1	1	1	1

(e) $X = \overline{\overline{(\overline{AB} + \overline{C})D}} + \overline{E}$

A	B	C	D	E	X	A	B	C	D	E	X
0	0	0	0	0	1	1	0	0	0	0	1
0	0	0	0	1	0	1	0	0	0	1	0
0	0	0	1	0	1	1	0	0	1	0	1
0	0	0	1	1	1	1	0	0	1	1	1
0	0	1	0	0	1	1	0	1	0	0	1
0	0	1	0	1	0	1	0	1	0	1	0
0	0	1	1	0	1	1	0	1	1	0	1
0	0	1	1	1	0	1	0	1	1	1	0
0	1	0	0	0	1	1	1	0	0	0	1
0	1	0	0	1	0	1	1	0	0	1	0
0	1	0	1	0	1	1	1	0	1	0	1
0	1	0	1	1	1	1	1	0	1	1	1
0	1	1	0	0	1	1	1	1	0	0	1
0	1	1	0	1	0	1	1	1	0	1	0
0	1	1	1	0	1	1	1	1	1	0	1
0	1	1	1	1	0	1	1	1	1	1	1

(f) $X = (\overline{\overline{AB}} + \overline{\overline{CD}})(\overline{\overline{EF}} + \overline{\overline{GH}})$

A	B	C	D	E	F	G	H	X
0	X	0	X	X	X	X	X	1
X	0	0	X	X	X	X	X	1
0	X	X	0	X	X	X	X	1
X	0	X	0	0	X	X	X	1
X	X	X	X	0	X	0	X	1
X	X	X	X	X	0	0	X	1
X	X	X	X	0	X	X	0	1
X	X	X	X	X	0	X	0	1

For all other entries $X = 0$

X = *don't care*

An abbreviated table is shown because there are 256 combinations

7. $X = \overline{A\overline{B} + \overline{A}B} = \overline{(A\overline{B})}\,\overline{(\overline{A}B)} = (\overline{A} + B)(A + \overline{B})$

8. See Figure 5-3.

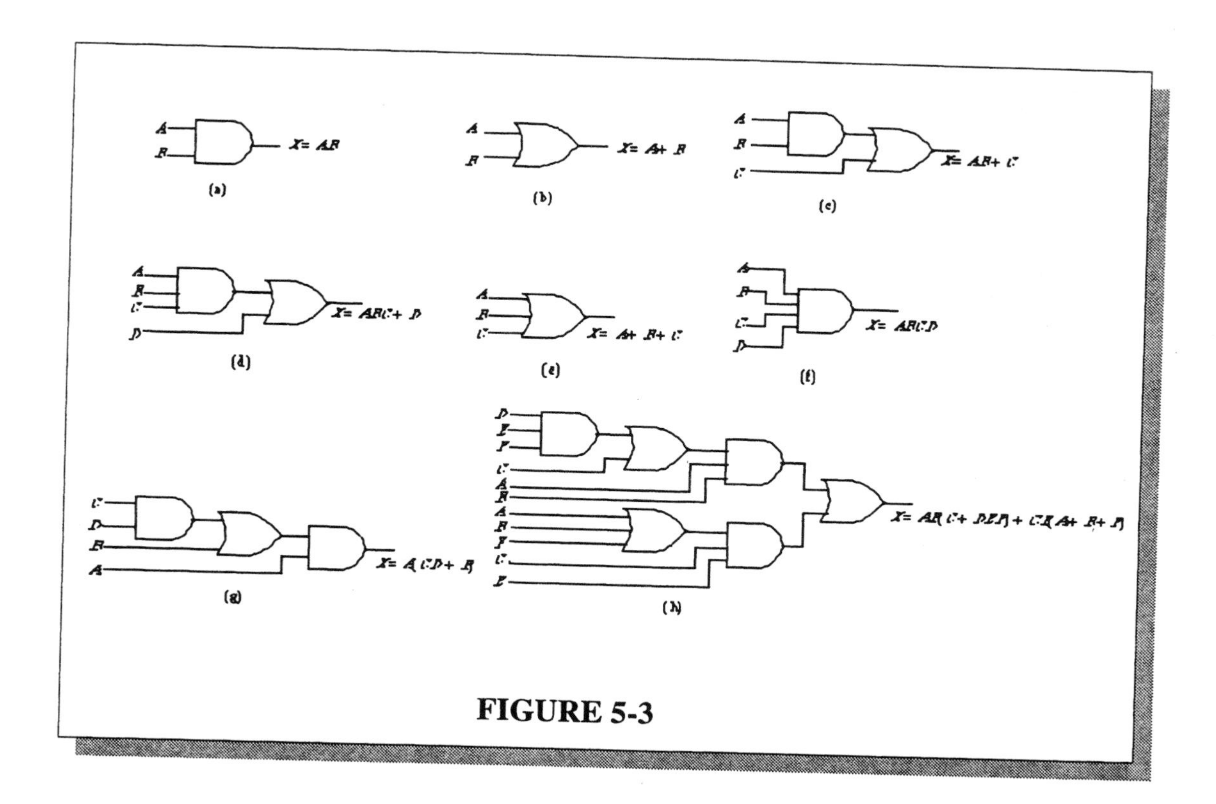

FIGURE 5-3

9. See Figure 5-4.

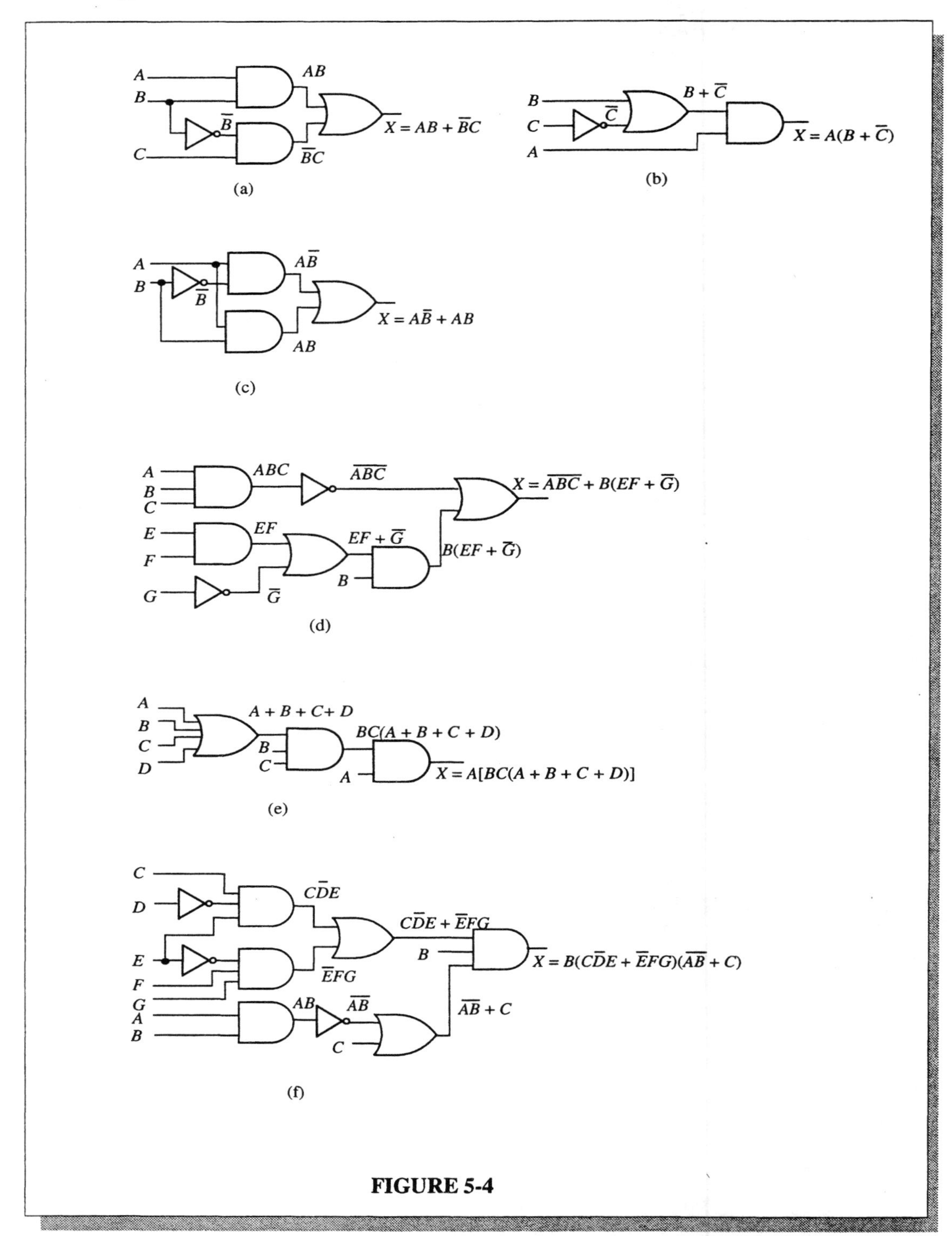

FIGURE 5-4

10. See Figure 5-5.

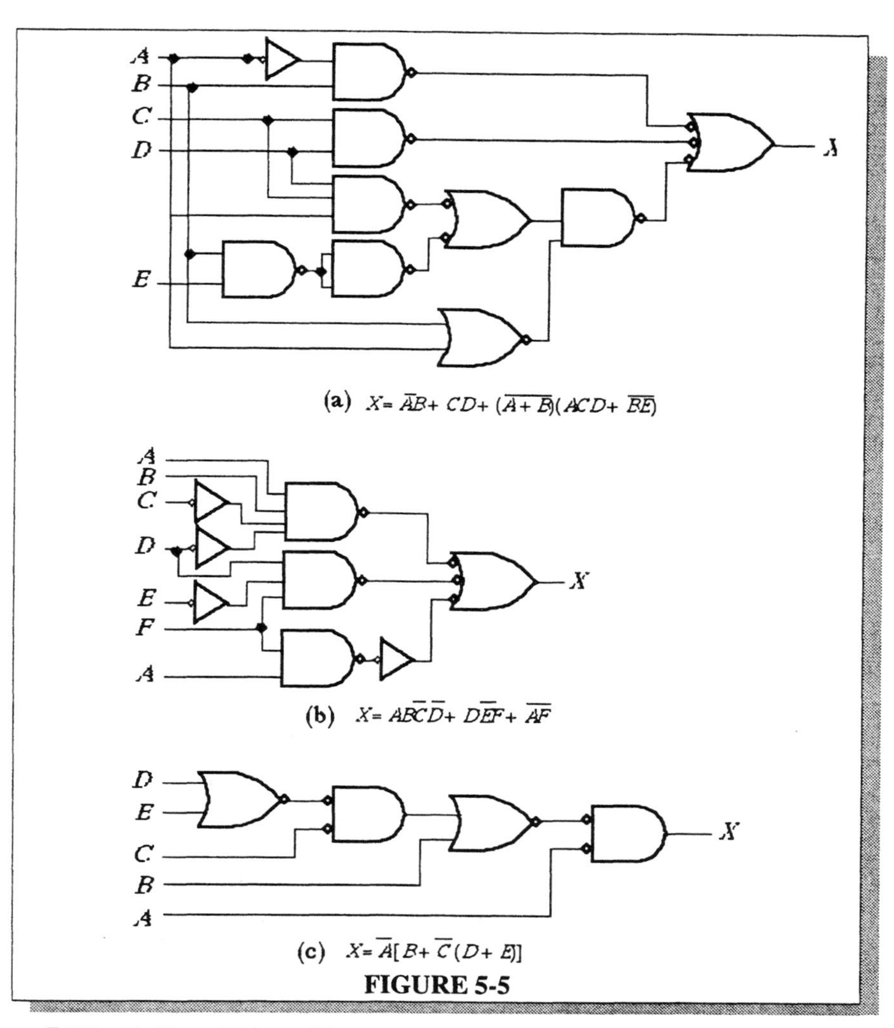

(a) $X = \overline{A}B + CD + (\overline{A + B})(ACD + \overline{BE})$

(b) $X = AB\overline{C}\,\overline{D} + D\overline{E}F + \overline{AF}$

(c) $X = \overline{A}[B + \overline{C}(D + E)]$

FIGURE 5-5

11. $X = \overline{A}\,\overline{B}\,\overline{C} + \overline{A}B\overline{C} + A\overline{B}\,\overline{C} + AB\overline{C} + ABC$

See Figure 5-6.

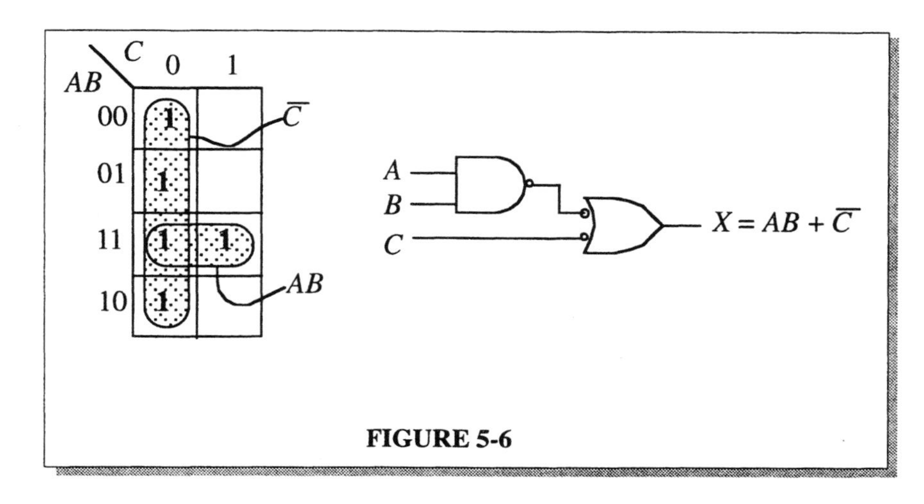

FIGURE 5-6

12. $X = \overline{A}\,\overline{B}C\overline{D} + \overline{A}\,\overline{B}CD + \overline{A}B\overline{C}\,\overline{D} + A\overline{B}\,\overline{C}\,\overline{D} + A\overline{B}\,\overline{C}D + A\overline{B}C\overline{D} + A\overline{B}CD + ABCD$

See Figure 5-7.

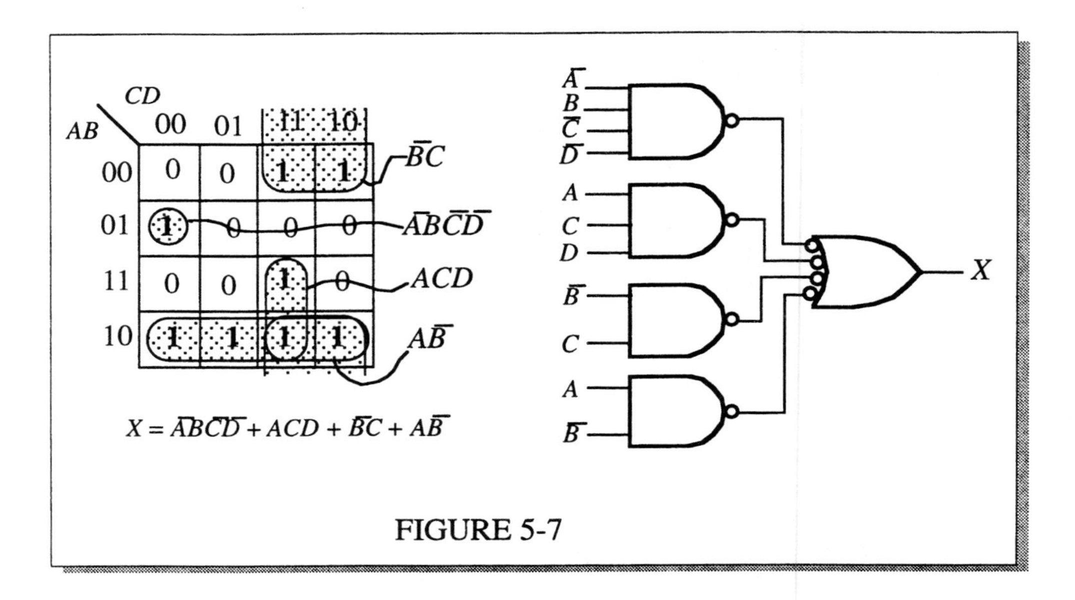

FIGURE 5-7

13. $X = AB + ABC = AB(1 + C) = \mathbf{AB}$

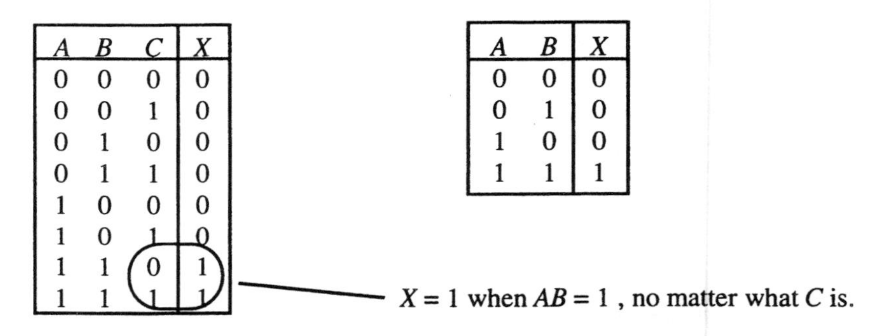

A	B	C	X
0	0	0	0
0	0	1	0
0	1	0	0
0	1	1	0
1	0	0	0
1	0	1	0
1	1	0	1
1	1	1	1

A	B	X
0	0	0
0	1	0
1	0	0
1	1	1

$X = 1$ when $AB = 1$, no matter what C is.

Since C is a don't care variable, the output depends only on A and B as shown by the two-variable truth table above which is implemented with the AND gate in Figure 5-8.

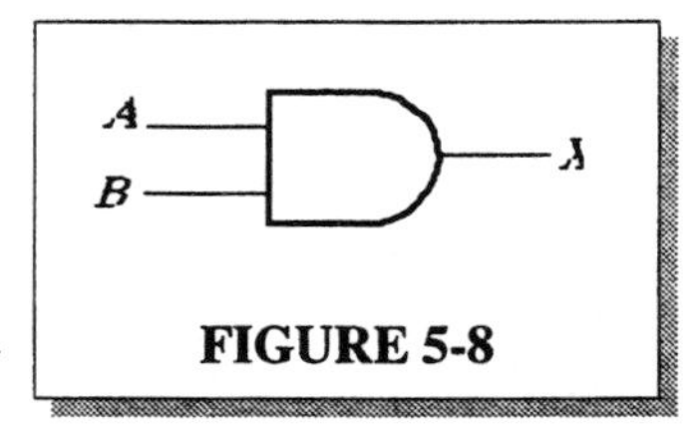

FIGURE 5-8

14. $X = \overline{\overline{(\overline{AB})(\overline{B+C})} + C} = \overline{\overline{(\overline{AB})(\overline{B+C})}}\,\overline{C} = (\overline{AB})(\overline{B+C})\overline{C} = (\overline{A} + \overline{B})(\overline{B}\,\overline{C})\overline{C}$

$= (\overline{A}\,\overline{B}\,\overline{C} + \overline{B}\,\overline{C})\overline{C} = \overline{A}\,\overline{B}\,\overline{C} + \overline{B}\,\overline{C} = \overline{B}\,\overline{C}(A + 1) = \mathbf{\overline{B}\,\overline{C}}$

See Figure 5-9.

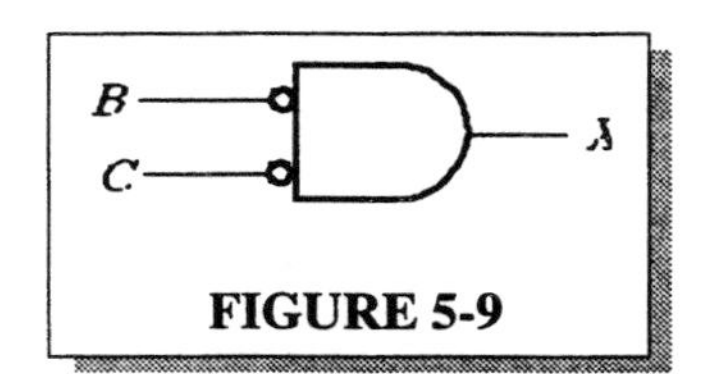

FIGURE 5-9

The output is dependent only on B and C. The value of A does not matter

A	B	C	X
0	0	0	1
0	0	1	1
0	1	0	1
0	1	1	0
1	0	0	1
1	0	1	1
1	1	0	1
1	1	1	0

15. (a) $X = AB + B\overline{C}$

No simplification. See Figure 5-10.

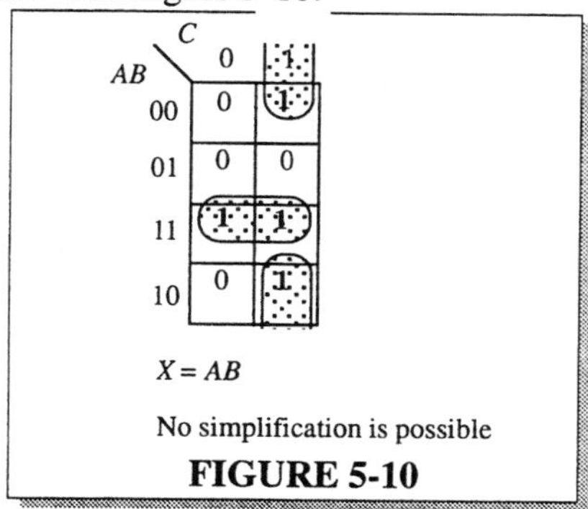

FIGURE 5-10

(b) $X = A(B + \overline{C}) = \boldsymbol{AB + A\overline{C}}$

See Figure 5-11.

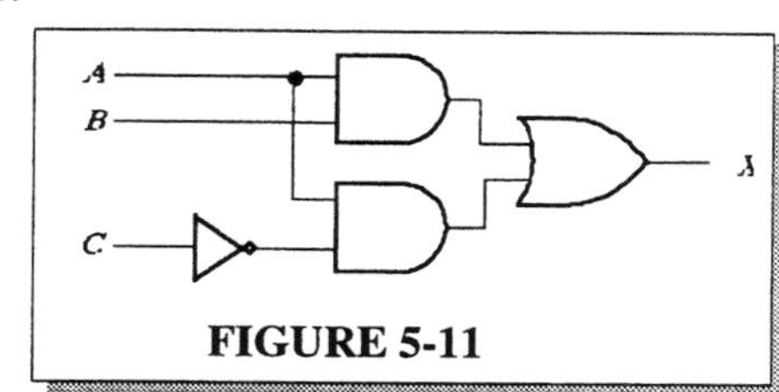

FIGURE 5-11

(c) $X = AB + A\overline{B} = A(B + \overline{B}) = \boldsymbol{A}$

A direct connection from input to output. No gates required.

(d) $X = \overline{ABC} + B(EF + \overline{G}) = \overline{A} + \overline{B} + \overline{C} + BEF + B\overline{G}$

$= \overline{A} + \overline{C} + BEF + \overline{B} + \overline{G} = \boldsymbol{\overline{A} + \overline{C} + \overline{B} + EF + \overline{G}}$

See Figure 5-12.

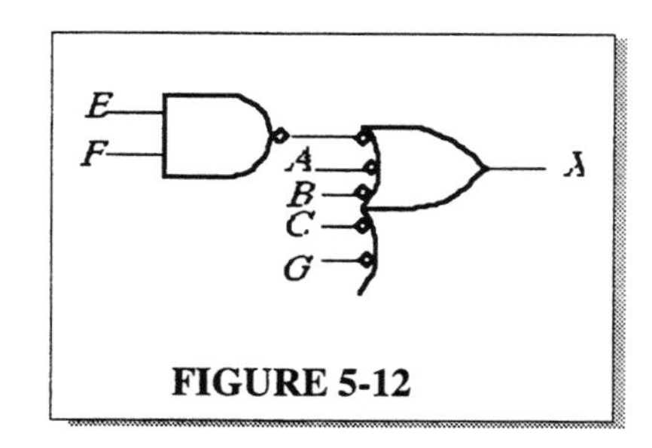

FIGURE 5-12

(e) $X = A(BC(A + B + C + D)) = ABCA + ABCB + ABCC + ABCD$

$= ABC + ABC + ABC + ABCD = ABC + ABC(1 + D)$

$= ABC + ABC = \boldsymbol{ABC}$

See Figure 5-13.

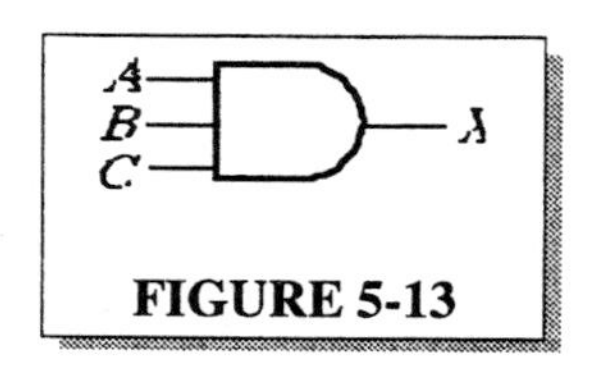

FIGURE 5-13

(f) $X = B(C\overline{D}E + \overline{E}FG)(\overline{AB} + C) = (BC\overline{D}E + B\overline{E}FG)(\overline{A} + \overline{B} + C)$

$= \overline{A}BC\overline{D}E + \overline{A}B\overline{E}FG + BC\overline{D}E + BC\overline{E}FG$

$= BC\overline{D}E(\overline{A} + 1) + \overline{A}B\overline{E}FG + BC\overline{E}FG$

$= \mathbf{BC\overline{D}E + \overline{A}B\overline{E}FG + BC\overline{E}FG}$

See Figure 5-14.

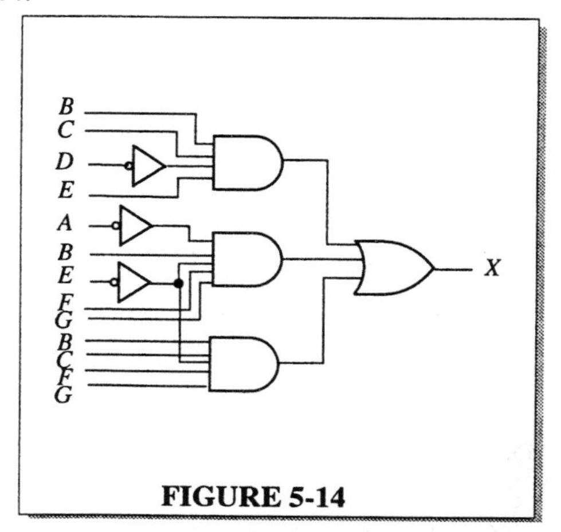

FIGURE 5-14

16. (a) $X = \overline{A}B + CD + (\overline{A + B})(ACD + \overline{BE}) = \overline{A}B + CD + \overline{A}\overline{B}(ACD + \overline{B} + \overline{E})$

$= \overline{A}B + CD + \overline{A}\overline{B} + \overline{A}\overline{B}\overline{E} = \overline{A}(B + \overline{B}) + CD + \overline{A}\overline{B}\overline{E}$

$= \overline{A} + \overline{A}\overline{B}\overline{E} + CD = \overline{A}(1 + \overline{B}\overline{E}) + CD = \mathbf{\overline{A} + CD}$

See Figure 5-15.

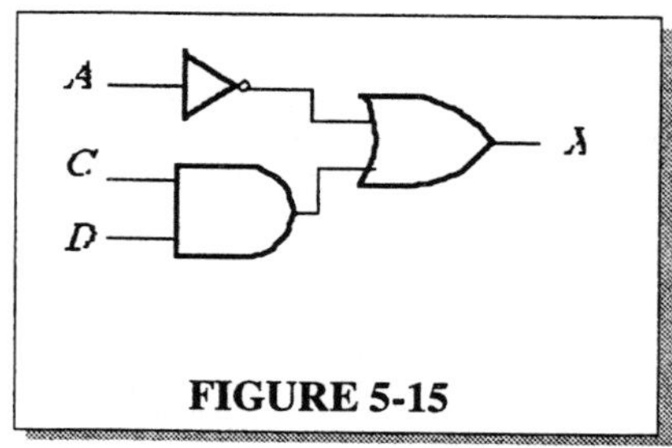

FIGURE 5-15

(b) $X = AB\overline{C}\overline{D} + D\overline{E}F + \overline{AF} = AB\overline{C}\overline{D} + D\overline{E}F + \overline{A} + \overline{F}$

$= \mathbf{\overline{A} + B\overline{C}\overline{D} + \overline{F} + D\overline{E}}$

See Figure 5-16.

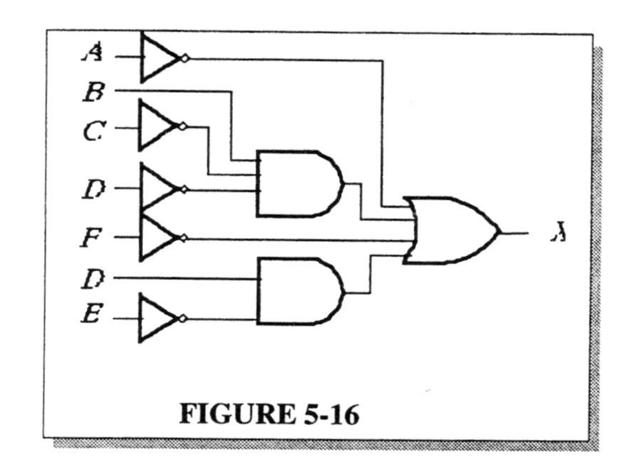

FIGURE 5-16

(c) $X = \overline{A}(B + \overline{C}(D + E)) = \overline{A}(B + \overline{C}D + \overline{C}E) = \mathbf{\overline{A}B + \overline{A}\overline{C}D + \overline{A}\overline{C}E}$

See Figure 5-17.

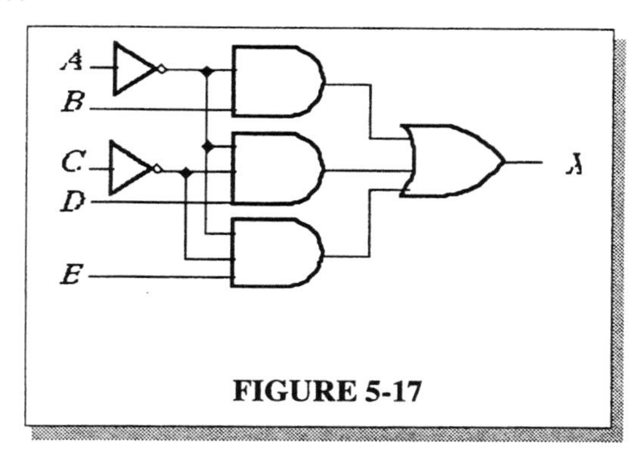

FIGURE 5-17

17. The SOP expressions are developed as follows and the resulting circuits are shown in Figure 5-18.

(a) $X = (A + B)(C + D) = AC + AD + BC + BD$

(b) $X = \overline{\overline{\overline{AB}C} + \overline{CD}} = (\overline{AB}C)(CD) = (\overline{A} + \overline{B})CCD = \overline{A}CD + \overline{B}CD$

(c) $X = (AB + C)D + E = ABD + CD + E$

(d) $X = \overline{\overline{(\overline{A} + B)}\,\overline{(\overline{BC})}} + D = \overline{\overline{(\overline{A} + B)}}\,\overline{\overline{(\overline{BC})}} + D = \overline{A} + B + BC + D$

(e) $X = \overline{\overline{(\overline{\overline{AB}} + \overline{C})D}} + \overline{E} = (AB + \overline{C})D + \overline{E} = ABD + \overline{C}D + \overline{E}$

(f) $X = (\overline{\overline{AB}} + \overline{\overline{CD}})(\overline{\overline{EF}} + \overline{\overline{GH}}) = \overline{(AB + CD)(EF + GH)} = \overline{(AB + CD)} + \overline{(EF + GHG)} = (\overline{AB})(\overline{CD}) + (\overline{EF})(\overline{GH})$
$(\overline{A} + \overline{B})(\overline{C} + \overline{D}) + (\overline{E} + \overline{F})(\overline{G} + \overline{H}) = \overline{A}\,\overline{C} + \overline{B}\,\overline{C} + \overline{A}\,\overline{D} + \overline{B}\,\overline{D} + \overline{E}\,\overline{G} + \overline{F}\,\overline{G} + \overline{E}\,\overline{H} + \overline{F}\,\overline{H}$

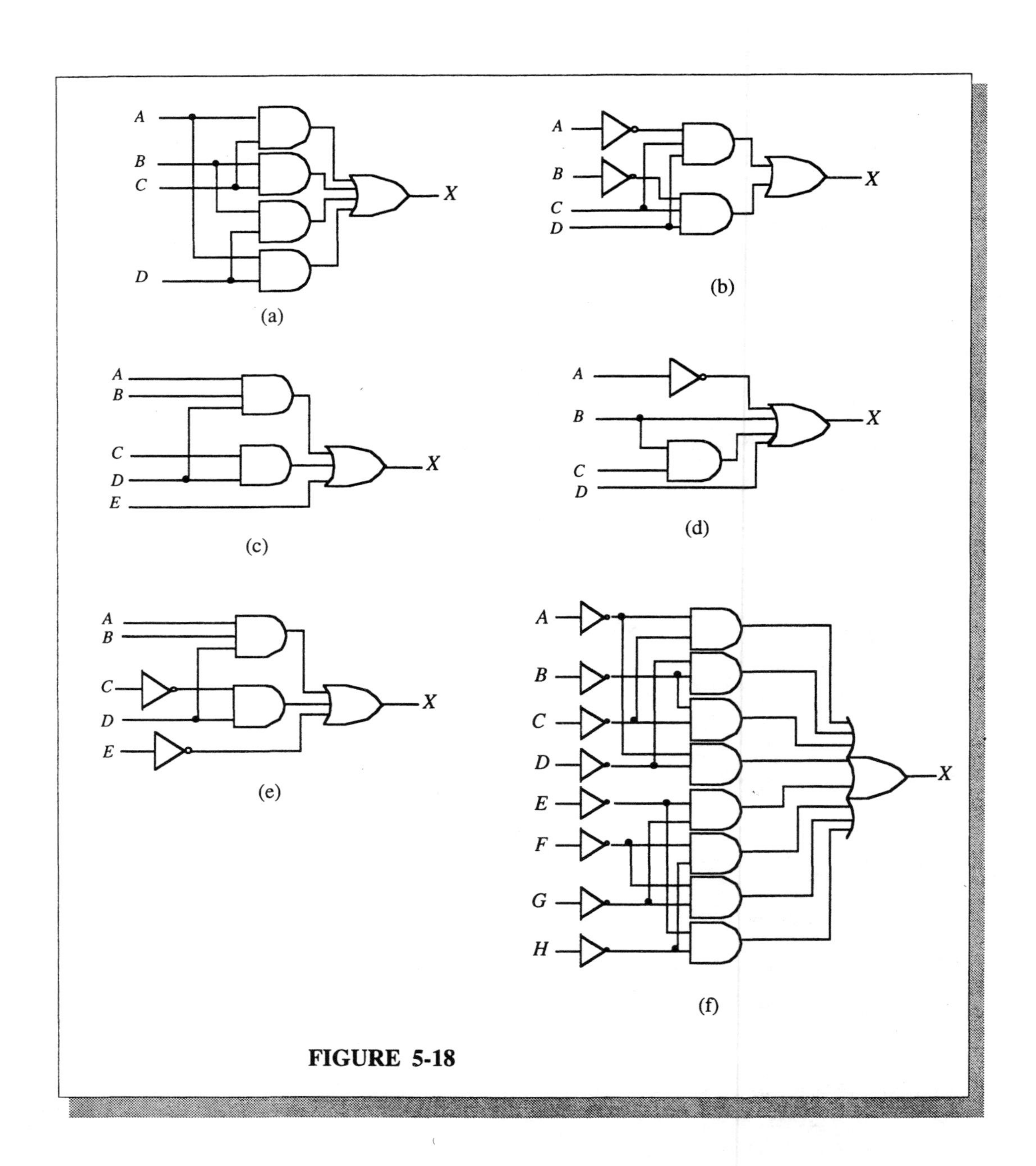

FIGURE 5-18

18. See Figure 5-19.

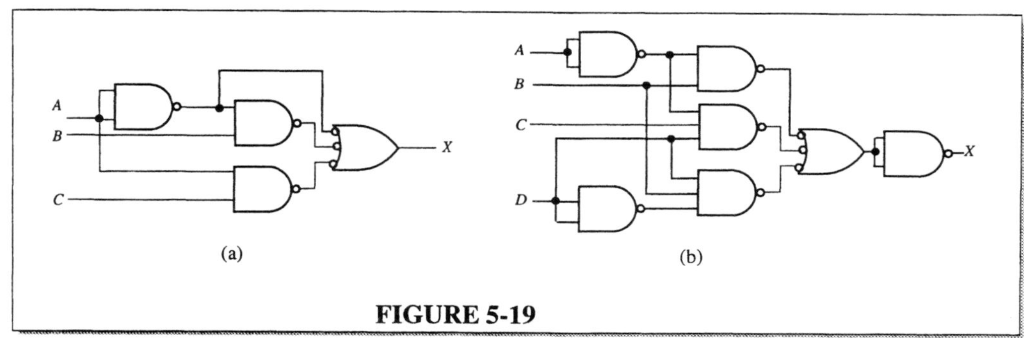

FIGURE 5-19

19. $X = \overline{\overline{(\overline{AB})(\overline{B + C})} + C}$

See Figure 5-20.

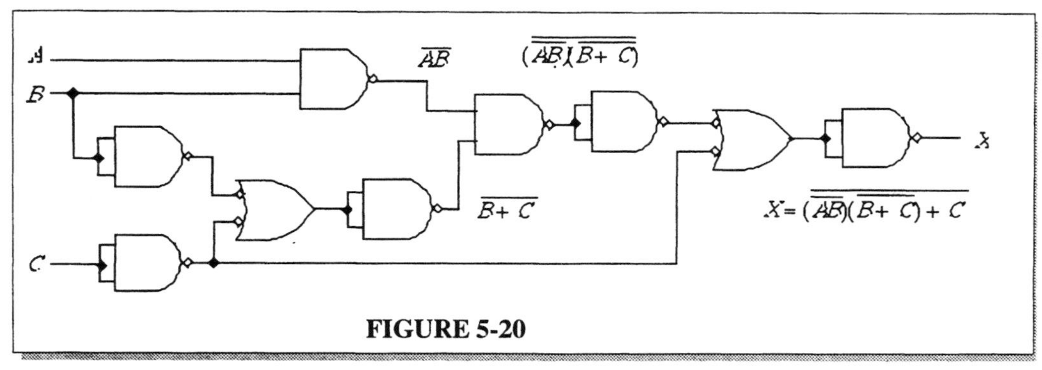

FIGURE 5-20

20. See Figure 5-21.

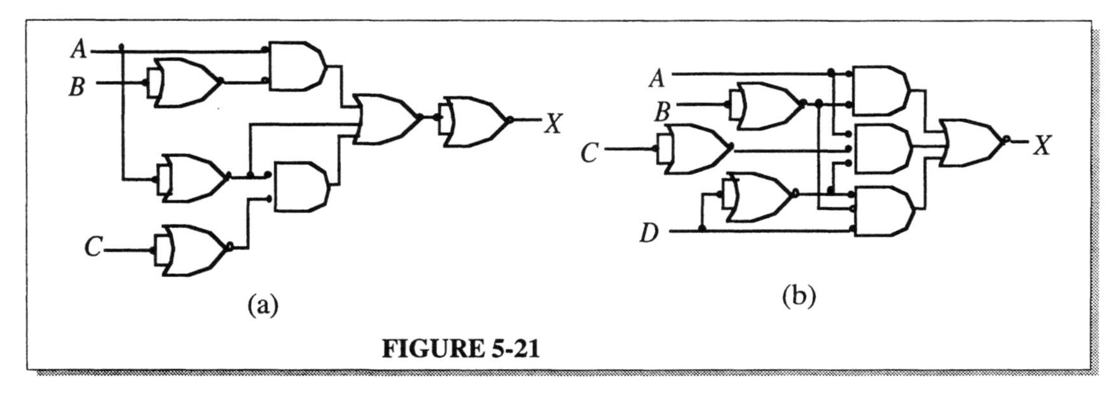

FIGURE 5-21

21. See Figure 5-22.

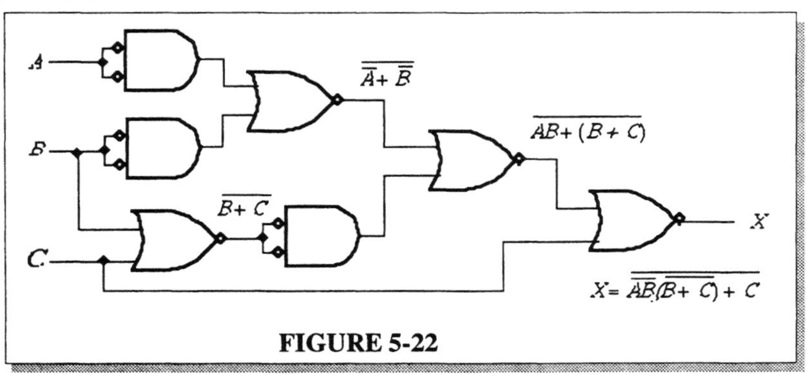

FIGURE 5-22

22. (a) $X = ABC$

See Figure 5-23.

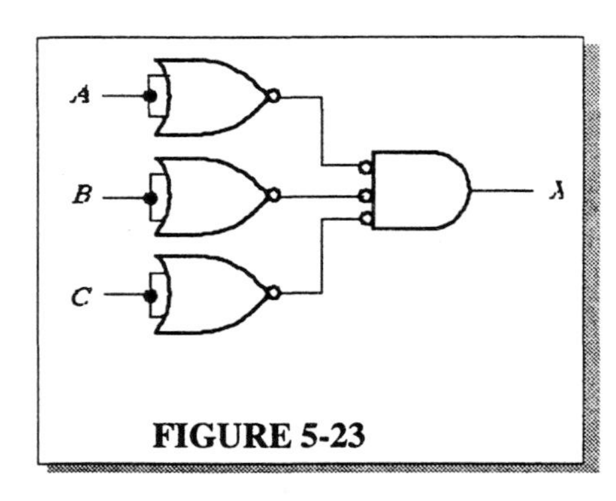

FIGURE 5-23

(b) $X = \overline{ABC}$

See Figure 5-24.

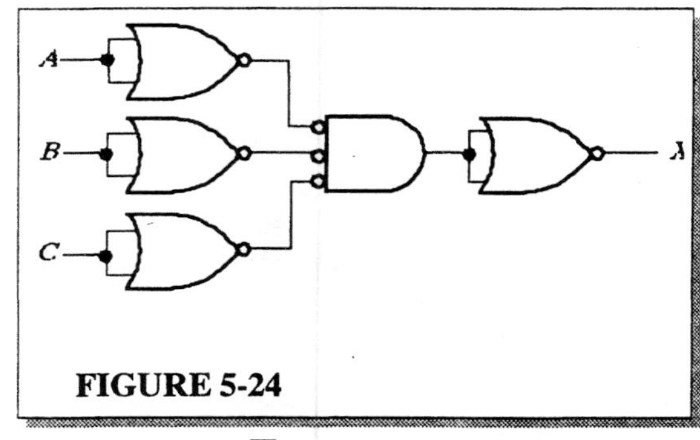

FIGURE 5-24

(c) $X = A + B$

See Figure 5-25.

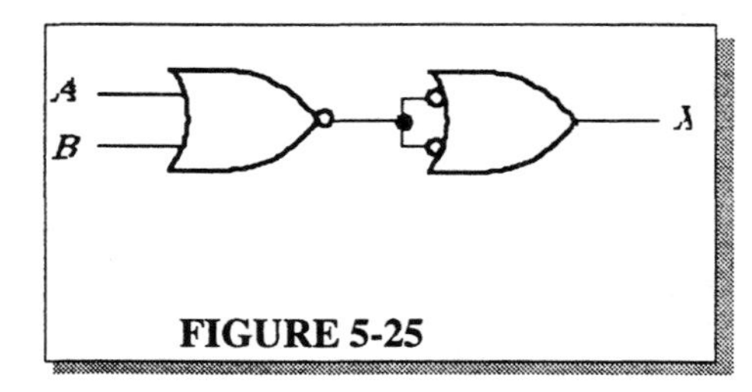

FIGURE 5-25

(d) $X = A + B + \overline{C}$

See Figure 5-26.

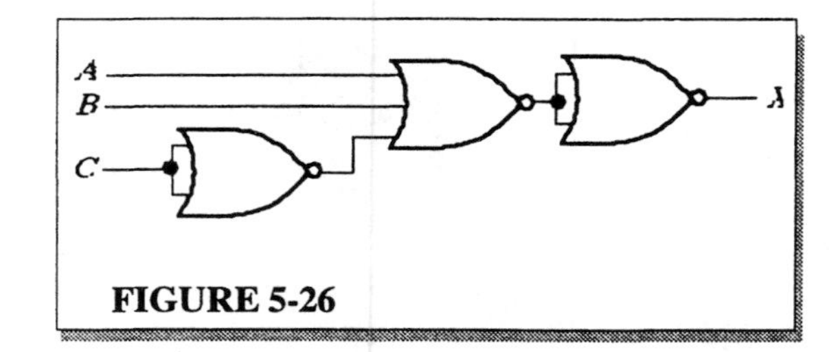

FIGURE 5-26

(e) $X = \overline{AB} + \overline{CD}$

See Figure 5-27.

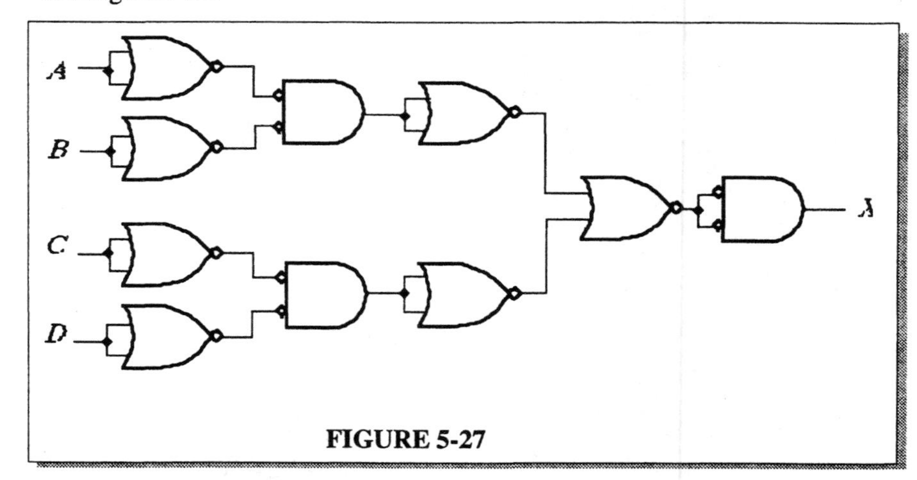

FIGURE 5-27

(f) $X = (A + B)(C + D)$

See Figure 5-28.

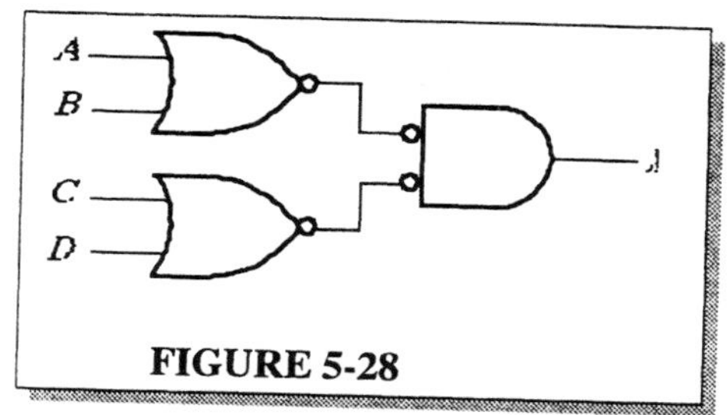

FIGURE 5-28

(g) $X = AB[C(\overline{DE} + \overline{AB}) + \overline{BCE}]$

See Figure 5-29.

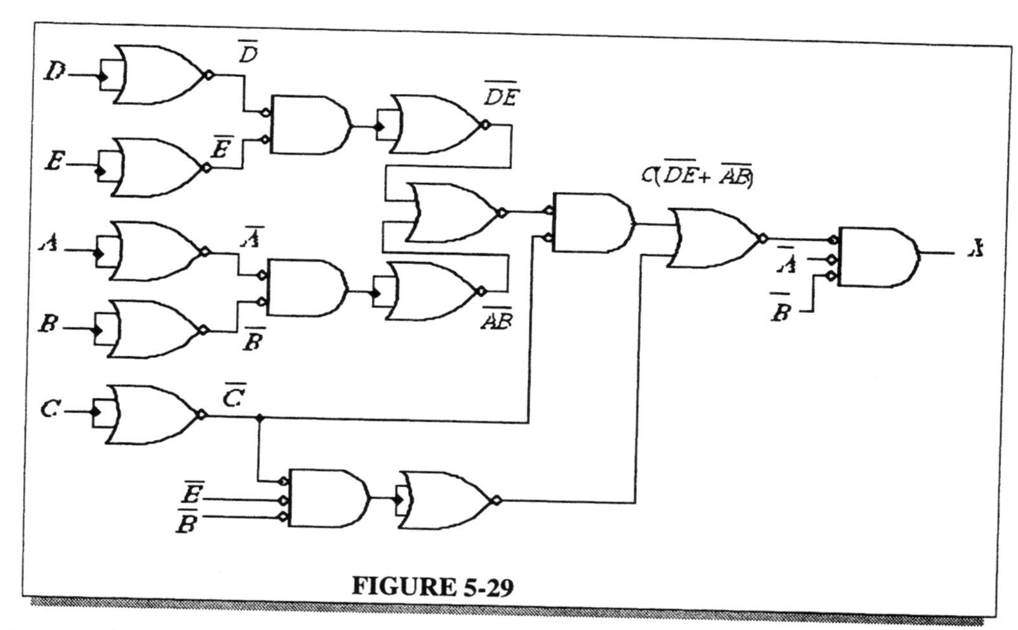

FIGURE 5-29

23. (a) $X = ABC$

See Figure 5-30.

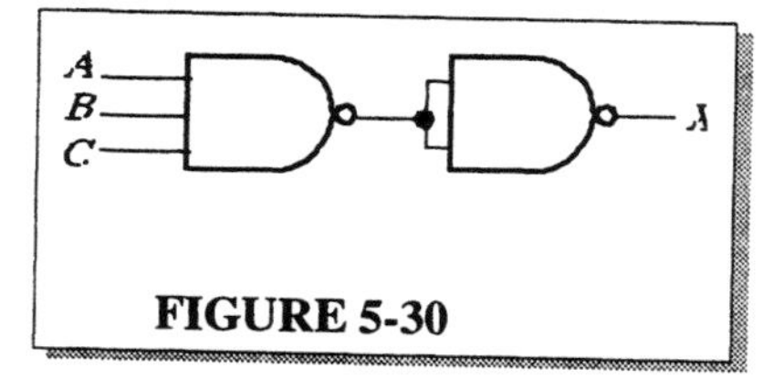

FIGURE 5-30

(b) $X = \overline{ABC}$

See Figure 5-31.

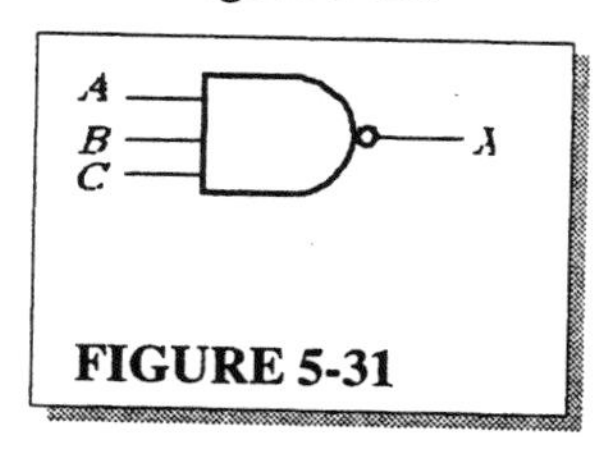

FIGURE 5-31

(c) $X = A + B$

See Figure 5-32.

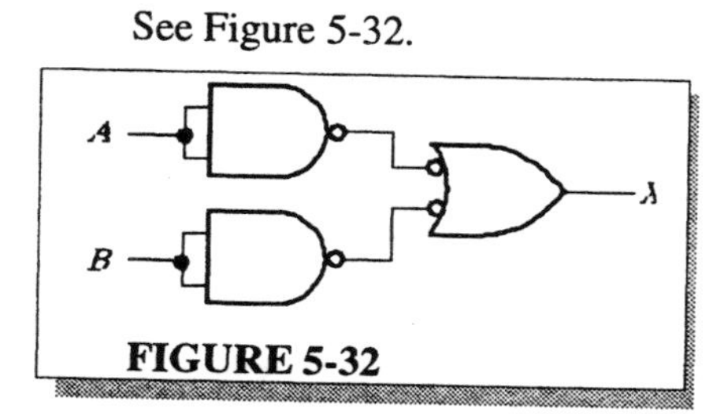

FIGURE 5-32

(d) $X = A + B + \overline{C}$

See Figure 5-33.

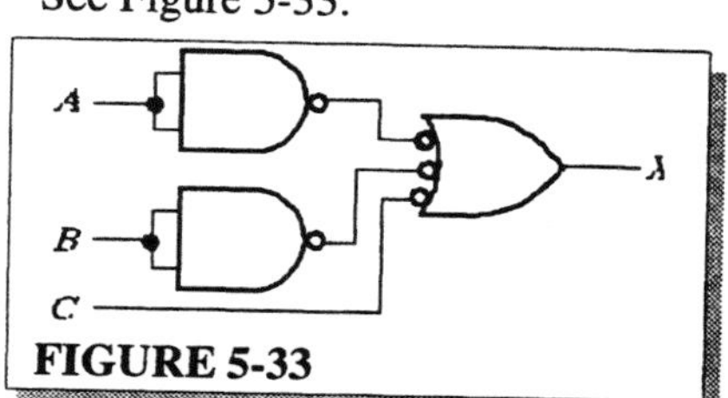

FIGURE 5-33

(e) $X = \overline{AB} + \overline{CD}$

See Figure 5-34.

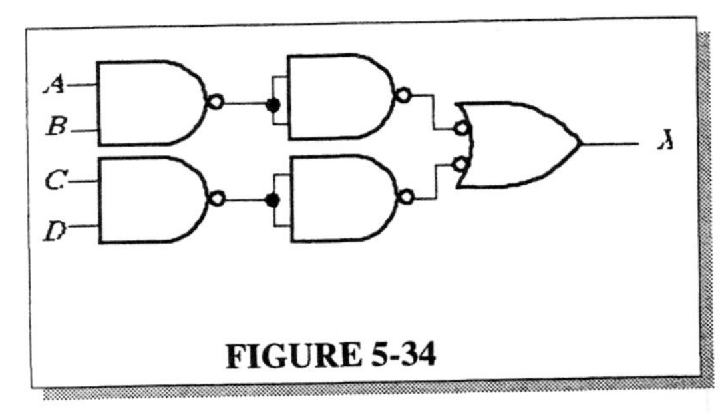

FIGURE 5-34

(f) $X = (A + B)(C + D)$

See Figure 5-35.

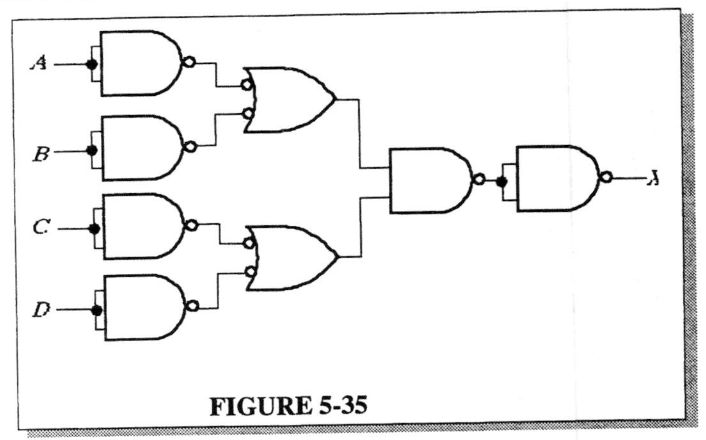

FIGURE 5-35

(g) $X = AB[C(\overline{DE} + \overline{AB}) + \overline{BCE}]$

See Figure 5-36.

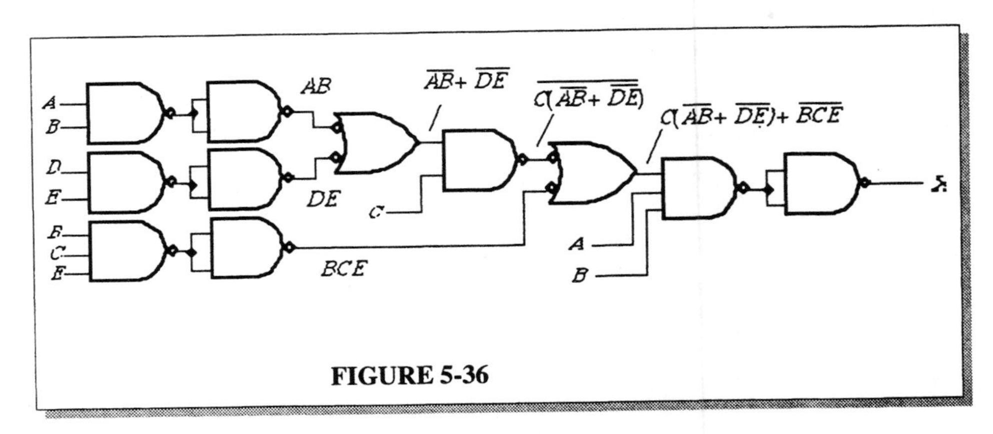

FIGURE 5-36

24. (a) $X = AB$

See Figure 5-37.

(b) $X = A + B$

See Figure 5-38.

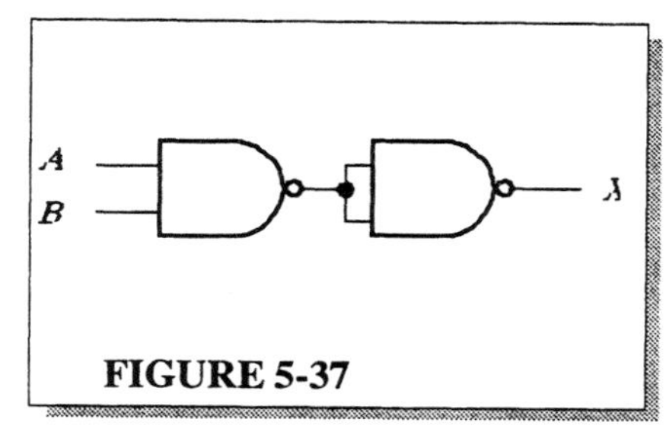

FIGURE 5-37

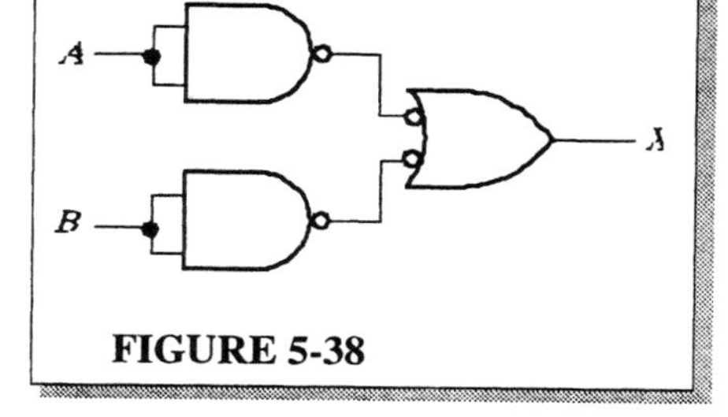

FIGURE 5-38

(c) $X = AB + C$

See Figure 5-39.

(d) $X = ABC + D$

See Figure 5-40.

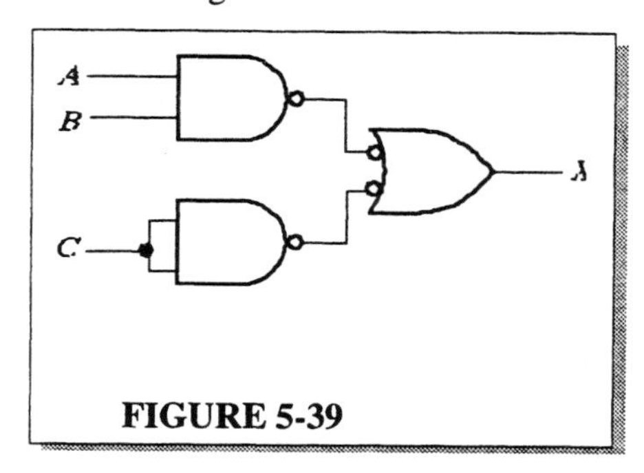

FIGURE 5-39

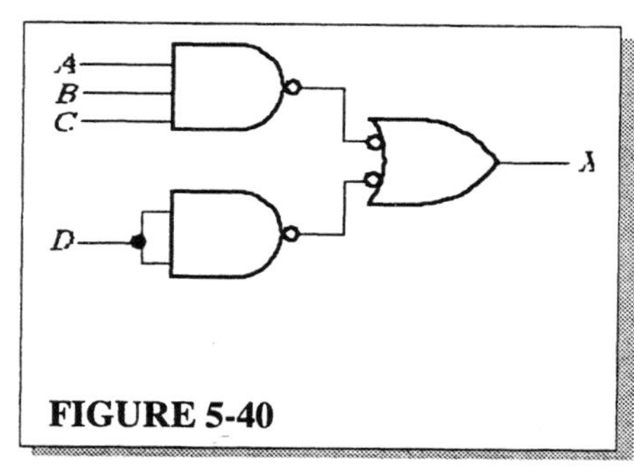

FIGURE 5-40

(e) $X = A + B + C$

See Figure 5-41.

(f) $X = ABCD$

See Figure 5-42.

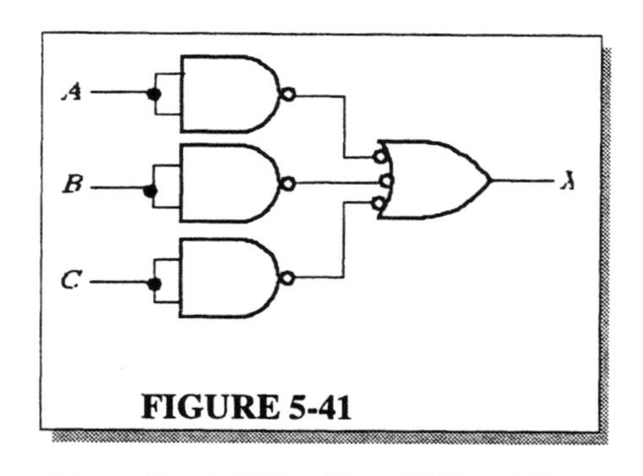

FIGURE 5-41

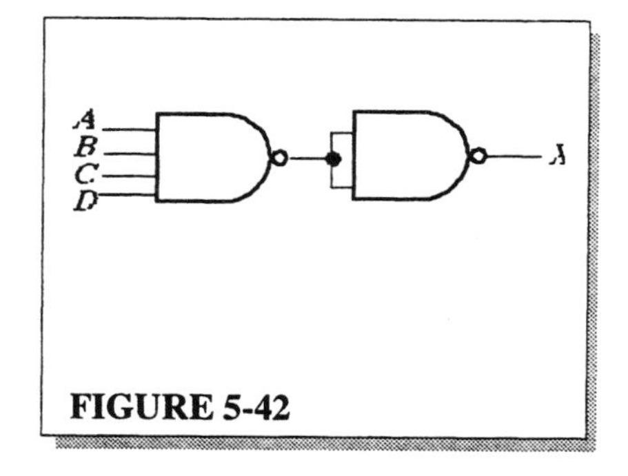

FIGURE 5-42

(g) $X = A(CD + B) = \mathbf{ACD + AB}$

See Figure 5-43.

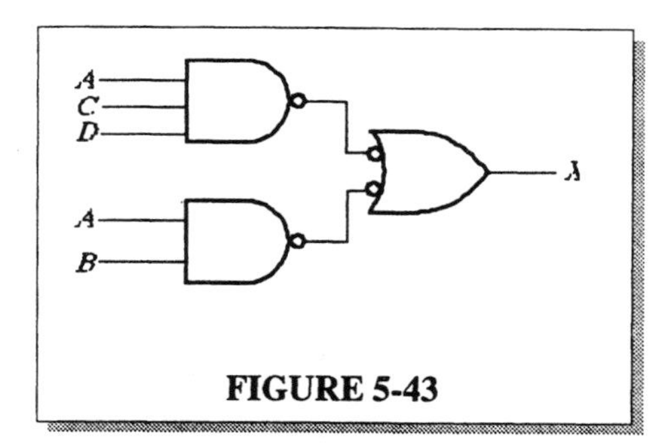

FIGURE 5-43

(h) $X = AB(C + DEF) + CE(A + B + F)$

$= \mathbf{ABC + ABDEF + CEA + CEB + CEF}$

See Figure 5-44.

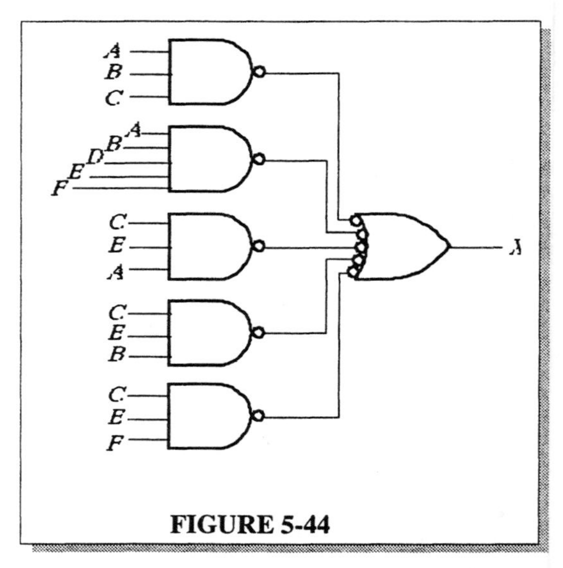

FIGURE 5-44

25. (a) $X = AB + \overline{B}C$

See Figure 5-45.

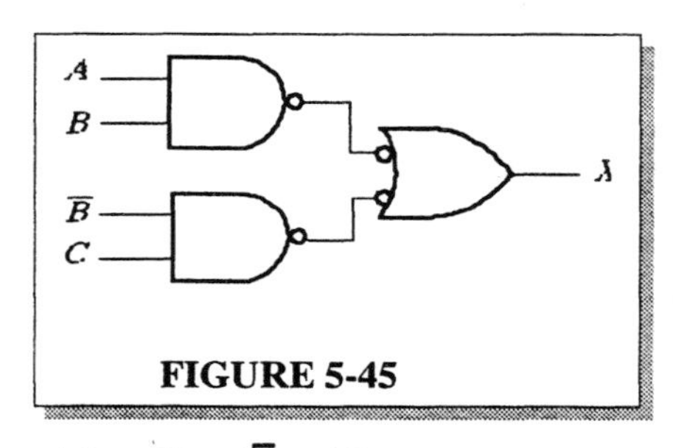

FIGURE 5-45

(b) $X = A(B + \overline{C}) = \mathbf{AB + A\overline{C}}$

See Figure 5-46.

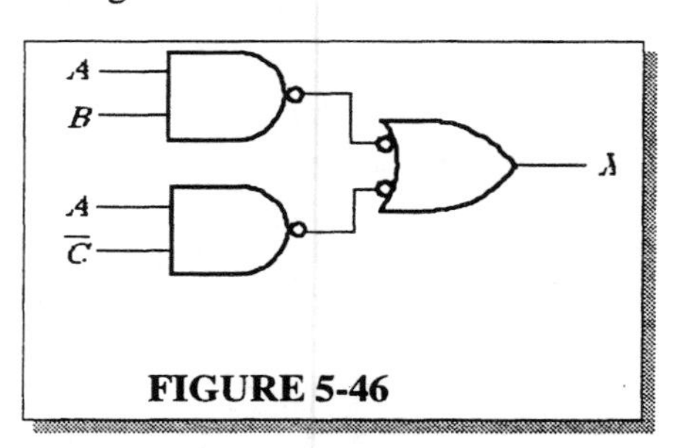

FIGURE 5-46

(c) $X = A\overline{B} + AB$

See Figure 5-47.

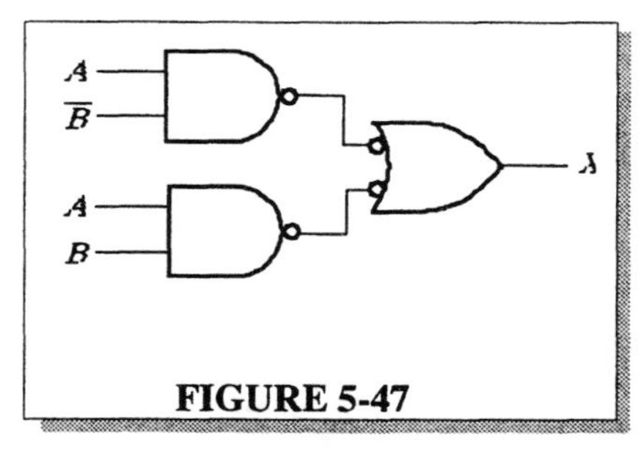

FIGURE 5-47

(d) $X = \overline{ABC} + B(EF + \overline{G}) = \mathbf{\overline{A} + \overline{B} + \overline{C} + BEF + B\overline{G}}$

See Figure 5-48.

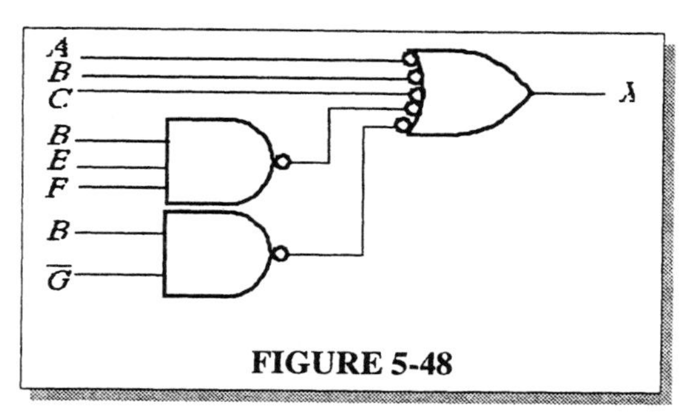

FIGURE 5-48

(e) $X = A[BC(A + B + C + D)] = ABCA + ABCB + ABCC + ABCD$

$= ABC + ABC + ABC + ABCD = ABC(1 + D) = \mathbf{ABC}$

See Figure 5-49.

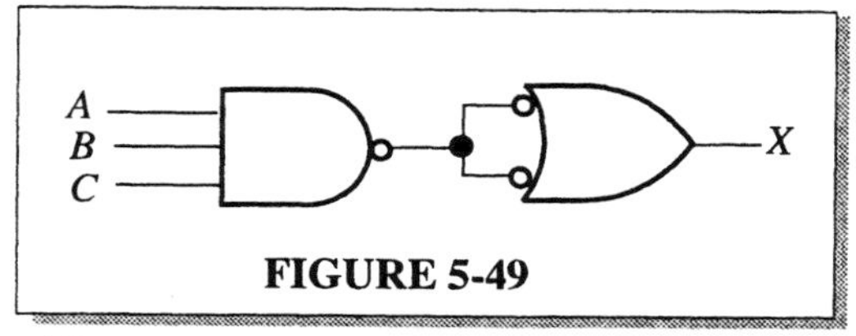

FIGURE 5-49

(f) $X = B(C\overline{D}E + \overline{E}FG)(\overline{AB} + C) = B(C\overline{D}E + \overline{E}FG)(\overline{A} + \overline{B} + C)$

$= B(\overline{A}C\overline{D}E + \overline{A}\overline{E}FG + \overline{B}C\overline{D}E + \overline{B}\overline{E}FG + C\overline{D}E + C\overline{E}FG)$

$= \overline{A}B\overline{E}FG + B\overline{B}\overline{E}FG + BC\overline{D}E + BC\overline{E}FG$

$= \mathbf{\overline{A}B\overline{E}FG + BC\overline{D}E + BC\overline{E}FG}$

See Figure 5-50.

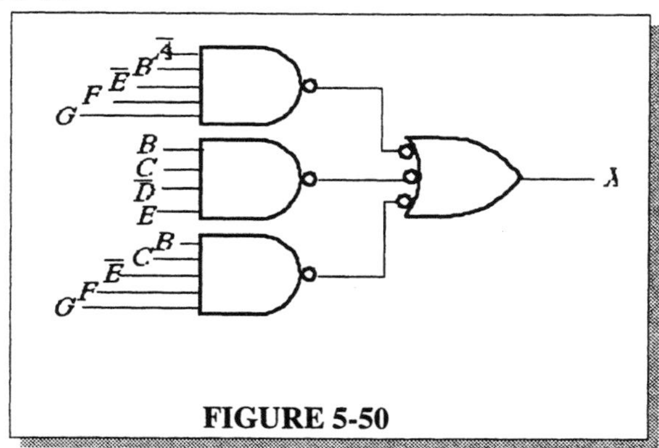

FIGURE 5-50

26. $X = \overline{\overline{A} + \overline{B} + B} = AB\overline{B} = 0$

The output X is always LOW.

27. $X = \overline{(\overline{A}B)B} = A + \overline{B} + \overline{B} = A + \mathbf{\overline{B}}$

See Figure 5-51.

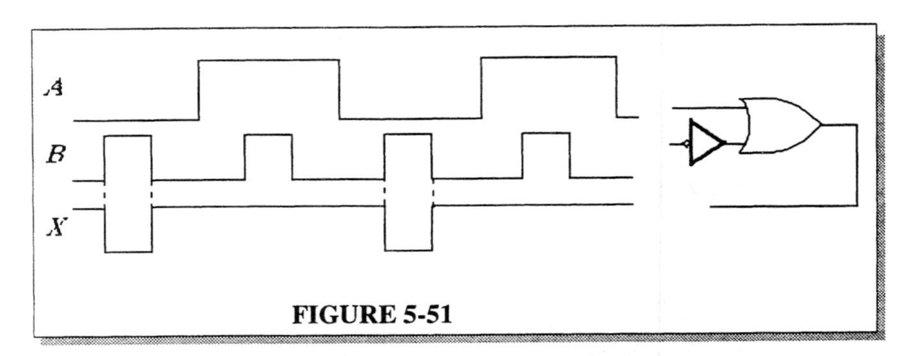

FIGURE 5-51

28. X is HIGH when ABC are all HIGH or when A is HIGH and B is LOW and C is LOW or when A is HIGH and B is LOW and C is HIGH.

$X = ABC + A\overline{B}\overline{C} + A\overline{B}C$

See Figure 5-52.

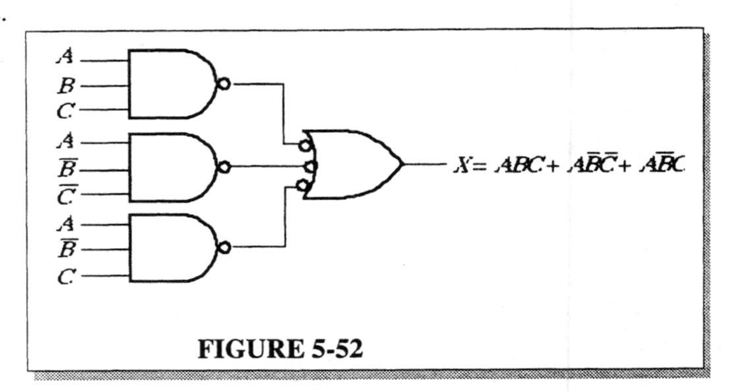

FIGURE 5-52

29. X is HIGH when A is HIGH, B is LOW, and C is LOW. We do not know if X is HIGH when all inputs are HIGH.

$X = A\overline{B}\overline{C}$

See Figure 5-53.

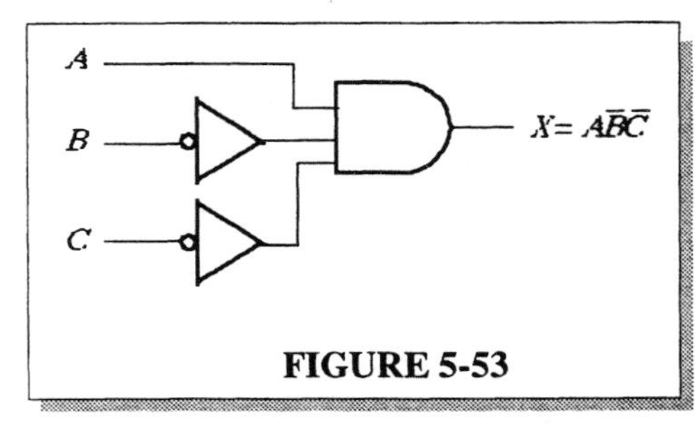

FIGURE 5-53

30. See Figure 5-54.

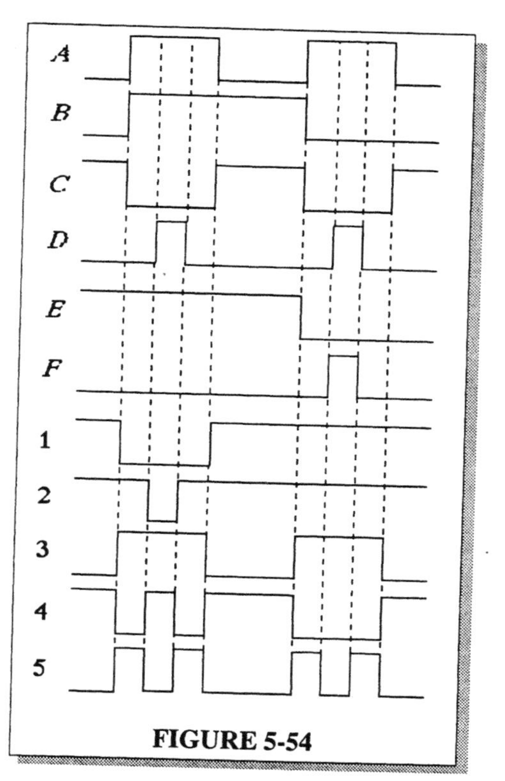

FIGURE 5-54

31. The output is sufficiently wide. It is greater than 25 ns. A maximum is not specified. See Figure 5-55.

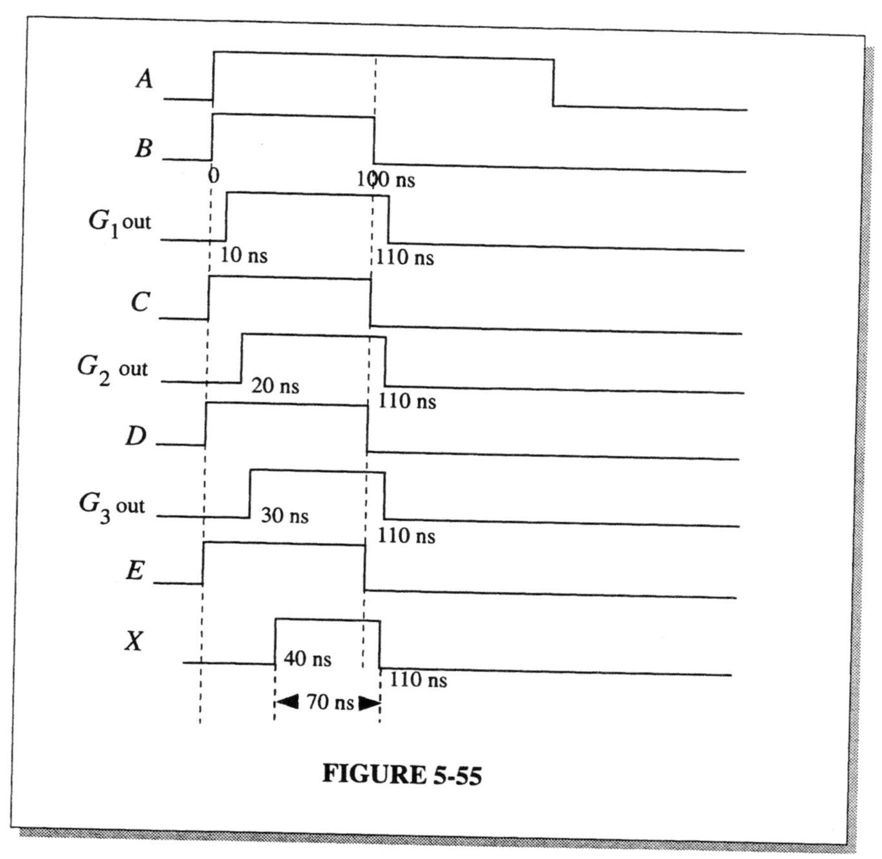

FIGURE 5-55

32. $X = \overline{\overline{AB} + \overline{CD}} = ABCD$

X is HIGH only when $ABCD$ are all HIGH. This does not occur in the waveforms, so X should remain LOW. **The output is incorrect**

33. $X = ABC + D\overline{E}$

Since X is the same as the G3 output, either G1 or G2 has failed with its output *stuck LOW.*

34. $X = AB + CD + EF$

X does no go HIGH when C and D are HIGH. G2 has failed with the output *open* or *stuck HIGH* or the corresponding input to G4 is *open.*

35. Gate G5 has an input *shorted* to ground because the current tracer indicates that all the current is into that input.

36. $X = \overline{\overline{AB} + \overline{CD} + \overline{EF}} = (\overline{\overline{AB}})(\overline{\overline{CD}})(\overline{\overline{EF}}) = (A + B)(C + D)(E + F)$

Since X does not go HIGH when C or D is HIGH the output of gate G2 must be *stuck* LOW.

37. (a) $X = (\overline{A} + \overline{B} + C)E + (C + \overline{D})E = \overline{A}E + \overline{B}E + CE + CE + \overline{D}E$

$= \overline{A}E + \overline{B}E + CE + \overline{D}E$

See Figure 5-56.

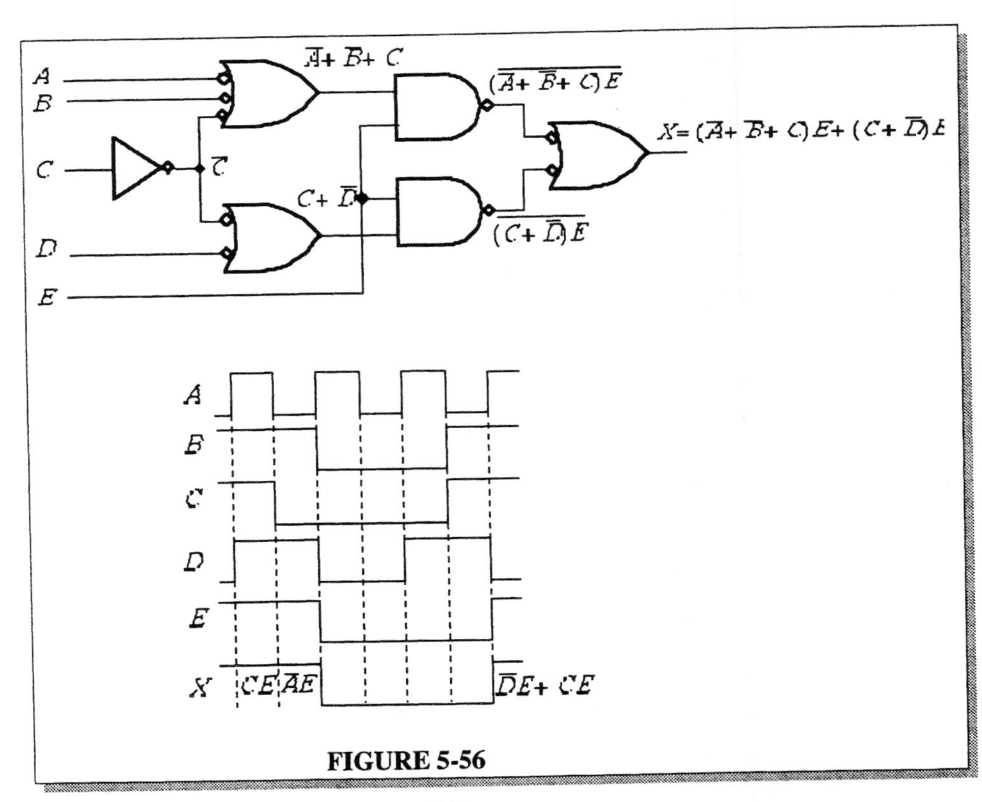

FIGURE 5-56

(b) $X = E + E(\overline{D} + C) = E(1 + \overline{D} + C) = E$

Waveform X is the same as waveform E, in Figure 5-56. Since this is the correct waveform, the open output of gate G3 does not show up for this *particular* set of input waveforms.

(c) $X = E + E(\overline{A} + \overline{B} + C) = E(1 + \overline{A} + \overline{B} + C) = E$

Again waveform X is the same as waveform E. As strange as it may seem, the shorted input to G5 does not affect the output for this *particular* set of input waveform.

Conclusion: the two faults are not indicated in the output waveform for these particular inputs.

38. $\text{TP} = \overline{\overline{A}\overline{B} + \overline{C}\overline{D}}$

The output of the $\overline{C}\overline{D}$ gate is *stuck LOW*. See Figure 57.

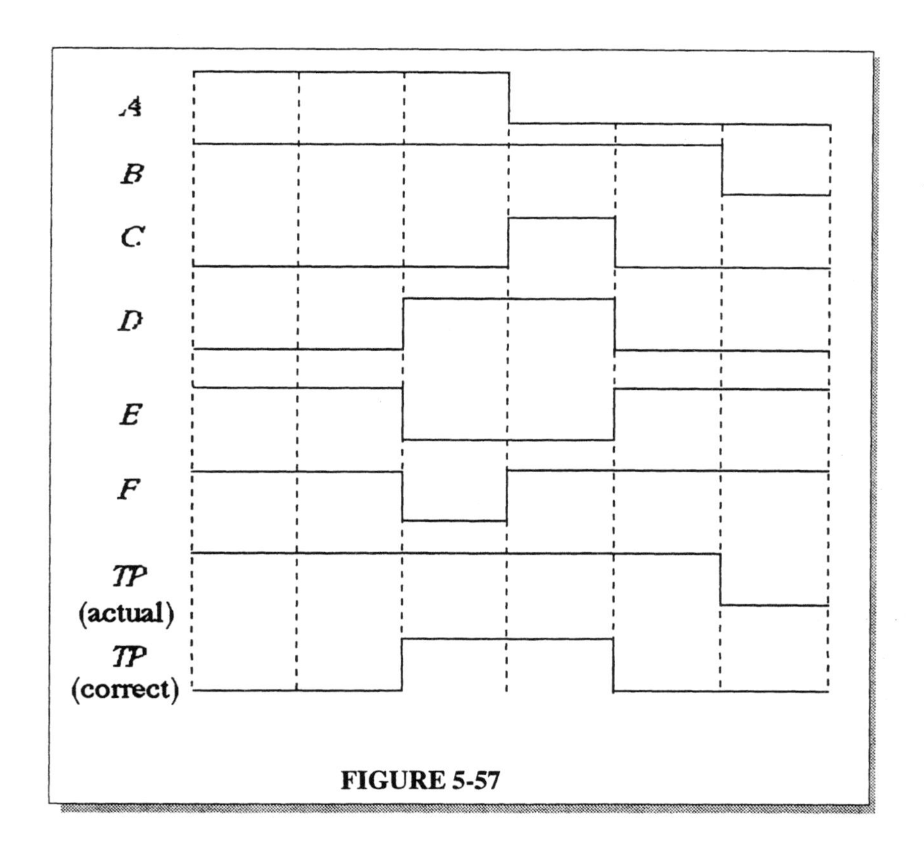

FIGURE 5-57

39. See Figure 5-58.

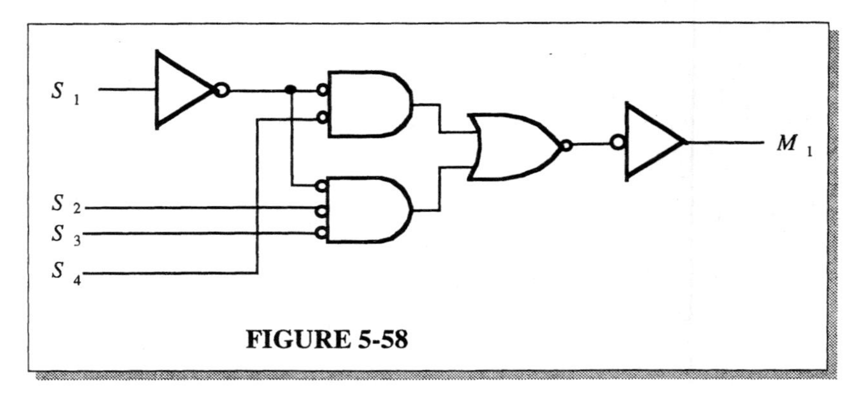

FIGURE 5-58

40. See Figure 5-59.

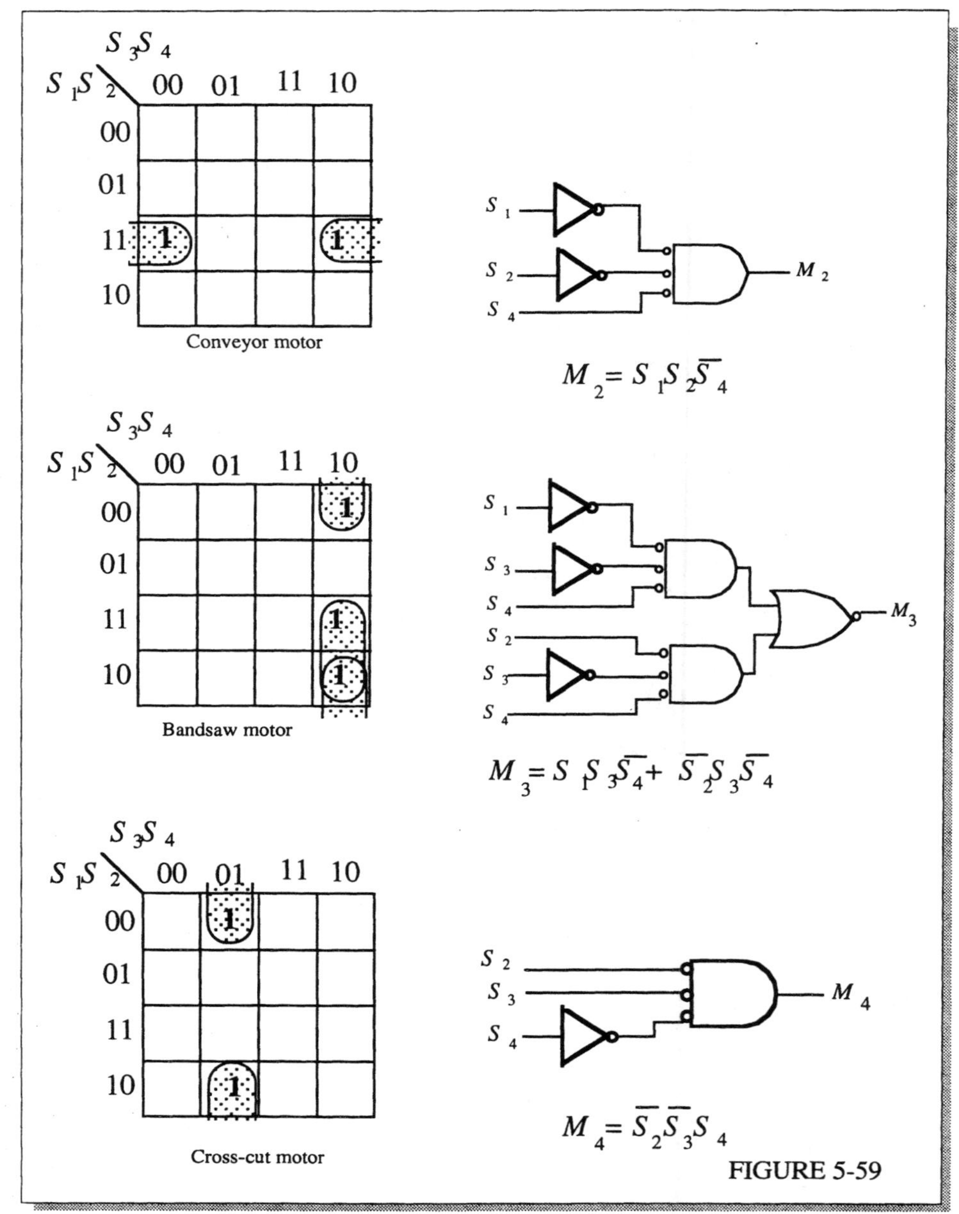

FIGURE 5-59

41. The logic for all but the cross-cut logic remains as shown in Figure 5-59. See Figure 5-60.

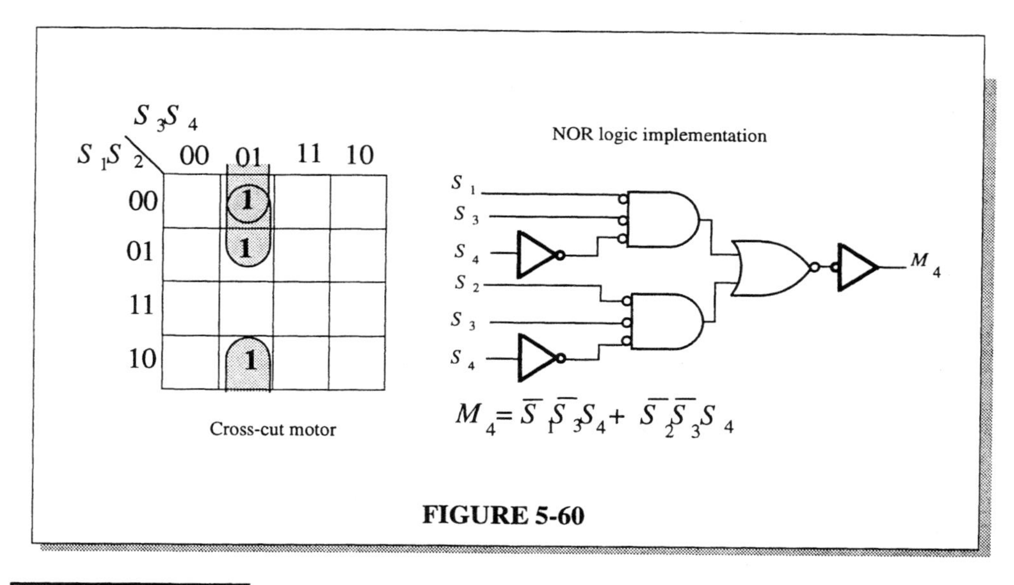

FIGURE 5-60

42.

A_3	A_2	A_1	A_0	X
0	0	0	0	1
0	0	0	1	1
0	0	1	0	1
0	0	1	1	0
0	1	0	0	0
0	1	0	1	0
0	1	1	0	0
0	1	1	1	0
1	0	0	0	0
1	0	0	1	0
1	0	1	0	0
1	0	1	1	0
1	1	0	0	0
1	1	0	1	1
1	1	1	0	1
1	1	1	1	1

See Figure 5-61.

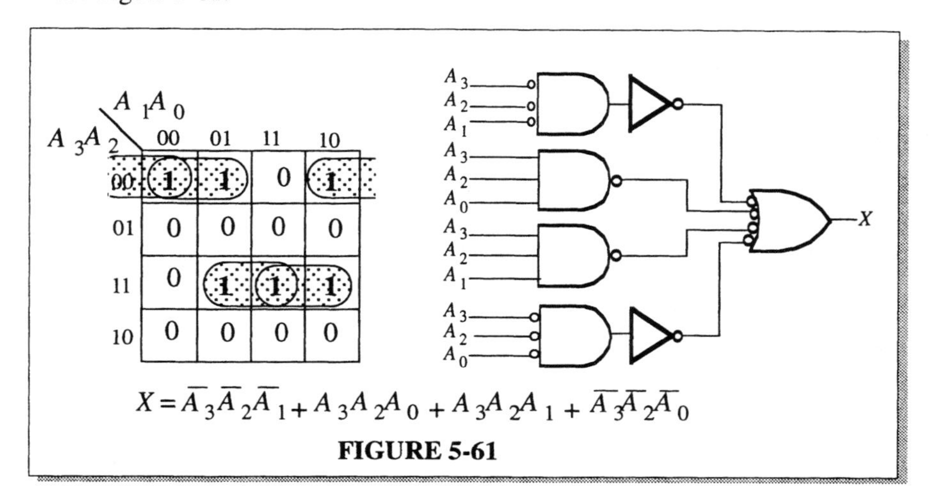

FIGURE 5-61

43. Let

X = Lamp on
A = Front door switch on
$\overline{A}$ = Front door switch off
B = Back door switch on
$\overline{B}$ = Back door switch off

$X = A\overline{B} + \overline{A}B$. This is an XOR operation.

See Figure 5-62.

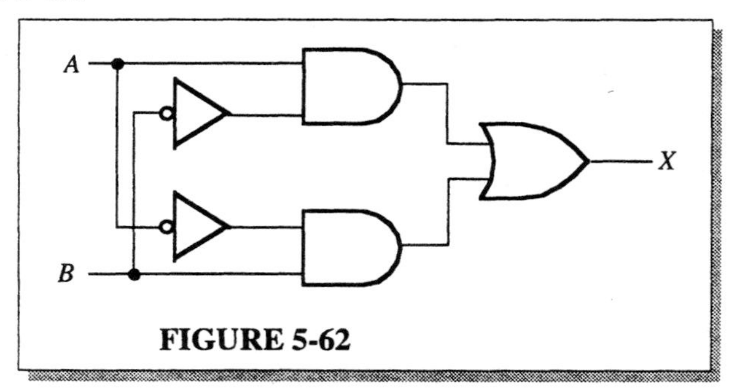

FIGURE 5-62

44. See Figure 5-63.

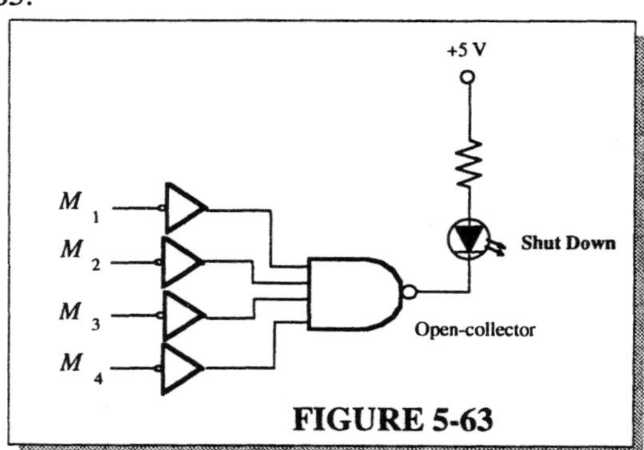

FIGURE 5-63

45. See Figure 5-64.

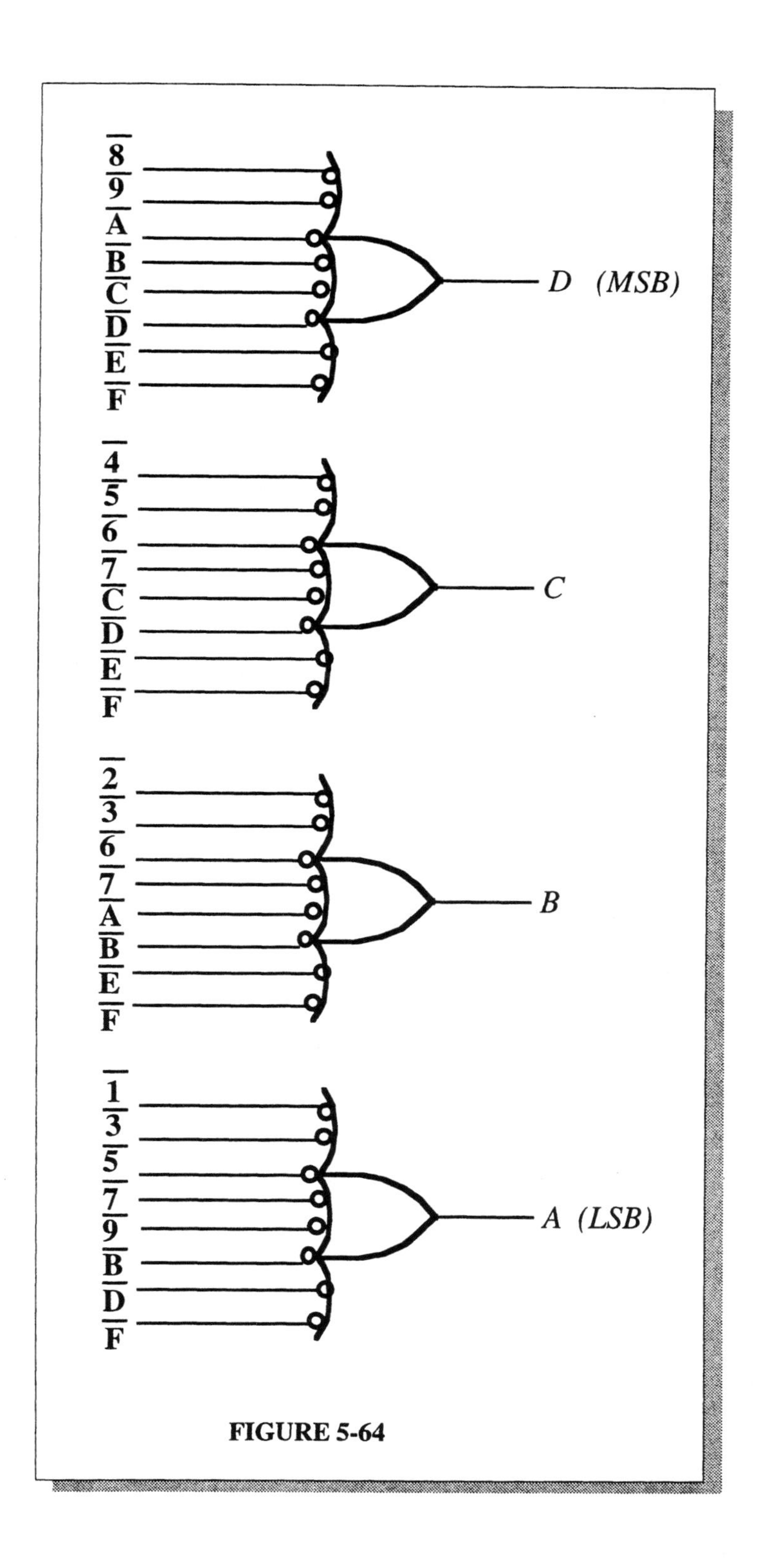

$\overline{8}$
$\overline{9}$
$\overline{A}$
$\overline{B}$
$\overline{C}$
$\overline{D}$
$\overline{E}$
$\overline{F}$
D (MSB)
$\overline{4}$
$\overline{5}$
$\overline{6}$
$\overline{7}$
$\overline{C}$
$\overline{D}$
$\overline{E}$
$\overline{F}$
C
$\overline{2}$
$\overline{3}$
$\overline{6}$
$\overline{7}$
$\overline{A}$
$\overline{B}$
$\overline{E}$
$\overline{F}$
B
$\overline{1}$
$\overline{3}$
$\overline{5}$
$\overline{7}$
$\overline{9}$
$\overline{B}$
$\overline{D}$
$\overline{F}$
A (LSB)

FIGURE 5-64

CHAPTER 6
FUNCTIONS OF COMBINATIONAL LOGIC

1.
(a) XOR (upper) output = 0, Sum output = 1, AND (upper) output = 0, AND (lower) output = 1, Carry output = 1
(b) XOR (upper) output = 1, Sum output = 0, AND (upper) output = 1, AND (lower) output = 0, Carry output = 1
(c) XOR (upper) output = 1, Sum output = 1, AND (upper) output = 0, AND (lower) output = 0, Carry output = 0

2.
(a) $A = 0, B = 0, C_{in} = 0$
(b) $A = 1, B = 0, C_{in} = 0$ or $A = 0, B = 1, C_{in} = 0$ or $A = 0, B = 0, C_{in} = 1$
(c) $A = 1, B = 1, C_{in} = 1$
(d) $A = 1, B = 1, C_{in} = 0$ or $A = 0, B = 1, C_{in} = 1$ or $A = 1, B = 0, C_{in} = 1$

3.
$$\Sigma = \bar{A}\bar{B}C_{in} + \bar{A}B\bar{C}_{in} + A\bar{B}\bar{C}_{in} + ABC_{in}$$

$$C_{out} = ABC_{in} + \bar{A}BC_{in} + A\bar{B}C_{in} + AB\bar{C}_{in}$$

See Figure 6-1.

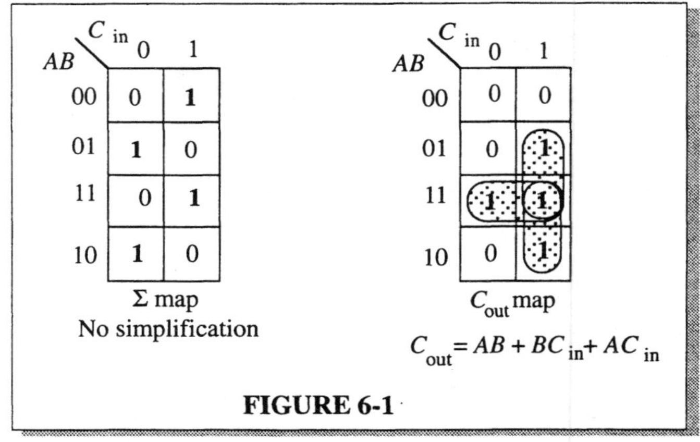

FIGURE 6-1

4.
111 See Figure 6-2.
101
1100

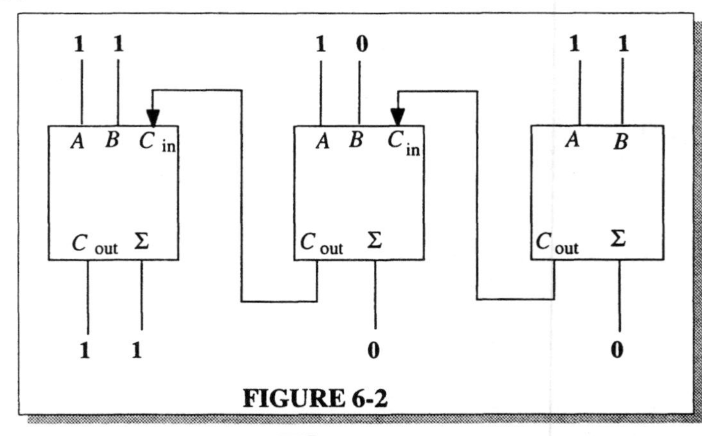

FIGURE 6-2

5.

10101 See Figure 6-3.
00111
11100

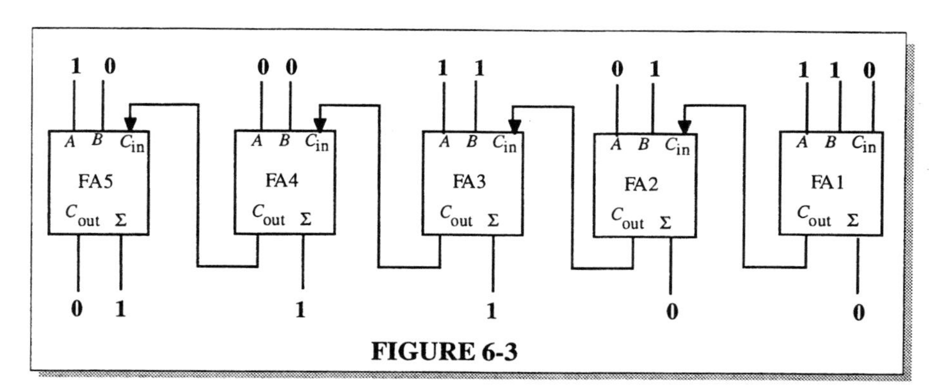

FIGURE 6-3

6. See Figure 6-4.

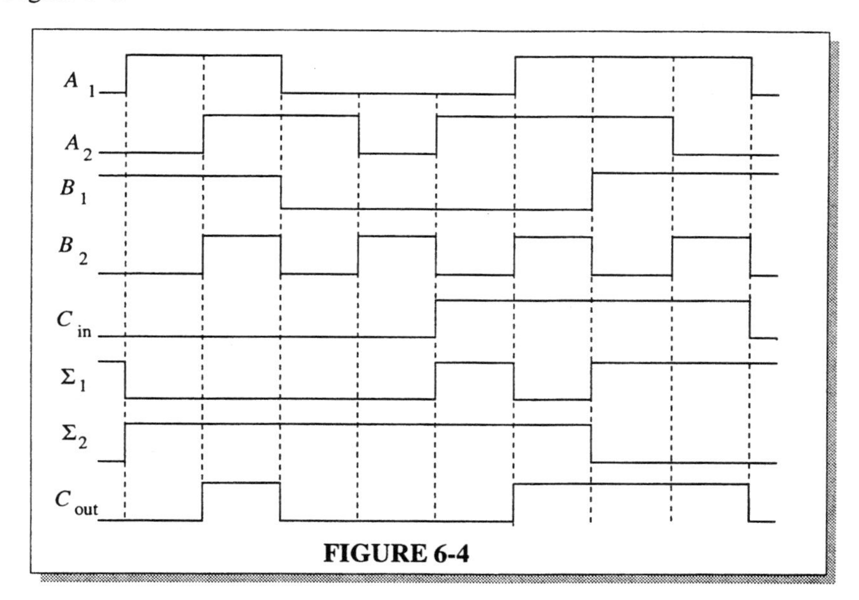

FIGURE 6-4

7.

A_4	A_3	A_2	A_1	B_3	B_2	B_1	B_0	Σ_5	Σ_4	Σ_3	Σ_2	Σ_1
0	0	0	0	0	0	0	0	0	0	0	0	0
1	1	0	1	0	1	0	0	1	0	0	0	1
0	0	0	1	1	0	1	0	0	1	0	1	1
1	1	1	0	0	1	1	1	1	0	1	0	1
1	0	0	1	0	0	0	1	0	1	0	1	0
1	0	1	0	1	1	0	1	1	0	1	1	1
0	0	1	0	0	0	1	1	0	0	1	0	1
1	0	1	1	0	1	1	1	1	0	0	1	0

$\Sigma_1 = 01101110$
$\Sigma_2 = 10110100$
$\Sigma_3 = 01101000$
$\Sigma_4 = 00010100$
$\Sigma_5 = 10101010$

8.

$$\begin{array}{r} 0100 \\ \underline{1110} \\ 10010 \end{array}$$

Σ outputs should be $C_{out}\Sigma_4\Sigma_3\Sigma_2\Sigma_1$ = 10010.

The Σ_3 output (pin 2) is HIGH and should be LOW.

See Figure 6-5.

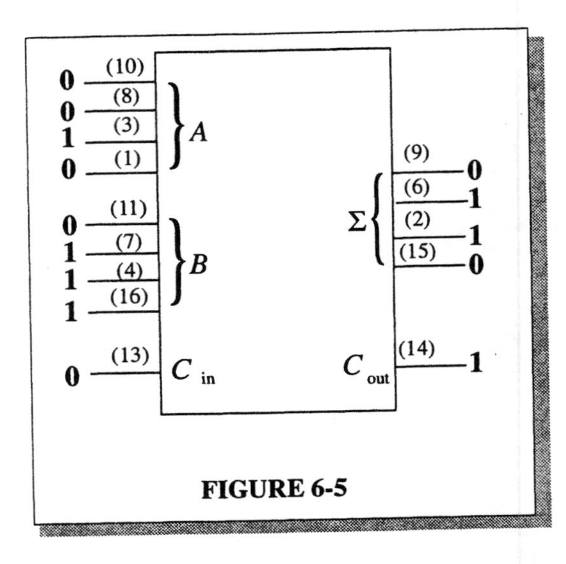

FIGURE 6-5

9. The $A = B$ output is HIGH when $A_0 = B_0$ and $A_1 = B_1$.

See Figure 6-6.

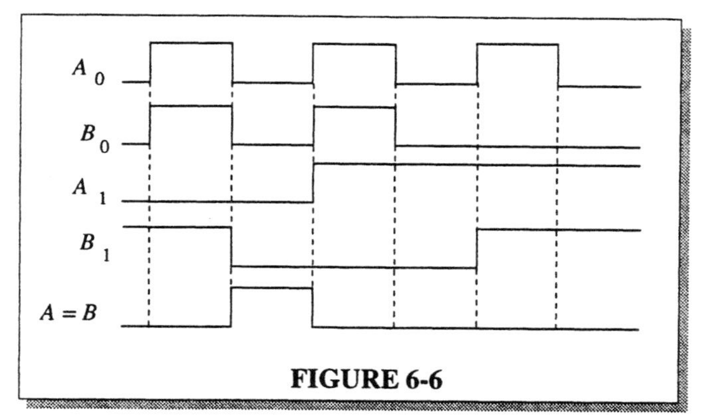

FIGURE 6-6

10. See Figure 6-7.

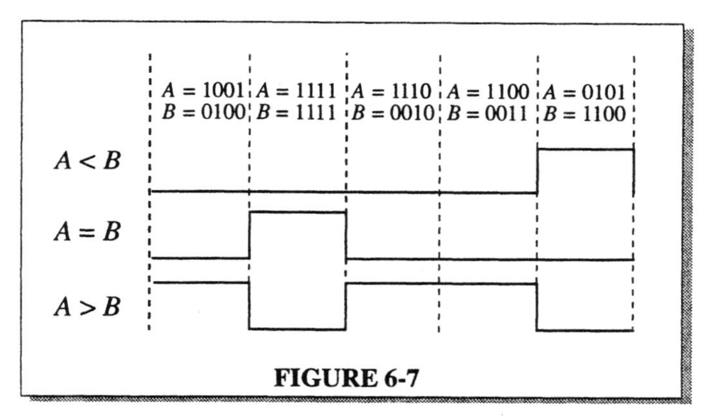

FIGURE 6-7

11. (a) A > B: 1, A = B: 0, A < B: 0

(b) A > B: 0, A = B: 0, A < B: 1

(c) A > B: 0, A = B: 1, A < B: 0

12. (a) $A_3A_2A_1A_0$ = **1110** (b) $A_3A_2A_1A_0$ = **1100**
(c) $A_3A_2A_1A_0$ = **1111** (d) $A_3A_2A_1A_0$ = **1000**

13. See Figure 6-8.

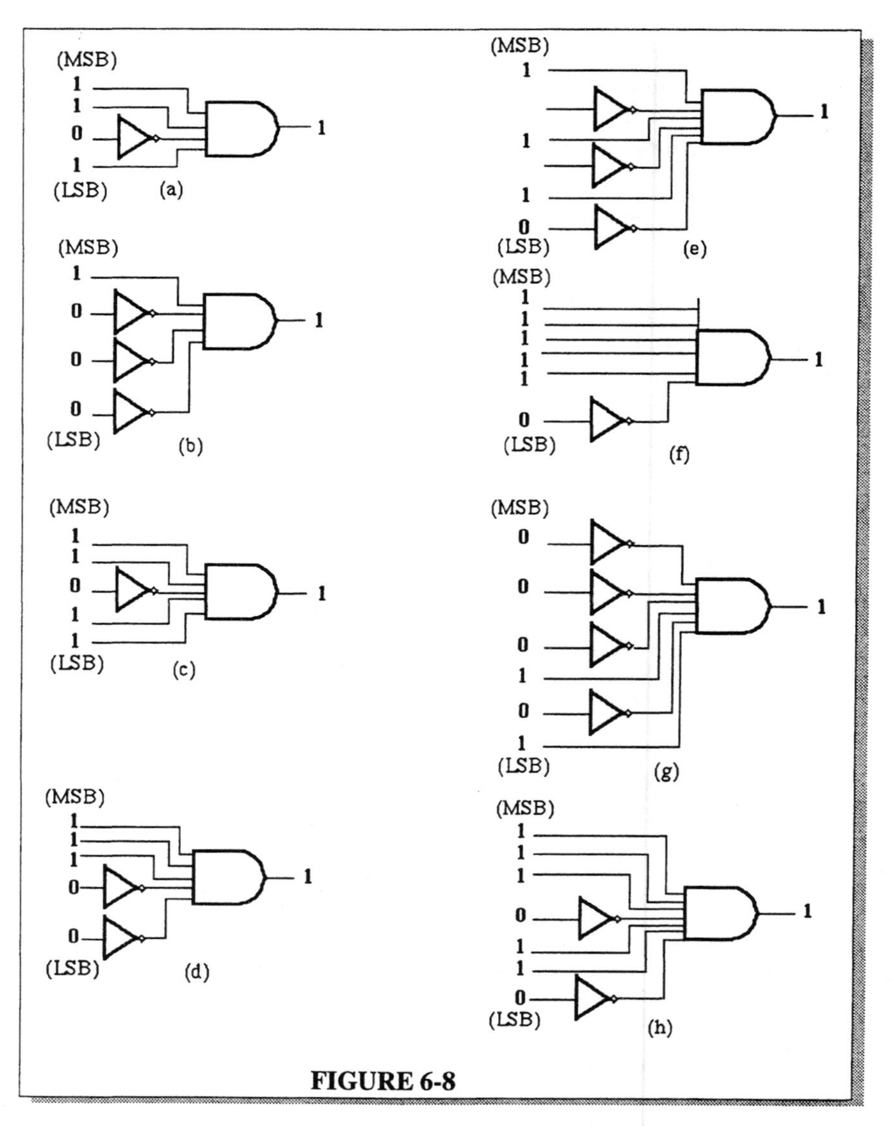

FIGURE 6-8

14. Change the AND gates to NAND gates in Figure 6-8.

15. $X = \overline{A}_3\overline{A}_2\overline{A}_1A_0 + A_3\overline{A}_2A_1\overline{A}_0 + A_3A_2\overline{A}_1\overline{A}_0 + A_3\overline{A}_2A_1A_0$

See Figure 6-9.

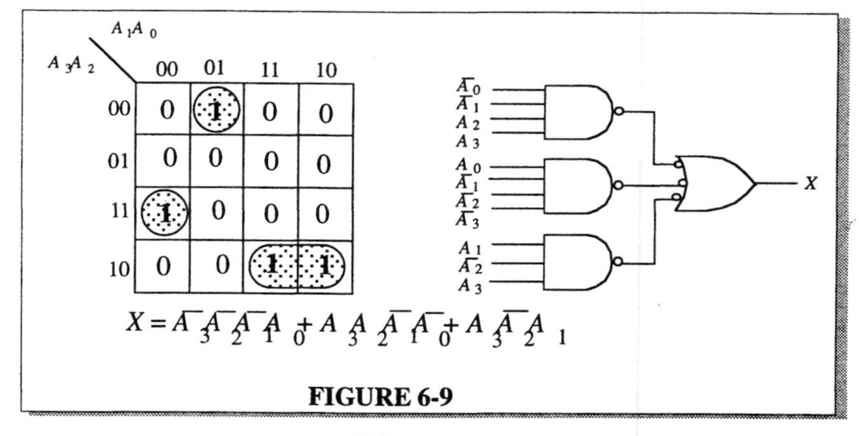

FIGURE 6-9

16. $Y = A_2A_1\overline{A}_0 + A_2\overline{A}_1A_0 + \overline{A}_2A_1\overline{A}_0$

See Figure 6-10.

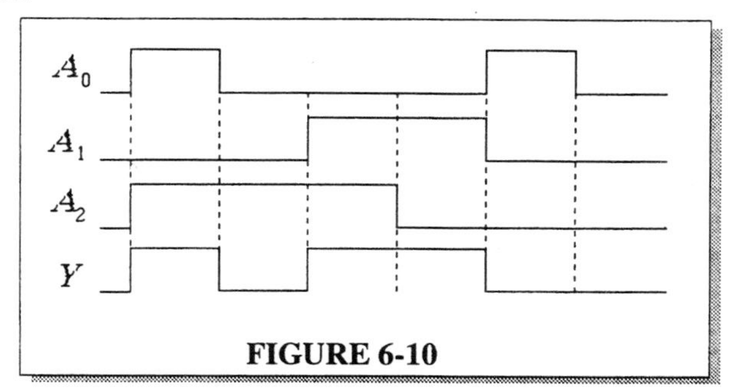

FIGURE 6-10

17. See Figure 6-11.

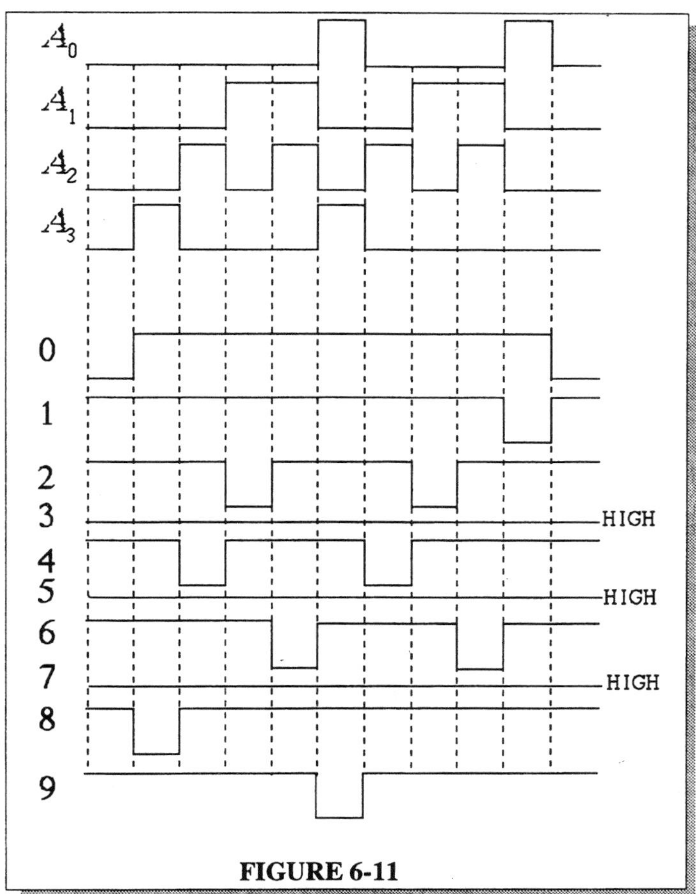

FIGURE 6-11

18. **0 1 6 9 4 4 4 8 0**

19. A_0, A_1, and A_3 are HIGH. $A_3A_2A_1A_0 = 1011$, which is an invalid BCD code.

20. Pin 2 is for decimal 5, pin 5 is for decimal 8, and pin 12 is for decimal 2. The highest priority input is pin 5.

The complemented outputs are: $\overline{A}_3\overline{A}_2\overline{A}_1\overline{A}_0 = 0111$, which is binary 8 (1000).

21. (a) 2_{10} = $\mathbf{0010}_{BCD}$ = $\mathbf{0010}_2$

(b) 8_{10} = $\mathbf{1000}_{BCD}$ = $\mathbf{1000}_2$

(c) 13_{10} = $\mathbf{00010011}_{BCD}$ = $\mathbf{1101}_2$

(d) 26_{10} = $\mathbf{00100110}_{BCD}$ = $\mathbf{11010}_2$

(e) 33_{10} = $\mathbf{00110011}_{BCD}$ = $\mathbf{100001}_2$

22.

(a) 1010101010 binary
1111111111 gray

(b) 1111100000 binary
1000010000 gray

(c) 0000001110 binary
0000001001 gray

(d) 1111111111 binary
1000000000 gray

See Figure 6-12.

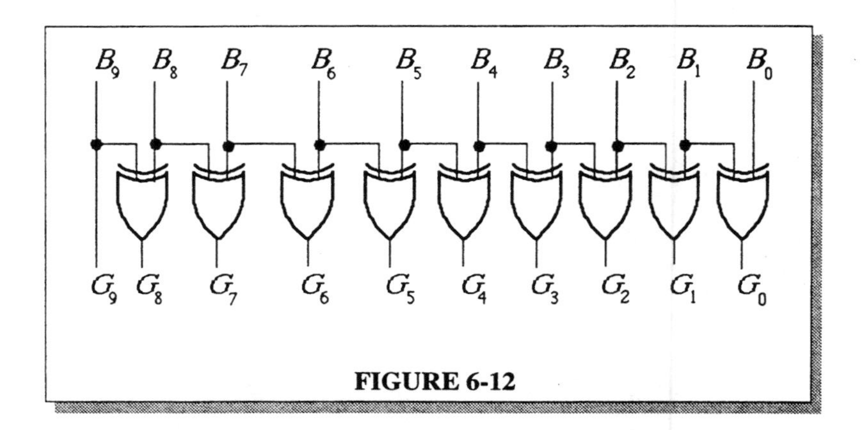

FIGURE 6-12

23.

(a) 1010000000 gray
1100000000 binary

(b) 0011001100 gray
0010001000 binary

(c) 1111000111 gray
1010000101 binary

(d) 0000000001 gray
0000000001 binary

See Figure 6-13.

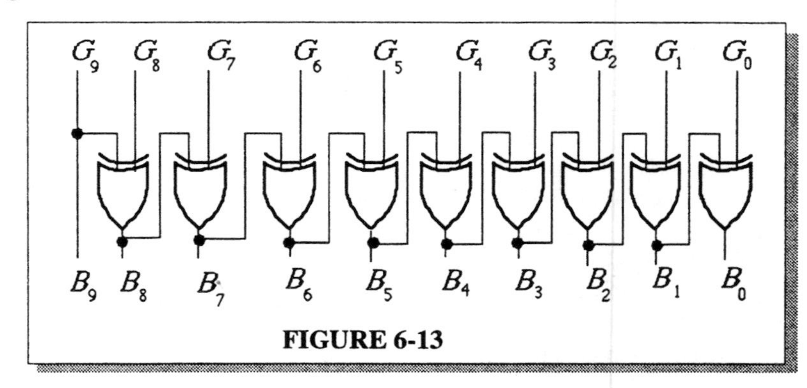

FIGURE 6-13

24. $S_1S_0 = 01$ selects D_1, therefore $Y = 1$.

25. See Figure 6-14.

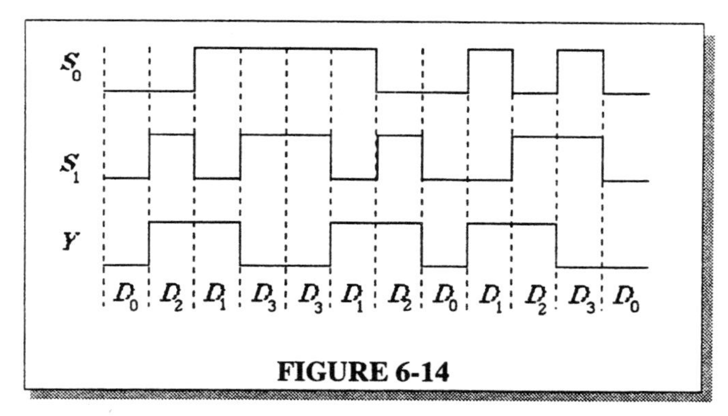

FIGURE 6-14

26. See Figure 6-15.

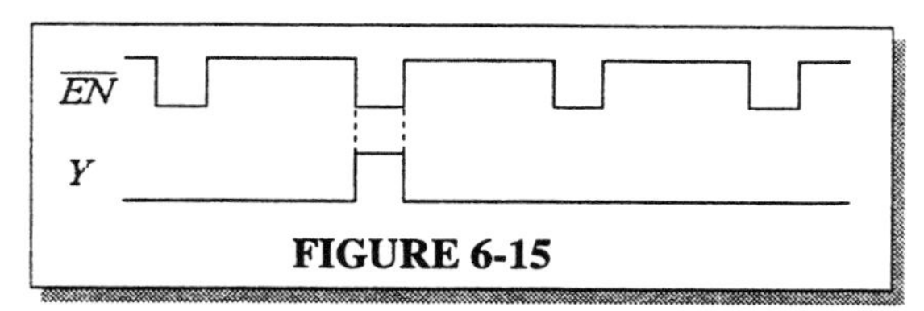

FIGURE 6-15

27. See Figure 6-16.

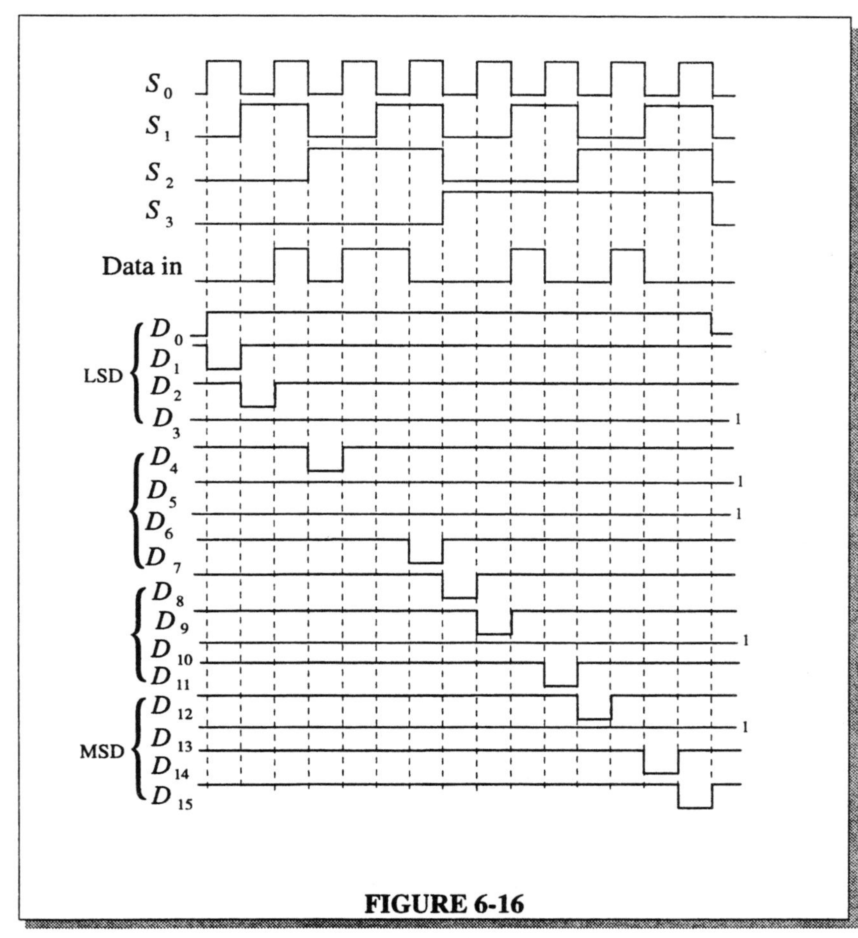

FIGURE 6-16

28. See Figure 6-17.

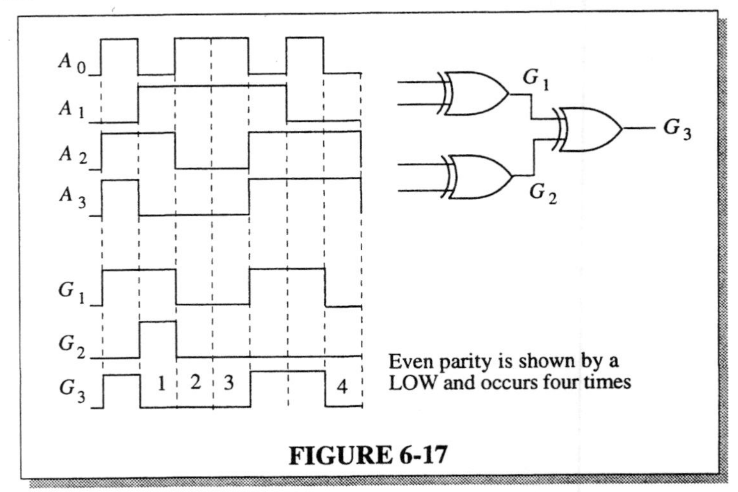

FIGURE 6-17

29. See Figure 6-18.

30. The outputs given in the problem are incorrect. By observation of these incorrect waveforms, we can conclude that the outputs of the device are not open or shorted because both waveforms are changing.

Observe that at the beginning of the timing diagram all inputs are 0 but the sum is 1. This indicates that an input is stuck HIGH. Start by assuming that C_{in} is stuck HIGH. This results in Σ and C_{out} output waveforms that match the waveforms given in the problem, indicating that C_{in} is indeed stuck HIGH, perhaps shorted to V_{CC}.

See Figure 6-19 for the correct output waveforms.

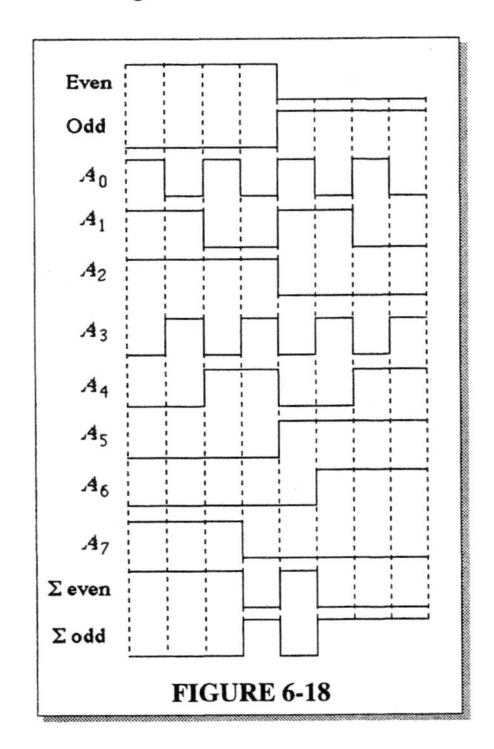

FIGURE 6-18

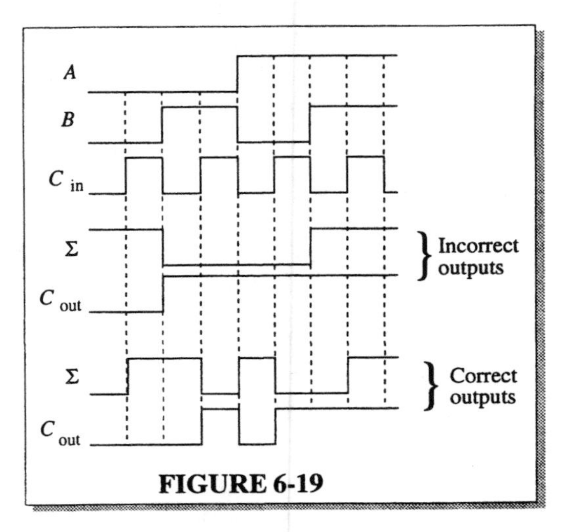

FIGURE 6-19

31. (a) OK (b) Segment *g* burned out; ourpur G open (c) Segment *b* output stuck LOW

32. *Step 1*: Verify that the supply voltage is applied.
Step 2: Go through the key sequence and verify the output code in Table 1.

Key	A_3	A_2	A_1	A_0
None	1	1	1	1
0	1	1	1	1
1	1	1	1	0
2	1	1	0	1
3	1	1	0	0
4	1	0	1	1
5	1	0	1	0
6	1	0	0	1
7	1	0	0	0
8	0	1	1	1
9	0	1	1	0

TABLE 1

Step 3: Check for proper priority operation by repeating the key sequence in Table 1 except that for each key closure, hold that key down and depress each lower valued key as specified in Table 2.

Hold down keys	Depress keys one at a time	A_3	A_2	A_1	A_0
1	0	1	1	1	0
2	1, 0	1	1	0	1
3	2, 1, 0	1	1	0	0
4	3, 2, 1, 0	1	0	1	1
5	4, 3, 2, 1, 0	1	0	1	0
6	5, 4, 3, 2, 1, 0	1	0	0	1
7	6, 5, 4, 3, 2, 1, 0	1	0	0	0
8	7, 6, 5, 4, 3, 2, 1, 0	0	1	1	1
9	8, 7, 6, 5, 4, 3, 2, 1, 0	0	1	1	0

TABLE 2

33. (a) Open A_1 input acts as a HIGH. All binary values corresponding to a BCD number having a 1's value of 0, 1, 4, 5, 8, or 9 will be off by 2. This will first be seen for a BCD value of 000000.

(b) Open C_{out} of top adder. All values not normally involving a carry out will be off by 32. This will first be seen for a BCD value of 000000.

(c) The $\Sigma 4$ output of top adder is shorted to ground. Same binary values above 15 will be short by 16. The first BCD value to indicate this will be 00011000.

(d) Σ_2 of bottom adder is shorted to ground. Every other set of 16 value starting with 16 will be short 16. The first BCD value to indicate this will be 00010110.

34.

(a) The 1Y1 output of the 74LS139 is *stuck HIGH* or *open*; B cathode open.
(b) No power; EN input to the 74LS139 is *open.*
(c) The f output of the 74LS48 is *stuck HIGH.*
(d) The frequency of the data select input is too *low.*

35.

1. Place a LOW on pin 7 (Enable).
2. Apply a HIGH to D_0 and a LOW to D_1 through D_7.
3. Go through the binary sequence on the select inputs and check Y and $\overline{Y}$ according to Table 3.

S_2	S_1	S_0	Y	$\overline{Y}$
0	0	0	1	0
0	0	1	0	1
0	1	0	0	1
0	1	1	0	1
1	0	0	0	1
1	0	1	0	1
1	1	0	0	1
1	1	1	0	1

TABLE 3

4. Repeat the binary sequence of select inputs for each set of data inputs listed in Table 4. A HIGH on the Y output should occur only for the corresponding combinations of select inputs shown.

D_0	D_1	D_2	D_3	D_4	D_5	D_6	D_7	Y	$\overline{Y}$	S_2	S_1	S_0
L	H	L	L	L	L	L	L	1	0	0	0	1
L	L	H	L	L	L	L	L	1	0	0	1	0
L	L	L	H	L	L	L	L	1	0	0	1	1
L	L	L	L	H	L	L	L	1	0	1	0	0
L	L	L	L	L	H	L	L	1	0	1	0	1
L	L	L	L	L	L	H	L	1	0	1	1	0
L	L	L	L	L	L	L	H	1	0	1	1	1

TABLE 4

36. The Σ EVEN output of the 74LS280 should be HIGH and the output of the error gate should be HIGH because of the error condition. Possible faults are:

1. Σ EVEN output of the 74LS280 *stuck LOW.*
2. Error gate faulty.
3. The ODD input to the 74LS280 is *open* thus acting as a HIGH.
4. The inverter going to the ODD input of the 74LS280 has an *open* output or the output is *stuck HIGH.*

37. Apply a HIGH in turn to each Data input, D_0 through D_7 with LOWs on all the other inputs. For each HIGH applied to a data input, sequence through all eight binary combinations of select inputs ($S_2S_1S_0$) and check for a HIGH on the corresponding data output and LOWs on all the other data outputs.

One possible approach to implementation is to decode the $S_2S_1S_0$ inputs and generate an inhibit pulse during any given bit time as determined by the settings of seven switches. The inhibit pulse effectively changes a LOW on the *Y* serial data line to a HIGH during the selected bit time(s), thus producing a bit error. A basic diagram of this approach is shown in Figure 6-20.

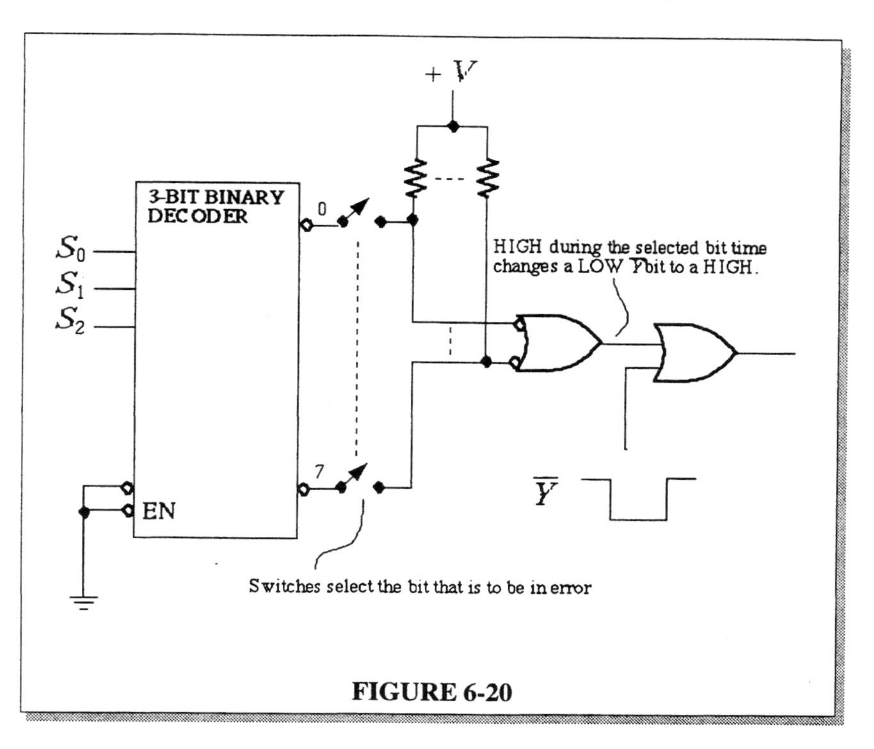

FIGURE 6-20

38. See Figure 6-21.

39. See Figure 6-22.

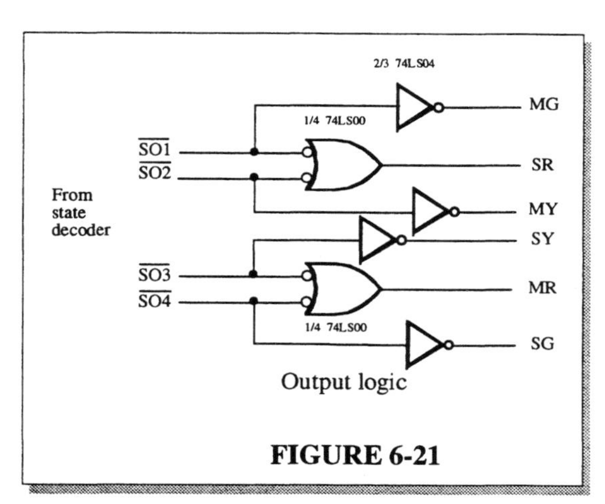

FIGURE 6-21

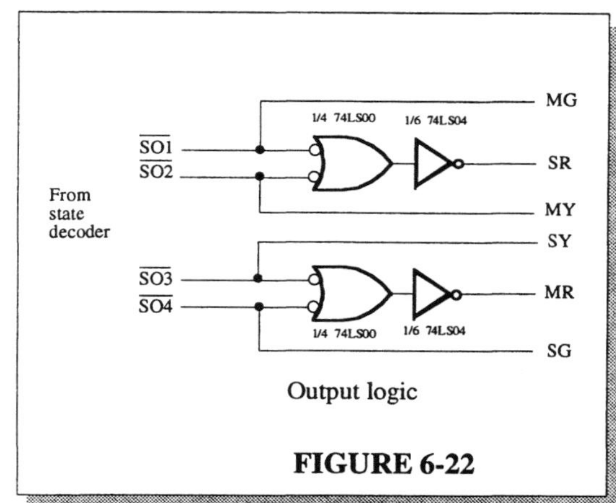

FIGURE 6-22

40. See Figure 6-23.

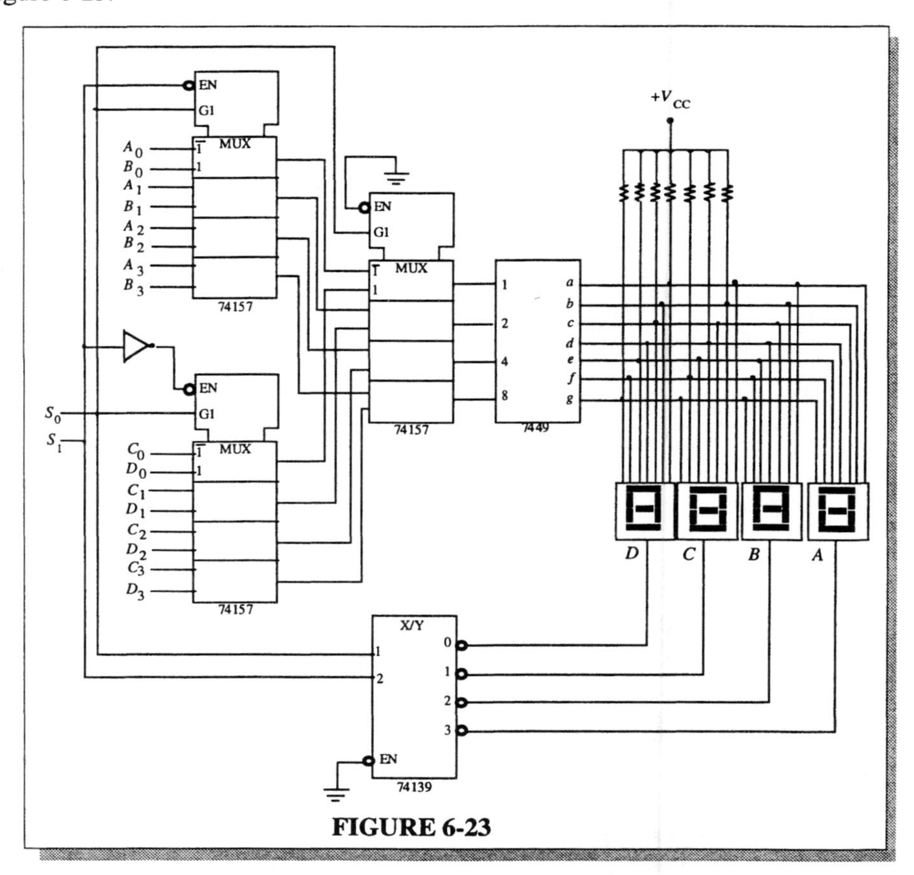

FIGURE 6-23

41. $Y = \overline{A}\,\overline{B}C_{in} + \overline{A}B\overline{C}_{in} + A\overline{B}\,\overline{C}_{in} + ABC_{in}$

See Figure 6-24.

A	B	C_{in}	Y
0	0	0	0
0	0	1	1
0	1	0	1
0	1	1	0
1	0	0	1
1	0	1	0
1	1	0	0
1	1	1	1

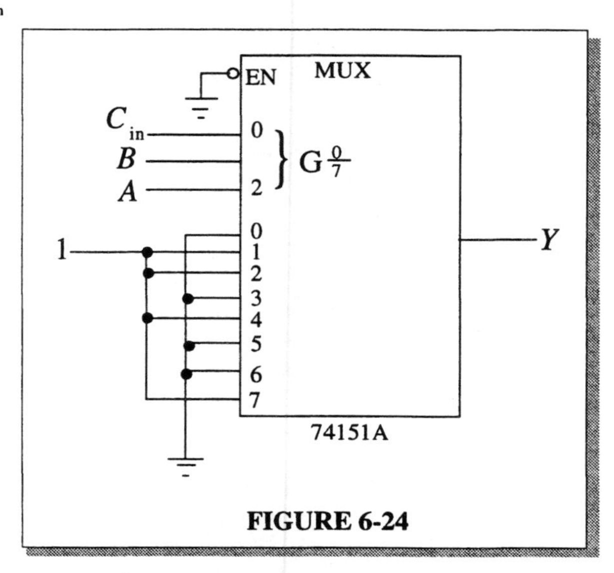

FIGURE 6-24

42. $Y = \overline{A_3}\,\overline{A_2}A_1\overline{A_0} + \overline{A_3}\,\overline{A_2}A_1A_0 + \overline{A_3}A_2A_1\overline{A_0} + \overline{A_3}A_2A_1A_0 + A_3\overline{A_2}\,\overline{A_1}\,\overline{A_0}$

$+ A_3\overline{A_2}A_1\overline{A_0} + A_3\overline{A_2}A_1A_0 + A_3A_2\overline{A_1}A_0 + A_3A_2A_1A_0$

See Figure 6-25.

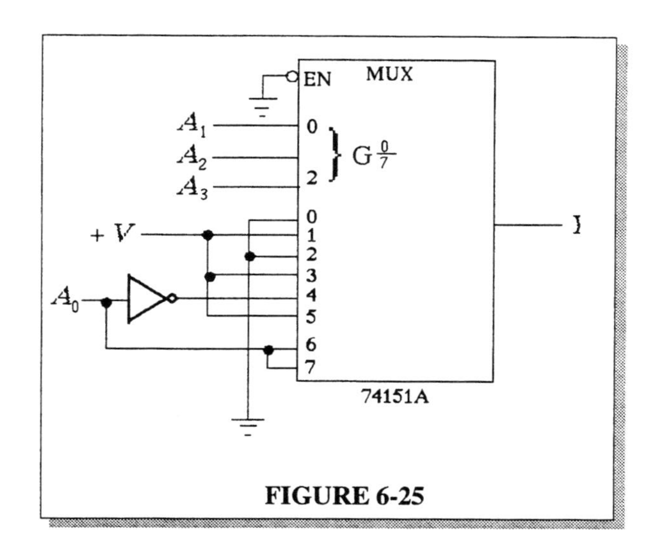

FIGURE 6-25

43. See Figure 6-26.

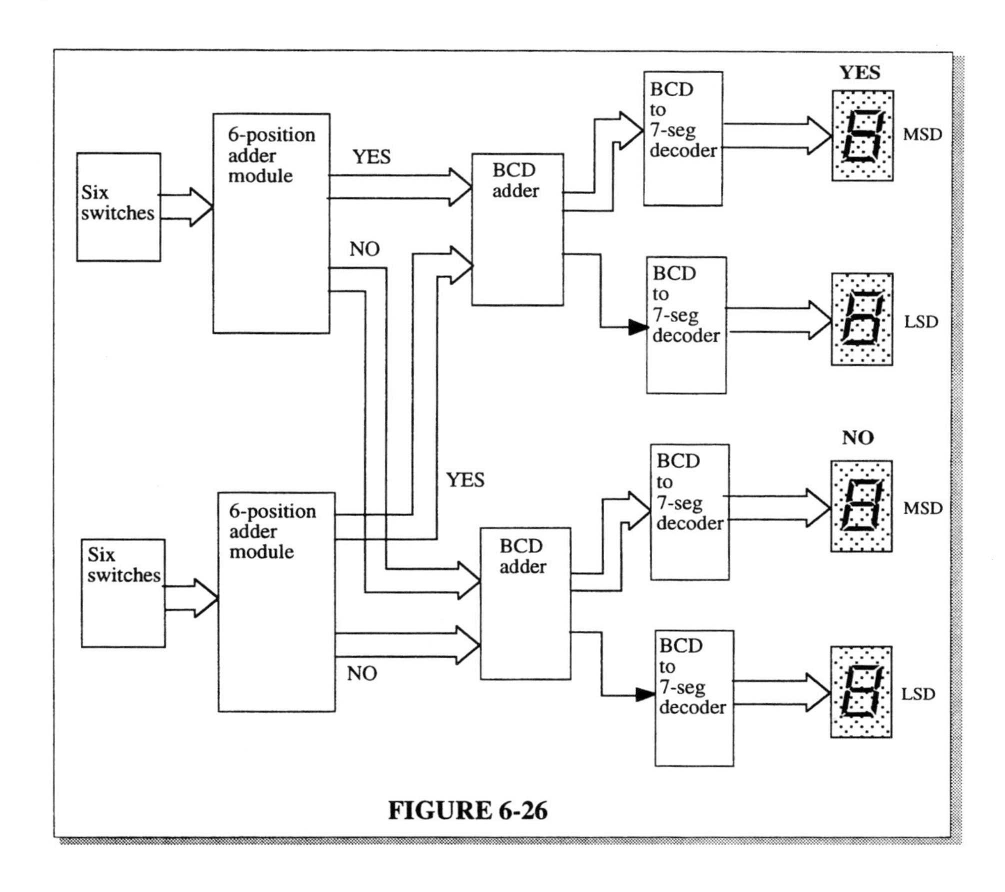

FIGURE 6-26

44. See Figure 6-27.

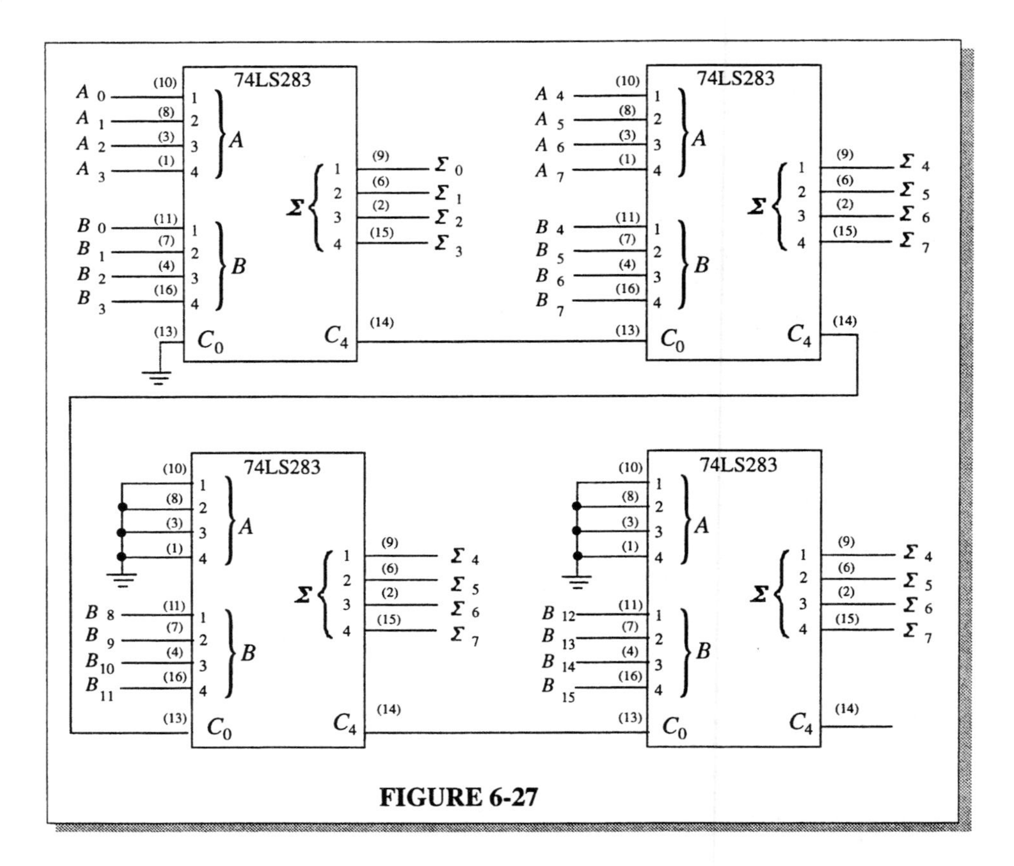

FIGURE 6-27

45. See Figure 6-28.

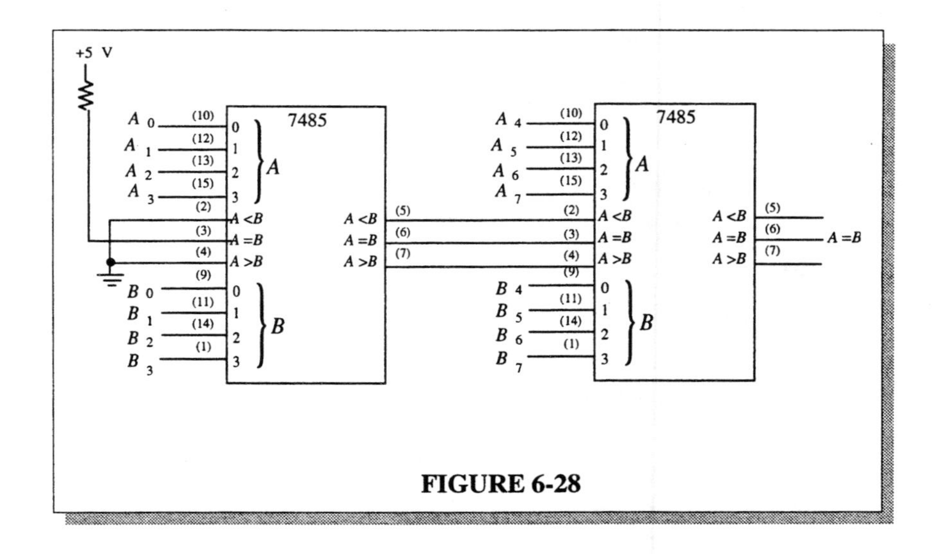

FIGURE 6-28

46. See Figure 6-29.

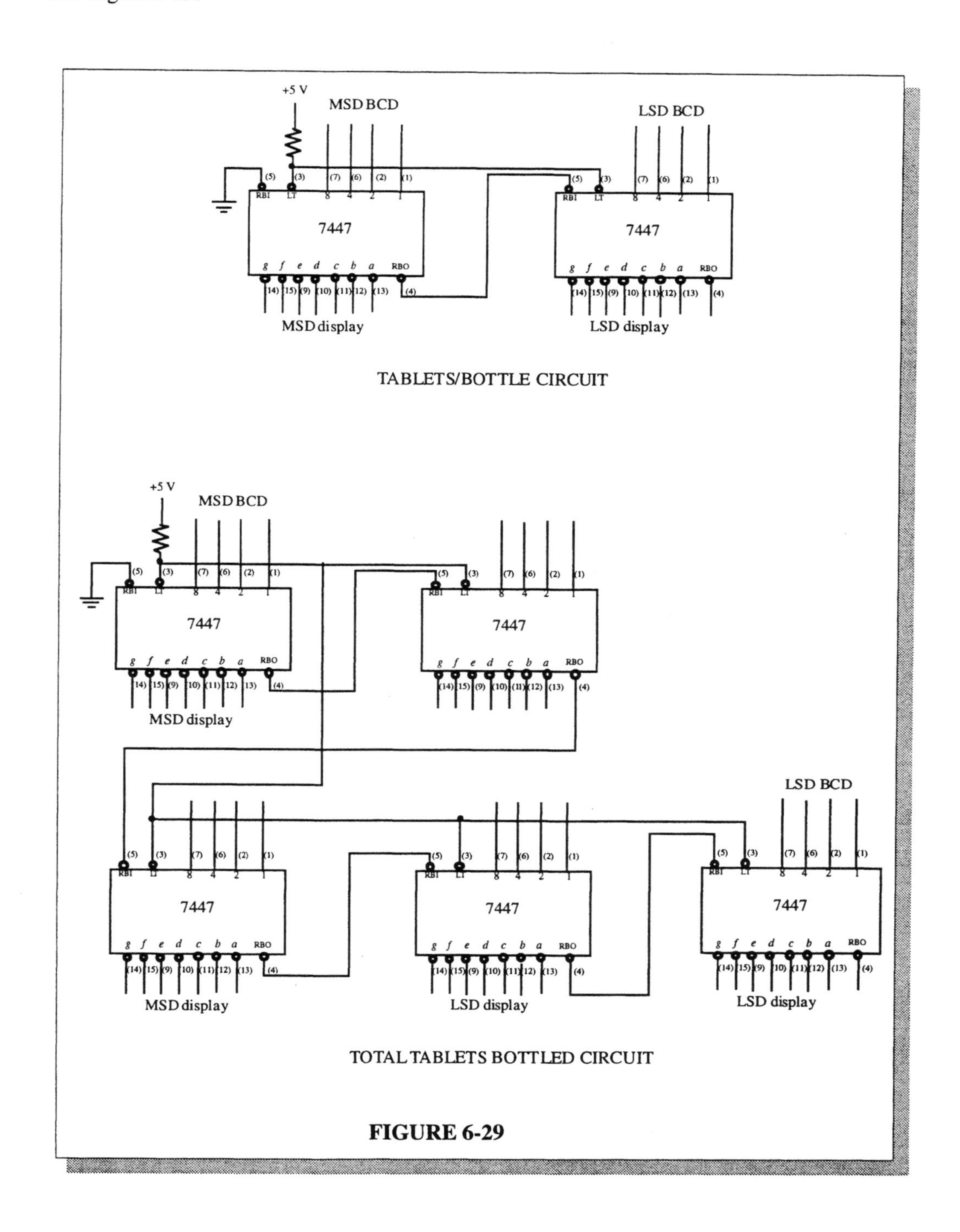

FIGURE 6-29

47. See Figure 6-30.

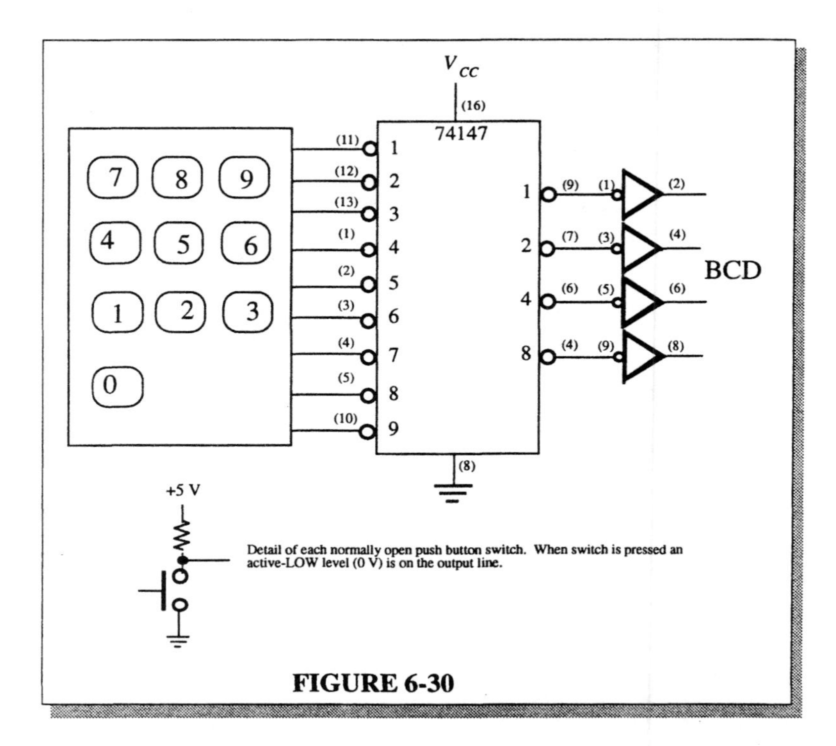

FIGURE 6-30

48. See Figure 6-31.

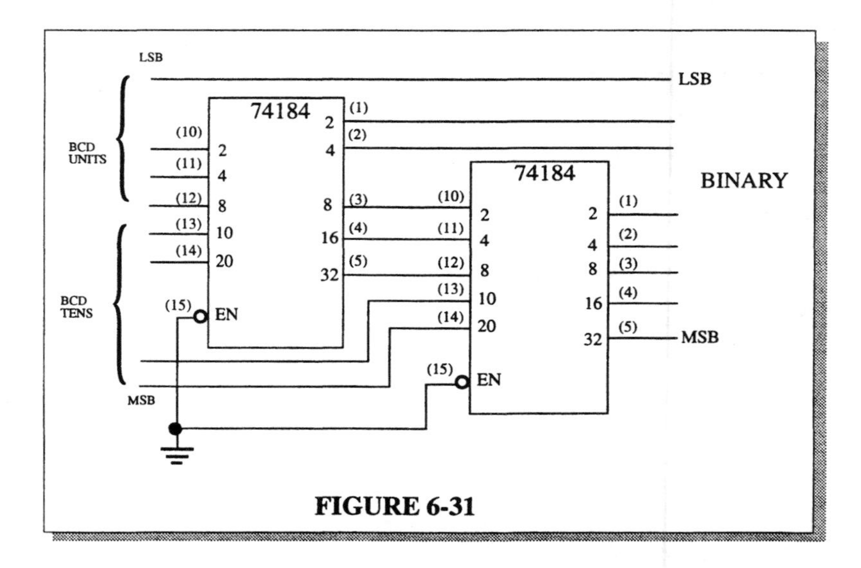

FIGURE 6-31

INTRODUCTION TO PROGRAMMABLE LOGIC DEVICES

1. Four types of PLDs are:

   Programmable read only memories (PROMs) with fixed AND and programmable OR arrays.
   Programmable logic arrays (PLAs) with programmable AND and programmable OR arrays.
   Programmable array logic (PAL) with programmable AND and fixed OR arrays.
   Generic array logic (GAL) with reprogrammable AND and fixed OR arrays.

2. $X_1 = \bar{A}B$
   $X_2 = \bar{A}\bar{B}$
   $X_3 = A\bar{B}$

3. $X_1 = \bar{A}BC$
   Row 1: blow A, B, $\bar{B}$, C, and $\bar{C}$ column fuses
   Row 2: blow A, $\bar{A}$, $\bar{B}$, C, and $\bar{C}$ column fuses
   Row 3: blow A, $\bar{A}$, B, $\bar{B}$, and $\bar{C}$ column fuses
   $X_2 = AB\bar{C}$
   Row 4: blow $\bar{A}$, B, $\bar{B}$, C, and $\bar{C}$ column fuses
   Row 5: blow A, $\bar{A}$, $\bar{B}$, C, and $\bar{C}$ column fuses
   Row 6: blow A, $\bar{A}$, B, $\bar{B}$, and C column fuses
   $X_3 = \bar{A}B\bar{C}$
   Row 7: blow A, B, $\bar{B}$, C, and $\bar{C}$ column fuses
   Row 8: blow A, $\bar{A}$, $\bar{B}$, C, and $\bar{C}$ column fuses
   Row 9: blow A, $\bar{A}$, B, $\bar{B}$, and C column fuses

4. $X = \bar{A}\bar{B}\bar{C} + \bar{A}B\bar{C} + A\bar{B}C$

5. (a) $Y = A\bar{B}C + \bar{A}B\bar{C} + ABC$. See Figure 7–1.

   (*b*) $Y = A\bar{B}C + \bar{A}\bar{B}C + A\bar{B}\bar{C} + \bar{A}BC = \bar{B}C + A\bar{B}\bar{C} + \bar{A}BC$
   See Figure 7–2.

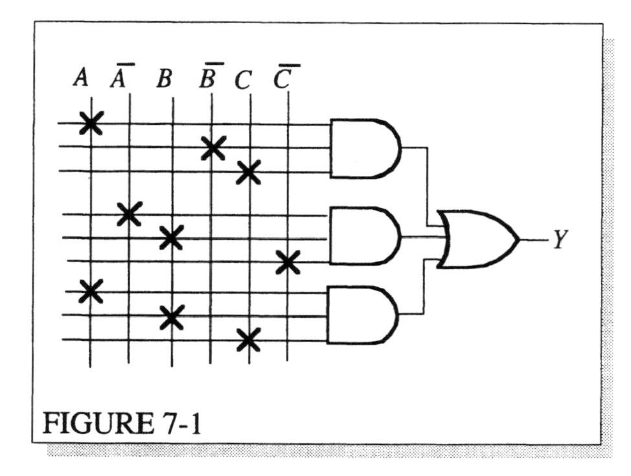

FIGURE 7-1

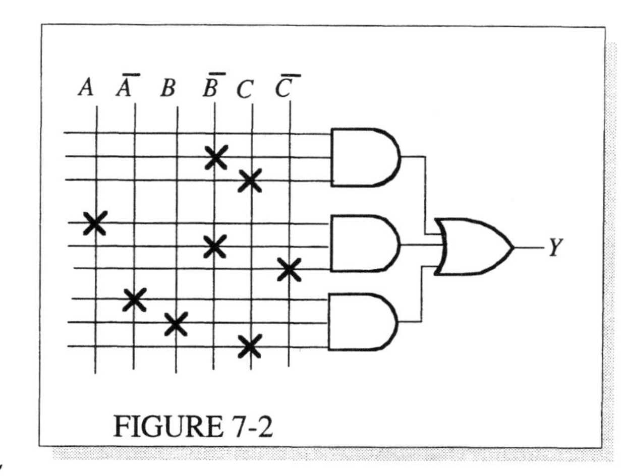

FIGURE 7-2

6. (a) PAL16H2: 16 inputs, 2 active-HIGH outputs
   (b) PAL12H6: 12 inputs, 6 active-HIGH outputs
   (c) PAL10P8: 10 inputs, 8 programmable-polarity outputs
   (d) PAL16R6: 16 inputs, 6 registered outputs

7. The programmed polarity output in a PAL works by selectively opening or closing a CMOS cell on one input of an XOR gate, causing the gate to invert or not invert the signal.

8. $X = A\bar{B}C + \bar{A}\bar{C} + B\bar{C} + \bar{B}$

9. (a) $X = A\bar{B}C + \bar{A}B\bar{C} + A\bar{B} + BC$. See Figure 7–3.

   (b) $X = (A + \bar{B} + \bar{C})(\bar{A} + B)$
   $= A\bar{A} + \bar{A}\bar{B} + \bar{A}\bar{C} + AB + B\bar{B} + B\bar{C}$
   $= \bar{A}\bar{B} + \bar{A}\bar{C} + AB + B\bar{C}$
   See Figure 7–4.

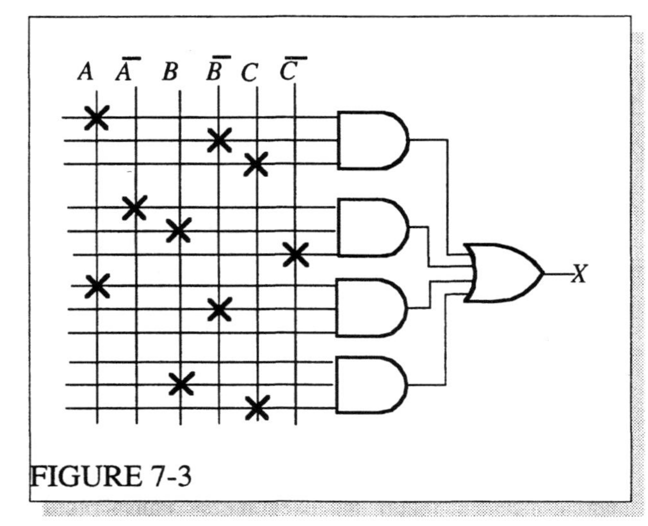

FIGURE 7-3

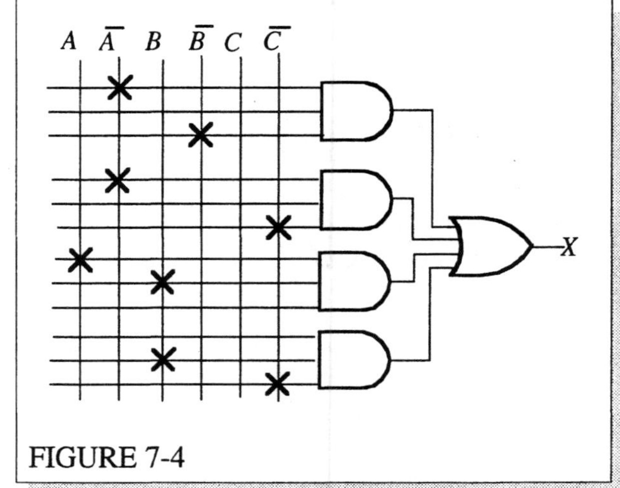

FIGURE 7-4

10. A GAL22V10 has
    (a) 12 dedicated inputs
    (b) 10 I/Os
    (c) 22 dedicated inputs plus I/Os.

11. If a GAL22V10 has 15 inputs, the maximum number of available outputs is 22-15 = 7.

12. The largest number of SOP terms that can be implemented with a GAL22V10 using one output is 16 with 21 input variables.

13. The select inputs to each OLMC in a GAL22V10 determine whether the associated pin is an in put or output, and whether the OLMC output is inverted or noninverted and combinational or registered.

14. $X = ABC\bar{D} + \bar{A}BCD + \bar{A}BC\bar{D} + \bar{A}B\bar{C}\bar{D} + ABCD + AB\bar{C}D$

15. A GAL16V8 has
    (a) 8 dedicated inputs
    (b) 8 I/Os
    (c) 16 dedicated inputs plus I/Os

16. If a GAL16V8 has 10 inputs, the maximum number of outputs is 16 - 10 = 6.

17. The largest SOP term using one output of a GAL16V8 has 8 product terms with 15 input variables.

18. PAL emulation with a GAL16V8 can be done with the simple mode, the complex mode, or the registered mode.

19. The items required to program a PLD are a computer, a compiler, and a programmer.

20. To program a PLD, the sequence to follow is:
    Edit a source file, compile the source file to a JEDEC file, download the JEDEC file to a programmer, and program the PLD. This assumes no errors requiring iterations in editing and simulating the design file.

21. An ispGAL22V10 differs from a GAL22V10 in that the isp version can be programmed after being inserted on a circuit board.

22. The ABEL symbols are:
    (a) NOT = !
    (b) AND = &
    (c) OR = #
    (d) XOR = $

23. (a) $\bar{A}\bar{B}\bar{C} \rightarrow$!A & !B & !C;
    (b) $A + B + \bar{C} \rightarrow$ A # B # !C;
    (c) $A(\overline{B + C}) \rightarrow$ A & !(B # C);

24. (a) W = X & Y & Z # !X & !Y & !Z; $\rightarrow W = XYZ + \bar{X}\bar{Y}\bar{Z}$
    (b) X = !(A & B) # !(!A & B & !C); $\rightarrow X = \overline{AB} + \overline{\bar{A}B\bar{C}}$
    (c) !Y = !(A # A & !B # !A & B); $\rightarrow \bar{Y} = \overline{A + A\bar{B} + \bar{A}B}$

25. (a) $X = A\bar{B}C\bar{D}E + A\bar{B}C + \bar{A}\bar{B} \rightarrow$ X = A & !B & C & !D & E # A & !B & C # !A & !B;
    (b) $Y = AB\bar{C} + DEF + \bar{G}H\bar{I} + J\bar{K}L$
    $\rightarrow$ Y = A & B & !C # D & E & F # !G & H &!I # J & !K & L;
    (c) $Z = (\bar{A} + \bar{B} + C + D)(E + \bar{F}) + G \rightarrow$ Z = (!A # !B # C # D) & (E # !F) # G;

26. The set declarations for the ABEL equations

```
X1 = A1& P1& Q1 # B1&!P1& Q1 # C1&P1&!Q1;
X2 = A2& P1& Q1 # B2&!P1& Q1 # C2&P1&!Q1;
X3 = A3& P1& Q1 # B3&!P1& Q1 # C3&P1&!Q1;
```

are

```
X = [X3, X2, X1];
A = [A3, A2, A1];
B = [B3, B2, B1];
C = [C3, C2, C1];
S = [Q1, P1];

X = A&(S= = 3)#B&(S = = 2)#C&(S = = 1);
```

27. An ABEL truth table for an exclusive-NOR is

```
TRUTH_TABLE  ([A, B] → [X])
                [0, 0] → [1];
                [0, 1] → [0];
                [1, 0] → [0];
                [1, 1] → [1];
```

28. An ABEL truth table for a BCD-to-decimal decoder with active-HIGH outputs is

```
TRUTH_TABLE  (BCD → [ a, b, c, d, e, f, g, h, i, j ])
               0 → [1, 0, 0, 0, 0, 0, 0, 0, 0, 0 ];
               1 → [0, 1, 0, 0, 0, 0, 0, 0, 0, 0 ];
               2 → [0, 0, 1, 0, 0, 0, 0, 0, 0, 0 ];
               3 → [0, 0, 0, 1, 0, 0, 0, 0, 0, 0 ];
               4 → [0, 0, 0, 0, 1, 0, 0, 0, 0, 0 ];
               5 → [0, 0, 0, 0, 0, 1, 0, 0, 0, 0 ];
               6 → [0, 0, 0, 0, 0, 0, 1, 0, 0, 0 ];
               7 → [0, 0, 0, 0, 0, 0, 0, 1, 0, 0 ];
               8 → [0, 0, 0, 0, 0, 0, 0, 0, 1, 0 ];
               9 → [0, 0, 0, 0, 0, 0, 0, 0, 0, 1 ];
```

29. The test vectors for the BCD-to-decomal decoder of probelm 28 is

```
TEST_VECTORS  (BCD → [ a, b, c, d, e, f, g, h, i, j ])
                      0 → [1, 0, 0, 0, 0, 0, 0, 0, 0, 0 ];
                      1 → [0, 1, 0, 0, 0, 0, 0, 0, 0, 0 ];
                      2 → [0, 0, 1, 0, 0, 0, 0, 0, 0, 0 ];
                      3 → [0, 0, 0, 1, 0, 0, 0, 0, 0, 0 ];
                      4 → [0, 0, 0, 0, 1, 0, 0, 0, 0, 0 ];
                      5 → [0, 0, 0, 0, 0, 1, 0, 0, 0, 0 ];
                      6 → [0, 0, 0, 0, 0, 0, 1, 0, 0, 0 ];
                      7 → [0, 0, 0, 0, 0, 0, 0, 1, 0, 0 ];
                      8 → [0, 0, 0, 0, 0, 0, 0, 0, 1, 0 ];
                      9 → [0, 0, 0, 0, 0, 0, 0, 0, 0, 1 ];
```

30. Changing the durations of the long and short time intervals will not affect the truth table for the state decoding and output logic.

31. A GAL16V8 can be used to implement the state decoding and output logic because the design requires only 2 input and 8 outputs.

32. A complete ABEL input file for a BCD-to-decimal decoder with active-HIGH outputs using a GAL22V10 is

```
Module      bcd_to_decimal_decoder
Title       'BCD-to-DECIMAL DECODER IN A GAL22V10'

        decoder       device        'P22V10';

        A0, A1, A2, A3              pin  1, 2, 3, 4;
        Y0, Y1, Y2, Y3, Y4          pin  23, 22, 21, 20, 19;
        Y5, Y6, Y7, Y8, Y9          pin  18, 17, 16, 15, 14:

Equations
        BCD = [A3, A2, A1, A0];
```

```
TRUTH_TABLE  (BCD → [Y0, Y1, Y2, Y3, Y4, Y5, Y6, Y7, Y8, Y9])
             0 → [ 1,  0,  0,  0,  0,  0,  0,  0,  0,  0 ];
             0 → [ 0,  1   0,  0,  0,  0,  0,  0,  0,  0 ];
             0 → [ 0,  0,  1,  0,  0,  0,  0,  0,  0,  0 ];
             0 → [ 0,  0,  0,  1,  0,  0,  0,  0,  0,  0 ];
             0 → [ 0,  0,  0,  0,  1,  0,  0,  0,  0,  0 ];
             0 → [ 0,  0,  0,  0,  0,  1,  0,  0,  0,  0 ];
             0 → [ 0,  0,  0,  0,  0,  0,  1,  0,  0,  0 ];
             0 → [ 0,  0,  0,  0,  0,  0,  0,  1,  0,  0 ];
             0 → [ 0,  0,  0,  0,  0,  0,  0,  0,  1,  0 ];
             0 → [ 0,  0,  0,  0,  0,  0,  0,  0,  0,  1 ];

TEST_VECTORS  (BCD → [Y0, Y1, Y2, Y3, Y4, Y5, Y6, Y7, Y8, Y9])
             0 → [ 1,  0,  0,  0,  0,  0,  0,  0,  0,  0 ];
             0 → [ 0,  1   0,  0,  0,  0,  0,  0,  0,  0 ];
             0 → [ 0,  0,  1,  0,  0,  0,  0,  0,  0,  0 ];
             0 → [ 0,  0,  0,  1,  0,  0,  0,  0,  0,  0 ];
             0 → [ 0,  0,  0,  0,  1,  0,  0,  0,  0,  0 ];
             0 → [ 0,  0,  0,  0,  0,  1,  0,  0,  0,  0 ];
             0 → [ 0,  0,  0,  0,  0,  0,  1,  0,  0,  0 ];
             0 → [ 0,  0,  0,  0,  0,  0,  0,  1,  0,  0 ];
             0 → [ 0,  0,  0,  0,  0,  0,  0,  0,  1,  0 ];
             0 → [ 0,  0,  0,  0,  0,  0,  0,  0,  0,  1 ];
```

33. An ABEL input file for a Decimal-to-BCD priority encoder with active-HIGH outputs using a GAL16V8 is

```
Module      decimal_to_bcd_encoder
Title       'DECIMAL-to-BCD PRIORITY ENCODER IN A GAL16V8'

        encoder_1     device        'P16V8';

        I0, I1, I2, I3, I4          pin  1, 2, 3, 4, 5;
        I5, I6, I7, I8, I9          pin  6, 7, 8, 9, 12;
        Q0, Q1, Q2, Q3              pin  16, 17, 18, 19;

Equations
        Q0 = I1&!I2&!I4&!I6&!I8 # I3&!I4&!I6&!I8 # I5&!I6&!I8 # I7&!I8 # I9;
        Q1 = I2&!I4&!I5&!I8&!I9 # I3&!I4&!I5&!I8&!I9 # I6&!I8&!I9 # I7&!I8&!I9;
        Q2 = I4&!I8&!I9 # I5&!I8&!I9 # I6&!I8&!I9 # I7&!I8&!I9;
        Q3 = I8 # I9;
```

Test vectors can be added, but for a priority encoder there are1024 input combinations. A partial testing can be considered to keep the test vectors to a reasonable size.

34. An ABEL input file for an 8-bit binary-to-BCD encoder using a GAL22V10 is

```
Module          binary_to_bcd_encoder
Title           '8-BIT BINARY-TO-BCD ENCODER IN A GAL22V10'

        encoder_2      device          'P22V10';

        A, B, C, D, E, F, G, H                   PIN;
        A0, A1, A2, A3, B0, B1, B2, B3           PIN;

"No pin assignments to maximize chance of design fit"

Equations
        BIN = [H,G,F,E,D,C,B,A];
        BCD1 = [A3,A2,A1,A0];
        BCD2 = [B3,B2,B1,B0];

        TRUTH_TABLE  (BIN → [BCD2, BCD1])
                              0 → [  0,     0  ];
                              1 → [  0,     1  ];
                              2 → [  0,     2  ];
                              •
                              •
                             61 → [  6,     1  ];
                             62 → [  6,     2  ];
                             63 → [  6,     3  ];
```

CHAPTER 8
FLIP-FLOPS AND RELATED DEVICES

1. See Figure 8-1.

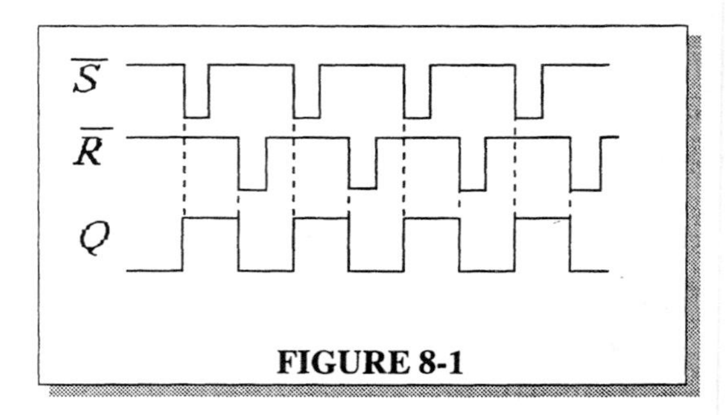

FIGURE 8-1

2. See Figure 8-2.

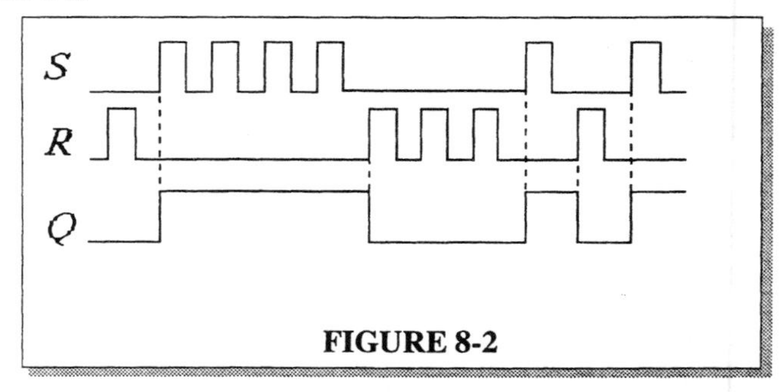

FIGURE 8-2

3. See Figure 8-3.

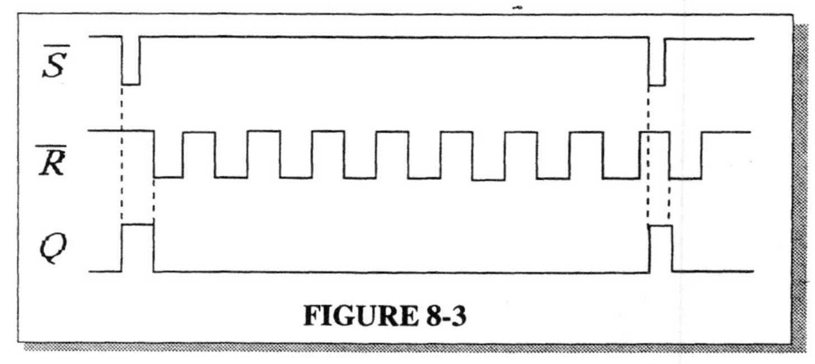

FIGURE 8-3

4. See Figure 8-4.

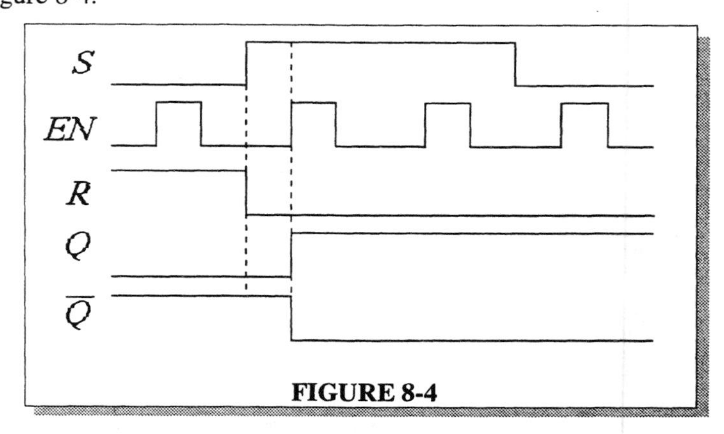

FIGURE 8-4

5. See Figure 8-5.

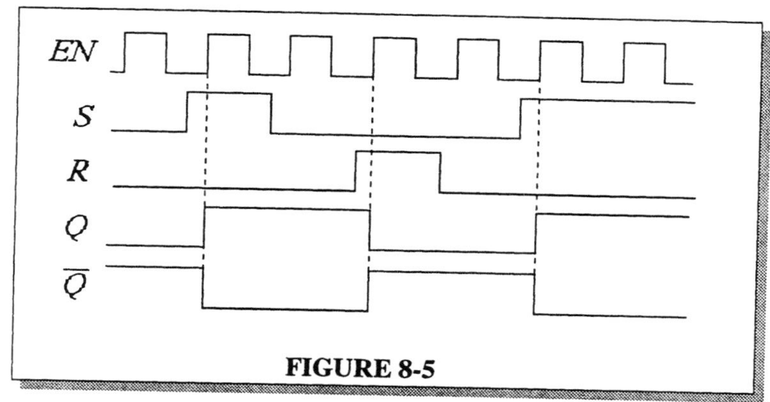

FIGURE 8-5

6. See Figure 8-6.

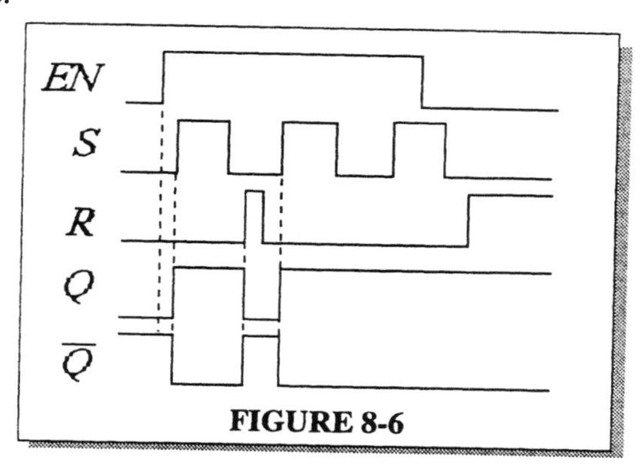

FIGURE 8-6

7. See Figure 8-7.

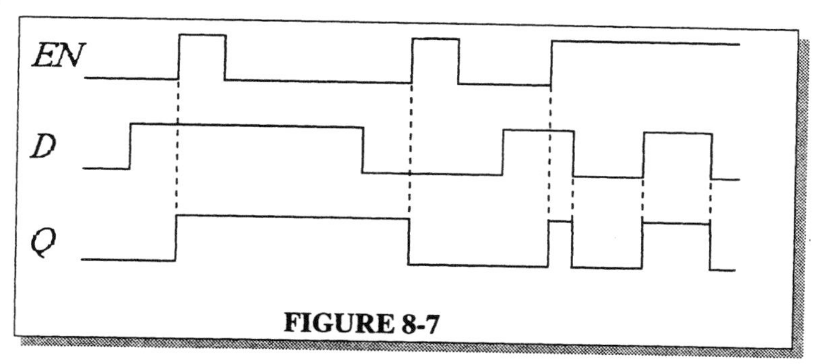

FIGURE 8-7

8. See Figure 8-8.

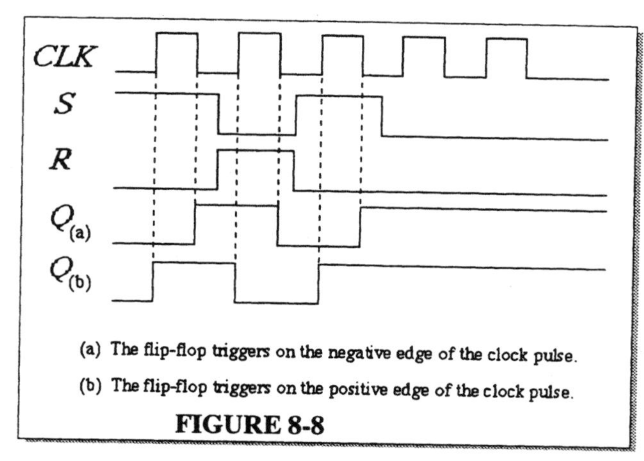

(a) The flip-flop triggers on the negative edge of the clock pulse.

(b) The flip-flop triggers on the positive edge of the clock pulse.

FIGURE 8-8

9. See Figure 8-9.

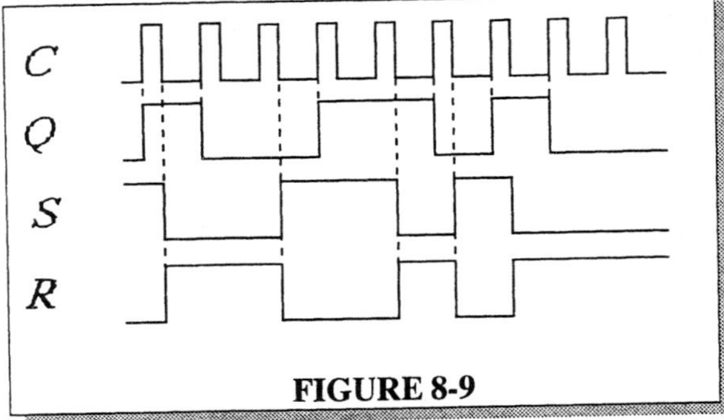

FIGURE 8-9

10. See Figure 8-10.

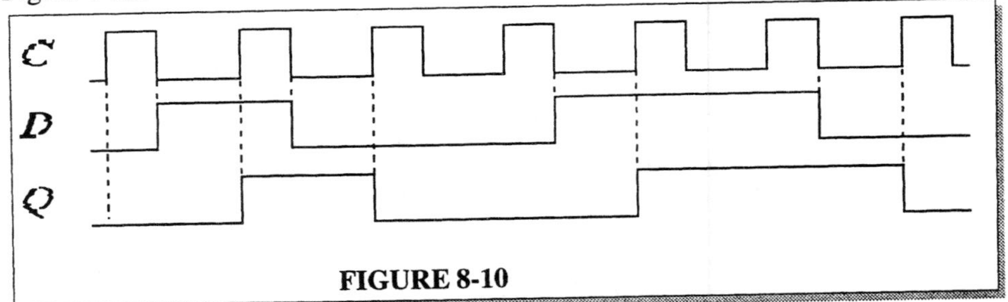

FIGURE 8-10

11. See Figure 8-11.

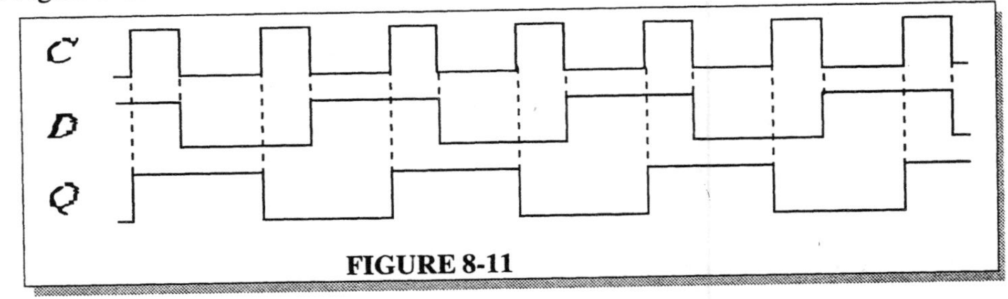

FIGURE 8-11

12. See Figure 8-12.

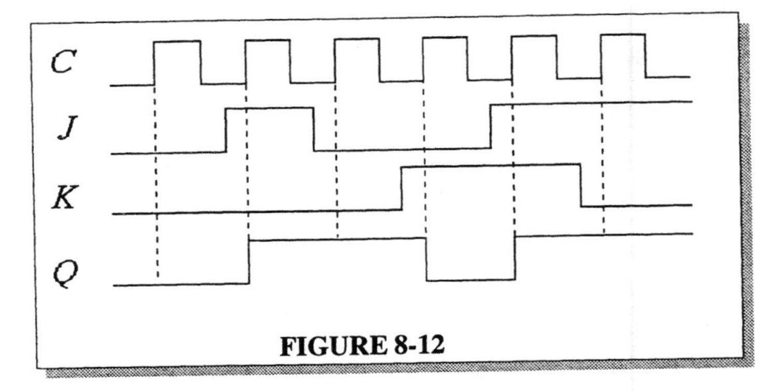

FIGURE 8-12

13. See Figure 8-13.

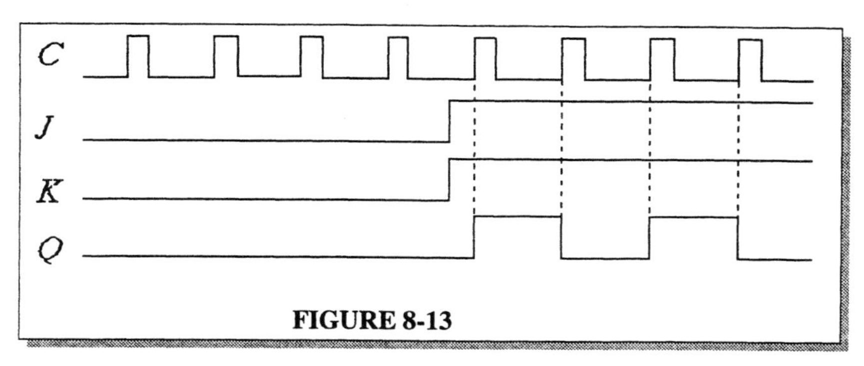

FIGURE 8-13

14. See Figure 8-14.

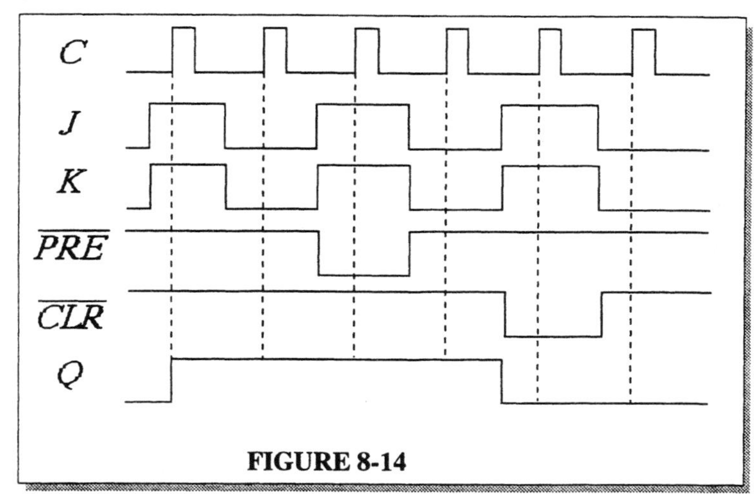

FIGURE 8-14

15. See Figure 8-15.

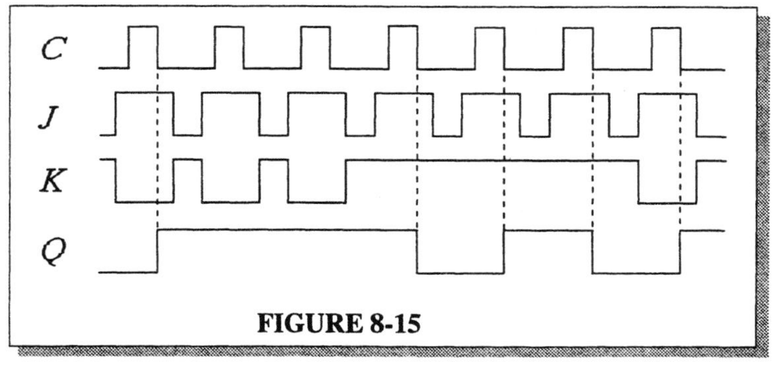

FIGURE 8-15

16. *J*: 0010000
K: 0000100
Q: 0011000

17. See Figure 8-16.

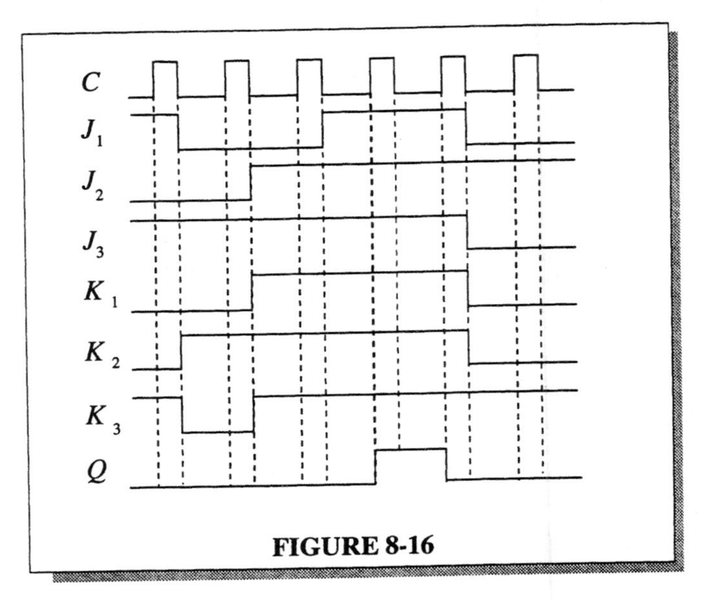

FIGURE 8-16

18. See Figure 8-17.

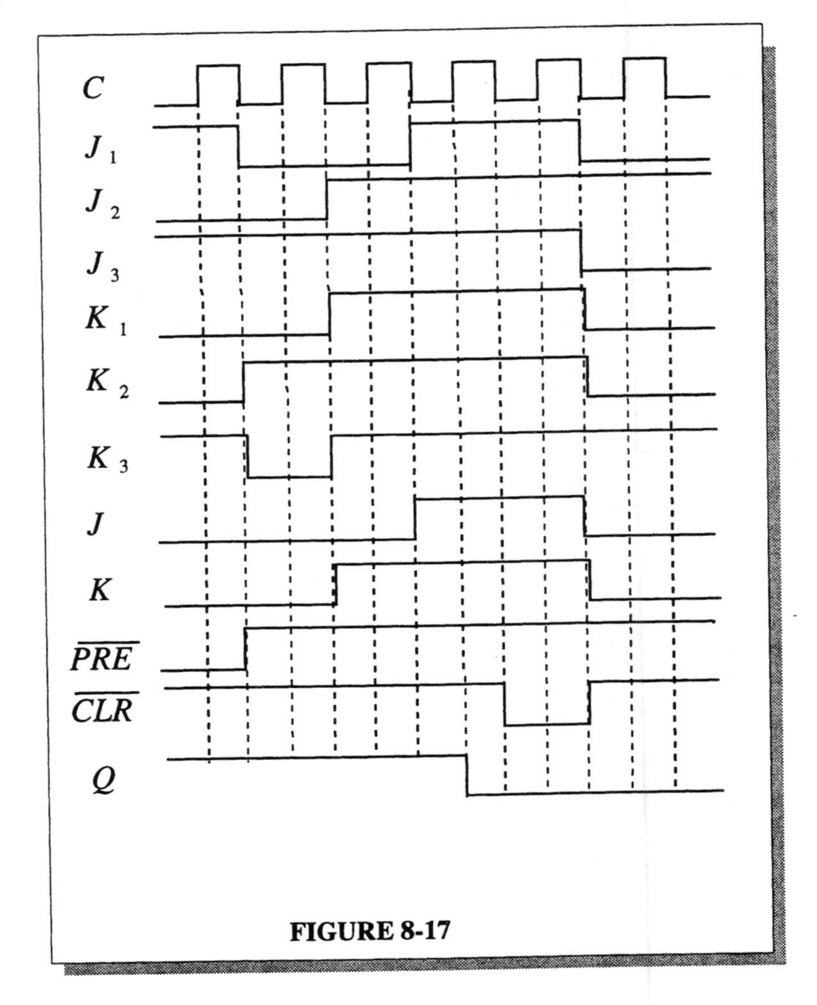

FIGURE 8-17

19. See Figure 8-18.

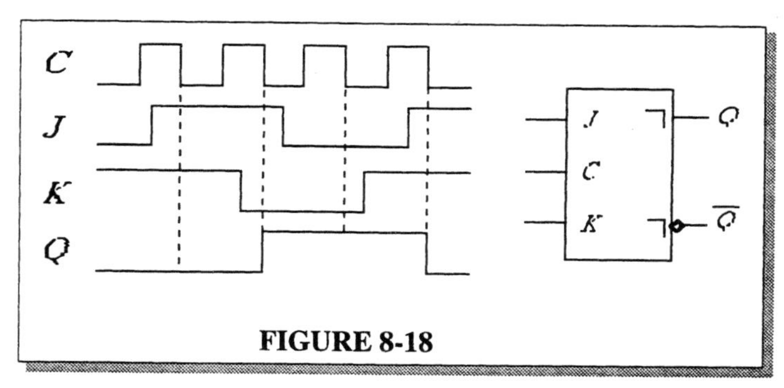

FIGURE 8-18

20. See Figure 8-19.

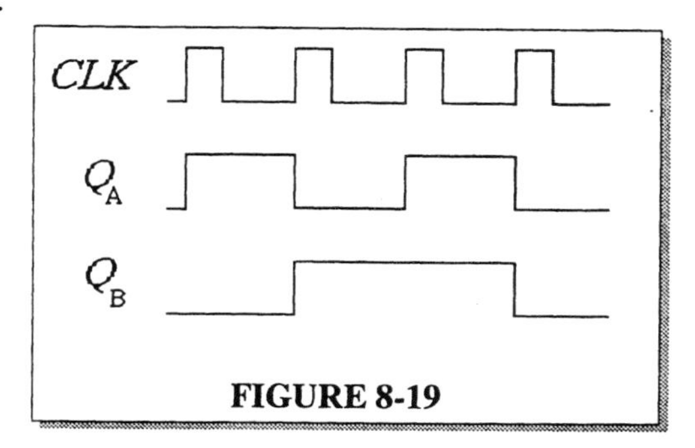

FIGURE 8-19

21. See Figure 8-20.

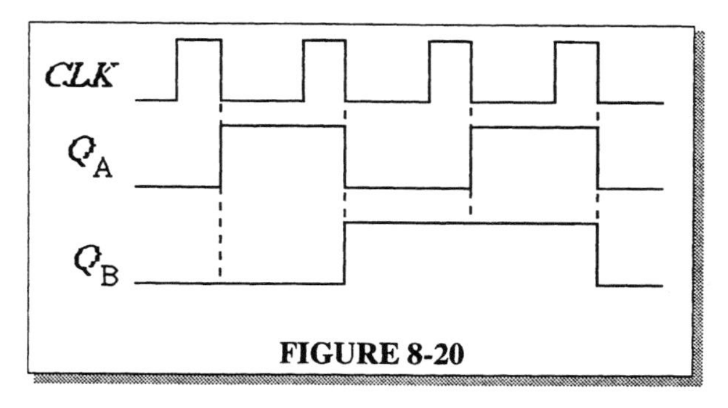

FIGURE 8-20

22. See Figure 8-21.

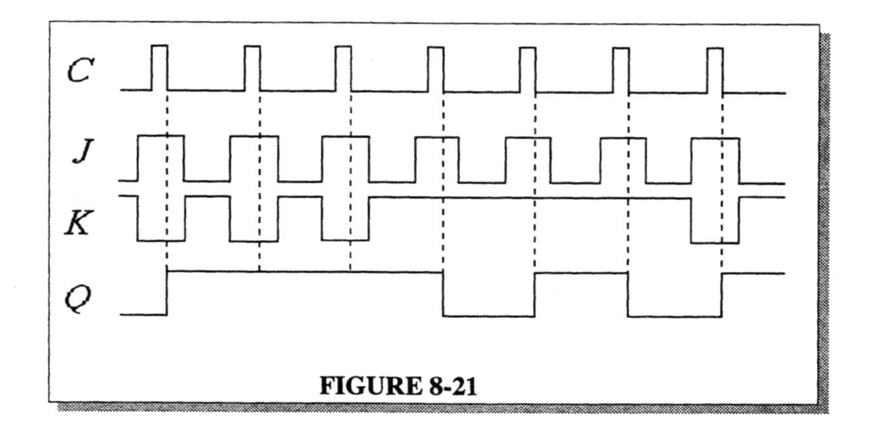

FIGURE 8-21

23. See Figure 8-22.

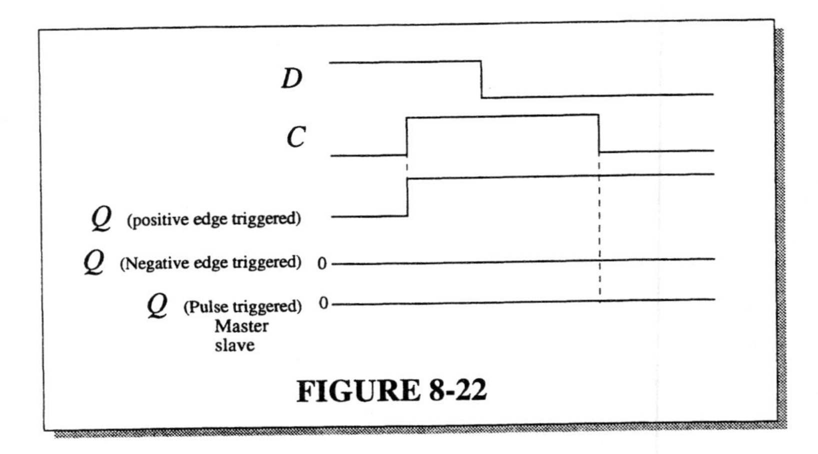

FIGURE 8-22

24. t_{pLH} (Clock to Q):
Time from triggering edge of clock to the LOW-to-HIGH transition of the Q output.

t_{pHL} (Clock to Q):
Time from triggering edge of clock to the HIGH-to-LOW transition of the Q output.

t_{pLH} ($\overline{\text{PRE}}$ to Q):
Time from assertion of the Preset input to the LOW-to-HIGH transition of the Q output.

t_{pHL} ($\overline{\text{CLR}}$ to Q):
Time from assertion of the clear input to the HIGH-to-LOW transition of the Q output.

25. $T_{min} = 30 \text{ ns} + 37 \text{ ns} = 67 \text{ ns}$

$$f_{max} = \frac{1}{T_{min}} = 14.9 \text{ MHz}$$

26. See Figure 8-23.

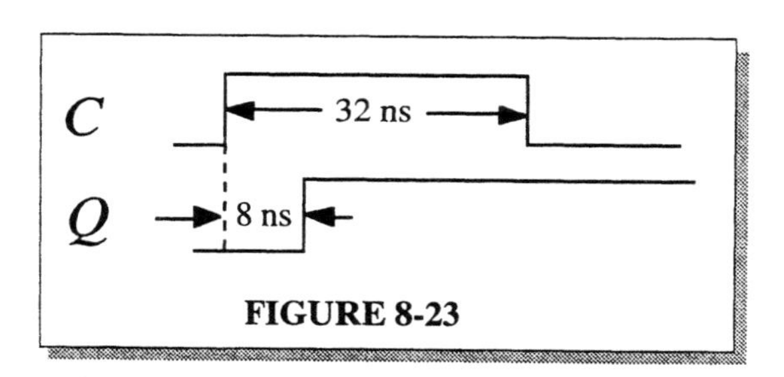

FIGURE 8-23

27. $I_T = 15(10\text{ mA}) = \mathbf{150\text{ mA}}$

$P_T = (5\text{ V})(150\text{ mA}) = \mathbf{750\text{ mW}}$

28. See Figure 8-24.

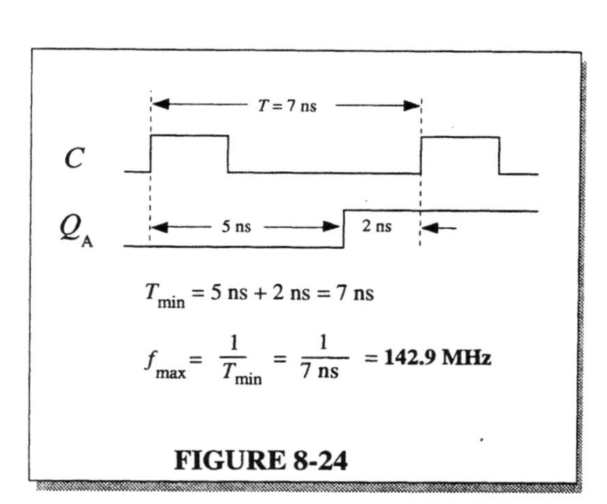

FIGURE 8-24

29. See Figure 8-25.

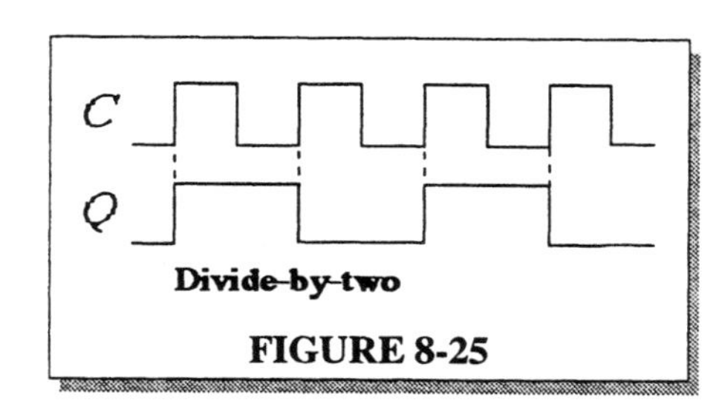

FIGURE 8-25

30. See Figure 8-26.

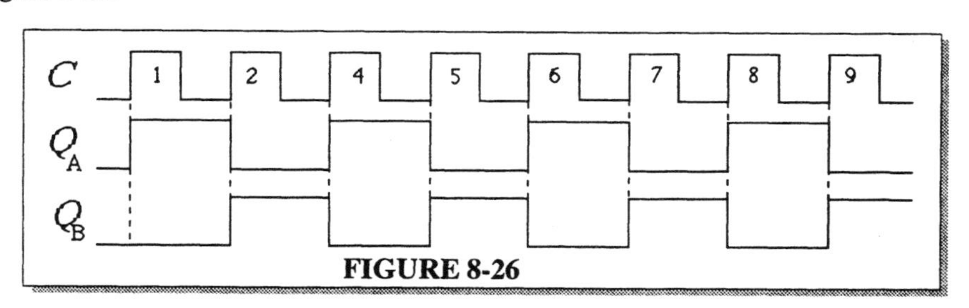

FIGURE 8-26

31. $t_w = 0.7RC_{EXT} = 0.7(3.3\text{ k}\Omega)(2000\text{ pF}) = \mathbf{4.62\ \mu s}$

32. $R_X = \frac{t_W}{RC_{EXT}} - 0.7 = \frac{5000\text{ ns}}{(0.32 \times 10{,}000\text{ pF})} - 0.7 = 1.56\text{ k}\Omega$

33. See Figure 8-27.

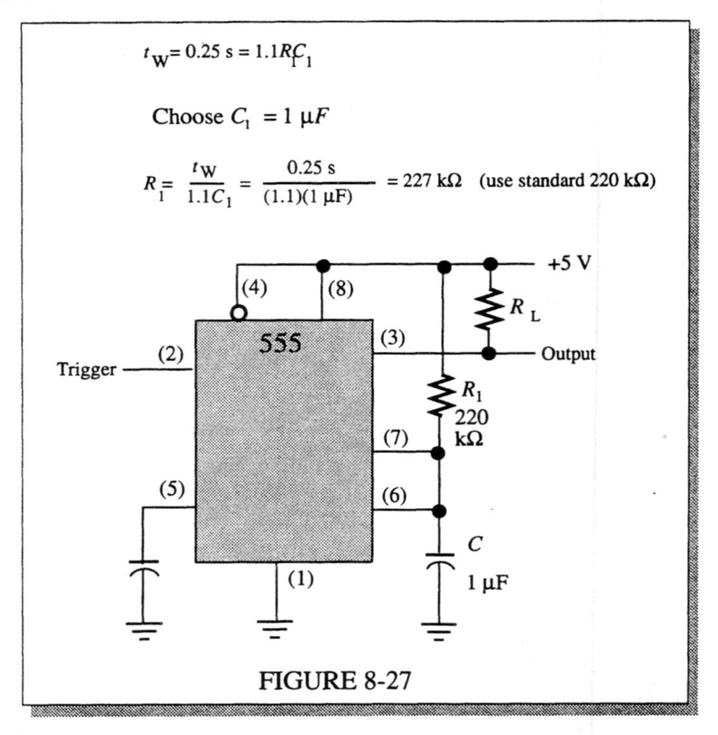

FIGURE 8-27

34.

$$f = \frac{1}{0.7(R_1 + 2R_2)C_2} = \frac{1}{0.7(1000\ \Omega + 4000\ \Omega)(0.01\ \mu F)} = 28.6\text{ kHz}$$

35.

$$T = \frac{1}{f} = \frac{1}{20\text{ kHz}} = 50\ \mu\text{s}$$

For a duty cycle of 75%:

$t_H = 37.5\ \mu\text{s}$ *and* $t_L = 12.5\ \mu\text{s}$

$$R_1 + R_2 = \frac{t_H}{0.7C} = \frac{37.5\ \mu\text{s}}{0.7(0.002\ \mu\text{F})} = 26{,}786\ \Omega$$

$$R_2 = \frac{t_L}{0.7C} = \frac{12.5\ \mu\text{s}}{0.7(0.002\ \mu\text{F})} = 98{,}929\ \Omega \text{ (use 9.1 k}\Omega\text{)}$$

$$R_1 = 26{,}786\ \Omega - R_2 = 26{,}786\ \Omega - 8{,}929\ \Omega = 17{,}857\ \Omega \text{ (use 18 k}\Omega\text{)}$$

36. The flip-flop in Figure 8-96 of the text has an internally open *J* input.

37. The wire from pin 6 to pin 10 and the gound wire are reversed. Pin 7 should be at ground and pin 6 connected to pin 10.

38. See Figure 8-28.

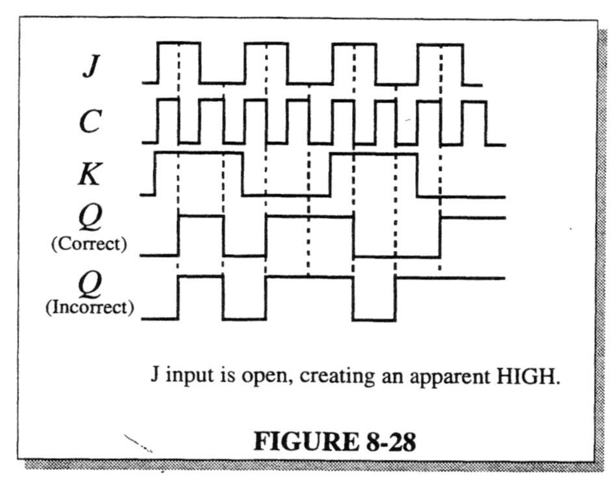

FIGURE 8-28

39. Since none of the flip-flops change, the problem must be a fault that affects all of them. The two functions common to all the flip-flops are the clock(CLK) and clear (CLR)inputs. One of these lines must be shorte to groundbecause a LOWon either one will prevent the flip-flops from chan ging state. Most likely, the $\overline{CLR}$ line is shorted to ground because if the clock line were shorted chances are that all of the flip-flops would not have ended up reset when the power was turned on unless an initial LOW was applied to the $\overline{CLR}$ at power on.

40. Small differences in the switching times of flip-flop A and flip-flop B due to propagation delay cause the glitches as shown in the expanded timing diagram in Figure 8-29. The delays are exaggerated greatly for purposes of illustration. Glitches are eliminated by strobing the output with the clock pulse.

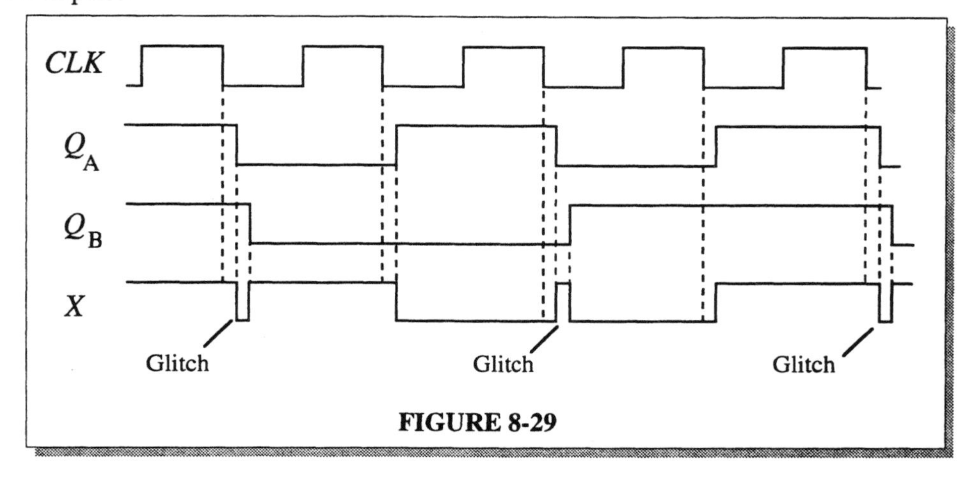

FIGURE 8-29

41. (a) See Figure 8-30.

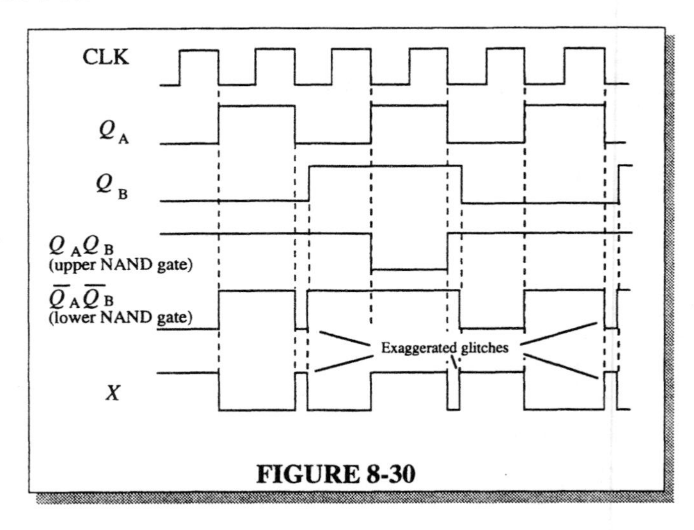

FIGURE 8-30

(b) K_B open acts as a HIGH and the operation is normal. The timing diagram is the same as Figure 8-30.

(c) See Figure 8-31.

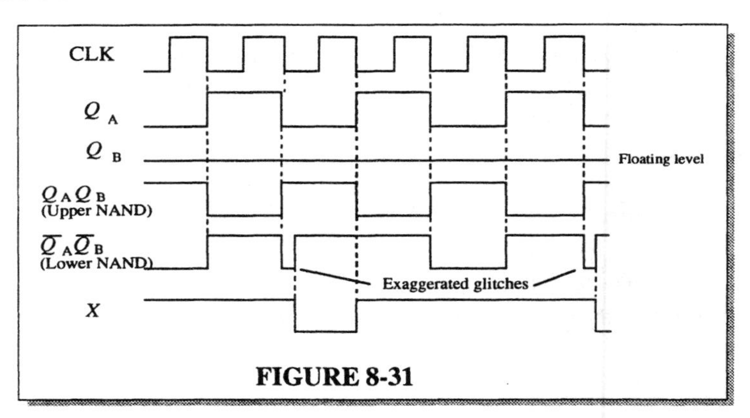

FIGURE 8-31

(d) X remains LOW if $Q_B = 1$ ($\overline{Q}_B = 0$). X follows $\overline{Q}_A$ if $Q_B = 0$ ($\overline{Q}_B = 1$).

(e) See Figure 8-32.

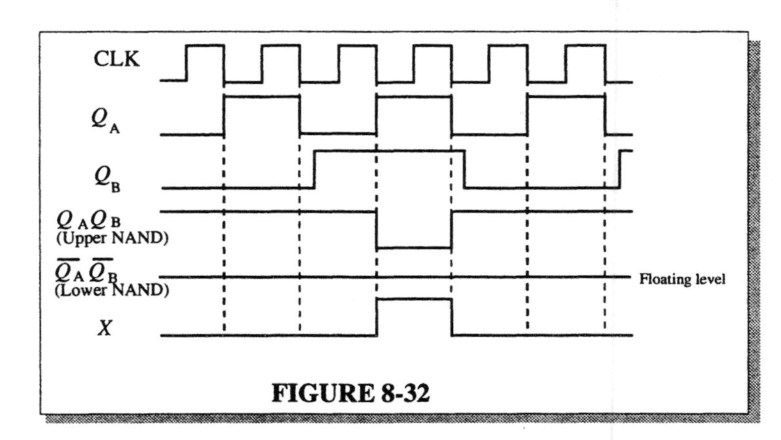

FIGURE 8-32

42. $t_w = 0.7RC_{EXT}$

One-shot A: $t_w = 0.7(0.22\ \mu F)(100\ k\Omega) = 15.4$ ms

One-shot B: $t_w = 0.7(0.1\ \mu F)(100\ k\Omega) = 7$ ms

The pulse width of one shot A is apparently not controlled by the external components and the one-shot is producing its minimum pulse width of about 40 ns. An *open pin 11* would cause this problem. See Figure 8-33.

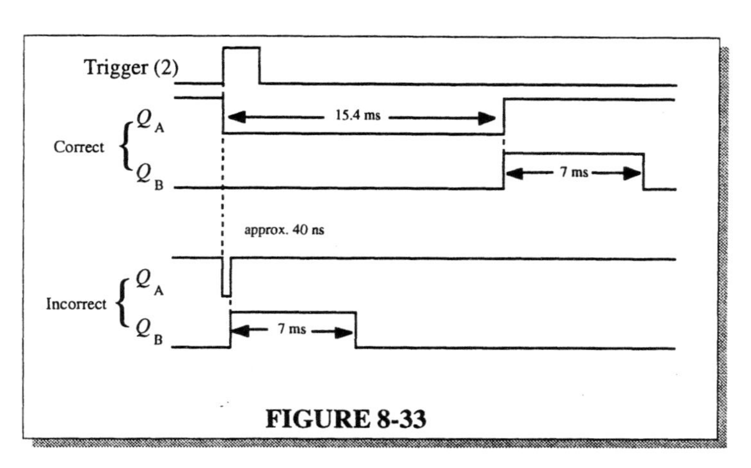

FIGURE 8-33

43. For the 4 s timer let $C_1 = 1\ \mu F$

$$R_1 = \frac{4\ s}{(1.1)(1\ \mu F)} = 3.63\ M\Omega \quad \text{(use 3.9 M}\Omega\text{)}$$

For the 25 s timer let $C_1 = 2.2\ \mu F$

$$R_1 = \frac{25\ s}{(1.1)(2.2\ \mu F)} = 10.3\ M\Omega \quad \text{(use 10 M}\Omega\text{)}$$

See Figure 8-34.

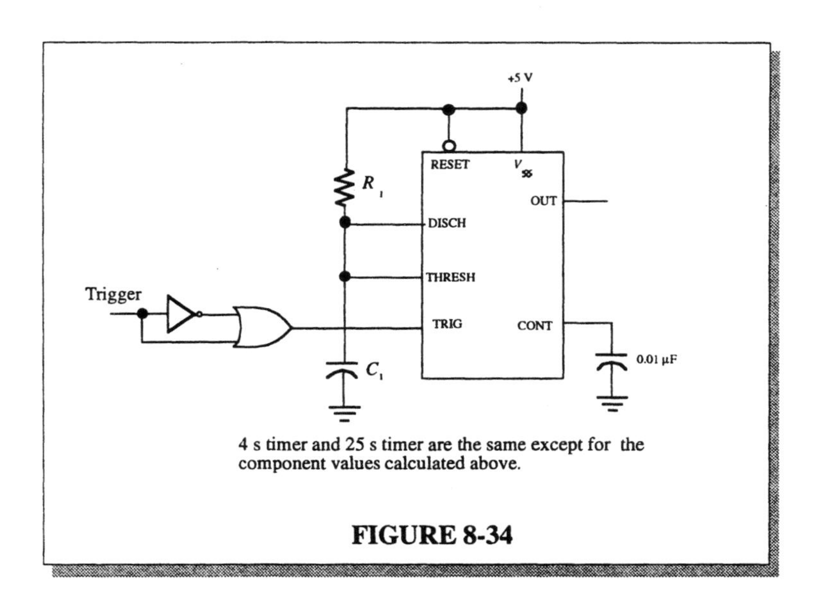

FIGURE 8-34

44. For a 10 kHz, 25% duty cycle 555 oscillator (see Figure 8-35):

$$10\text{ kHz} = \frac{1.44}{(R_1 + 2R_2)C_1}$$

If $C_1 = 0.1\ \mu\text{F}$,

$$R_1 + 2R_2 = \frac{1.44}{(10\text{ kHz})(0.1\ \mu\text{F})} = 1.44\text{ k}\Omega$$

$$R_1 = 1.44\text{ k}\Omega - 2R_2$$

$$\text{Also, } 0.25 = \frac{R_1}{R_1 + R_2} = \frac{1.44\text{ k}\Omega - 2R_2}{1.44\text{ k}\Omega - 2R_2 + R_2} = \frac{1.44\text{ k}\Omega - 2R_2}{1.44\text{ k}\Omega - R_2}$$

$$360\ \Omega - 0.25\ R_2 = 1.44\text{ k}\Omega - 2R_2$$

$$1.75R_2 = 1.08\text{ k}\Omega$$

$$R_2 = 617\ \Omega \quad (\text{use } 620\ \Omega)$$

$$R_1 = 1.44\text{ k}\Omega - 2(620\ \Omega) = 200\ \Omega$$

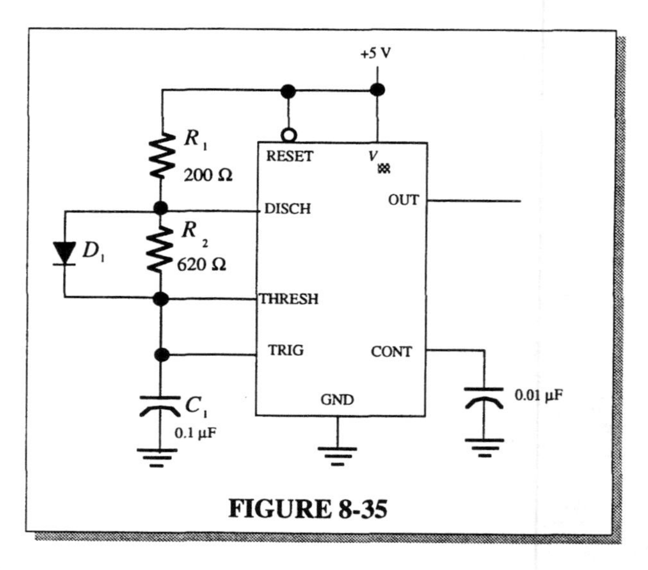

FIGURE 8-35

45. See Figure 8-36.

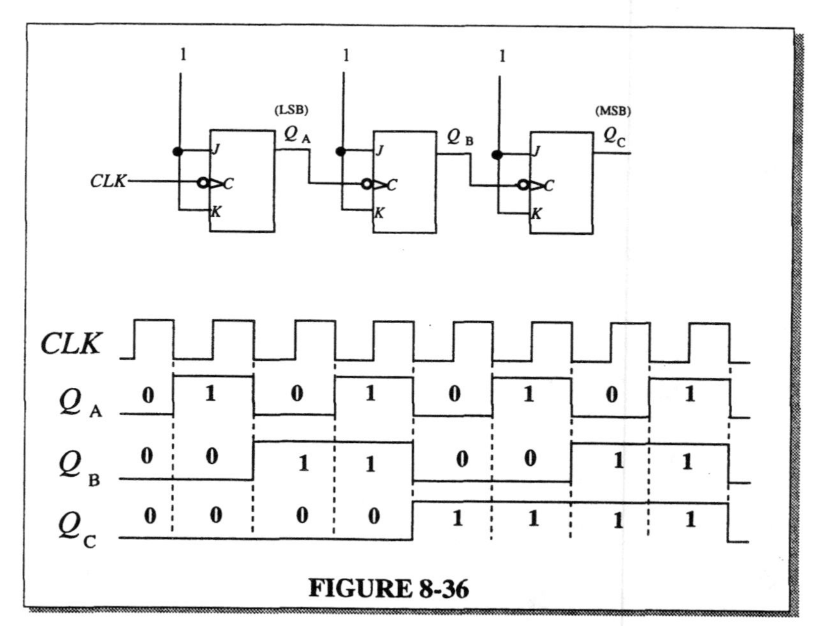

FIGURE 8-36

47. See Figure 8-37 for one possibility.

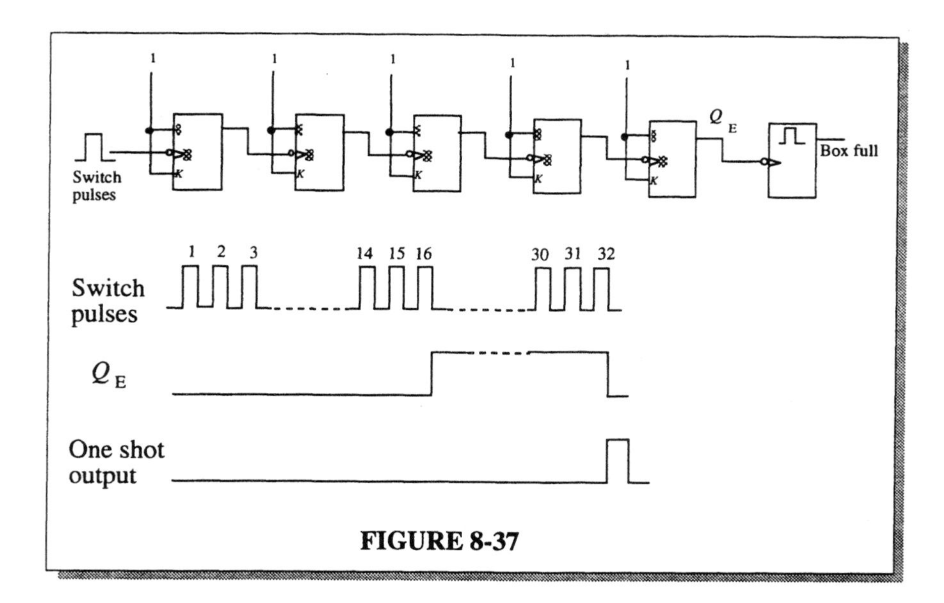

FIGURE 8-37

47. Changes required for the system to incorporate a 15 s left turn signal on main:

1. Change the 2-bit gray code sequence to a 3-bit sequence.
2. Add decoding logic to the State Decoder to decode the turn signal state.
3. Change the Output Logic to incorporate the turn signal output.
4. Change the Trigger Logic incorporate a trigger output for the turn signal. timer.
5. Add a 15 second timer.

See Figure 8-38.

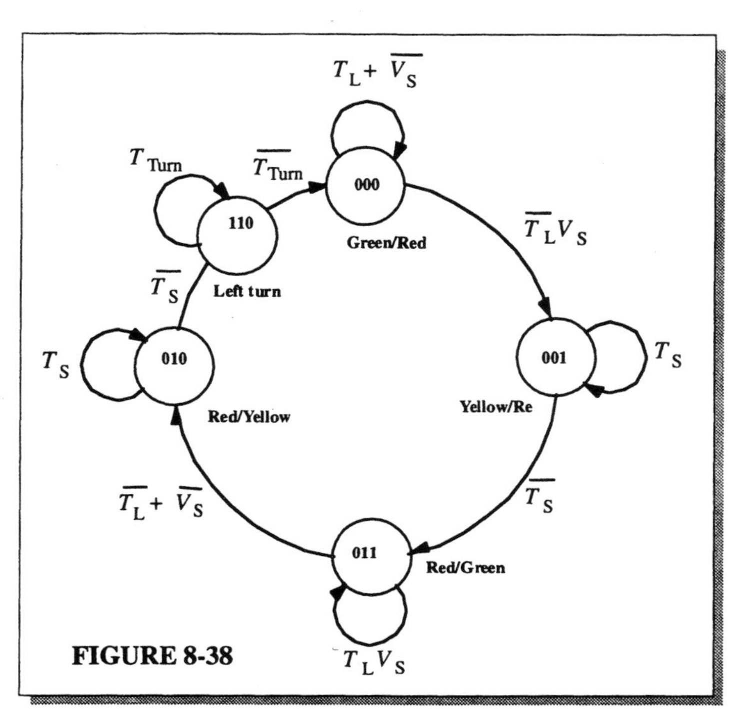

FIGURE 8-38

CHAPTER 9
COUNTERS

1. See Figure 9-1.

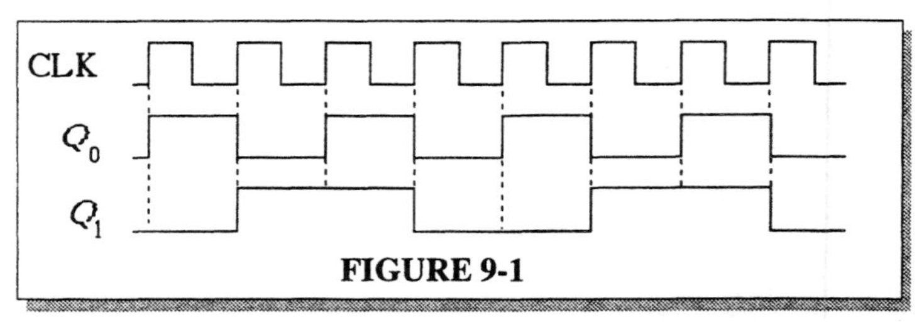

FIGURE 9-1

2. See Figure 9-2.

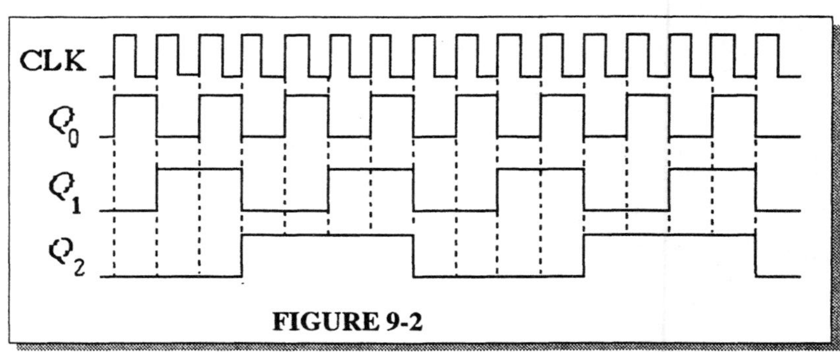

FIGURE 9-2

3. $t_{p(max)} = 3(8\text{ ns}) = \mathbf{24\ ns}$

Worst-case delay occurs when all flip-flops change state from 011 to 100 or from 111 to 000.

4. See Figure 9-3.

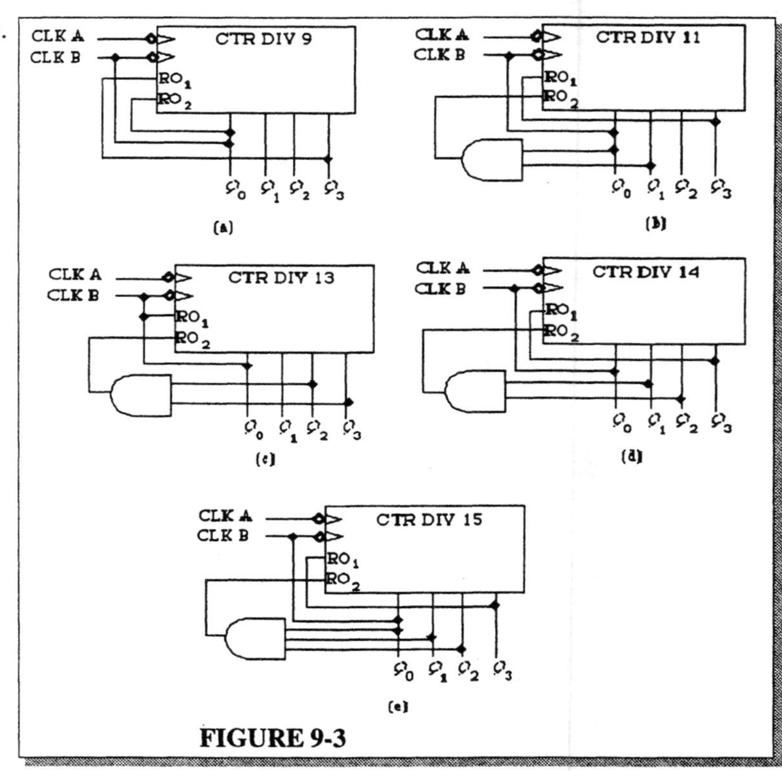

FIGURE 9-3

5. **8 ns**, the time it takes one flip-flop to change state.

6. See Figure 9-4.

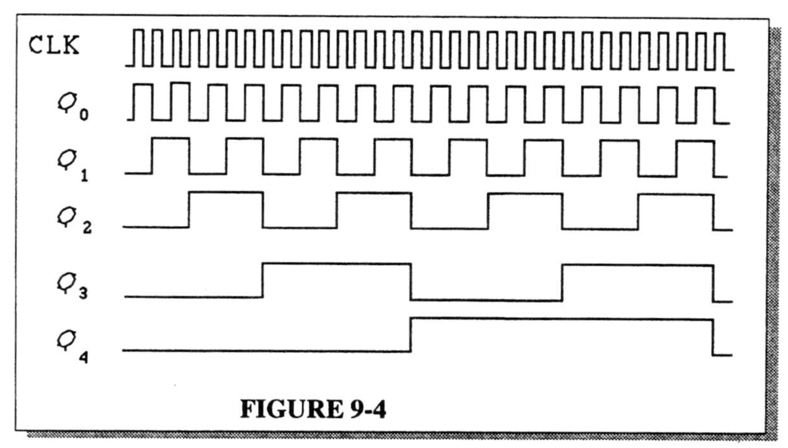

FIGURE 9-4

7. Each flip-flop is initially reset.

CLK	J_0K_0	J_1K_1	J_2K_2	J_3K_3	Q_0	Q_1	Q_2	Q_3
1	1	0	0	0	1	0	0	0
2	1	1	0	0	0	1	0	0
3	1	0	0	0	1	1	0	0
4	1	1	1	0	0	0	1	0
5	1	0	0	0	1	0	1	0
6	1	1	0	0	0	1	1	0
7	1	0	0	0	1	1	1	0
8	1	1	1	1	0	0	0	1
9	1	0	0	0	1	0	0	1
10	1	0	0	1	0	0	0	0

8. See Figure 9-5.

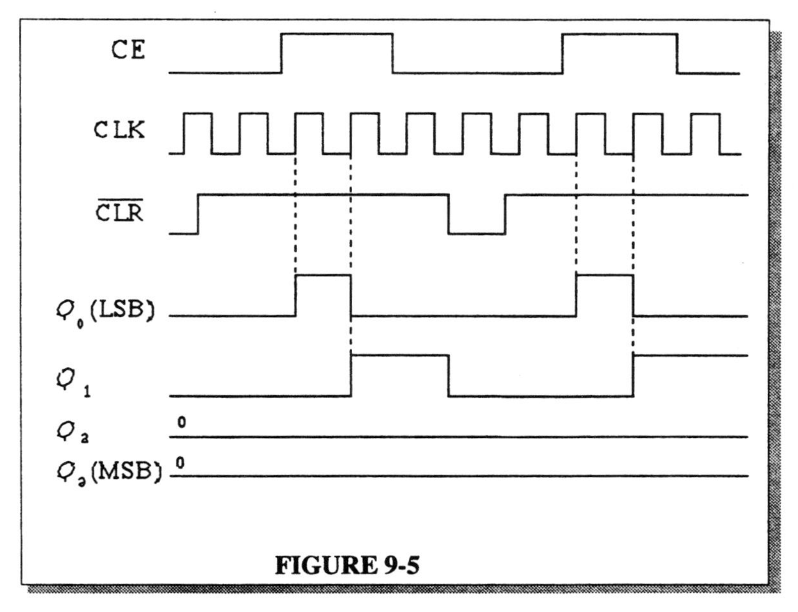

FIGURE 9-5

9. See Figure 9-6.

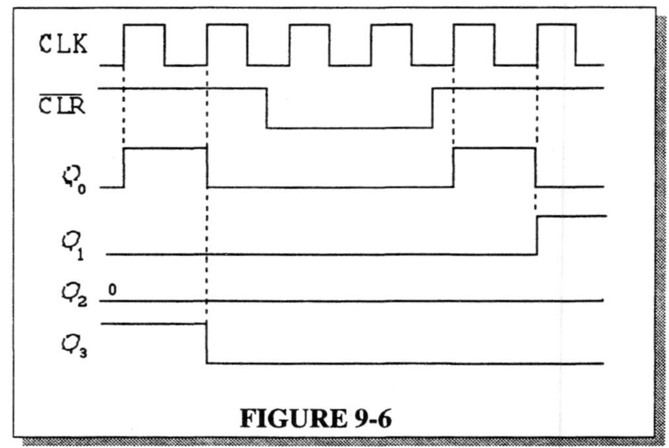

FIGURE 9-6

10. See Figure 9-7.

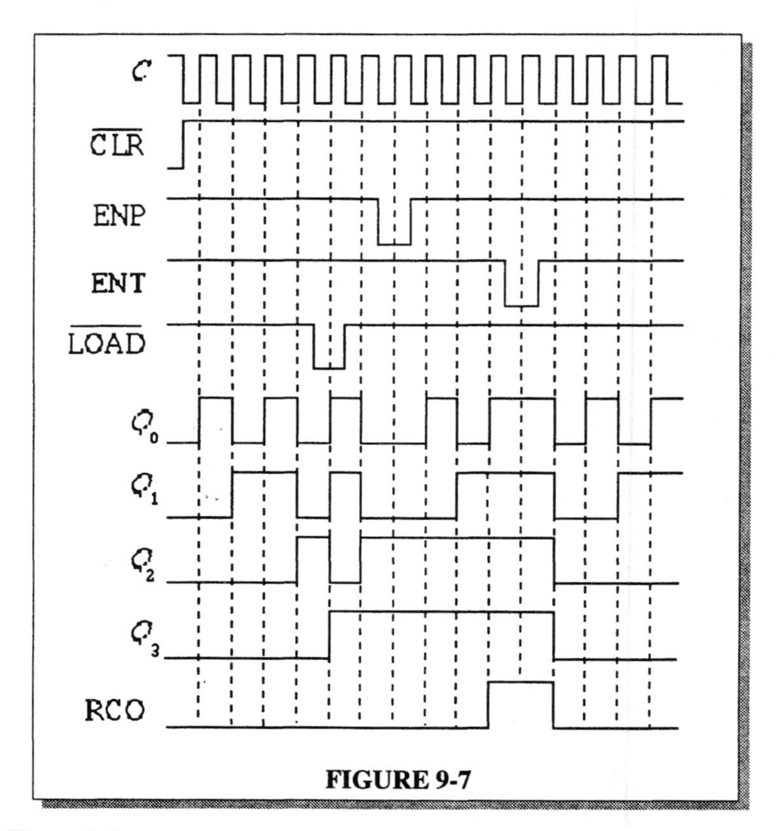

FIGURE 9-7

11. See Figure 9-8.

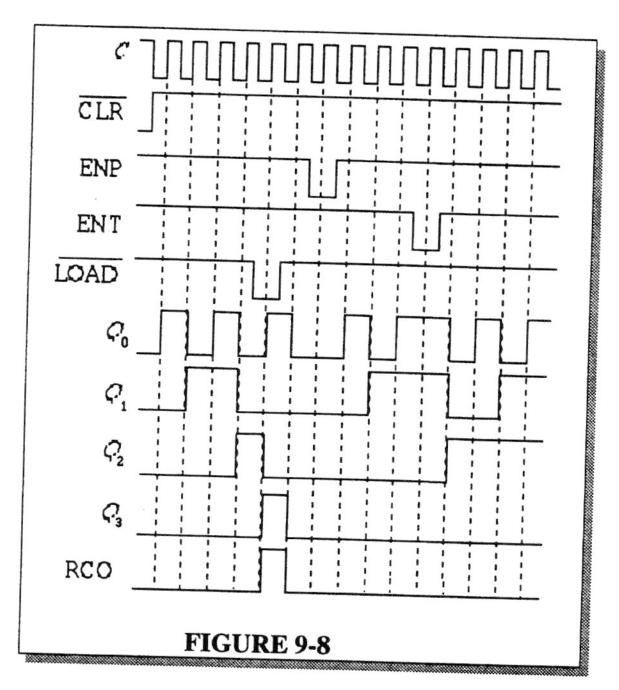

FIGURE 9-8

12. See Figure 9-9.

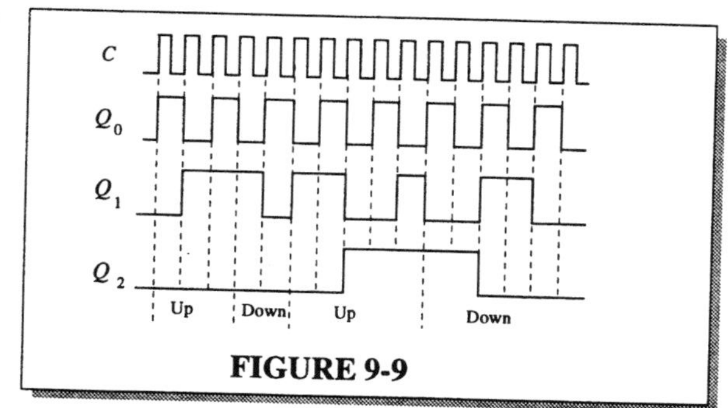

FIGURE 9-9

13. See Figure 9-10.

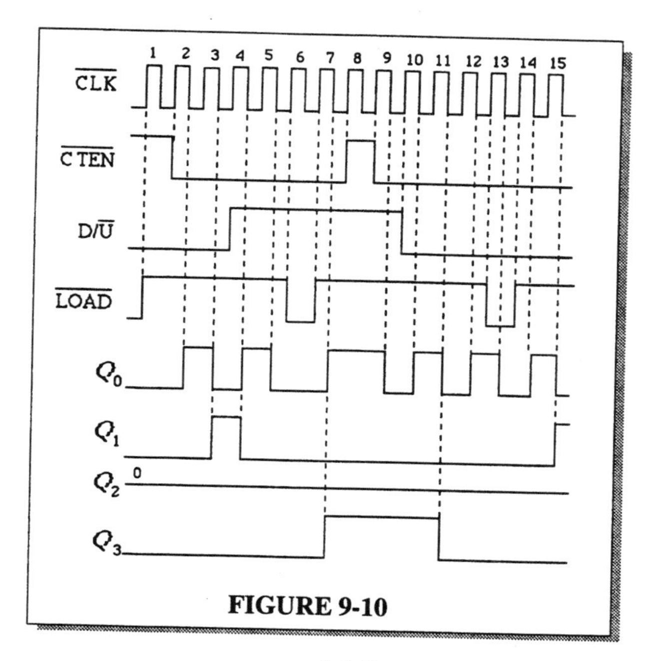

FIGURE 9-10

14.

	Q_2	Q_1	Q_0	D_2	D_1	D_0
Initially	0	0	0	0	0	1
At CLK 1	0	0	1	0	1	1
At CLK 2	0	1	1	1	1	1
At CLK 3	1	1	1	1	1	0
At CLK 4	1	1	0	1	0	0
At CLK 5	1	0	0	0	0	1
At CLK 6	0	0	1	0	1	1

The sequence is 000 to 001 to 011 to 111 to 110 to 100 and back to 001, etc.

15.

	FF3	FF2	FF1	FF0	Q_3	Q_2	Q_1	Q_0
Initially	Tog	Tog	Tog	Tog	0	0	0	0
After CLK 1	NC	NC	NC	Tog	1	1	1	1
After CLK 2	NC	NC	Tog	Tog	1	1	1	0
After CLK 3	NC	Tog	Tog	Tog	1	1	0	1
After CLK 4	Tog	Tog	Tog	Tog	1	0	1	0
After CLK 5	Tog	Tog	Tog	Tog	0	1	0	1

Tog = toggle, NC = no change

The counter locks ujp in the 1010 and 0101 states, alternating between them.

16.

NEXT STATE TABLE

Present State		Next State	
Q_1	Q_0	Q_1	Q_0
0	0	1	0
1	0	0	1
0	1	1	1
1	1	0	0

TRANSITION TABLE

Output State Transitions (Present state to next state)		Flip-flop inputs			
Q_1	Q_0	J_1	K_1	J_0	K_0
0 to 1	0 to 0	1	X	0	X
1 to 0	0 to 1	X	1	1	X
0 to 1	1 to 1	1	X	X	0
1 to 0	1 to 0	X	1	X	1

See Figure 9-11.

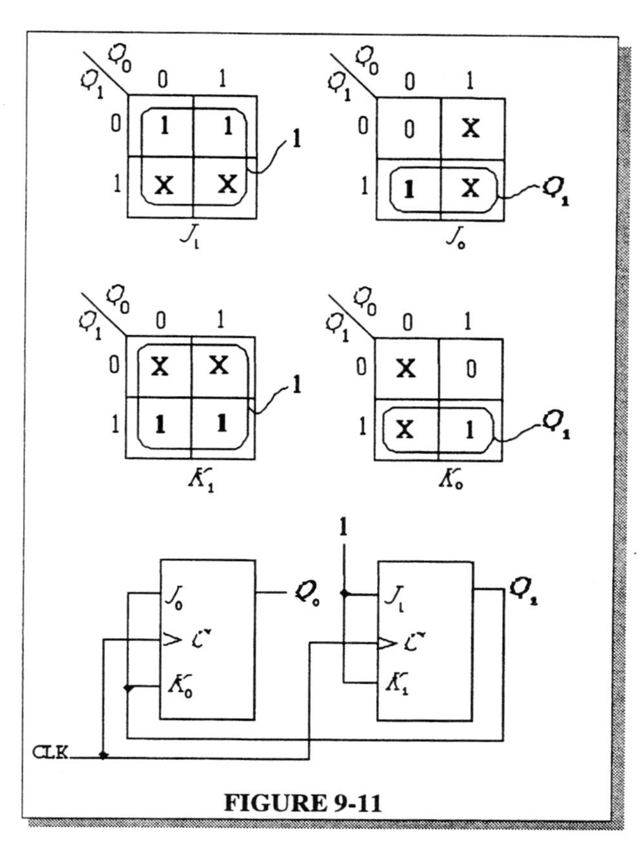

FIGURE 9-11

17. NEXT STATE TABLE

Present State Q_2	Q_1	Q_0	Next State Q_2	Q_1	Q_0
0	0	1	1	0	0
1	0	0	0	1	1
0	1	1	1	0	1
1	0	1	1	1	1
1	1	1	1	1	0
1	1	0	0	1	0
0	1	0	0	0	1

TRANSITION TABLE

Output State Transitions (Present state to next state) Q_2	Q_1	Q_0	Flip-flop inputs J_2	K_2	J_1	K_1	J_0	K_0
0 to 1	0 to 0	1 to 0	1	X	0	X	X	1
1 to 0	0 to 1	0 to 1	X	1	1	X	1	X
0 to 1	1 to 0	1 to 1	1	X	X	1	X	0
1 to 1	0 to 1	1 to 1	X	0	1	X	X	0
1 to 1	1 to 1	1 to 0	X	0	X	0	X	1
0 to 0	1 to 0	0 to 1	0	X	X	1	1	X
1 to 0`	1 to 1	0 to 0	X	1	X	0	0	X

See Figure 9-12.

$Q_2Q_1 \backslash Q_0$	0	1
00	X	1
01	0	1
11	X	X
10	X	X

$J_2 = Q_0$

$Q_2Q_1 \backslash Q_0$	0	1
00	X	0
01	X	X
11	X	X
10	1	1

$J_1 = Q_2$

$Q_2Q_1 \backslash Q_0$	0	1
00	X	X
01	1	X
11	0	X
10	1	X

$J_0 = \overline{Q}_1 + \overline{Q}_2$

$Q_2Q_1 \backslash Q_0$	0	1
00	X	X
01	X	X
11	1	0
10	1	0

$K_2 = \overline{Q}_0$

$Q_2Q_1 \backslash Q_0$	0	1
00	X	X
01	1	1
11	0	0
10	X	X

$K_1 = \overline{Q}_2$

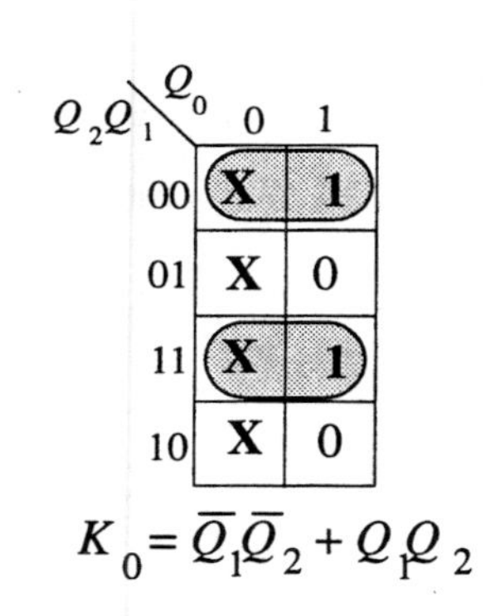

$Q_2Q_1 \backslash Q_0$	0	1
00	X	1
01	X	0
11	X	1
10	X	0

$K_0 = \overline{Q}_1\overline{Q}_2 + Q_1Q_2$

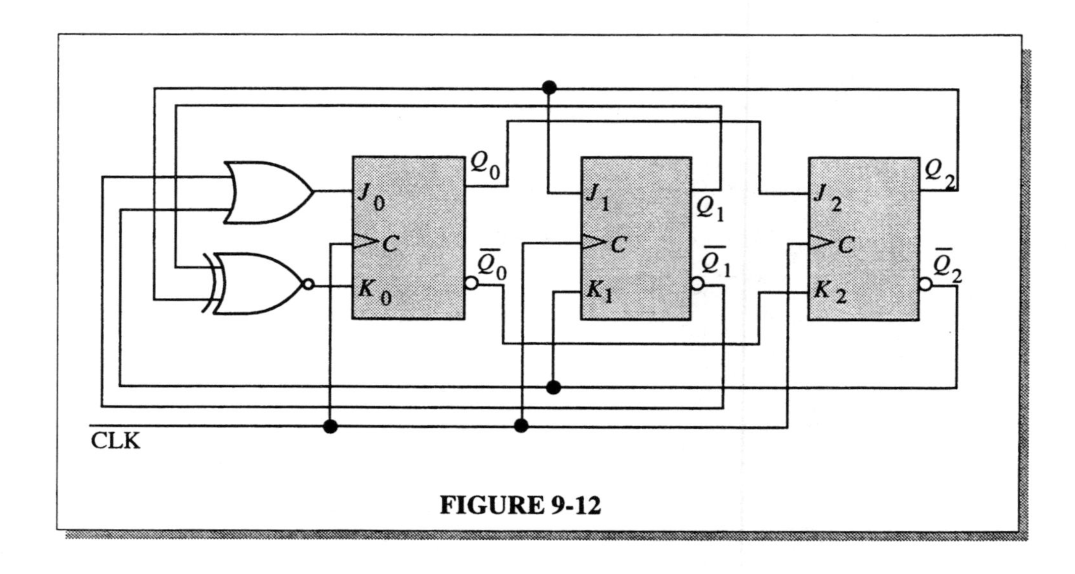

FIGURE 9-12

18. NEXT STATE TABLE

Present state				Next state			
Q_3	Q_2	Q_1	Q_0	Q_3	Q_2	Q_1	Q_0
0	0	0	0	1	0	0	1
1	0	0	1	0	0	0	1
0	0	0	1	1	0	0	0
1	0	0	0	0	0	1	0
0	0	1	0	0	1	1	1
0	1	1	1	0	0	1	1
0	0	1	1	0	1	1	0
0	1	1	0	0	1	0	0
0	1	0	0	0	1	0	1
0	1	0	1	0	0	0	0

TRANSITION TABLE

Output State Transitions (Present state to next state)				Flip-flop inputs							
Q_3	Q_2	Q_1	Q_0	J_3	K_3	J_2	K_2	J_1	K_1	J_0	K_0
0 to 1	0 to 0	0 to 0	0 to 1	1	X	0	X	0	X	1	X
1 to 0	0 to 0	0 to 0	0 to 1	X	1	0	X	0	X	X	0
0 to 1	0 to 0	0 to 0	1 to 0	1	X	0	X	0	X	X	1
1 to 0	0 to 0	0 to 1	0 to 0	X	1	0	X	1	X	0	X
0 to 0	0 to 1	1 to 1	0 to 1	0	X	1	X	X	0	1	X
0 to 0	1 to 0	1 to 1	1 to 1	0	X	X	1	X	0	X	0
0 to 0	0 to 1	1 to 1	1 to 0	0	X	1	X	X	0	X	1
0 to 0	1 to 1	1 to 0	0 to 0	0	X	X	0	X	1	0	X
0 to 0	1 to 1	0 to 0	0 to 1	0	X	X	0	0	X	1	X
0 to 0	1 to 0	0 to 0	1 to 0	0	X	X	1	0	X	X	1

Binary states for 10, 11, 12, 13, 14, and 15 are unallowed and can be represented by don't cares.

See Figure 9-13. Counter implementation is straightforward from input expressions.

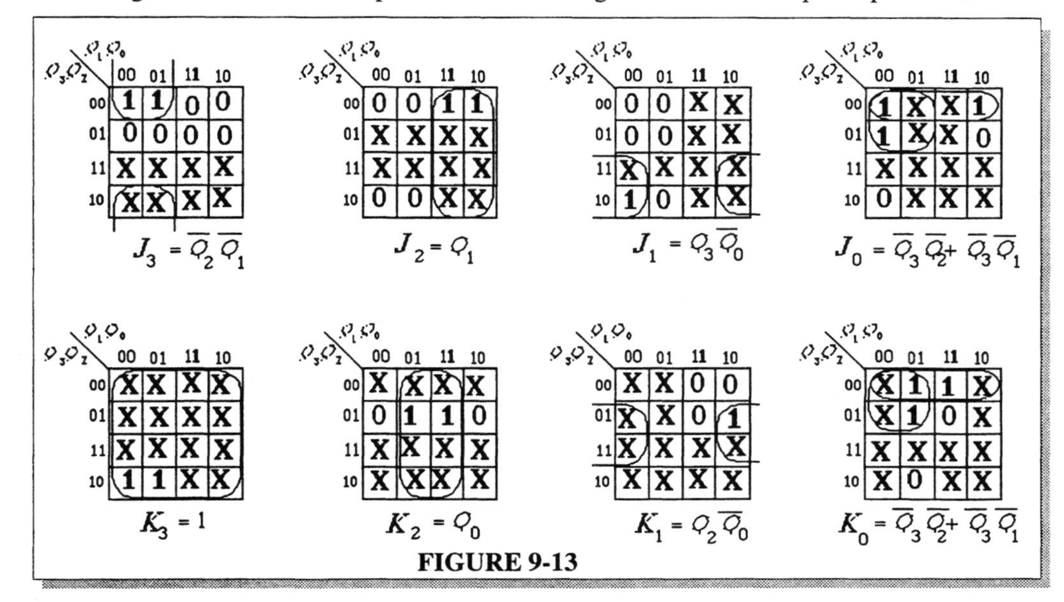

FIGURE 9-13

19. NEXT STATE TABLE

Present state				Next State							
				$Y = 1$ (Up)				$Y = 0$ (Down)			
Q_3	Q_2	Q_1	Q_0	Q_3	Q_2	Q_1	Q_0	Q_3	Q_2	Q_1	Q_0
0	0	0	0	0	0	1	1	1	0	1	1
0	0	1	1	0	1	0	1	0	0	0	0
0	1	0	1	0	1	1	1	0	0	1	1
0	1	1	1	1	0	0	1	0	1	0	1
1	0	0	1	1	0	1	1	0	1	1	1
1	0	1	1	0	0	0	0	1	0	0	1

TRANSITION TABLE

Output State Transistions (Present state to next state)				Y	Flip-flop inputs			
Q_3	Q_2	Q_1	Q_0		J_3K_3	J_2K_2	J_1K_1	J_0K_0
0 to 1	0 to 0	0 to 1	0 to 1	0	1X	0X	1X	1X
0 to 0	0 to 0	0 to 1	0 to 1	1	0X	0X	1X	1X
0 to 0	0 to 0	1 to 0	1 to 0	0	0X	0X	X1	X1
0 to 0	0 to 1	1 to 0	1 to 1	1	0X	1X	X1	X0
0 to 0	1 to 0	0 to 1	1 to 1	0	0X	X1	1X	X0
0 to 0	1 to 1	0 to 1	1 to 1	1	0X	X0	1X	X0
0 to 0	1 to 1	1 to 0	1 to 1	0	0X	X0	X1	X0
0 to 1	1 to 0	1 to 0	1 to 1	1	1X	X1	X1	X0
1 to 0	0 to 1	0 to 1	1 to 1	0	X1	1X	1X	X0
1 to 1	0 to 0	0 to 1	1 to 1	1	X0	0X	1X	X0
1 to 1	0 to 0	1 to 0	1 to 1	0	X0	0X	X1	X0
1 to 0	0 to 0	1 to 0	1 to 0	1	X1	0X	X1	X1

See Figure 9-14.

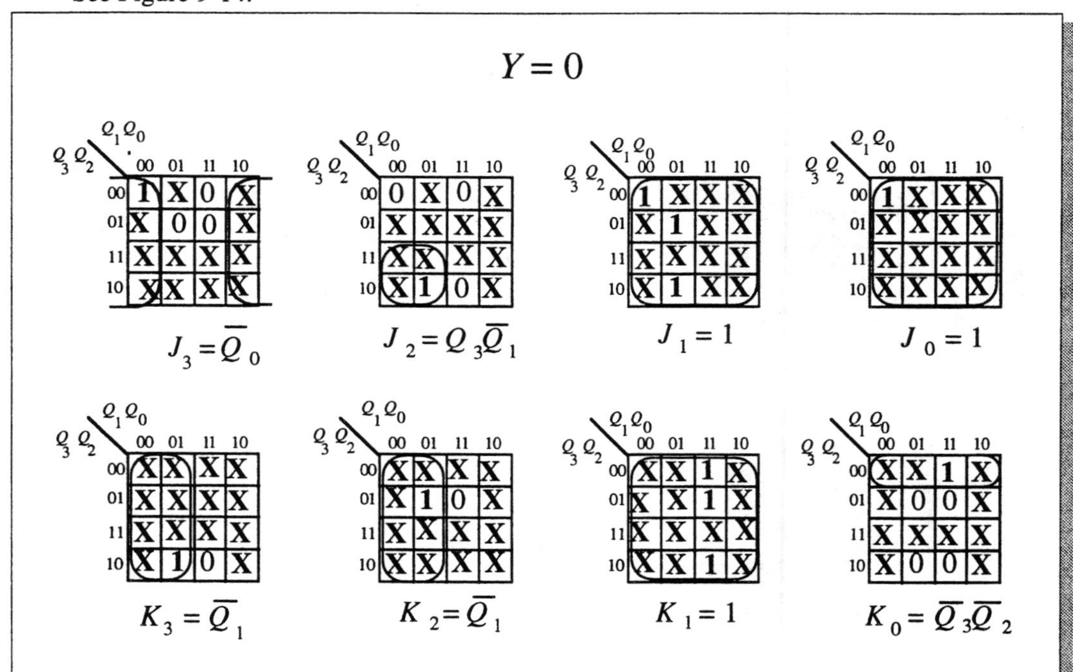

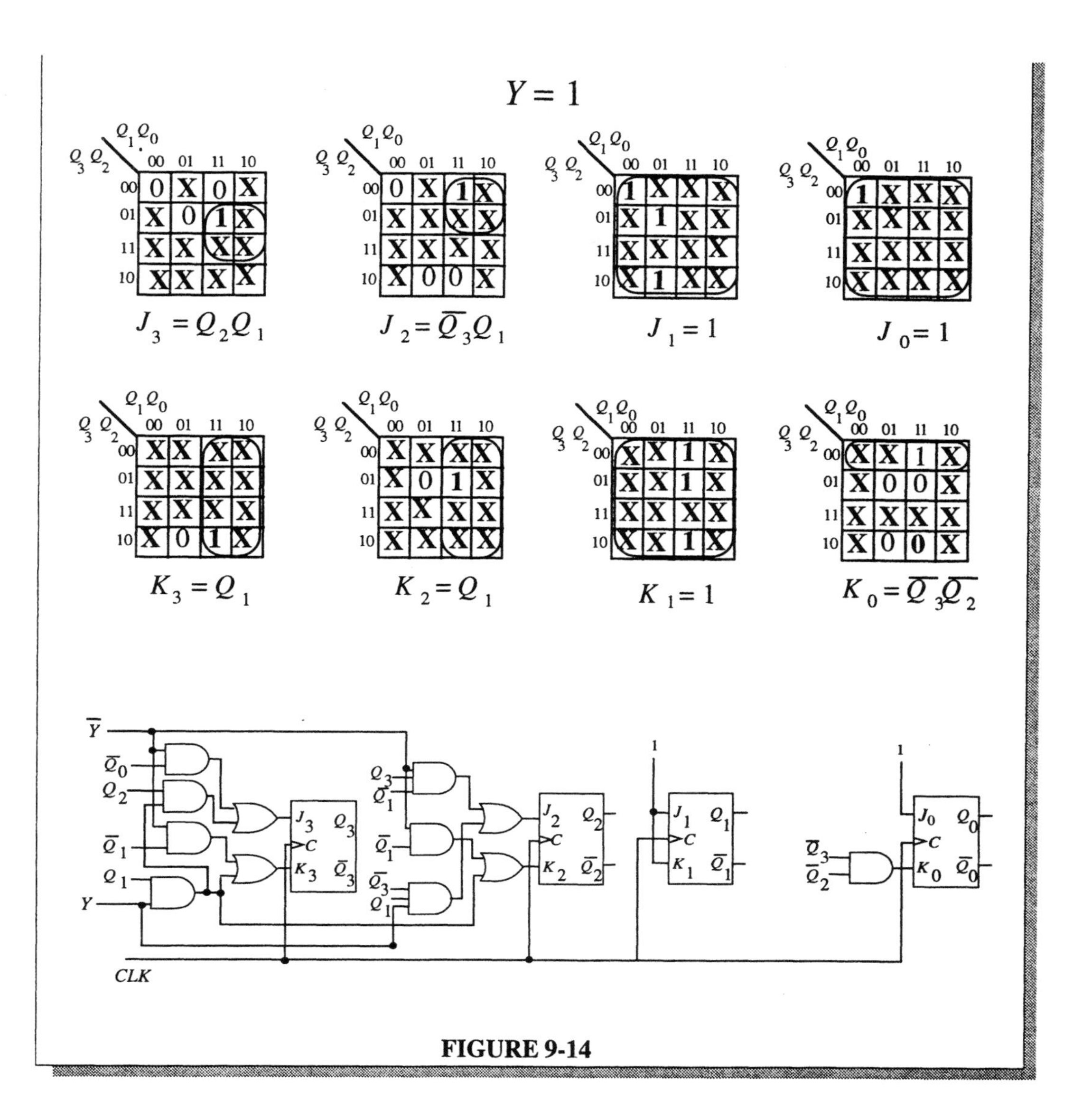

FIGURE 9-14

20. (a) Modulus = 4X8x2 = **64**

$$f_1 = \frac{1\text{ kHz}}{4} = 250\ \mathbf{Hz}$$

$$f_2 = \frac{250\text{ Hz}}{8} = \mathbf{31.25\ Hz}$$

$$f_3 = \frac{31.25\text{ Hz}}{2} = \mathbf{15.625Hz}$$

(b) Modulus = 10X10X10X2 = **2000**

$$f_1 = \frac{100 \text{ kHz}}{10} = 10 \text{ kHz}$$
$$f_2 = \frac{10 \text{ kHz}}{10} = 1 \text{ kHz}$$
$$f_3 = \frac{1 \text{ kHz}}{10} = 100 \text{ Hz}$$
$$f_4 = \frac{100 \text{ Hz}}{2} = 50 \text{ Hz}$$

(c) Modulus = 3X6X8X10X10 = **14400**

$$f_1 = \frac{21 \text{ MHz}}{3} = 7 \text{ MHz}$$
$$f_2 = \frac{7 \text{ MHz}}{6} = 1.167 \text{ MHz}$$
$$f_3 = \frac{1.167 \text{ MHz}}{8} = 145.875 \text{ MHz}$$
$$f_4 = \frac{145.875 \text{ MHz}}{10} = 14.588 \text{ kHz}$$
$$f_5 = \frac{14.588 \text{ kHz}}{10} = 1.459 \text{ kHz}$$

(d) Modulus = 2X4X6X8X16 = **6144**

$$f_1 = \frac{39.4 \text{ kHz}}{2} = 19.7 \text{ kHz}$$
$$f_2 = \frac{19.7 \text{ kHz}}{4} = 4.925 \text{ kHz}$$
$$f_3 = \frac{4.925 \text{ kHz}}{6} = 820.83 \text{ Hz}$$
$$f_4 = \frac{820.683}{8} = 102.6 \text{ Hz}$$
$$f_5 = \frac{102.6 \text{ Hz}}{16} = 6.41 \text{ Hz}$$

21. See Figure 9-15.

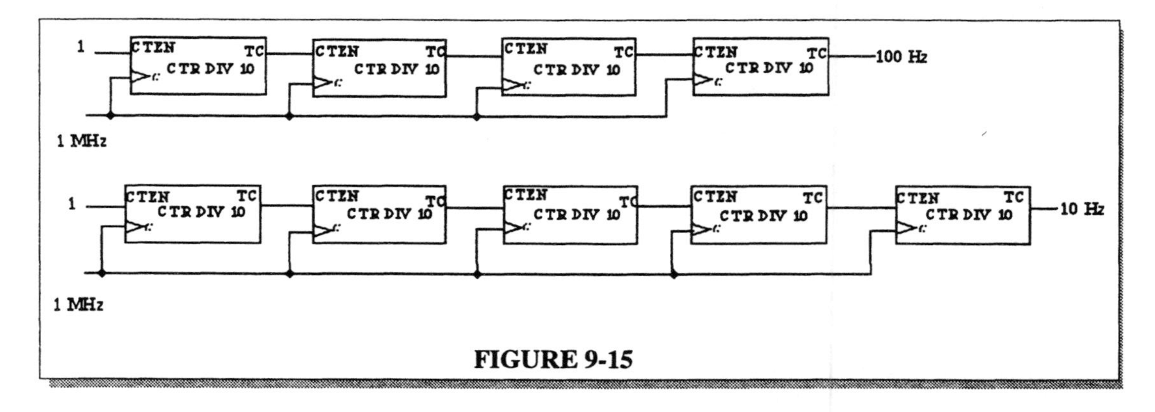

FIGURE 9-15

22. See Figure 9-16.

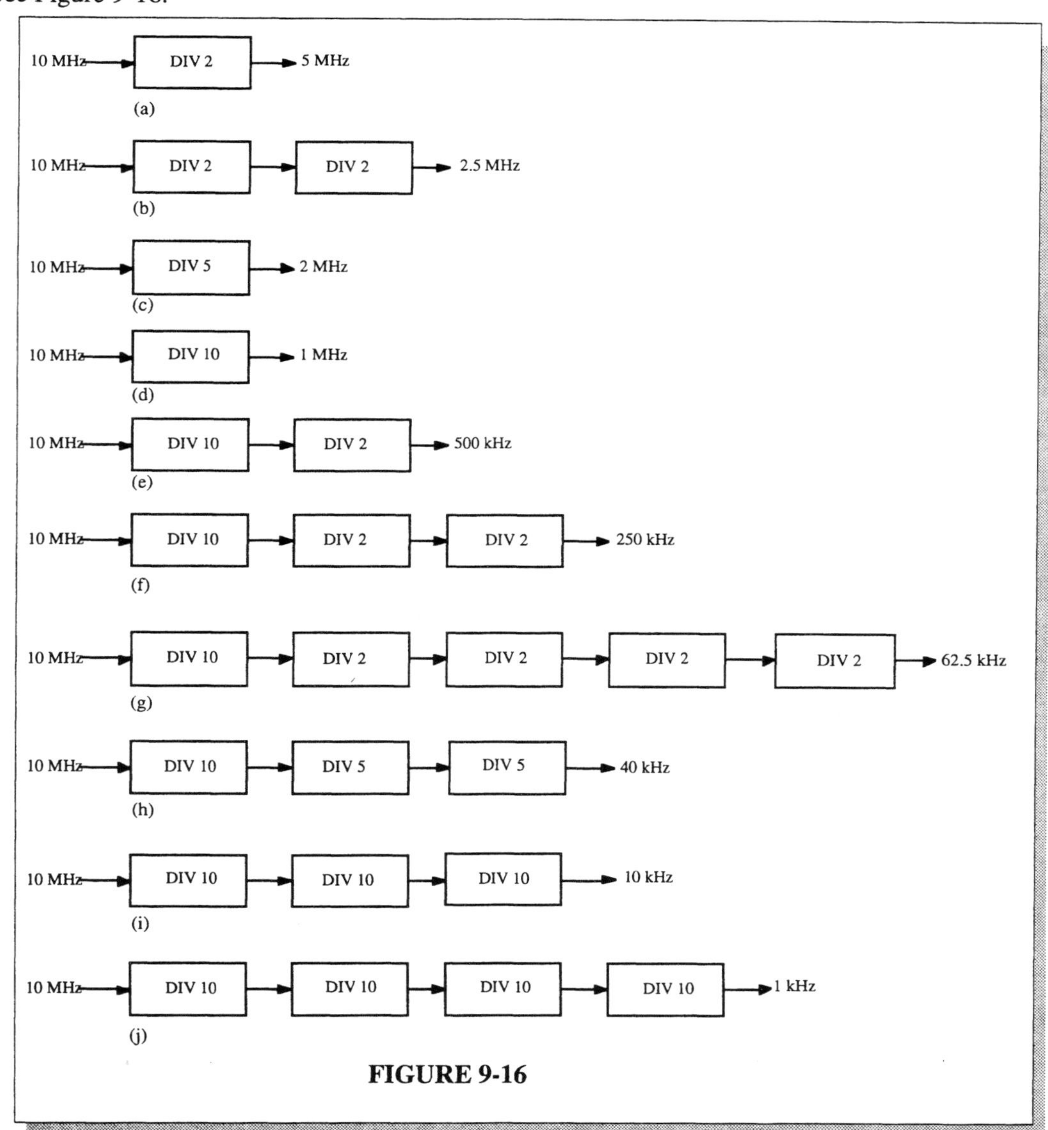

FIGURE 9-16

23. See Figure 9-17.

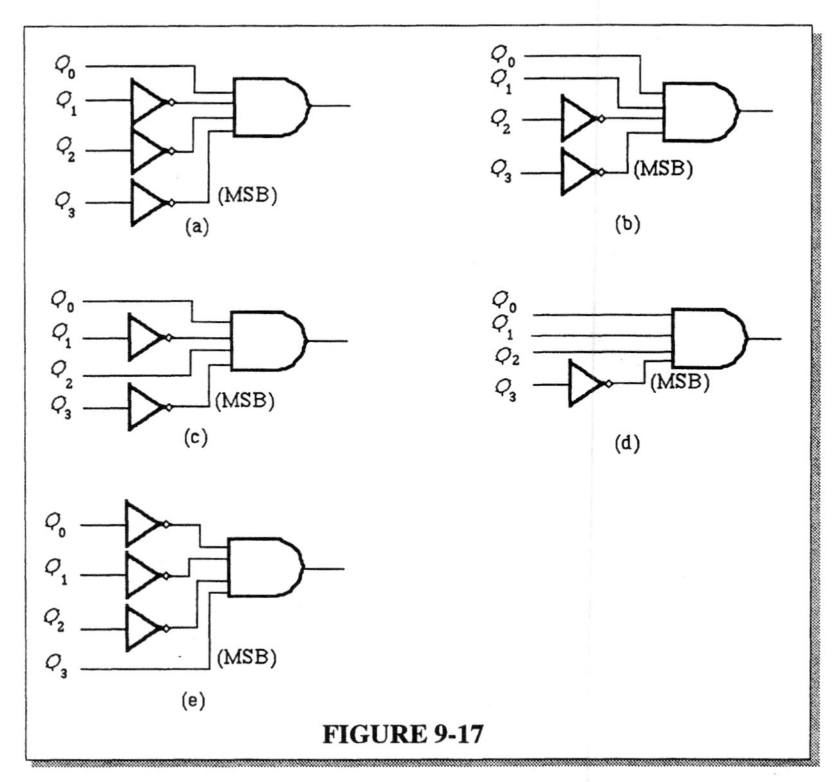

FIGURE 9-17

24. See Figure 9-18.

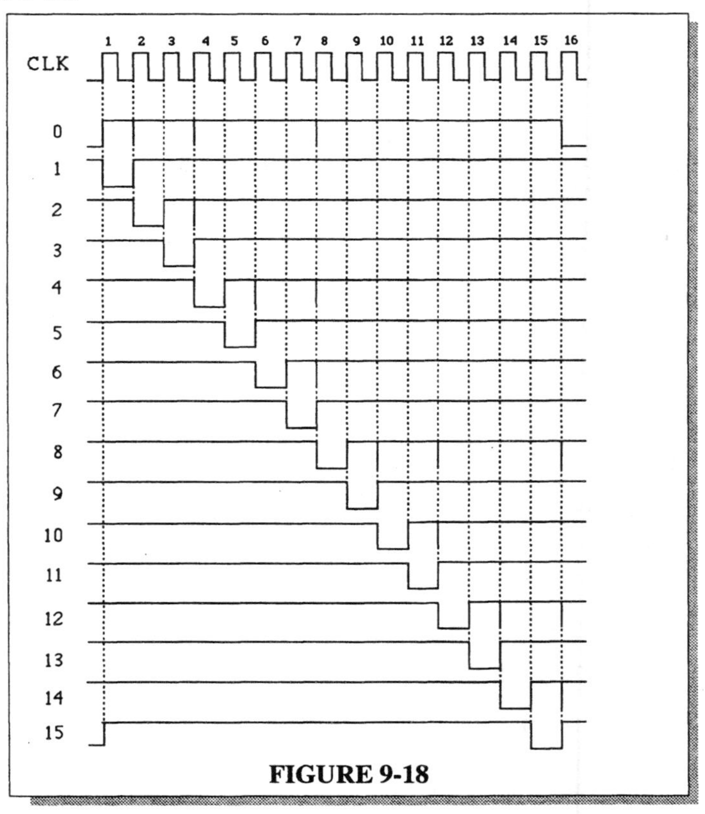

FIGURE 9-18

25. The states with an asterik are the transition states that produce glitches on the decoder outputs. The glitches are indicated on the waveforms in Figure 9-18 (Problem 9-24) by vertical lines.

Initial	0000
CLK 1	0001
CLK 2	0000 *
	0010
CLK 3	0011
CLK 4	0010 *
	0000 *
	0100
CLK 5	0100
CLK 6	0100 *
	0110
CLK 7	0111
CLK 8	0110 *
	0100 *
	0000 *
	1000
CLK 9	1001
CLK 10	1000 *
	1010
CLK 11	1011
CLK 12	1010 *
	1000 *
	1100
CLK 13	1101
CLK 14	1100 *
	1110
CLK 15	1111
CLK 16	1110 *
	1100 *
	1000 *
	0000

26. See Figure 9-19.

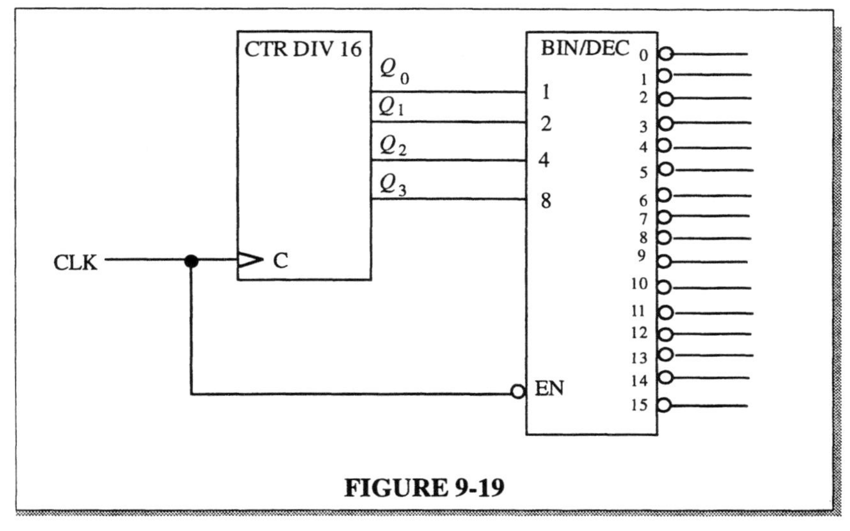

FIGURE 9-19

27. See Figure 9-20.

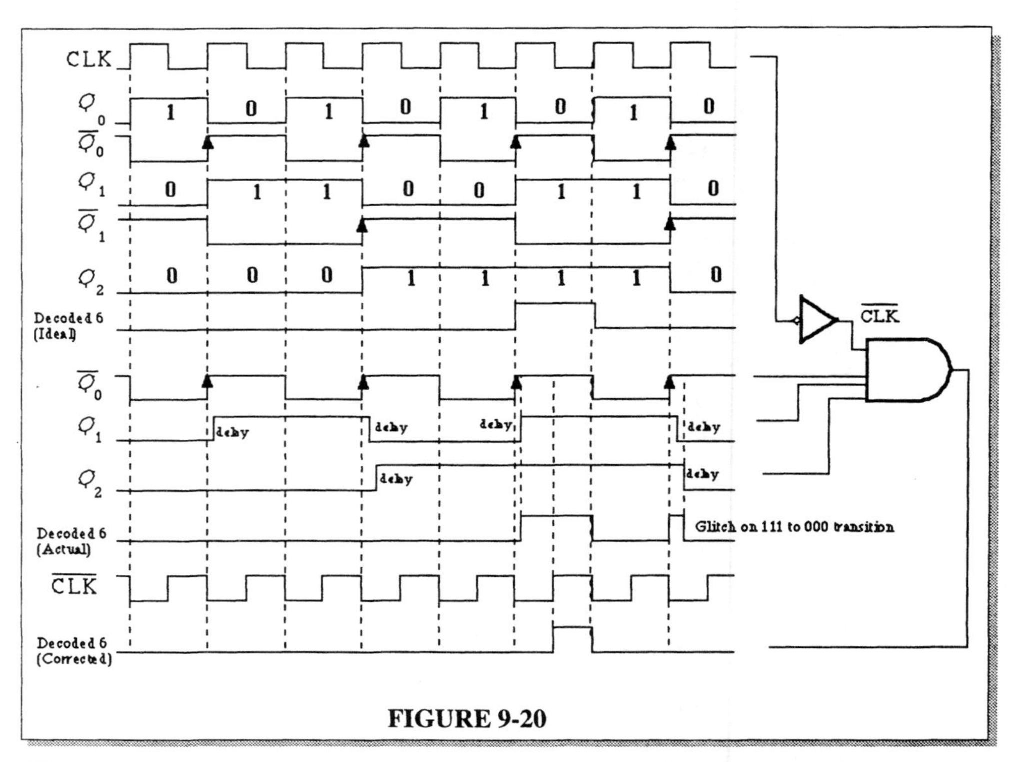

FIGURE 9-20

28. ① There is a possibility of a glitch on decode 2 at the positive-going edge ofCLK 4 if the propagation delay of FF0 is less than FF1 or FF2.

② There is a possibility of a glitch on decode 7 at the positive-going edge ofCLK 4 if the propagation delay of FF2 is less than FF0 and FF1.

③ There is a possibility of a glitch on decode 7 at the positive-going edge ofCLK 6 if the propagation delay of FF1 is less than FF0.

See the timing diagram in Figure 9-21 which is expanded to show the delays.

Any glitches can be prevented by using CLK as an input to both decode gates.

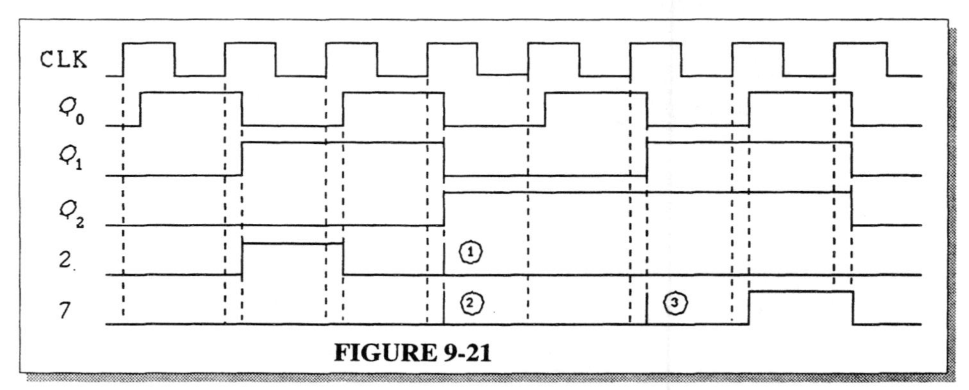

FIGURE 9-21

29. For the digital clock in Figure 9-51 of the text reset to 12:00:00, the binary state of each counter after 62 60-Hz pulses are:

Hours, tens: **0001**
Hours, units: **0010**
Minutes, tens: **0000**
Minutes, units: **0001**
Seconds, tens: **0000**
Seconds, units: **0010**

30. For the digital clock, the counter output frequencies are:
Divide-by-60 input counter:

$$\frac{60\text{ Hz}}{60} = 1\text{ Hz}$$

Seconds counter:

$$\frac{1\text{ Hz}}{60} = 16.7\text{ mHz}$$

Minutes counter:

$$\frac{16.7\text{ mHz}}{60} = 278\ \mu\text{Hz}$$

Hours counter:

$$\frac{278\ \mu\text{Hz}}{12} = 23.1\ \mu\text{Hz}$$

31. 53 + 37 - 26 = 64

32. See Figure 9-22.

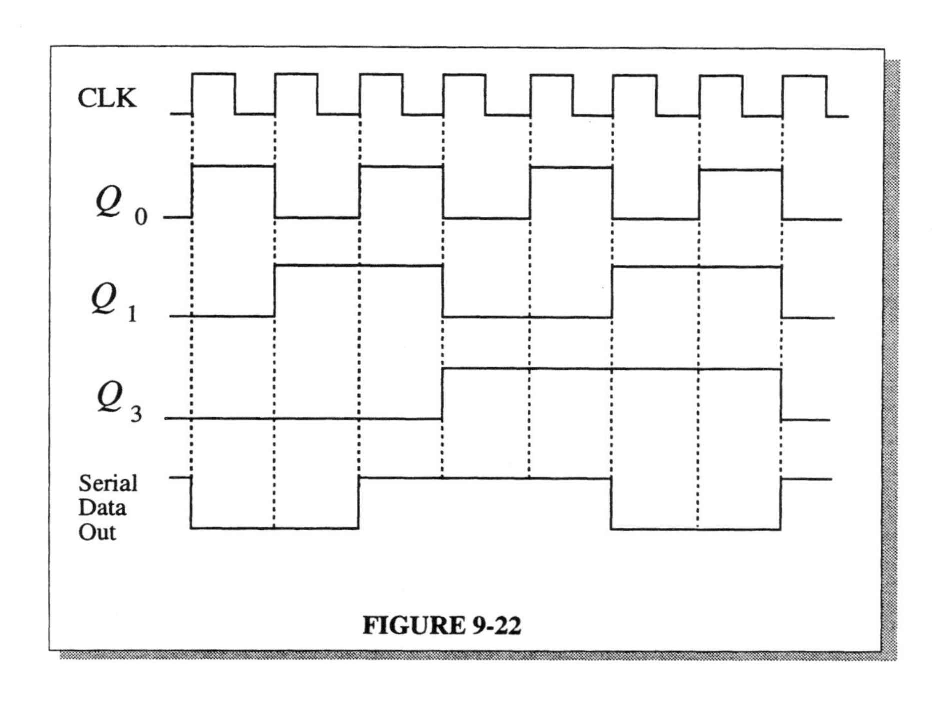

FIGURE 9-22

33. (a) Q_0 and Q_1 will not change due to the clock shorted to ground at FF0.

(b) Q_0 being open does not affect normal operation. See Figure 9-23.

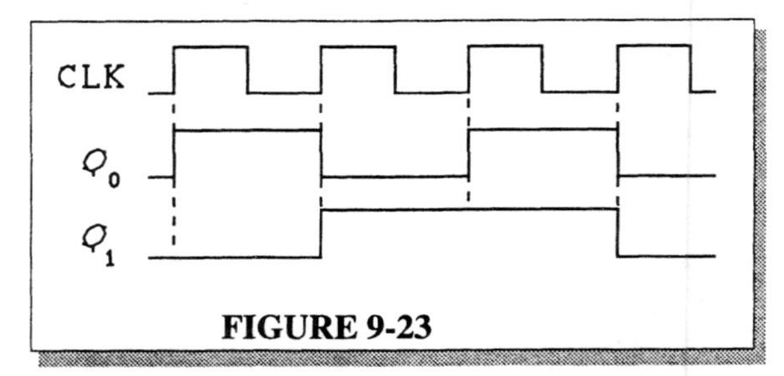

FIGURE 9-23

(c) See Figure 9-24.

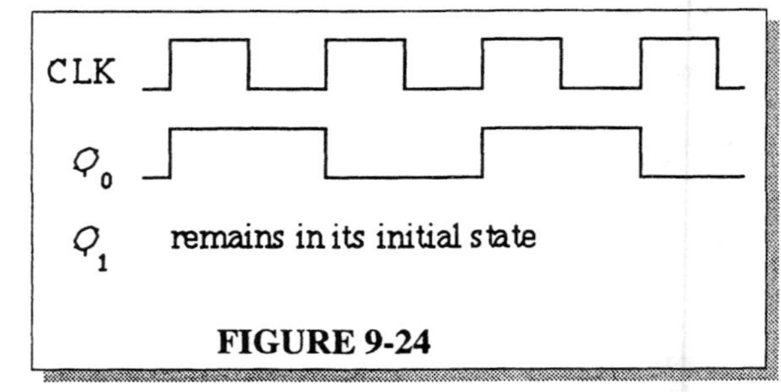

FIGURE 9-24

(d) Normal operation because an open J input acts as a HIGH.

(e) A shorted K input will pull all J and K inputs LOW and the counter will not change from its initial state.

34. (a) Q_0 and Q_1 will not change from initial states.

(b) See Figure 9-25.

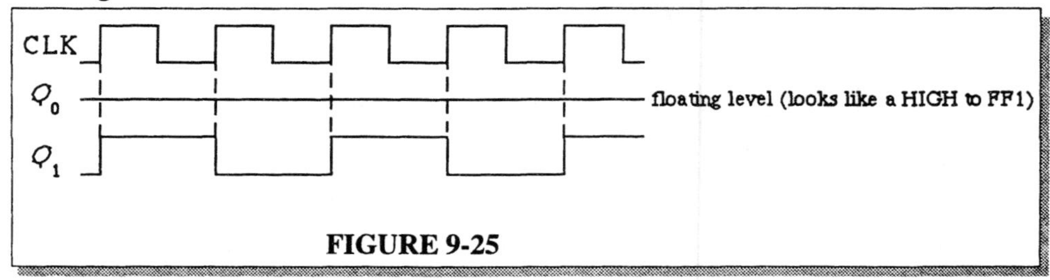

FIGURE 9-25

(c) See Figure 9-26.

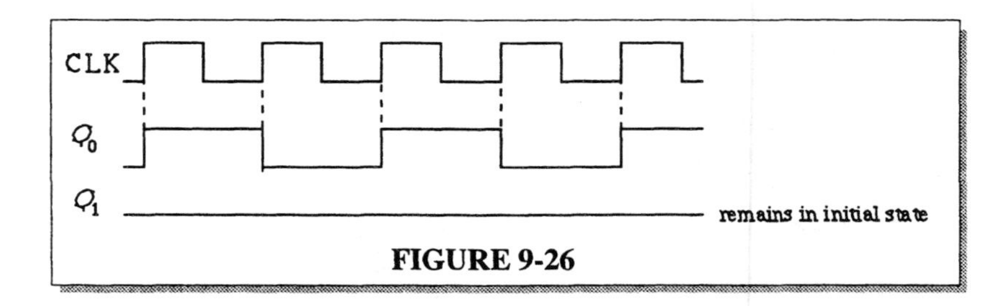

FIGURE 9-26

(d) Normal operation. See Figure 9-27.

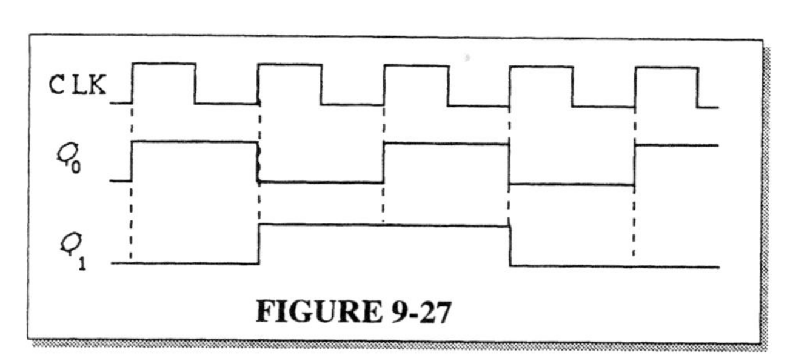

FIGURE 9-27

(e) Both J and K of FF1 arepulled LOW if K is grounded, producing a no-change condition. Q_0 also grounded. See Figure 9-28.

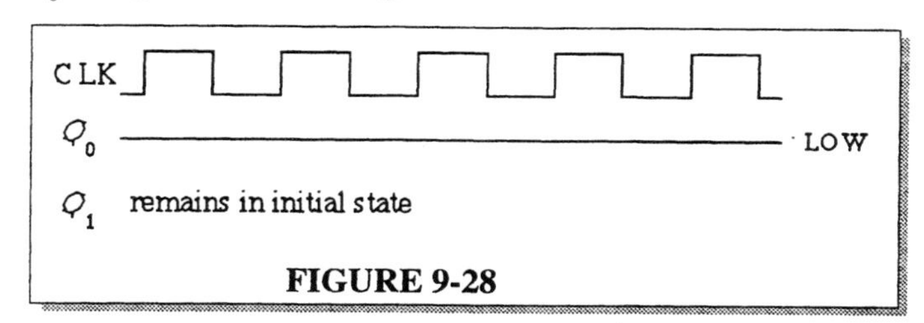

FIGURE 9-28

35. First, determine the correct waveforms and observe that Q_0 is correct but Q_1 and Q_2 are incorrect in Figure 8-83 in the text. See Figure 9-29.

Since Q_1 goes HIGH on the first clock transition and stays HIGH, FF1 must be in the SET state ($J = 1$, $K = 0$). There must be a wiring error at the J and K inputs to FF1; K must be connected to ground rather than to the J input.

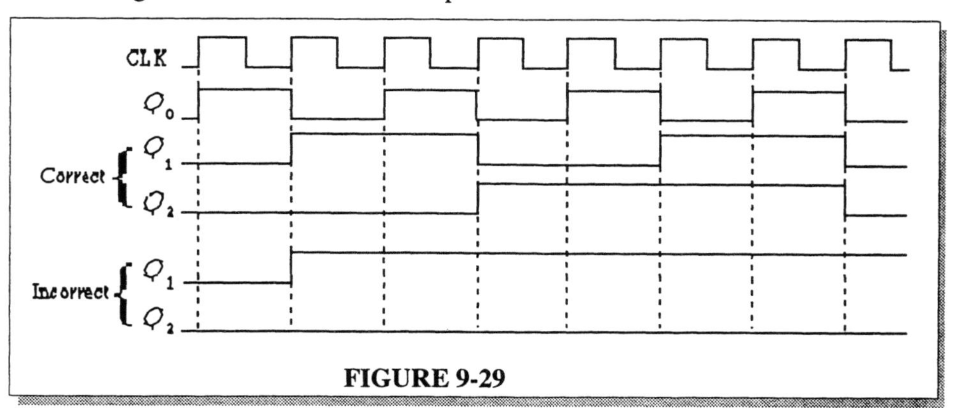

FIGURE 9-29

36. Since Q_2 toggles on each clock pulse, its J and K inputs must be constantly HIGH. The most probable fault is the AND gate's output is *open*.

37. If the Q_0 input to the AND gate is *open*, the JK inputs to FF2 is as shown in Figure 9-30.

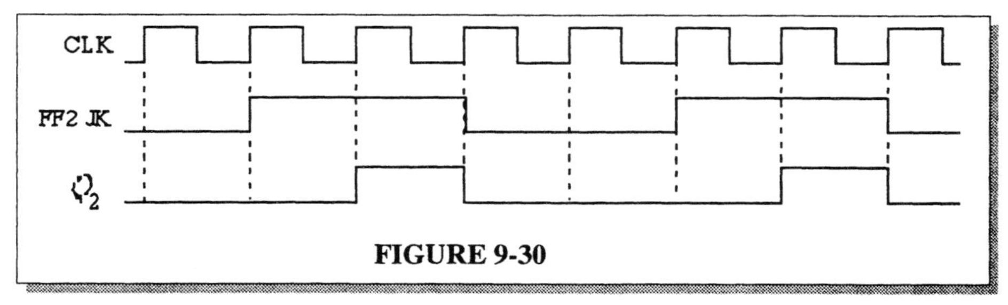

FIGURE 9-30

.38.

Number of states = 40,000

$f_{out} = \frac{5\text{ MHz}}{40{,}000} = 125\text{ Hz}$

76.2939 Hz is not correct. The faulty division factor is

$\frac{5\text{ MHz}}{76.2939\text{ Hz}} = 65{,}536$

Obviously, the counter is going through all of its states. This means that the $63C0_{16}$ on its parallel inputs is not being loaded. Possible faults are:

- Inverter output is stuck HIGH or open.
- RCO output of last counter is stuck LOW.

39.

Stage	Open	D_x	Loaded Count	f_{out}
1	0		63C1	250.006 Hz
1	1		63C2	250.012 Hz
1	2		63C4	250.025 Hz
1	3		63C8	250.050 Hz
2	0		63D0	250.100 Hz
2	1		63E0	250.200 Hz
2	2		63C0	250 Hz
2	3		63C0	250 Hz
3	0		63C0	250 Hz
3	1		63C0	250 Hz
3	2		67C0	256.568 Hz
3	3		6BC0	263.491 Hz
4	0		73C0	278.520 Hz
4	1		63C0	250 Hz
4	2		63C0	250 Hz
4	3		E3C0	1.383 kHz

40.

- The flip-flop output is stuck HIGH or open
- The least significant BCD/7-segment input is open.

See Figure 9-31.

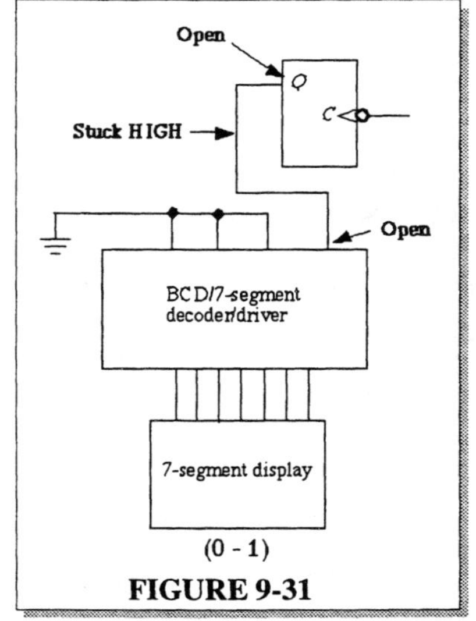

FIGURE 9-31

41. The DIV 6 is the tens of minutes counter. Q_1 open causes a continuous apparent HIGH output to the decode 6 gate and to the BCD/7-segment decoder/driver.

The apparent counter sequence is shown in the table.

Actual State of Ctr.	Apparent state			
	Q_3	Q_2	Q_1	Q_0
0	0	0	1	0
1	0	0	1	1
2	0	0	1	0
3	0	0	1	1
4	0	1	1	0

The decode 6 gate interprets count 4 as a 6 (0110) and clears the counter back to 0 (actually 0010). Thus the apparent (not actual) sequence is as shown in the table.

42. There are several possible causes of the malfunction. First check power to all units. Other possible faults are listed below.

- Sensor Latch
  Action: disconnect entrance sensor and pulse sensor input.
  Observation: latch should SET.
  Conclusion: if latch does not SET replace it.

- NOR gate
  Action: pulse sensor input.
  Observation: pulse on gate output.
  Conclusion: if there is no pulse replace gate.

- Counter
  Action: pulse sensor input.
  Observation: counter should advance.
  Conclusion: if counter does not advance replace it.

- Output Interface
  Action: pulse sensor intput until terminal count is reached.
  Observation: Full indication and gate lowered.
  Conclusion: No full indication of if gate does not lower replace interface.

- Sensor/Cable
  Action: try to activate sensor.
  Observation: if all previous checks are OK, sensor or cable is faulty.
  Conclusion: replace sensor or cable.

43. The expressions for the D_0 and the D_1 flip-flop inputs in the sequential logic portion of the system were developed for Test Bench 1, activities 1 and 2 and are restated below. Figure 9-32 shows the NAND implementation.

$$D_0 = \overline{Q_1}Q_0 + \overline{Q_1}\,\overline{T_L}V_S + Q_0T_LV_S$$

$$D_1 = Q_0\overline{T_S} + Q_1T_S$$

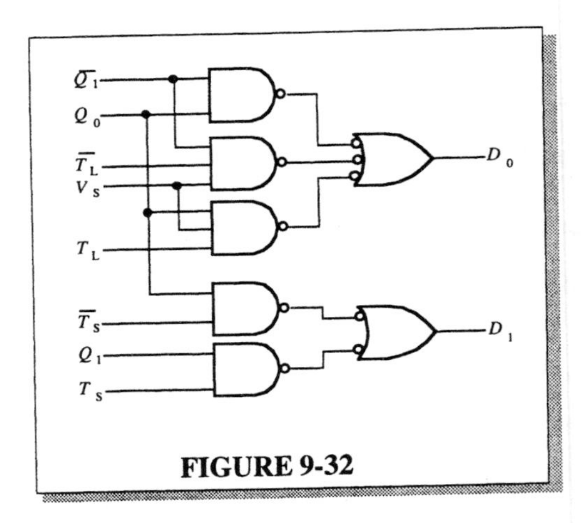

FIGURE 9-32

44. See Figure 9-33.

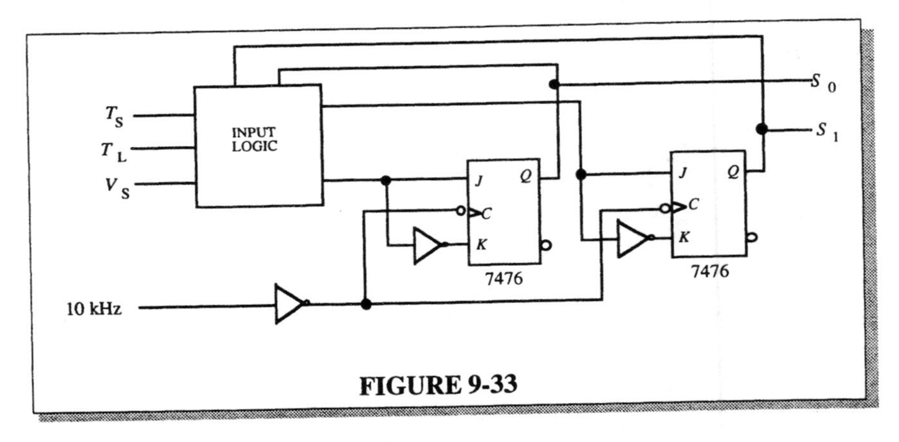

FIGURE 9-33

45. The time interval for the green light can be increased from 25 s to 60 s by increasing the value of either the resistor or the capacitor value by

$$\frac{60\text{ s}}{25\text{ s}} = 2.4\text{ times}$$

46. See Figure 9-34.

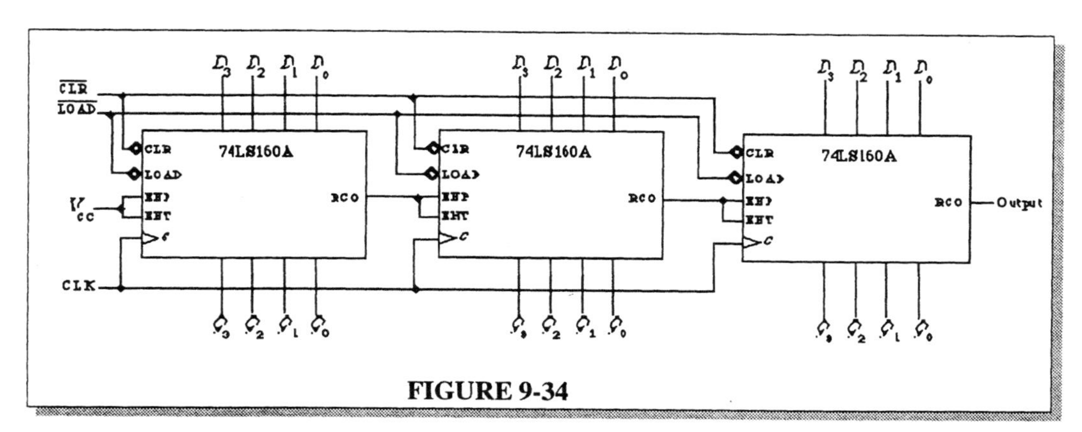

FIGURE 9-34

47. 65,536 - 30,000 = 35,536

Preset the counter to 35,536 so that it counts from 35,536 up to 65,536 on each full cycle, thus producing a sequence of 30,000 states (modulus 30,000).

$35{,}536 = 1000101011010000_2 = 8AD0_{16}$

See Figure 9-35.

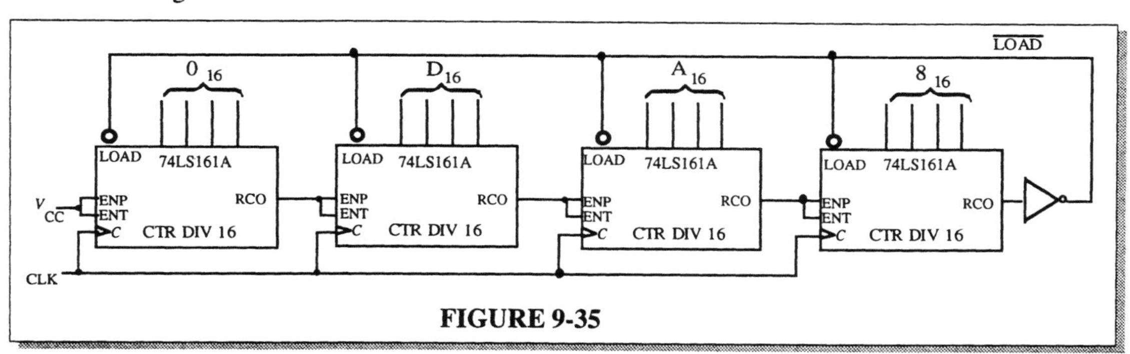

FIGURE 9-35

48. 65,536 - 50,000 = 15,536

Preset the counter to 15,536 so that it counts from 15,536 up to 65,536 on each full cycle, thus producing a sequence of 50,000 states (modulus 50,000).

$15{,}536 = 11110010110000_2 = \mathbf{3CB0_{16}}$

See Figure 9-36.

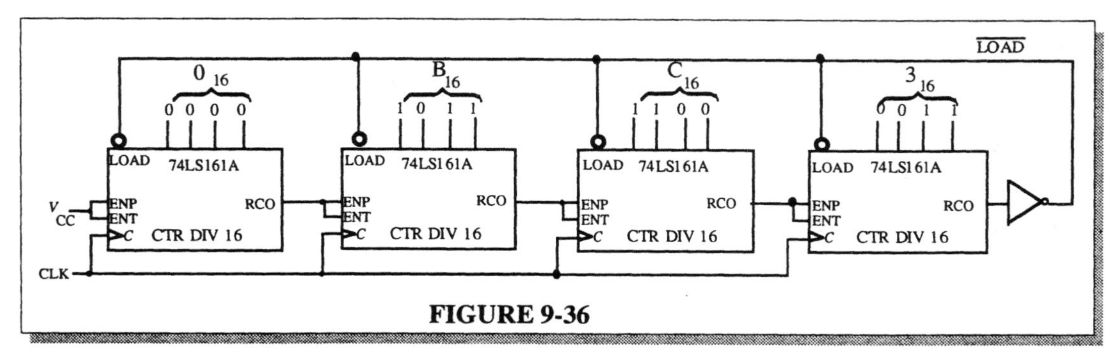

FIGURE 9-36

49. The approach is to preset the hours and minutes counters independently, each with a fast or slow preset mode. The seconds counter is not preset. One possible implementation is shown in Figure 9-37.

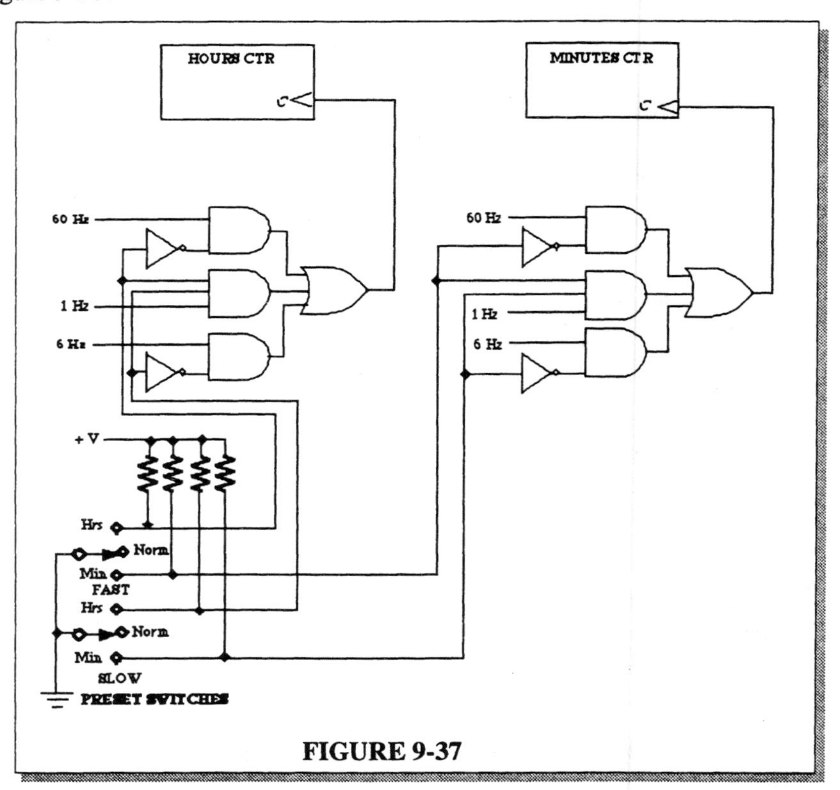

FIGURE 9-37

50. See Figure 9-38.

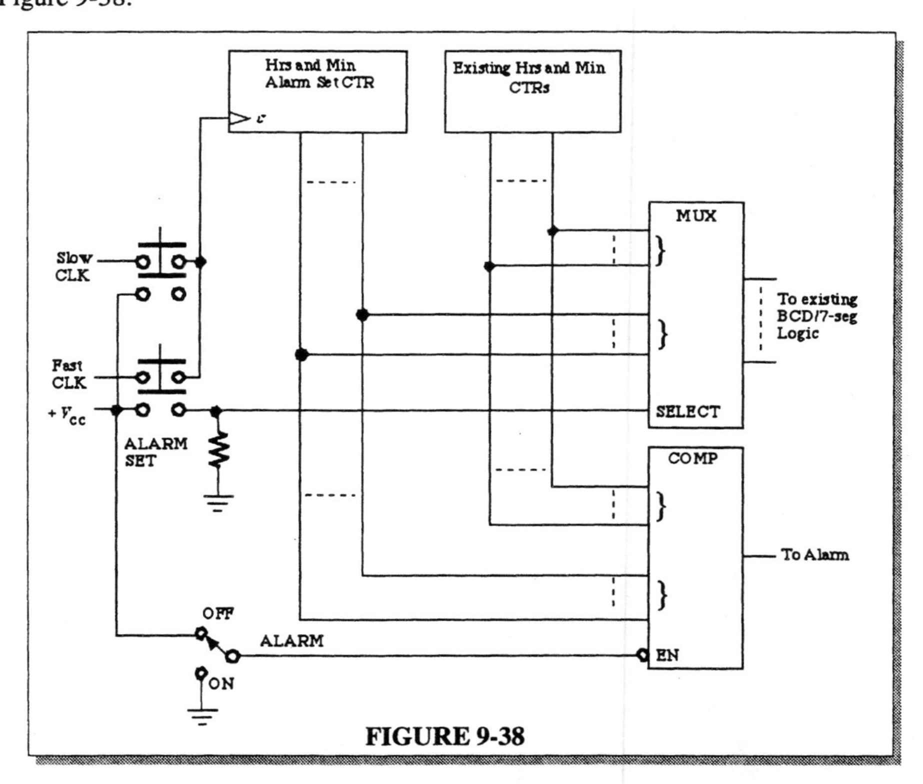

FIGURE 9-38

51. See Figure 9-39.

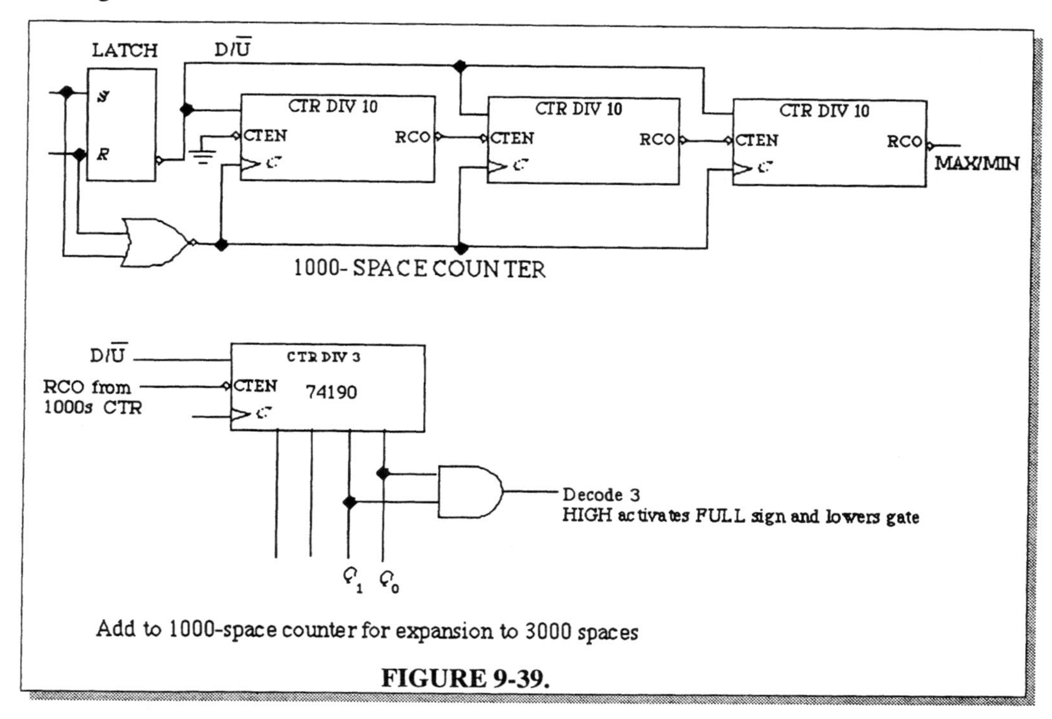

FIGURE 9-39.

52. See Figure 9-40.

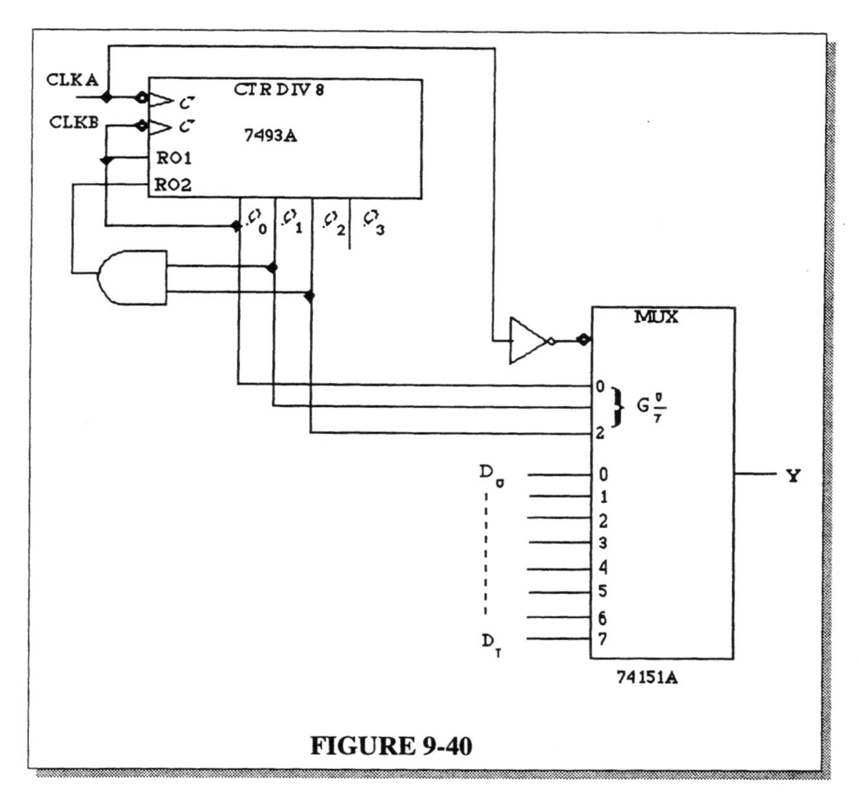

FIGURE 9-40

53. NEXT STATE TABLE

Present State				Next State			
Q_3	Q_2	Q_1	Q_0	Q_3	Q_2	Q_1	Q_0
0	0	0	0	1	1	1	1
1	1	1	1	1	1	1	0
1	1	1	0	1	1	0	1
1	1	0	1	1	0	1	0
1	0	1	0	0	1	0	1
0	1	0	1	0	0	0	0

TRANSITION TABLE

Output State Transistions				Flip-Flop inputs			
Q_3	Q_2	Q_1	Q_0	J_3K_3	J_2K_2	J_1K_1	J_0K_0
0 to 1	0 to 1	0 to 1	0 to 1	1X	1X	1X	1X
1 to 1	1 to 1	1 to 1	1 to 0	X0	X0	X0	X1
1 to 1	1 to 1	1 to 0	0 to 1	X0	X0	X1	1X
1 to 1	1 to 0	0 tol	1 to 0	X0	X1	1X	X1
1 to 0	0 to 1	1 to 0	0 to 1	X1	1X	X1	1X
0 to 0	1 to 0	0 to0	1 to 0	0X	X1	0X	X1

See Figure 9-41.

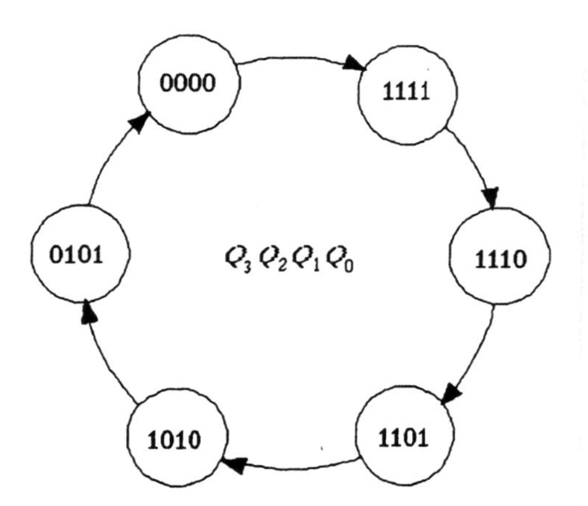

The desired sequence

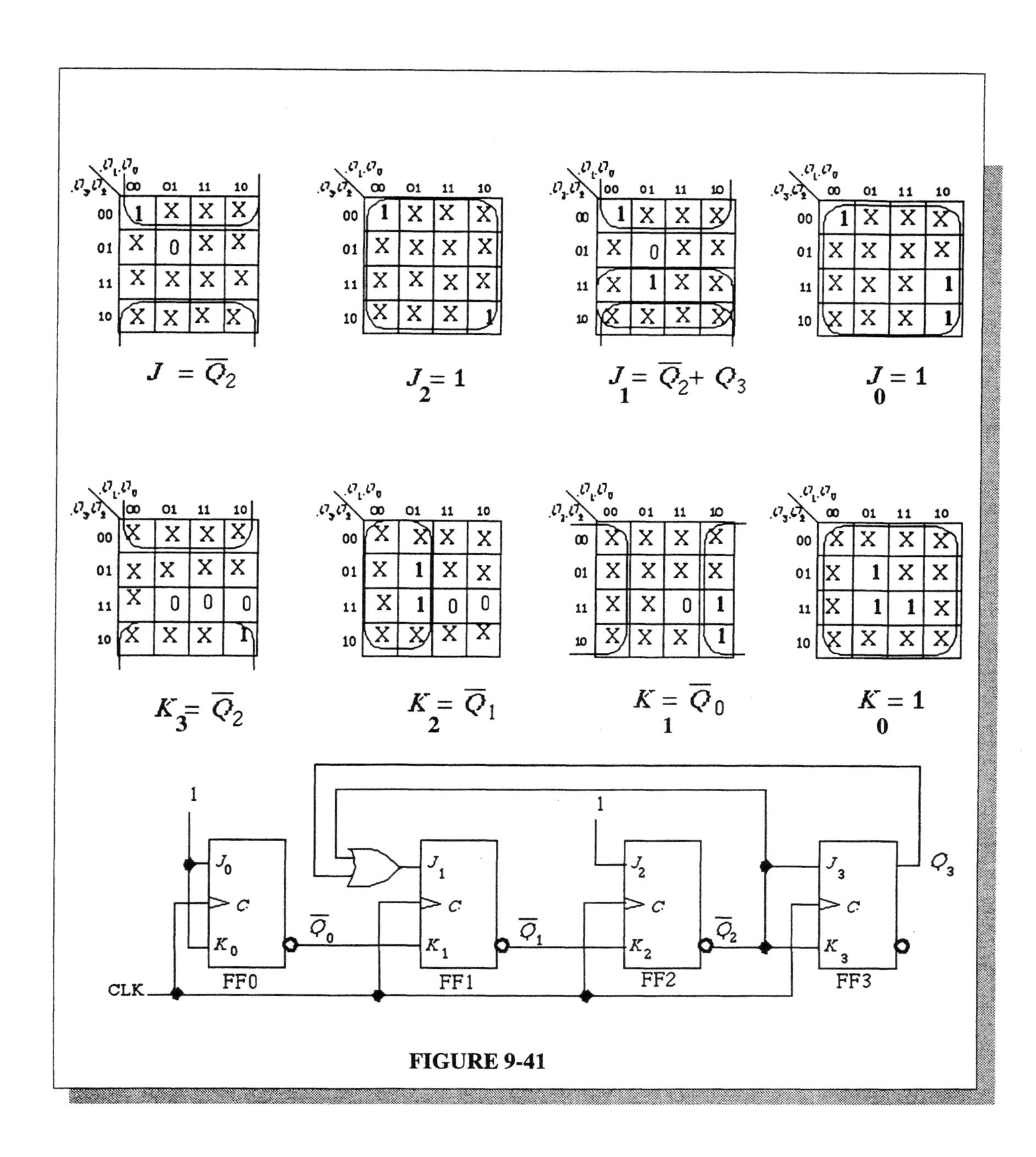

$J_3 = \overline{Q}_2$
$J_2 = 1$
$J_1 = \overline{Q}_2 + Q_3$
$J_0 = 1$
$K_3 = \overline{Q}_2$
$K_2 = \overline{Q}_1$
$K_1 = \overline{Q}_0$
$K_0 = 1$
CLK
FF0
FF1
FF2
FF3

FIGURE 9-41

54. See Figure 9-42.

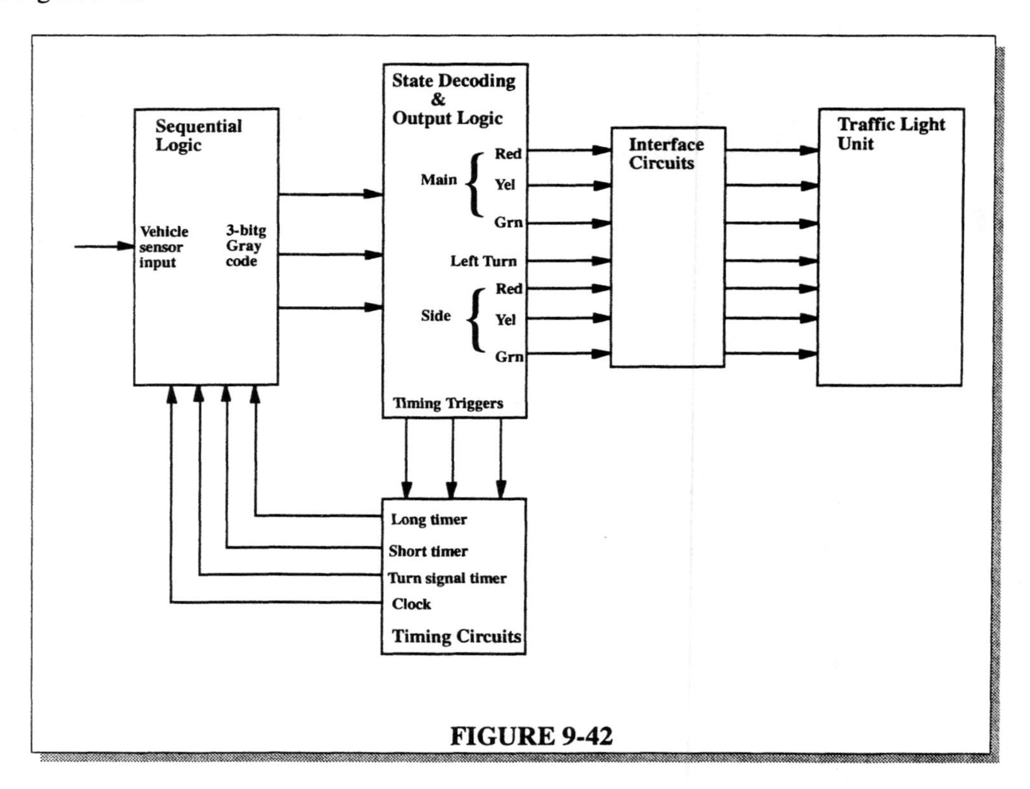

FIGURE 9-42

CHAPTER 10
SHIFT REGISTERS

1. Shift registers store binary data in a series of flip-flops or other storage elements.

2. 1 byte = **8 bits**; 2 bytes = **16 bits**

3. See Figure 10-1.

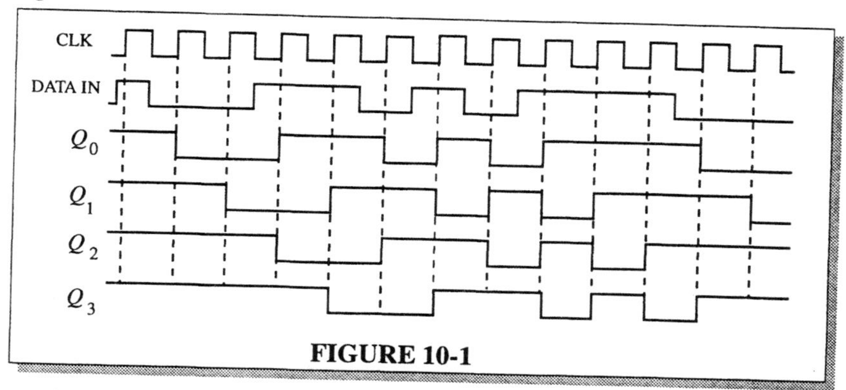

FIGURE 10-1

4. See Figure 10-2.

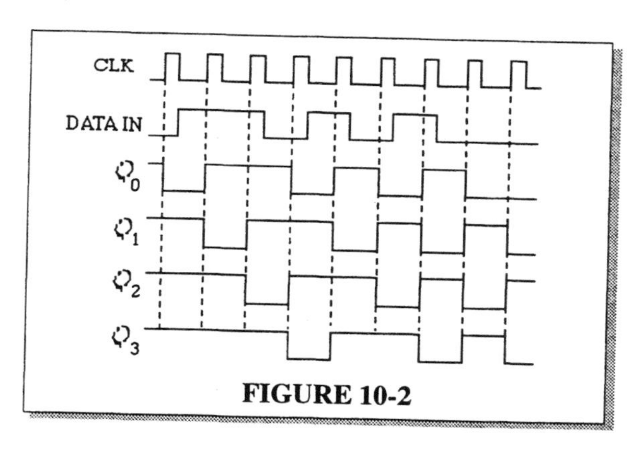

FIGURE 10-2

5.

Initially	101001111000
CLK 1	010100111100
CLK 2	001010011110
CLK 3	000101001111
CLK 4	000010100111
CLK 5	100001010011
CLK 6	110000101001
CLK 7	111000010100
CLK 8	011100001010
CLK 9	001110000101
CLK 10	000111000010
CLK 11	100011100001
CLK 12	110001110000

6. See Figure 10-3.

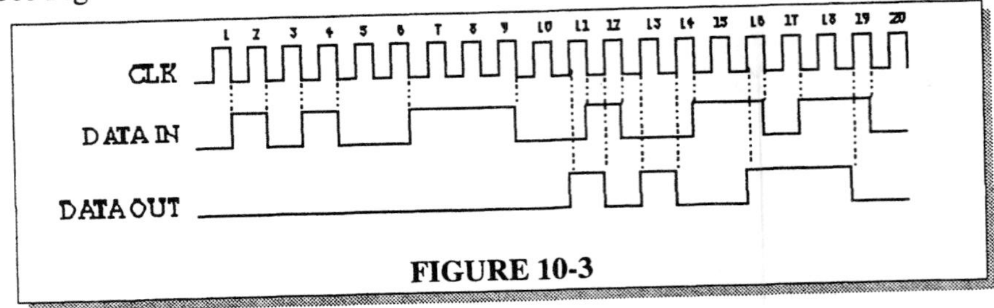

FIGURE 10-3

7. See Figure 10-4.

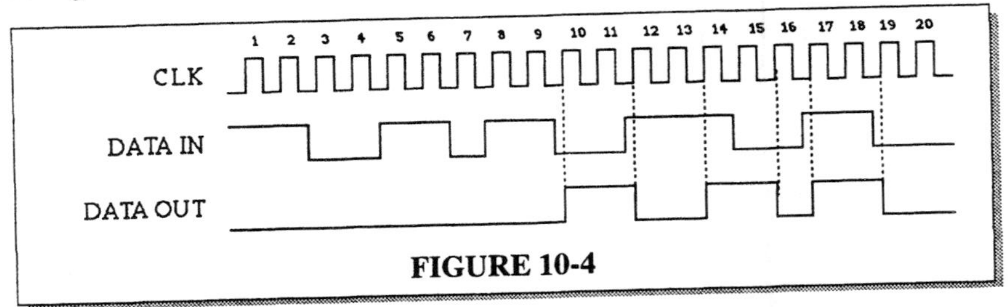

FIGURE 10-4

8. See Figure 10-5.

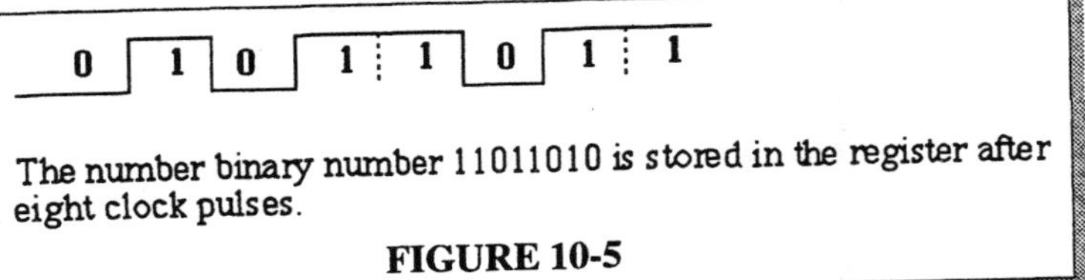

FIGURE 10-5

9. See Figure 10-6.

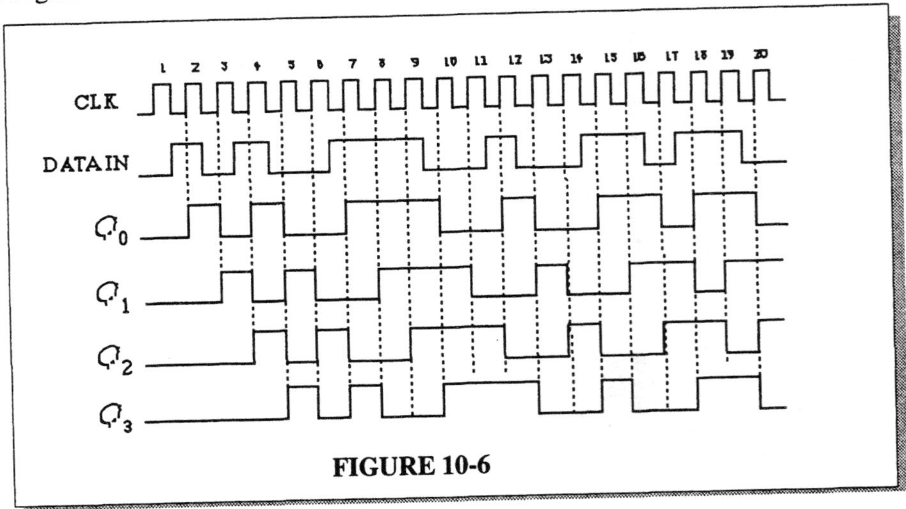

FIGURE 10-6

10. See Figure 10-7.

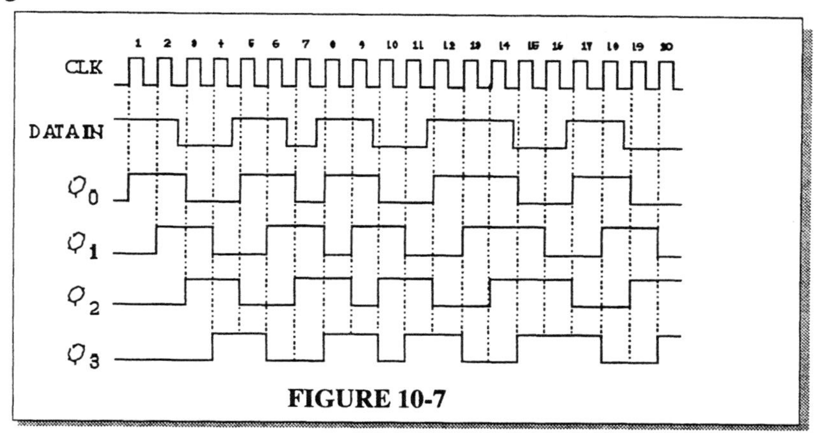

FIGURE 10-7

11. See Figure 10-8.

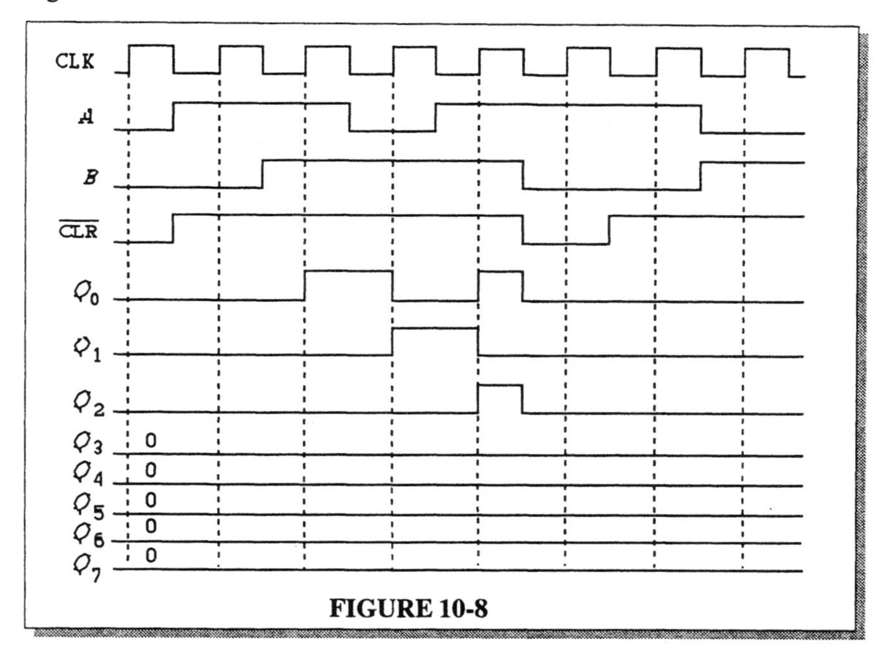

FIGURE 10-8

12. See Figure 10-9.

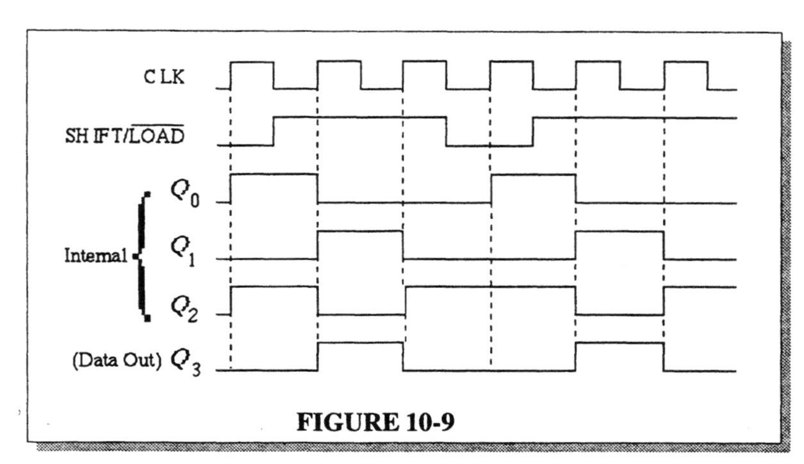

FIGURE 10-9

13. See Figure 10-10.

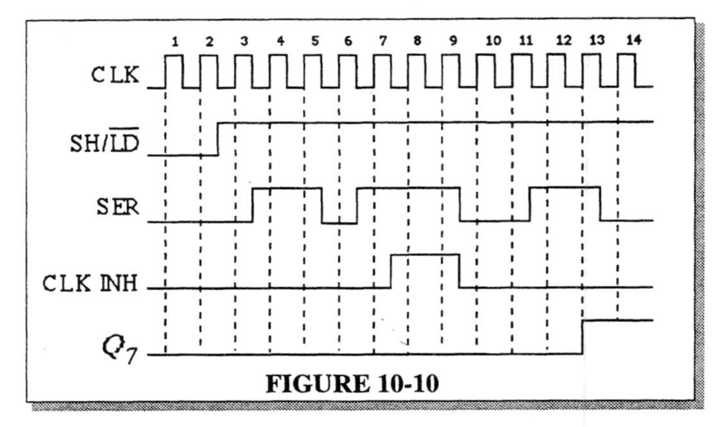

FIGURE 10-10

14. See Figure 10-11.

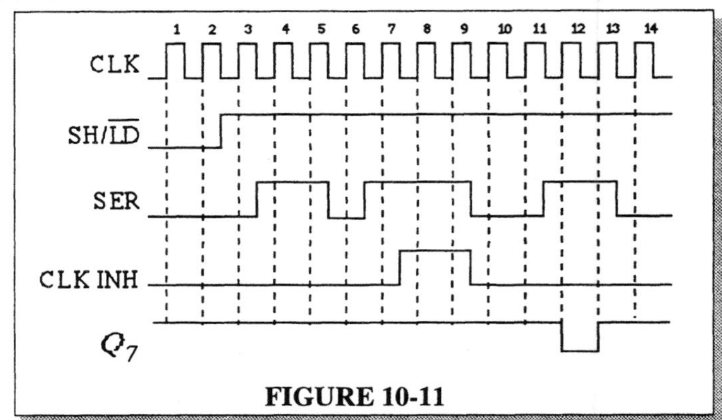

FIGURE 10-11

15. See Figure 10-12.

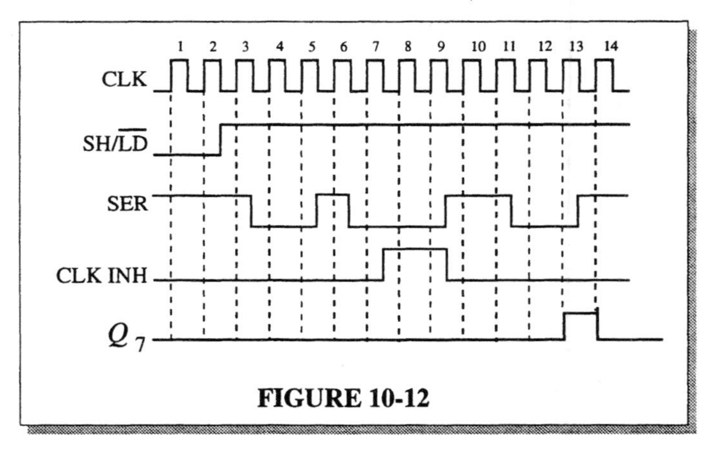

FIGURE 10-12

16. See Figure 10-13.

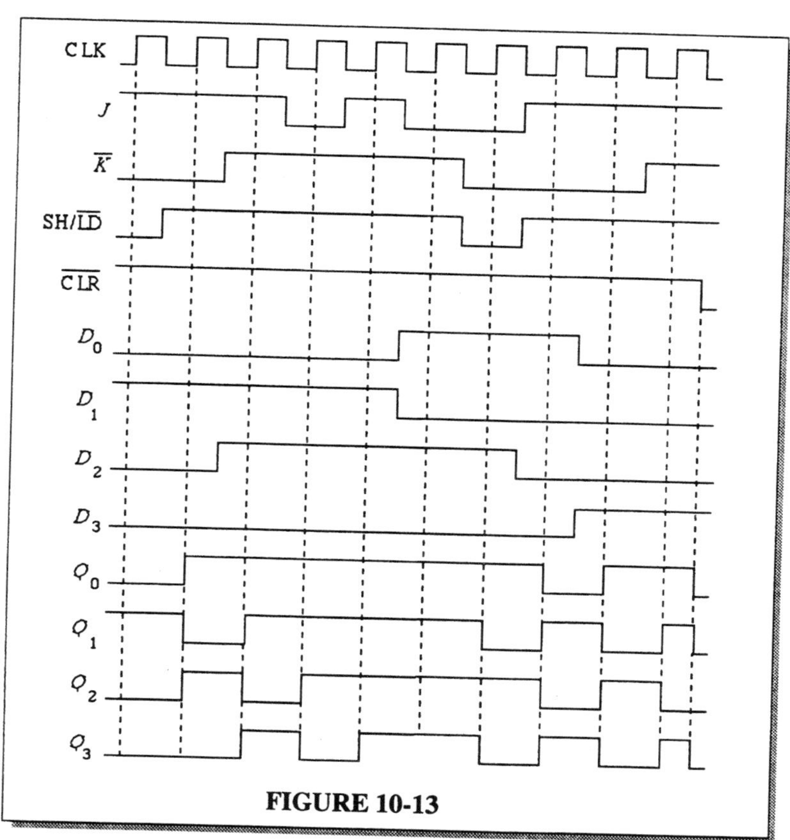

FIGURE 10-13

17. See Figure 10-14.

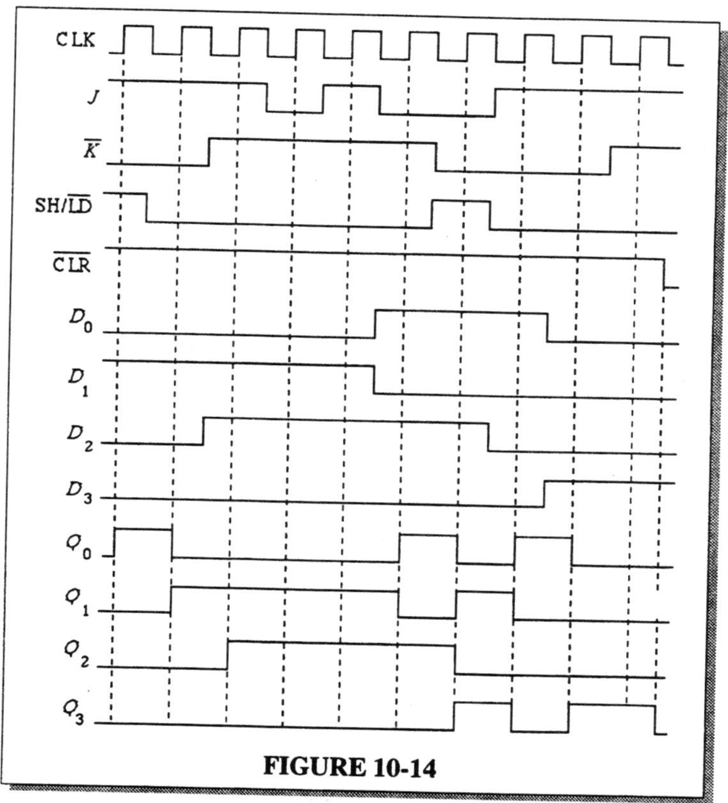

FIGURE 10-14

18. See Figure 10-15.

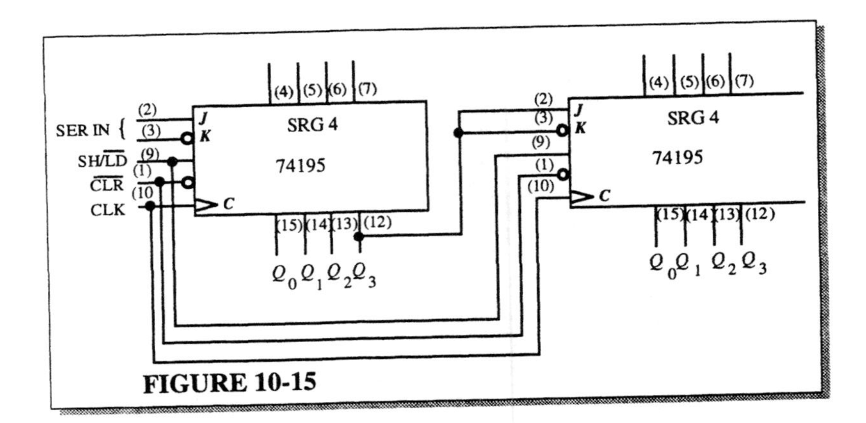

FIGURE 10-15

19.

Initially (76)	01001100	
CLK 1	10011000	Shift left
CLK 2	01001100	Shift right
CLK 3	00100110	Shift right
CLK 4	00010011	Shift right
CLK 5	00100110	Shift left
CLK 6	01001100	Shift left
CLK 7	00100110	Shift right
CLK 8	01001100	Shift left
CLK 9	00100110	Shift right
CLK 10	01001100	Shift left
CLK 11	10011000	Shift left

20.

Initially (76)	01001100	
CLK 1	00100110	Shift right
CLK 2	00010011	Shift right
CLK 3	00001001	Shift right
CLK 4	00010010	Shift left
CLK 5	00100100	Shift left
CLK 6	01001000	Shift left
CLK 7	00100100	Shift right
CLK 8	01001000	Shift left
CLK 9	10010000	Shift left
CLK 10	00100000	Shift left
CLK 11	00010000	Shift right
CLK 12	00001000	Shift right

21. See Figure 10-16.

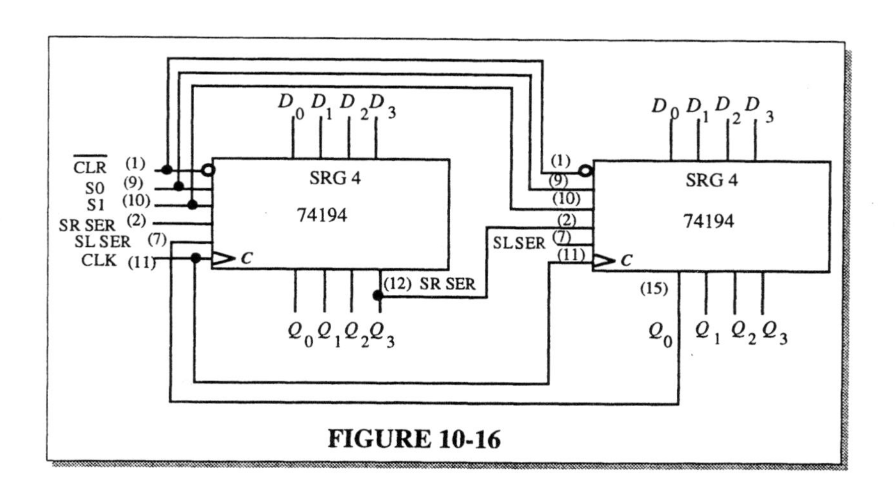

FIGURE 10-16

22. See Figure 10-17.

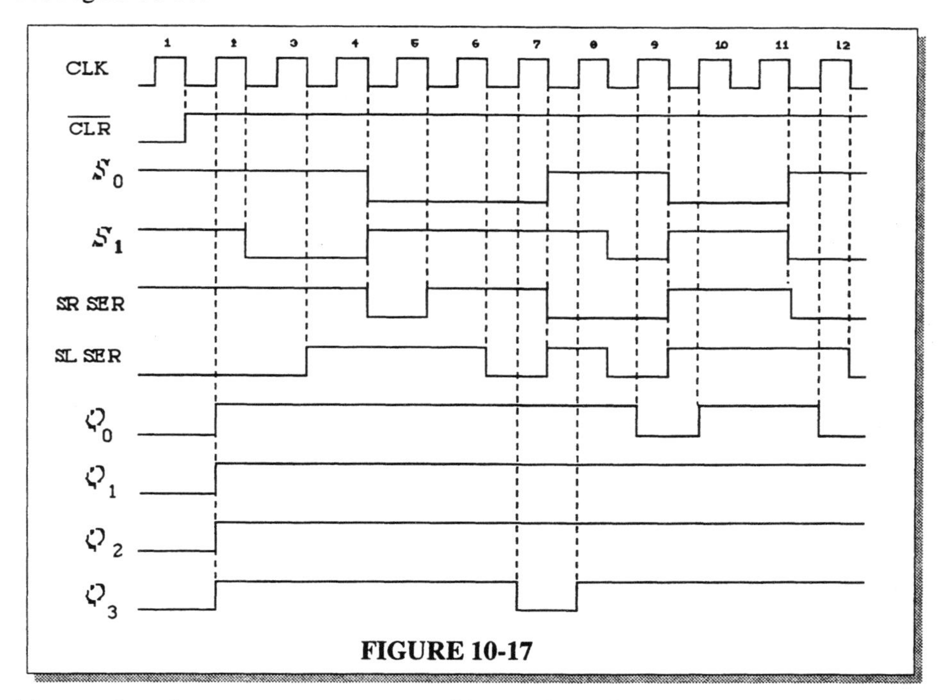

FIGURE 10-17

23.

(a) $2n = 6$
$n = \mathbf{3}$

(b) $2n = 10$
$n = \mathbf{5}$

(c) $2n = 14$
$n = \mathbf{7}$

(d) $2n = 16$
$n = \mathbf{8}$

24. $2n = 18$; $n = \mathbf{9}$ flip-flops

Q_0	Q_1	Q_2	Q_3	Q_4	Q_5	Q_6	Q_7	Q_8	Q_9
0	0	0	0	0	0	0	0	0	0
1	0	0	0	0	0	0	0	0	0
1	1	0	0	0	0	0	0	0	0
1	1	1	0	0	0	0	0	0	0
1	1	1	1	0	0	0	0	0	0
1	1	1	1	1	0	0	0	0	0
1	1	1	1	1	1	0	0	0	0
1	1	1	1	1	1	1	0	0	0
1	1	1	1	1	1	1	1	0	0
1	1	1	1	1	1	1	1	1	1
0	1	1	1	1	1	1	1	1	1
0	0	1	1	1	1	1	1	1	1
0	0	0	1	1	1	1	1	1	1
0	0	0	0	1	1	1	1	1	1
0	0	0	0	0	1	1	1	1	1
0	0	0	0	0	0	1	1	1	1
0	0	0	0	0	0	0	1	1	1
0	0	0	0	0	0	0	0	1	1
0	0	0	0	0	0	0	0	0	1
0	0	0	0	0	0	0	0	0	0

See Figure 10-18.

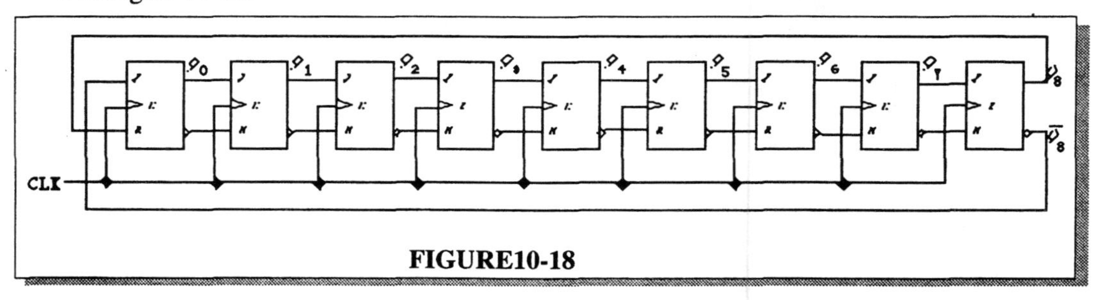

FIGURE10-18

25. See Figure 10-19.

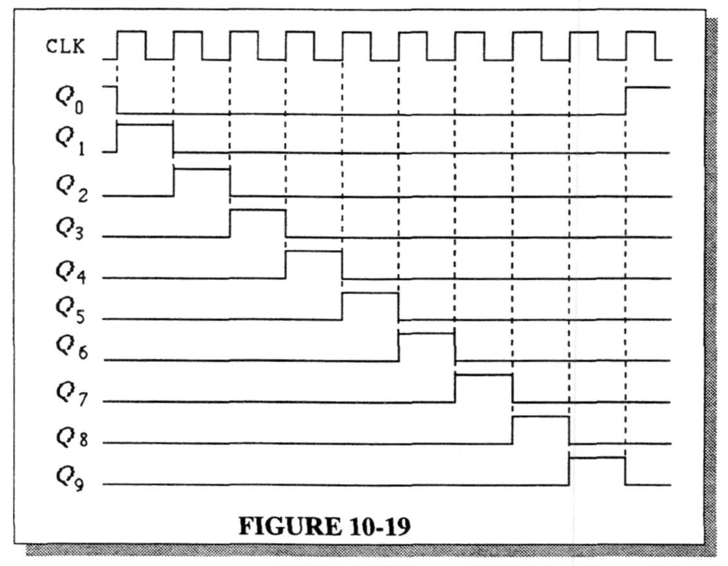

FIGURE 10-19

26. A 15-bit ring counter with stages **3**, **7**, and **12** SET and the remaining stages RESET. See Figure 10-20.

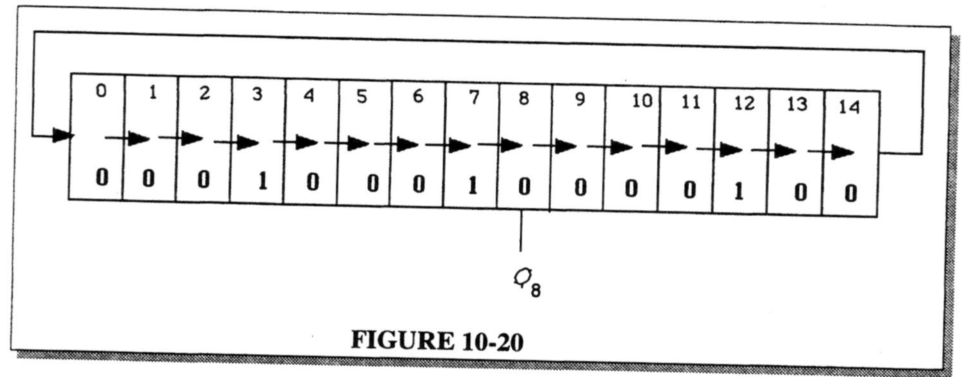

FIGURE 10-20

27. See Figure 10-21.

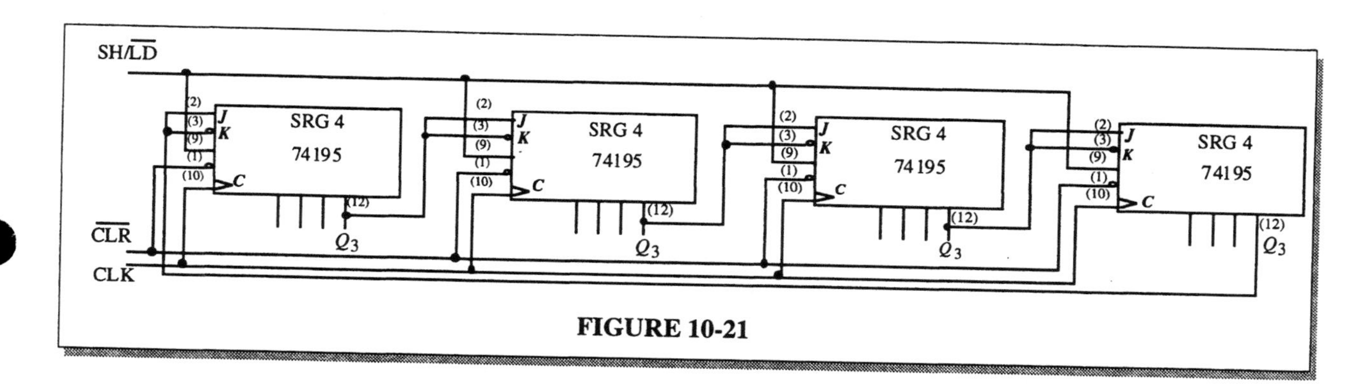

FIGURE 10-21

28. The power-on $\overline{LOAD}$ input provides a momentary LOW to parallel load the ring counter when power is turned on.

29. An incorrect code may be produced.

30. Q_2 goes HIGH on the first clock pulse indicating that the D input is open. See Figure 10-22.

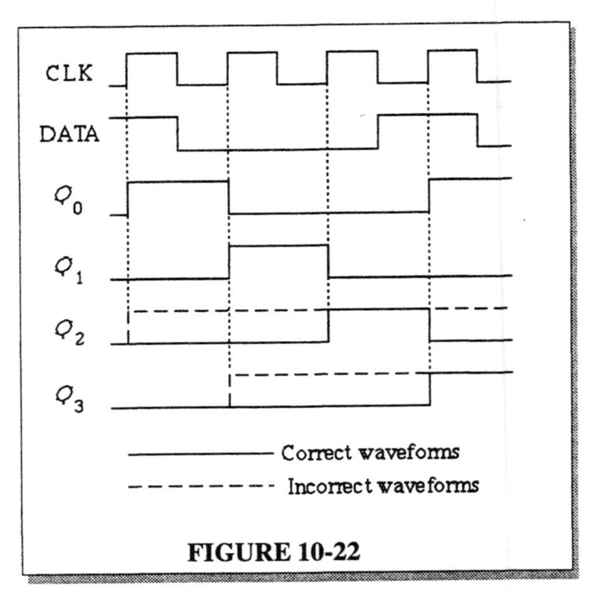

FIGURE 10-22

31. Since the LSB flip-flop works during serial shift, the problem is most likely in gate G3. An open D_3 input at G3 will cause the observed waveform. See Figure 10-23.

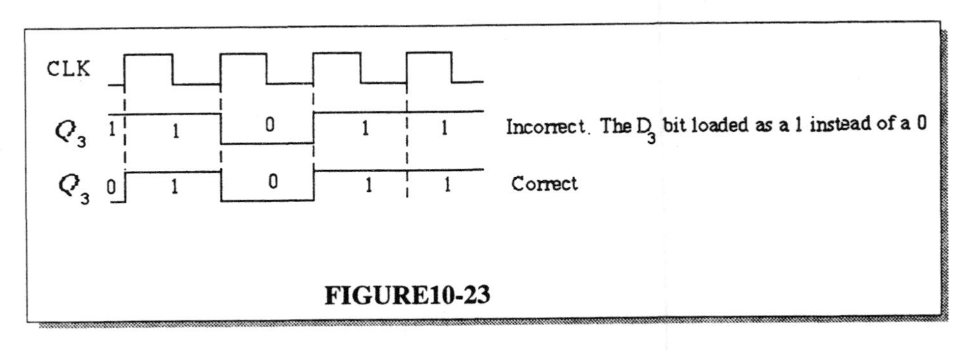

FIGURE10-23

32. It takes a LOW on the RIGHT/LEFT input to shift data left. An open inverter input will keep the inverter output LOW thus disabling all of the shift-left control gates G5, G6, G7, and G8.

33. (a) No clock at switch closure due to faulty NAND gate or one-shot; open clock input to key code register; open SH/$\overline{LD}$ input to key code register.

(b) The diode in the third row is open; Q_5 output of ring counter is open.

(c) The NAND (negative-OR) gate input connected to the first column is shorted to ground or open, preventing a switch closure transition.

(d) The "2" input to the column encoder is open.

34. *1.* Number the switches in the matrix according to the following format:

Top row								
	1	2	3	4	5	6	7	8
	9	10	11	12	13	14	15	16
	17	18	19	20	21	22	23	24
	25	26	27	28	29	30	31	32
	33	34	35	36	37	38	39	40
	41	42	43	44	45	46	47	48
	49	50	51	52	53	54	55	56
	57	58	59	60	61	62	63	64

2. Depress switches one at a time and observe the key code output according to the following Table 1.

Switch number	Key Code Register Q_0	Q_1	Q_2	Q_3	Q_4	Q_5
1	0	1	1	0	1	1
2	0	1	1	1	0	1
3	0	1	1	0	0	1
4	0	1	1	1	1	0
5	0	1	1	0	1	0
6	0	1	1	1	0	0
7	0	1	1	0	0	0
8	0	1	1	1	1	1
9	1	0	1	0	1	1
10	1	0	1	1	0	1
11	1	0	1	0	0	1
12	1	0	1	1	1	0
13	1	0	1	0	1	0
14	1	0	1	1	0	0
15	1	0	1	0	0	0
16	1	0	1	1	1	1
17	0	0	1	0	1	1
18	0	0	1	1	0	1
19	0	0	1	0	0	1
20	0	0	1	1	1	0
21	0	0	1	0	1	0
22	0	0	1	1	0	0
23	0	0	1	0	0	0
24	0	0	1	1	1	1

25	1	1	0	0	1	1
26	1	1	0	1	0	1
27	1	1	0	0	0	1
28	1	1	0	1	1	0
29	1	1	0	0	1	0
30	1	1	0	1	0	0
31	1	1	0	0	0	0
32	1	1	0	1	1	1
33	0	1	0	0	1	1
34	0	1	0	1	0	1
35	0	1	0	0	0	1
36	0	1	0	1	1	0
37	0	1	0	0	1	0
38	0	1	0	1	0	0
39	0	1	0	0	0	0
40	0	1	0	1	1	1
41	1	0	0	0	1	1
42	1	0	0	1	0	1
43	1	0	0	0	0	1
44	1	0	0	1	1	0
45	1	0	0	0	1	0
46	1	0	0	1	0	0
47	1	0	0	0	0	0
48	1	0	0	1	1	1
49	0	0	0	0	1	1
50	0	0	0	1	0	1
51	0	0	0	0	0	1
52	0	0	0	1	1	0
53	0	0	0	0	1	0
54	0	0	0	1	0	0
55	0	0	0	0	0	0
56	0	0	0	1	1	1
57	1	1	1	0	1	1
58	1	1	1	1	0	1
59	1	1	1	0	0	1
60	1	1	1	1	1	0
61	1	1	1	0	1	0
62	1	1	1	1	0	0
63	1	1	1	0	0	0
64	1	1	1	1	1	1

TABLE 1

35.

(a) Contents of Data Output Register remain constant.
(b) Contents of both registers do not change.
(c) Third stage output of Data Output Register remains HIGH.
(d) Clock generator is disabled after each pulse by the flip-flop being continuously SET and then RESET.

36. See Figure 10-24.

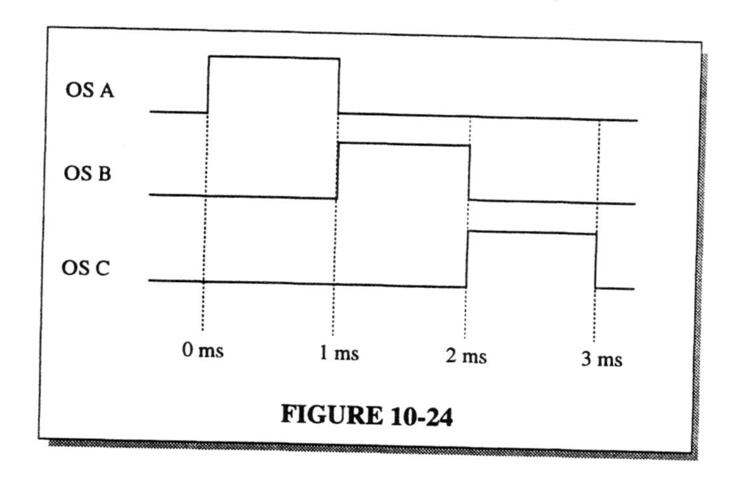

FIGURE 10-24

37. The states of shift registers A and B after two correct key closures are:

Shift register A : 10001
Shift register B: 11100

38. The states of shift registers A and B after each key closurewhen entering 7645 are:

After key 7 is pressed:
Shift register A contains 0111
Shift register B contains 11000
After key 6 is pressed:
Shift register A contains 0110
Shift register B contains 11100
After key 4 is pressed:
Shift register A contains 0100
Shift register B contains 11110
After key 5 (an incorrect entry) is pressed:
Shift register A contains 0000
Shift register B contains 10000

39. If one-shot A never triggers, the output of gate G_1 could be open or shorted, any input of gate G_1 could be open, the one-shot could have an open output, or the one-shot could have an open trigger input.

40. Possible faults are:

1. Open reset or clock input to shift register A.
2. Open reset or clock input to the address counter.
3. Open input to gate G_1.
4. Open input to gate G_6.
5. Faulty memory device.
6. Shorted serial in put to shift register B.
7. Open serial output of shift register B.

41. See Figure 10-25.

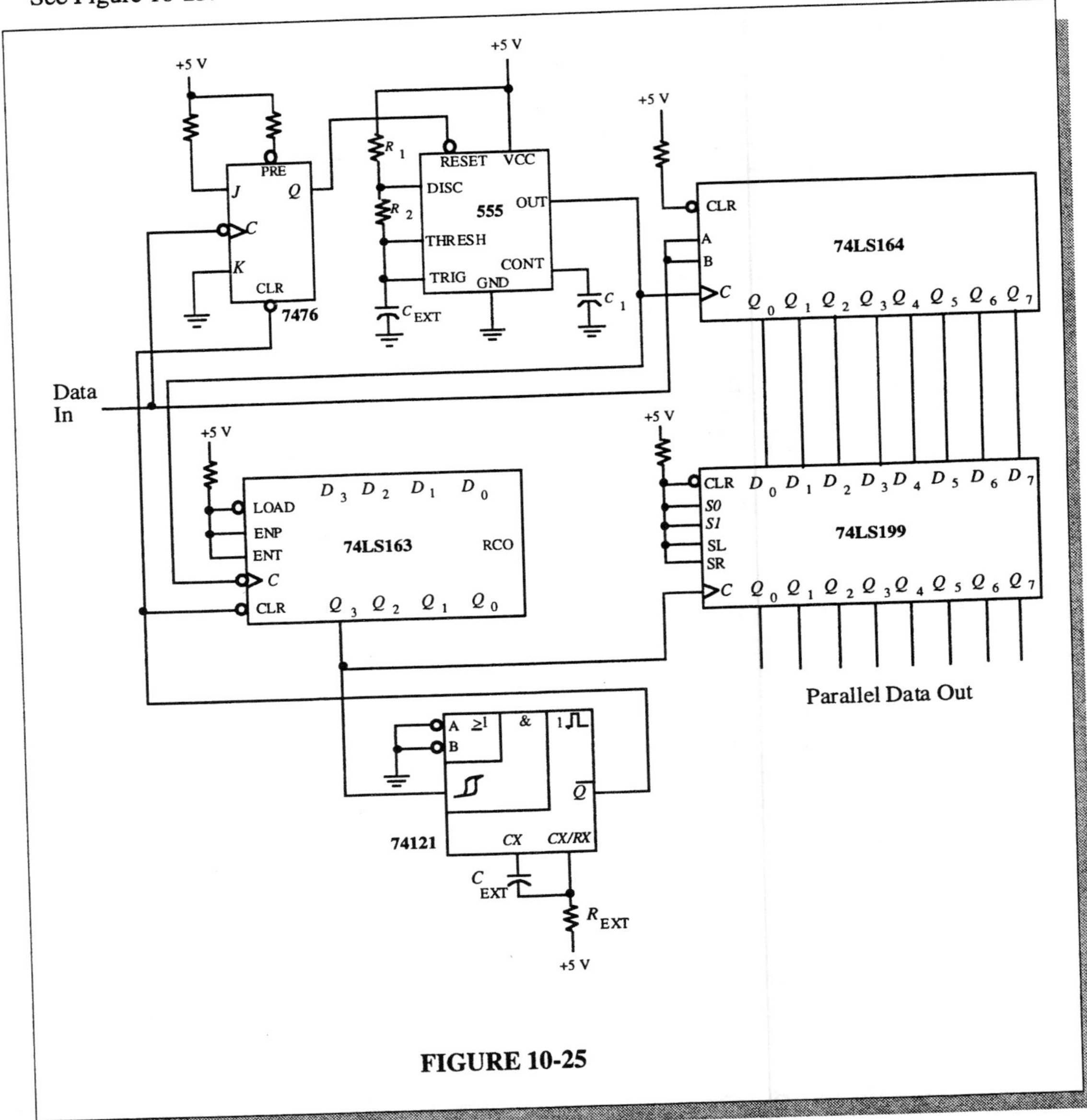

FIGURE 10-25

42. Figure 10-26 shows only the 74LS164, 74LS199, and 74LS163 portions of the circuit that require modification for 16-bit conversion.

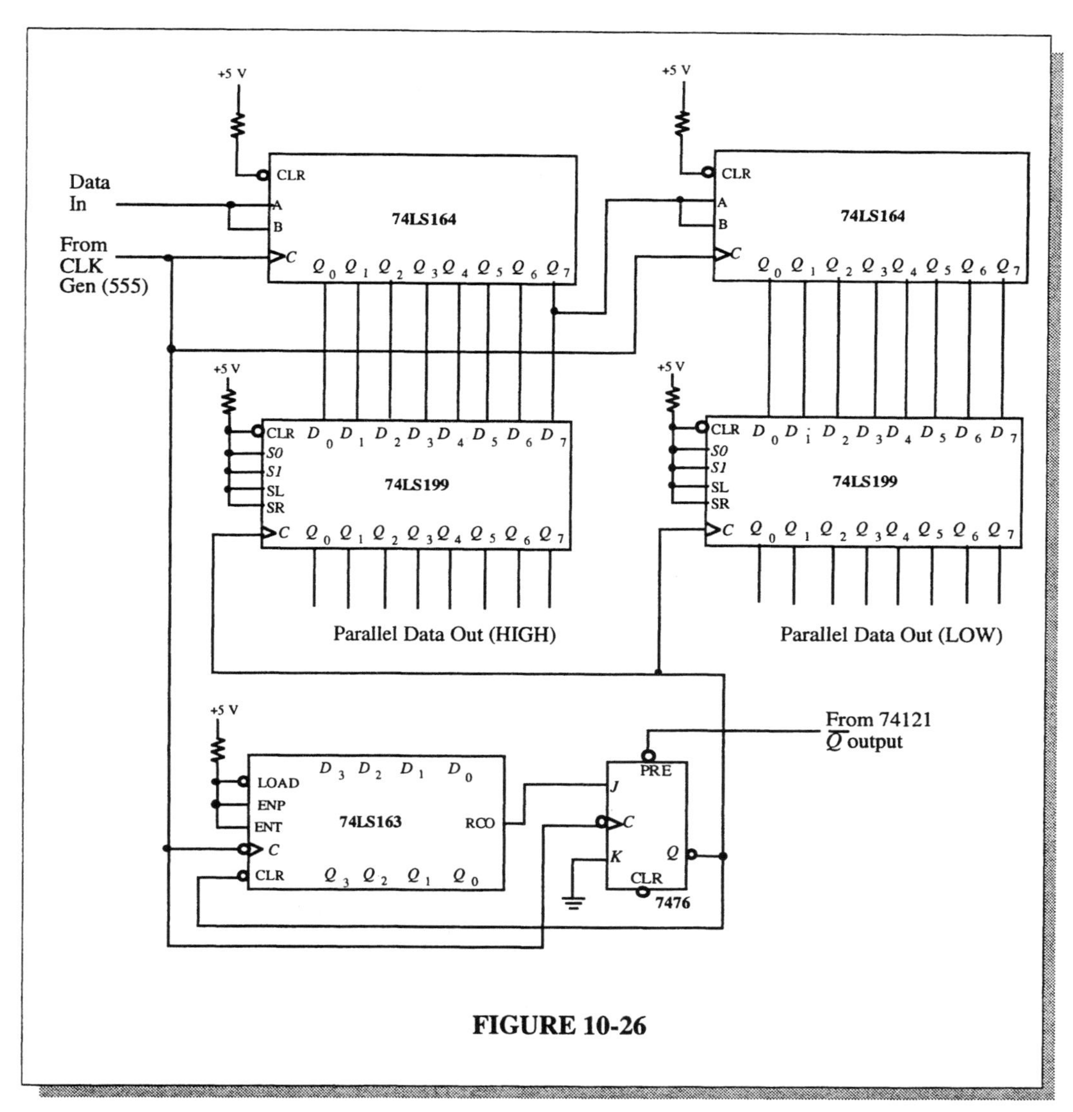

FIGURE 10-26

43. See Figure 10-27 for one possible implementation.

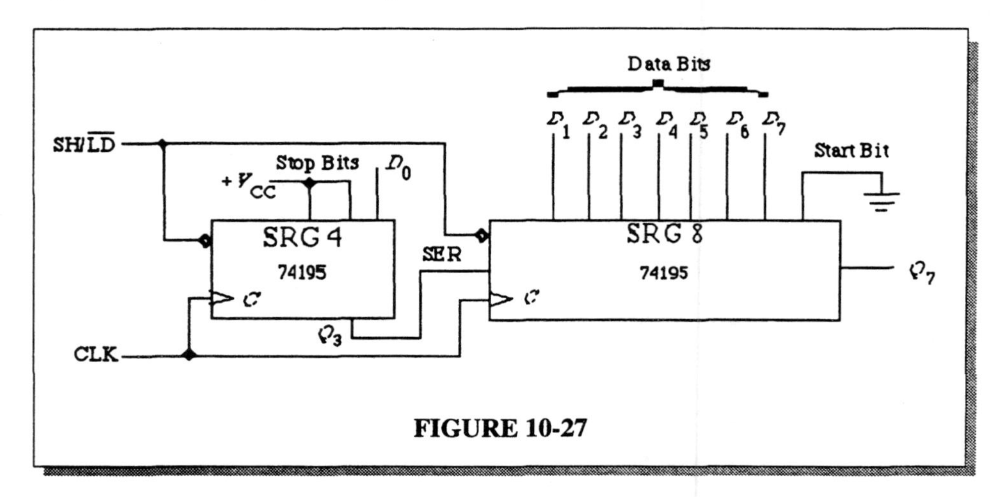

FIGURE 10-27

44. One possible approach is shown in Figure 10-28.

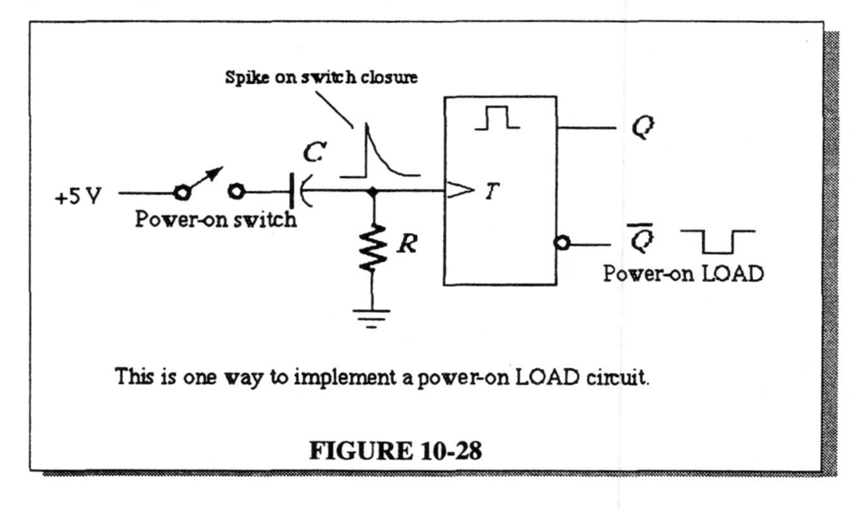

FIGURE 10-28

45. See Figure 10-29.

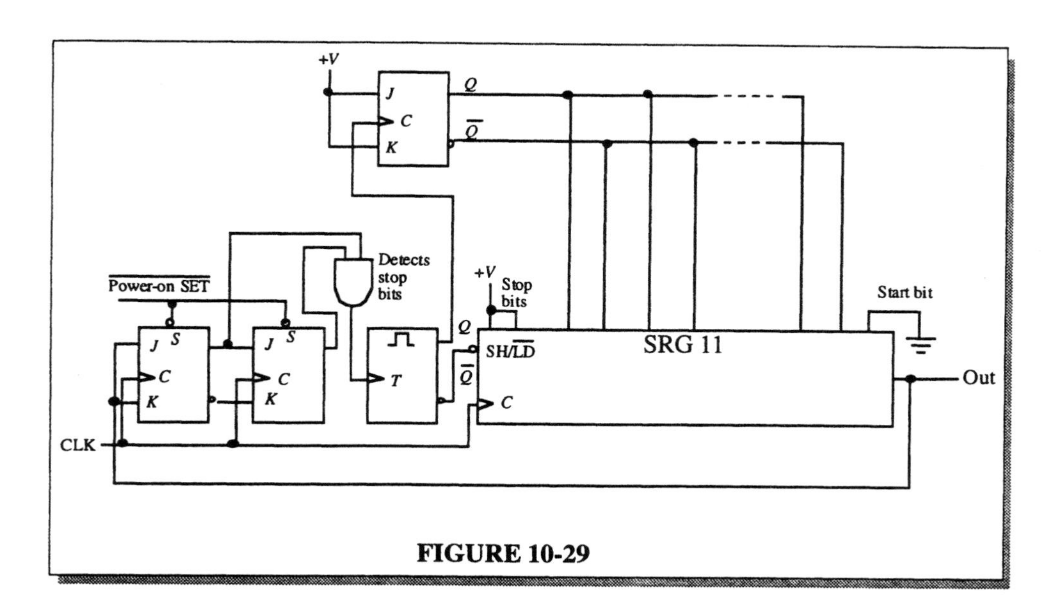

FIGURE 10-29

46. Register A requires 8 bits and can be implemented with one 74199. Register B requires 16 bits and can be implemented with two 74199s.

CHAPTER 11

SEQUENTIAL LOGIC APPLICATIONS OF PLDs

1. The four GAL22V10 OLMC configurations are:
   1. Combinational with active-LOW outputs
   2. Combinational with active-HIGH outputs
   3. Registered mode with active-LOW outputs
   4. Registered mode with active-HIGH outputs

2. For the cell values specified, the GAL16V8 output configurations are:

   (a) SYN=0, ACO=1, XOR=1: registered, active-LOW
   (b) SYN=1, ACO=0, XOR=0: combinational, active-HIGH

3. In addition to the flip-flop and logic gates the OLMC uses multiplexers.

4. For the bit combinations specified, the GAL22V10 modes are:
   (a) $S_1S_0 = 10$: combinational active-LOW
   (b) $S_1S_0 = 00$: registered active-LOW
   (c) $S_1S_0 = 11$: combinational active-HIGH
   (d) $S_1S_0 = 01$: registered active-HIGH

5. For the ABEL output pin declarations are described as follows:
   (a) X pin 15 ISTYPE 'com";
   Pin 15 is a combinational output associated with the output variable X.
   (b) Q2 pin 20 ISTYPE 'reg";
   Pin 20 is a registered output associated with the output variable Q2.
   (c) A pin 18 ISTYPE 'com, invert";
   Pin 18 is an active-LOW combinational output associated with the output variableA.
   (d) Q1 pin 21 ISTYPE 'reg, buffer";
   Pin 21 is an active-HIGH registered output associated with the output variable Q1.

6. For the given expressions, the output types are:
   (a) Y = A# B; combinational
   (b) X := A & !B; registered
   (c) Q1 := D1 # D0; registered
   (d) Q2 = !C; combinational

7. A clock equation to accompany the expression Q1 := (A&B)#C; is

   Q1.CLK = Clock (where Clock is the clock pin variable)

8. An ABEL input file to implement an 8-bit SISO shift register is a GAL22V10 is

```
Module          8-bit_SISO_shift_register
Title           '8-bit SISO shift register in a GAL22V10'
"Device declaration
        Register        DEVICE        'P22V10'
"Pin declarations
        Clock, Clear      PIN  1,2;
        Din               PIN  3;
        Q0,Q1,Q2,Q3       PIN  16,17,18,19  ISTYPE 'reg, buffer';
        Q4,Q5,Q6,Qout     PIN  20,21,22,23  ISTYPE 'reg, buffer';
Equations
        [Q0,Q1,Q2,Q3,Q4,Q5,Q6,Qout].CLK = Clock;
        [Q0,Q1,Q2,Q3,Q4,Q5,Q6,Qout].AR = !Clear;
        Q0 := Din;
        Q1 := Q0.FB;
        Q2 := Q1.FB;
        Q3 := Q2.FB;
        Q4 := Q3.FB;
        Q5 := Q4.FB;
        Q6 := Q5.FB;
        Qout := Q6.FB;
END
```

9. The test vector section for the file of Problem 8 is:

```
Test_vector  ([Clock,Clear,Din] → [Q0,Q1,Q2,Q3,Q4,Q5,Q6,Qout])
             [ .x.,  0,  .x. ] → [ 0,  0,  0,  0,  0,  0,  0,  0 ];
             [ .c.,  1,  1 ] → [ 1,  0,  0,  0,  0,  0,  0,  0 ];
             [ .c.,  1,  0 ] → [ 0,  1,  0,  0,  0,  0,  0,  0 ];
             [ .c.,  1,  1 ] → [ 1,  0,  1,  0,  0,  0,  0,  0 ];
             [ .c.,  1,  0 ] → [ 0,  1,  0,  1,  0,  0,  0,  0 ];
             [ .c.,  1,  1 ] → [ 1,  0,  1,  0,  1,  0,  0,  0 ];
             [ .c.,  1,  0 ] → [ 0,  1,  0,  1,  0,  1,  0,  0 ];
             [ .c.,  1,  1 ] → [ 1,  0,  1,  0,  1,  0,  1,  0 ];
             [ .c.,  1,  0 ] → [ 0,  1,  0,  1,  0,  1,  0,  1 ];
```

10. An ABEL input file to implement the circuit in Problem 8 using a GAL16V8 is:

```
Module          8-bit_SISO_shift_register
Title           '8-bit SISO shift register in a GAL16V8'
"Device declaration
        Register        DEVICE          'P16V8'
"Pin declarations
        Clock, Clear    PIN  1,11;
        Din             PIN  3;
        Q0,Q1,Q2,Q3     PIN  12,13,14,15  ISTYPE 'reg, buffer';
        Q4,Q5,Q6,Qout   PIN  16,17,18,19  ISTYPE 'reg, buffer';
Equations
        [Q0,Q1,Q2,Q3,Q4,Q5,Q6,Qout].CLK = Clock;
        [Q0,Q1,Q2,Q3,Q4,Q5,Q6,Qout].AR = !Clear;
        Q0 := Din;
        Q1 := Q0.FB;
        Q2 := Q1.FB;
        Q3 := Q2.FB;
        Q4 := Q3.FB;
        Q5 := Q4.FB;
        Q6 := Q5.FB;
```

11. The test vectors for the GAL16V8 in Problem 10 are identical to those in Problem 9:

```
Test_vectors  ([Clock,Clear,Din] → [Q0,Q1,Q2,Q3,Q4,Q5,Q6,Qout])
              [ .x.,   0,  .x. ] → [ 0,  0,  0,  0,  0,  0,  0,   0 ];
              [ .c.,   1,   1  ] → [ 1,  0,  0,  0,  0,  0,  0,   0 ];
              [ .c.,   1,   0  ] → [ 0,  1,  0,  0,  0,  0,  0,   0 ];
              [ .c.,   1,   1  ] → [ 1,  0,  1,  0,  0,  0,  0,   0 ];
              [ .c.,   1,   0  ] → [ 0,  1,  0,  1,  0,  0,  0,   0 ];
              [ .c.,   1,   1  ] → [ 1,  0,  1,  0,  1,  0,  0,   0 ];
              [ .c.,   1,   0  ] → [ 0,  1,  0,  1,  0,  1,  0,   0 ];
              [ .c.,   1,   1  ] → [ 1,  0,  1,  0,  1,  0,  1,   0 ];
              [ .c.,   1,   0  ] → [ 0,  1,  0,  1,  0,  1,  0,   1 ];
```

12. An ABEL input file to expand the 4-bit PISO shift register in Figue 11-23 to an 8-bit PISO using a GAL22V10 is

```
Module          8-bit_PISO_shift_register
Title           '8-bit PISO shift register in a GAL22V10'
"Device declaration
        Register        DEVICE          'P22V10'
"Pin declarations
        Clock, Clear    PIN  1,2;
        SHLD            PIN  3;
        D0,D1,D2,D3     PIN  4,5,6,7;
        D4,D5,D6,D7     PIN  8,9,10,11;
        Q0,Q1,Q2,Q3     PIN  16,17,18,19  ISTYPE "reg, buffer';
        Q4,Q5,Q6,Q7     PIN  20,23,24,25  ISTYPE 'reg, buffer';
        BININ   = [D7,D6,D5,D4,D3,D2,D1,D0];
        BINOUT = [Q7,Q6,Q5,Q4,Q3,Q2,Q1,Q0];
```

(CONT)

```
Equations
        BININ.CLK = Clock;
        BINOUT.AR = !Clear;
        Q0 := D0;
        Q1 := Q0.FB&SHLD # D1&!SHLD;
        Q2 := Q1.FB&SHLD # D2&!SHLD;
        Q3 := Q2.FB&SHLD # D3&!SHLD;
        Q4 := Q3.FB&SHLD # D4&!SHLD;
        Q5 := Q4.FB&SHLD # D5&!SHLD;
        Q6 := Q5.FB&SHLD # D6&!SHLD;
        Q7 := Q6.FB&SHLD # D7&!SHLD;

Test_vectors ([Clock,Clear,SHLD,   BININ   ]→[ BINOUT ])
             [ .x.,  0,  .x.,      .x.     ]→ [^B00000000];
             [ .c.,  1,  0,  ^B01010101]→ [^B01010101];
             [ .c.,  1,  0,  ^B10101010]→ [^B10101010];
             [ .c.,  1,  0,  ^B00000000]→ [^B01010101];
             [ .c.,  1,  0,  ^B00000000]→ [^B00101010];
             [ .c.,  1,  0,  ^B00000000]→ [^B00010101];
             [ .c.,  1,  0,  ^B00000000]→ [^B00001010];
             [ .c.,  1,  0,  ^B00000000]→ [^B00000101];
             [ .c.,  1,  0,  ^B00000000]→ [^B00000010];
            .[ .c.,  1,  0,  ^B00000000]→ [^B00000001];
            .[ .c.,  1,  0,  ^B00000000]→ [^B00000000];
            .[ .c.,  1,  0,  ^B00000001]→ [^B10000000];
            .[ .c.,  1,  0,  ^B00000000]→ [^B11000000];
            .[ .c.,  1,  0,  ^B00000000]→ [^B11100000];
            .[ .c.,  1,  0,  ^B00000000]→ [^B11110000];
            .[ .c.,  1,  0,  ^B00000000]→ [^B11111000];
            .[ .c.,  1,  0,  ^B00000000]→ [^B11111100];
            .[ .c.,  1,  0,  ^B00000000]→ [^B11111110];
            .[ .c.,  1,  0,  ^B00000000]→ [^B11111111];

END
```

13. A GAL16V8 has an insufficient number of inputs available to implement the 8-bit register in Problem 12.

14. The difference between a Mealy machine and a Moore machine is that in a Moore machine the outputs depend on the states only while in a Mealy machine the outputs are dependent on both the states and the inputs.

15. The largest counter that can be implemented in a single GAL16V8 is an 8-bit counter.

16. The largest counter that can be implemented in a single GAL22V10 is a 10-bit counter.

17. For an up-only Gray code counter, the ABEL logic equations are:

```
Q0 = !Q2&!Q1&!Q0 # !Q2&!Q1&Q0 # Q2&Q1&!Q0 # Q2&Q1&Q0
   => !Q2&!Q1 # Q2&Q1
Q1 = !Q2&!Q1&Q0 # !Q2&Q1&Q0 # !Q2&Q1&!Q0 # Q2&Q1&!Q0
   => !Q2&Q0 # Q1&!Q0
Q2 = !Q2&Q1&!Q0 # Q2&Q1&!Q0 # Q2&Q1&Q0 # Q2&!Q1&Q0
   => Q1&!Q0 # Q2&Q0
```

An ABEL input file for a GAL22V10 circuit is:

```
Module          3-bit_Gray_code_up_counter
Title           '3-bit Gray code up counter in a GAL22V10'
"Device declaration
        Counter         DEVICE          'P22V10'
"Pin declarations
        Clock, Clear    PIN  1,2;
        Q0,Q1,Q2        PIN  21,22,23  ISTYPE 'reg, buffer';
Equations
        [Q0,Q1,Q2].CLK = Clock;
        [Q0,Q1,Q2].AR = !Clear;
        Q0 := !Q2.FB&!Q1.FB # Q2.FB&Q1.FB;
        Q1 := !Q2.FB&Q0.FB # Q1.FB&!Q0.FB;
        Q2 := Q1.FB&Q0.FB # Q2.FB&Q0.FB;

Test_vectors  ([Clock,Clear] → [Q2,Q1,Q0])
              [ .x.,   0  ] → [ 0,  0,  0 ];
              [ .c.,   1  ] → [ 0,  0,  1 ];
              [ .c.,   1  ] → [ 0,  1,  1 ];
              [ .c.,   1  ] → [ 0,  1,  0 ];
              [ .c.,   1  ] → [ 1,  1,  0 ];
              [ .c.,   1  ] → [ 1,  1,  1 ];
              [ .c.,   1  ] → [ 1,  0,  1 ];
              [ .c.,   1  ] → [ 1,  0,  0 ];
              [ .c.,   1  ] → [ 0,  0,  0 ];

END
```

18. An ABEL input file for the counter of Problem 17 using truth table entry is:

```
Module          3-bit_Gray_code_up_counter
Title           '3-bit Gray code up counter in a GAL22V10'
        "Device declaration
                Counter         DEVICE          'P22V10'
        "Pin declarations
                Clock, Clear    PIN  1,2;
                Q0,Q1,Q2        PIN  21,22,23  ISTYPE 'reg, buffer';
        Equations
                [Q0,Q1,Q2].CLK = Clock;
                [Q0,Q1,Q2].AR = !Clear;
```

(CONT)

```
Truth_table  ([Clock,Q2,Q1,Q0] : > [Q2,Q1,Q0])
             [ .c.,  0, 0, 0 ] → [ 0, 0, 1 ];
             [ .c.,  0, 0, 1 ] → [ 0, 1, 1 ];
             [ .c.,  0, 1, 1 ] → [ 0, 1, 0 ];
             [ .c.,  0, 1, 0 ] → [ 1, 1, 0 ];
             [ .c.,  1, 1, 0 ] → [ 1, 1, 1 ];
             [ .c.,  1, 1, 1 ] → [ 1, 0, 1 ];
             [ .c.,  1, 0, 1 ] → [ 1, 0, 0 ];
             [ .c.,  1, 0, 0 ] → [ 0, 0, 0 ];

Test_vectors ([Clock,Clear] → [Q2,Q1,Q0])
             [ .x.,   0  ] → [ 0, 0, 0 ];
             [ .c.,   1  ] → [ 0, 0, 1 ];
             [ .c.,   1  ] → [ 0, 1, 1 ];
             [ .c.,   1  ] → [ 0, 1, 0 ];
             [ .c.,   1  ] → [ 1, 1, 0 ];
             [ .c.,   1  ] → [ 1, 1, 1 ];
             [ .c.,   1  ] → [ 1, 0, 1 ];
             [ .c.,   1  ] → [ 1, 0, 0 ];
             [ .c.,   1  ] → [ 0, 0, 0 ];

END
```

19. An ABEL input file for the Gray code counter in Problem 17 using state diagram entry is:

```
Module          3-bit_Gray_code_up_counter
Title           '3-bit Gray code up counter in a GAL22V10'
"Device declaration
        Counter         DEVICE          'P22V10'
"Pin declarations
        Clock, Clear    PIN  1,2;
        Q0,Q1,Q2        PIN  21,22,23  ISTYPE 'reg, buffer';

        COUNT = [Q2,Q1,Q0];
        A       = ^B000;
        B       = ^B001;
        C       = ^B011;
        D       = ^B010;
        E       = ^B110;
        F       = ^B111;
        G       = ^B101;
        H       = ^B100;
Equations
        COUNT.CLK = Clock;
        COUNT.AR = !Clear;

State_diagram  COUNT
        State A  : GOTO B;
        State B  : GOTO C;
        State C  : GOTO D;
        State D  : GOTO E;
        State E  : GOTO F;
        State F  : GOTO G;
        State G  : GOTO H;
        State H  : GOTO A;
```

```
Test_vectors ([Clock,Clear] → [Q2,Q1,Q0])
        [ .x.,   0  ]→[ 0,  0,  0 ];
        [ .c.,   1  ]→[ 0,  0,  1 ];
        [ .c.,   1  ]→[ 0,  1,  1 ];
        [ .c.,   1  ]→[ 0,  1,  0 ];
        [ .c.,   1  ]→[ 1,  1,  0 ];
        [ .c.,   1  ]→[ 1,  1,  1 ];
        [ .c.,   1  ]→[ 1,  0,  1 ];
        [ .c.,   1  ]→[ 1,  0,  0 ];
        [ .c.,   1  ]→[ 0,  0,  0 ];

END
```

20. The input file of Problem 17 modified for a GAL16V8 target device is:

```
Module          3-bit_Gray_code_up_counter
Title           '3-bit Gray code up counter in a GAL16V8'
"Device declaration
        Counter         DEVICE          'P16V8'
"Pin declarations
        Clock, Clear    PIN  1,11;
        Q0,Q1,Q2        PIN  17,18,19  ISTYPE 'reg, buffer';
Equations
        [Q0,Q1,Q2].CLK = Clock;
        [Q0,Q1,Q2].AR = !Clear;
        Q0 := !Q2.FB&!Q1.FB # Q2.FB&Q1.FB;
        Q1 := !Q2.FB&Q0.FB # Q1.FB&!Q0.FB;
        Q2 := Q1.FB&Q0.FB # Q2.FB&Q0.FB;

Test_vectors ([Clock,Clear] → [Q2,Q1,Q0])
        [ .x.,   0  ]→[ 0,  0,  0 ];
        [ .c.,   1  ]→[ 0,  0,  1 ];
        [ .c.,   1  ]→[ 0,  1,  1 ];
        [ .c.,   1  ]→[ 0,  1,  0 ];
        [ .c.,   1  ]→[ 1,  1,  0 ];
        [ .c.,   1  ]→[ 1,  1,  1 ];
        [ .c.,   1  ]→[ 1,  0,  1 ];
        [ .c.,   1  ]→[ 1,  0,  0 ];
        [ .c.,   1  ]→[ 0,  0,  0 ];

END
```

21. The input file from probme 18 modified for a GAL16V8 target device is:

```
Module          3-bit_Gray_code_up_counter
Title           '3-bit Gray code up counter in a GAL16V8'
        "Device declaration
                Counter         DEVICE          'P16V8'
        "Pin declarations
                Clock, Clear    PIN  1,11;
                Q0,Q1,Q2        PIN  17,18,19  ISTYPE 'reg, buffer';
        Equations
                [Q0,Q1,Q2].CLK = Clock;
                [Q0,Q1,Q2].AR = !Clear;

    Truth_table ([Clock,Q2,Q1,Q0] : > [Q2,Q1,Q0])
                [ .c.,   0,  0,  0 ]→[ 0,  0,  1 ];
                [ .c.,   0,  0,  1 ]→[ 0,  1,  1 ];
                [ .c.,   0,  1,  1 ]→[ 0,  1,  0 ];
                [ .c.,   0,  1,  0 ]→[ 1,  1,  0 ];
                [ .c.,   1,  1,  0 ]→[ 1,  1,  1 ];
                [ .c.,   1,  1,  1 ]→[ 1,  0,  1 ];
                [ .c.,   1,  0,  1 ]→[ 1,  0,  0 ];
                [ .c.,   1,  0,  0 ]→[ 0,  0,  0 ];

    Test_vectors ([Clock,Clear] → [Q2,Q1,Q0])
                [ .x.,    0   ]→[ 0,  0,  0 ];
                [ .c.,    1   ]→[ 0,  0,  1 ];
                [ .c.,    1   ]→[ 0,  1,  1 ];
                [ .c.,    1   ]→[ 0,  1,  0 ];
                [ .c.,    1   ]→[ 1,  1,  0 ];
                [ .c.,    1   ]→[ 1,  1,  1 ];
                [ .c.,    1   ]→[ 1,  0,  1 ];
                [ .c.,    1   ]→[ 1,  0,  0 ];
                [ .c.,    1   ]→[ 0,  0,  0 ];

END
```

22. The input file from Problem 19 modified for a GAL16V8 is:

```
        Module          3-bit_Gray_code_up_counter
        Title           '3-bit Gray code up counter in a GAL16V8'
        "Device declaration
                Counter         DEVICE          'P16V8'
        "Pin declarations
                Clock, Clear    PIN  1,11;
                Q0,Q1,Q2        PIN  17,18,19  ISTYPE 'reg, buffer';
                COUNT = [Q2,Q1,Q0];
                A       = ^B000;
                B       = ^B001;
                C       = ^B011;
                D       = ^B010;
                E       = ^B110;
                F       = ^B111;
                G       = ^B101;
                H       = ^B100;
```

```
Equations
        COUNT.CLK = Clock;
        COUNT.AR = !Clear;

State_diagram  COUNT
        State A  : GOTO B;
        State B  : GOTO C;
        State C  : GOTO D;
        State D  : GOTO E;
        State E  : GOTO F;
        State F  : GOTO G;
        State G  : GOTO H;
        State H  : GOTO A;
Test_vectors ([Clock,Clear] → [Q2,Q1,Q0])
            [ .x.,   0  ] → [ 0,  0,  0 ];
            [ .c.,   1  ] → [ 0,  0,  1 ];
            [ .c.,   1  ] → [ 0,  1,  1 ];
            [ .c.,   1  ] → [ 0,  1,  0 ];
            [ .c.,   1  ] → [ 1,  1,  0 ];
            [ .c.,   1  ] → [ 1,  1,  1 ];
            [ .c.,   1  ] → [ 1,  0,  1 ];
            [ .c.,   1  ] → [ 1,  0,  0 ];
            [ .c.,   1  ] → [ 0,  0,  0 ];

END
```

23. For the state diagram of Figure 11-25, if the OPEN button is pressed in the following states the results are:
    (a) CLOSE 1 state: The system goes to the REST 1 state.
    (b) REST1 state: The system stays in the REST1 state.
    (c) UP state: The system stays in the UP state until the ARRIVE signal..
    (d) REST2 state: The system stays in the REST2 state.
    (e) CLOSE2 state: The system goes to the REST2 state.
    (f) DOWN state: The system stays in the DOWN state until the ARRIVE signal.

24. (a) FLOOR1 button pressed:
    CLOSE1 state: System stays in the CLOSE1 state.
    REST1 state: The system goes to the CLOSE1 state.
    UP state: The system stays in the UP state until the ARRIVE signal..
    REST2 state: The system goes to the CLOSE2 state.
    CLOSE2 state: The system goes to the DOWN state.
    DOWN state: The system stays in the DOWN state until the ARRIVE signal.

    (b) FLOOR2 button pressed:
    CLOSE1 state: System goes to the UP state.
    REST1 state: The system goes to the CLOSE1 state.
    UP state: The system stays in the UP state until the ARRIVE signal..
    REST2 state: The system goes to the CLOSE2 state.
    CLOSE2 state: The system stays in the CLOSE2 state.
    DOWN state: The system stays in the DOWN state until the ARRIVE signal.

(c) REQUEST1 button pressed:

CLOSE1 state: System goes to the REST1 state.
REST1 state: The system stays in the REST1 state.
UP state: The system stays in the UP state until the ARRIVE signal..
REST2 state: The system goes to the CLOSE2 state.
CLOSE2 state: The system stays goes to the DOWN state.
DOWN state: The system stays in the DOWN state until the ARRIVE signal.

(d) REQUEST2 button pressed:

CLOSE1 state: System goes to the UP state.
REST1 state: The system goes to the CLOSE1 state.
UP state: The system stays in the UP state until the ARRIVE signal..
REST2 state: The system stays in the REST2 state.
CLOSE2 state: The system stays in the REST2 state.
DOWN state: The system stays in the DOWN state until the ARRIVE signal.

25. The input file for the elevator control logic targeted for a GAL22V10 instead of a GAL16V8 is:

```
Module          Elevator_state_control
Title           'Control logic for a two-story elevator system'
"Device declaration
        Elevator_cont           DEVICE          'P22V10'
"Pin declarations
        Clock, !OE                              PIN  1,13;
        REQ1,REQ2                               PIN  2,3;
        FLOOR2,FLOOR1,OPEN,ARRIVE               PIN  4,5,6,7;
        L1PB,L1PB_                              PIN  20,21    ISTYPE 'com, buffer';
        L2PB,L2PB_                              PIN  22,23    ISTYPE 'com, buffer';
        DOOR,MOTION,DIR                         PIN  14,15,16   ISTYPE 'reg, buffer';

"State definitions
        CONSTATE = [DOOR,MOTION,DIR];
        REST1       = ^B000;
        OPEN1       = ^B100;
        UP          = ^B110;
        REST2       = ^B001;
        CLOSE2      = ^B101;
        DOWN        = ^B111;

Equations
        CONSTATE.CLK = CLK;
        CONSTATE.AR = OE;
        L1PB   = REQ1 # !DIR.FB&OPEN # DIR.FB&FLOOR1 # !L1PB_;
        L1PB_  = !DOOR.FB&!MOTION.FB&DIR.FB # !L1PB;
        L2PB   = REQ2 # DIR.FB&OPEN # !DIR.FB&FLOOR2 # !L2PB_;
        L2PB_  = !DOOR.FB&!MOTION.FB&DIR.FB # !L2PB;

State_diagram CONSTATE
        State RESET1    : if (L2PB) then UP else CLOSE1;
        State CLOSE1    : if (L2PB) then UP else if (L1PB) then RESET1 elseCLOSE1;
        State UP        : if (ARRIVE) then REST2 else UP;
        State RESET2    : if (L1PB) then DOWN else CLOSE2;
        State CLOSE2    : if (L1PB) then DOWN else if (L2PB) then RESET2 elseCLOSE2;
        State DOWN      : if (ARRIVE) then REST1 else DOWN;
```

```
Test_vectors ([CLK,REQ1,REQ2,FLOOR2,FLOOR1,OPEN,ARRIVE] → [DOOR,MOTION,DIR])
    [ .c., 0, 0, 0, 0, 0, 1 ] → [ .x., .x., .x. ];
    [ .c., 1, 0, 0, 0, 0, 1 ] → [ .x., .x., .x. ];
    [ .c., 1, 0, 0, 0, 0, 1 ] → [ 0, 0, 0 ];
    [ .c., 0, 0, 0, 0, 0, 0 ] → [ 1, 0, 0 ];
    [ .c., 1, 0, 0, 0, 0, 0 ] → [ 0, 0, 0 ];
    [ .c., 0, 0, 0, 0, 0, 1 ] → [ 1, 0, 0 ];
    [ .c., 0, 0, 0, 0, 1, 0 ] → [ 0, 0, 0 ];
    [ .c., 0, 1, 0, 0, 0, 0 ] → [ 1, 1, 0 ];
    [ .c., 0, 0, 0, 0, 0, 1 ] → [ 0, 0, 1 ];
    [ .c., 0, 0, 0, 0, 0, 0 ] → [ 1, 0, 1 ];
    [ .c., 0, 1, 0, 0, 0, 1 ] → [ 0, 0, 1 ];
    [ .c., 0, 0, 0, 0, 0, 0 ] → [ 1, 0, 1 ];
    [ .c., 0, 0, 0, 0, 1, 0 ] → [ 0, 0, 1 ];
    [ .c., 1, 0, 0, 0, 0, 0 ] → [ 1, 1, 1 ];
    [ .c., 0, 0, 0, 0, 0, 1 ] → [ 0, 0, 0 ];
    [ .c., 0, 0, 1, 0, 0, 0 ] → [ 1, 1, 0 ];
    [ .c., 0, 0, 0, 0, 0, 1 ] → [ 0, 0, 1 ];
    [ .c., 0, 0, 0, 1, 0, 0 ] → [ 1, 1, 1 ];
    [ .c., 0, 0, 0, 0, 0, 1 ] → [ 0, 0, 0 ];
    [ .c., 0, 0, 0, 0, 0, 0 ] → [ 1, 0, 0 ];
    [ .c., 0, 1, 0, 0, 0, 1 ] → [ 1, 1, 0 ];
    [ .c., 0, 0, 0, 0, 0, 1 ] → [ 0, 0, 1 ];
    [ .c., 0, 0, 0, 0, 0, 0 ] → [ 1, 0, 1 ];
    [ .c., 1, 0, 0, 0, 0, 0 ] → [ 1, 1, 1 ];
    [ .c., 0, 0, 0, 0, 0, 1 ] → [ 0, 0, 0 ];
    [ .c., 0, 0, 0, 0, 0, 0 ] → [ 1, 0, 0 ];
    [ .c., 0, 0, 1, 0, 0, 0 ] → [ 1, 1, 0 ];
    [ .c., 0, 0, 0, 0, 0, 1 ] → [ 0, 0, 1 ];
    [ .c., 0, 0, 0, 0, 0, 0 ] → [ 1, 0, 1 ];
    [ .c., 0, 0, 0, 1, 0, 0 ] → [ 1, 1, 1 ];
    [ .c., 0, 0, 0, 0, 0, 1 ] → [ 0, 0, 0 ];

END
```

26. The circuit file for the elevator display logic targeted for a GAL22V10 is:

```
Module          Elevator_display_logic
Title           'Display logic for a two-story elevator system'
"Device declaration
        Elevator_disp           DEVICE          'P22V10'
"Pin declarations
        MOTION, DIR             PIN 2,3;
        SEGA,SEGB,SEGC          PIN 12,13,14    ISTYPE 'com, invert';
        SEGD,SEBE,SEGG          PIN 15,16,17    ISTYPE 'com, invert';
        UPARR,DWNARR            PIN 18,19       ISTYPE 'com, invert';
Equations
        UPARR = MOTION&!DIR;
        DWNARR = MOTION&DIR;
        SEGA = !MOTION&DIR;
        SEGB = !MOTION;
        SEGC = !MOTION&!DIR;
        SEGD = !MOTION&DIR;
        SEGE = !MOTION&DIR;
        SEGG = !MOTION&DIR;
```

```
Test_vectors ([MOTION,DIR] → [SEGA,SEGB,SEGC,SEGD,SEGE,SEGG,UPARR,DWNARR])
    [ 0,  0 ] → [ 1, 0, 0, 1, 1, 1, 1, 1 ];
    [ 0,  1 ] → [ 0, 0, 1, 0, 0, 0, 1, 1 ];
    [ 1,  0 ] → [ 1, 1, 1, 1, 1, 1, 0, 1 ];
    [ 1,  1 ] → [ 1, 0, 0, 1, 1, 1, 1, 0 ];

END
```

27. The traffic light control system cannot be implemented in a single GAL16V8 because 10 outputs are required and the device has only 8.

28. If T_S in the traffic light sequential logic fails so it is always HIGH, the circuit will eventually be stuck in the second state with the main ligh yellow and the side light red, or in the fourth state with the main light red and the side light yellow.

29. If T_L in the traffic light sequential logic fails so it is always LOW, the third state will always be skipped and the first state skipped if a vehicle is on the side street.

30. If V_S is always HIGH in the traffic light sequential circuit, the light will always sequence through the side street light states (second, third, and fourth).

31. The circuit file for an 8-bit SISO shift register using a GAL16V8 as the taget device is:

```
Module        8-bit_SISO_shift_register
Title         '8-bit SISO shift register in a GAL16V8'
"Device declaration
        Register        DEVICE        'P16V8'
"Pin declarations
        Clock, Clear    PIN 1,11;
        Din             PIN 2;
        Q0,Q1,Q2,Q3     PIN 12,13,14,15  ISTYPE 'reg, buffer';
        Q4,Q5,Q6,Qout   PIN 16,17,18,19  ISTYPE 'reg, buffer';
Equations
        [Q0,Q1,Q2,Q3,Q4,Q5,Q6,Qout].CLK = Clock;
        [Q0,Q1,Q2,Q3,Q4,Q5,Q6,Qout].AR = !Clear;
        Q0 := Din;
        Q1 := Q0.FB;
        Q2 := Q1.FB;
        Q3 := Q1.FB;
        Q4 := Q3.FB;
        Q5 := Q4.FB;
        Q6 := Q5.FB;
        Qout := Q6.FB;
Test_vectors ([Clock,Clear,Din] → [Q0,Q1,Q2,Q3,Q4,Q5,Q6,Qout])
        [ .x., 0, .x. ] → [ 0, 0, 0 0, 0, 0, 0, 0 ];
        [ .c., 1, 1 ] → [ 1, 0, 0 0, 0, 0, 0, 0 ];
        [ .c., 1, 0 ] → [ 0, 1, 0 0, 0, 0, 0, 0 ];
        [ .c., 1, 1 ] → [ 1, 0, 1 0, 0, 0, 0, 0 ];
        [ .c., 1, 0 ] → [ 0, 1, 0 1, 0, 0, 0, 0 ];
        [ .c., 1, 1 ] → [ 1, 0, 1 0, 1, 0, 0, 0 ];
        [ .c., 1, 0 ] → [ 0, 1, 0 1, 0, 1, 0, 0 ];
        [ .c., 1, 1 ] → [ 1, 0, 1 0, 1, 0, 1, 0 ];
        [ .c., 1, 0 ] → [ 0, 1, 0 1, 0, 1, 0, 1 ];
END
```

32. Since it is not possible to implement a 16-bit shift register in a single GAL16V8 or GAL22V10, the solution is to program two GAL16V8s using the ABEL input file from Problem 31 and then cascade the two programmed devices to form a 16-bit shift register.

33. An ABEL input file for a 4-bit Excess-3 code counter using a GAL22V10 is:

```
Module          4-bit_Excess-3_code_counter
Title           '4-bit Excess-3 code counter in a GAL22V10'
        "Device declaration
                Counter         DEVICE          'P22V10'
        "Pin declarations
                Clock, Reset    PIN  1,2;
                Q0,Q1,Q2,Q3     PIN  12,13,14,15   ISTYPE 'reg, buffer';
                COUNT = [Q3,Q2,Q1,Q0];
        Equations
                COUNT.CLK = Clock;

 Truth_table  ([Clock,Reset,COUNT] : > [COUNT])
            [ .x.,    0,     .x.   ] : > [^B0011 ];
            [ .c.,    1,  ^B0000] : > [^B0011 ];
            [ .c.,    1,  ^B0001] : > [^B0011 ];
            [ .c.,    1,  ^B0010] : > [^B0011 ];
            [ .c.,    1,  ^B0011] : > [^B0100 ];
            [ .c.,    1,  ^B0100] : > [^B0101 ];
            [ .c.,    1,  ^B0101] : > [^B0110 ];
            [ .c.,    1,  ^B0110] : > [^B0111 ];
            [ .c.,    1,  ^B0111] : > [^B1000 ];
            [ .c.,    1,  ^B1000] : > [^B1001 ];
            [ .c.,    1,  ^B1001] : > [^B1010 ];
            [ .c.,    1,  ^B1010] : > [^B1011 ];
            [ .c.,    1,  ^B1011] : > [^B1100 ];
            [ .c.,    1,  ^B1100] : > [^B0000 ];
            [ .c.,    1,  ^B1101] : > [^B0011 ];
            [ .c.,    1,  ^B1110] : > [^B0011 ];
            [ .c.,    1,  ^B1111] : > [^B0011 ];

 Test_vectors  ([Clock,Reset] → [COUNT])
            [ .c.,    0   ] → [^B0011 ];
            [ .c.,    1   ] → [^B0100 ];
            [ .c.,    1   ] → [^B0101 ];
            [ .c.,    1   ] → [^B0110 ];
            [ .c.,    1   ] → [^B0111 ];
            [ .c.,    1   ] → [^B1000 ];
            [ .c.,    1   ] → [^B1001 ];
            [ .c.,    1   ] → [^B1010 ];
            [ .c.,    1   ] → [^B1011 ];
            [ .c.,    1   ] → [^B1100 ];
            [ .c.,    1   ] → [^B0011 ];

 END
```

34. The input file from Problem 33 modified to allow preloading any 4-bit Excess-3 code into the counter is:

```
Module          4-bit_Excess-3_code_counter
Title           '4-bit Excess-3 code counter in a GAL22V10'
        "Device declaration
                Counter         DEVICE          'P22V10'
        "Pin declarations
                Clock, Load     PIN  1,2;
                D0,D1,D2,D3     PIN  3,4,5,6;
                Q0,Q1,Q2,Q3     PIN  12,13,14,15   ISTYPE 'reg, buffer';
                DATA = [D3,D2,D1,D0];
                COUNT = [Q3,Q2,Q1,Q0];
        Equations
                COUNT.CLK = Clock;

        Truth_table  ([Clock,Load,  DATA,COUNT] : > [COUNT])
                [ .c.,   0,  ^B0000,     .x.   ] : > [^B0011];
                [ .c.,   0,  ^B0001,     .x.   ] : > [^B0011];
                [ .c.,   0,  ^B0010,     .x.   ] : > [^B0011];
                [ .c.,   0,  ^B0011,     .x.   ] : > [^B0011];
                [ .c.,   0,  ^B0100,     .x.   ] : > [^B0100];
                [ .c.,   0,  ^B0110,     .x.   ] : > [^B0110];
                [ .c.,   0,  ^B0111,     .x.   ] : > [^B0111];
                [ .c.,   0,  ^B1000,     .x.   ] : > [^B1000];
                [ .c.,   0,  ^B1001,     .x.   ] : > [^B1001];
                [ .c.,   0,  ^B1010,     .x.   ] : > [^B1010];
                [ .c.,   0,  ^B1011,     .x.   ] : > [^B1011];
                [ .c.,   0,  ^B1100,     .x.   ] : > [^B1100];
                [ .c.,   0,  ^B1101,     .x.   ] : > [^B0011];
                [ .c.,   0,  ^B1110,     .x.   ] : > [^B0011];
                [ .c.,   0,  ^B1111,     .x.   ] : > [^B0011];
                [ .c.,   1,      .x.     ^B0000] : > [^B0011];
                [ .c.,   1,      .x.     ^B0001] : > [^B0011];
                [ .c.,   1,      .x.     ^B0010] : > [^B0011];
                [ .c.,   1,      .x.     ^B0011] : > [^B0011];
                [ .c.,   1,      .x.     ^B0100] : > [^B0101];
                [ .c.,   1,      .x.     ^B0101] : > [^B0110];
                [ .c.,   1,      .x.     ^B0110] : > [^B0111];
                [ .c.,   1,      .x.     ^B0111] : > [^B1001];
                [ .c.,   1,      .x.     ^B1001] : > [^B1010];
                [ .c.,   1,      .x.     ^B1010] : > [^B1011];
                [ .c.,   1,      .x.     ^B1011] : > [^B1100];
                [ .c.,   1,      .x.     ^B1100] : > [^B1101];
                [ .c.,   1,      .x.     ^B1101] : > [^B0011];
                [ .c.,   1,      .x.     ^B1110] : > [^B0011];
                [ .c.,   1,      .x.     ^B0111] : > [^B0011];
```

```
Test_Vectors  ([Clock,Load, DATA] → [COUNT])
              [ .c.,  0,  ^B0000] → [ ^B0011 ];
              [ .c.,  0,  ^B0001] → [ ^B0011 ];
              [ .c.,  0,  ^B0010] → [ ^B0011 ];
              [ .c.,  0,  ^B0011] → [ ^B0011 ];
              [ .c.,  0,  ^B0100] → [ ^B0100 ];
              [ .c.,  0,  ^B0101] → [ ^B0101 ];
              [ .c.,  0,  ^B0110] → [ ^B0110 ];
              [ .c.,  0,  ^B0111] → [ ^B0111 ];
              [ .c.,  0,  ^B1000] → [ ^B1000 ];
              [ .c.,  0,  ^B1001] → [ ^B1001 ];
              [ .c.,  0,  ^B1010] → [ ^B1010 ];
              [ .c.,  0,  ^B1011] → [ ^B1011 ];
              [ .c.,  0,  ^B1100] → [ ^B1100 ];
              [ .c.,  0,  ^B1101] → [ ^B0011 ];
              [ .c.,  0,  ^B1110] → [ ^B0011 ];
              [ .c.,  0,  ^B1111] → [ ^B0011 ];
              [ .c.,  1,     .x.  ] → [ ^B0100 ];
              [ .c.,  1,     .x.  ] → [ ^B0101 ];
              [ .c.,  1,     .x.  ] → [ ^B0110 ];
              [ .c.,  1,     .x.  ] → [ ^B0111 ];
              [ .c.,  1,     .x.  ] → [ ^B1000 ];
              [ .c.,  1,     .x.  ] → [ ^B1001 ];
              [ .c.,  1,     .x.  ] → [ ^B1010 ];
              [ .c.,  1,     .x.  ] → [ ^B1011 ];
              [ .c.,  1,     .x.  ] → [ ^B1100 ];
              [ .c.,  1,     .x.  ] → [ ^B0011 ];

END
```

35. The state diagram for a 3-story elevator system will vary depending on the designer One approach is as follows:

The transition conditions appearing in the state diagram of Figure 11-1 are:

1. $\overline{L2PB}\ \overline{L3PB}$
2. REQ1+OPEN
3. ARV1
4. L2PB
5. L2PB
6. ARV2
7. L1PB
8. L1PB
9. $\overline{L1PB}\ \overline{L3PB}$
10. REQ2+OPEN
11. ARV2
12. L1PB
13. L1PB
14. L3PB
15. L3PB
16. L3PB
17. L3PB
18. ARV3
19. L2PB
20. L2PB
21 $\overline{L1PB}\ \overline{L2PB}$
22. REQ3+OPENJ

Explanation of these conditions and primary inputs are:

REQ1: Elevator button on 1st floor.
RFEQ2: Elevator button on 2nd floor.
REQ3: Elevator button on 3rd floor.
FLR1: 1st floor button inside elevator.
FLR2: 2nd floor button inside elevator.
FLR3: 3rd floor button inside elevator.
OPEN: Door open button inside elevator.
ARV1: 1st floor arrival switch.
ARV2: 2nd floor arrival switch.
ARV#: 3rd floor arrival switch.
L1PB: 1st floor request latch set.
L1PB_: 1st floow request latch reset.
L2PB: 2nd floor request latch set.
L2PB_: 2nd floow request latch reset.
L3PB: 3rd floor request latch set.
L3PB_: 3rd floow request latch reset.

States	DOOR	MOTION	DIR	RANGE
REST1	0	0	0	0
CLOSE1	1	0	0	0
UP2	1	1	0	0
UP3	1	1	0	1
REST2	0	0	0	1
CLOSE2	1	0	0	1
DOWN1	1	1	1	0
DOWN2	1	1	1	0
REST3	0	0	1	1
CLOSE3	1	0	1	1

DOOR: 0=OPEN, 1=CLOSED
MOTION: 0=STOPPED, 1=MOVING
DIR: 0=UP, 1= DOWN
RANGE: 0=1 FLOOR, 1=2 FLOORS

$$\text{L1PB} = \text{REQ1} + \text{FLR1} + \overline{\text{L1PB\_}}$$
$$\text{L1PB\_} = \overline{\text{MOTION}} \bullet \overline{\text{DIR}} \bullet \overline{\text{RANGE}} + \overline{\text{L1PB}}$$
$$\text{L2PB} = \text{REQ2} + \text{FLR2} + \overline{\text{L2PB\_}}$$
$$\text{L2PB\_} = \overline{\text{MOTION}} \bullet \overline{\text{DIR}} \bullet \text{RANGE} + \overline{\text{L2PB}}$$
$$\text{L3PB} = \text{REQ3} + \text{FLR3} + \overline{\text{L3PB\_}}$$
$$\text{L1PB\_} = \overline{\text{MOTION}} \bullet \text{DIR} \bullet \text{RANGE} + \overline{\text{L3PB}}$$

Although other designs are possible, the states and inputs shown attempt to decrease the possibility of some elevator requests being starved out. See the state diagram of theABEL input file in Problem 36 to see how floor requests are prioritized.

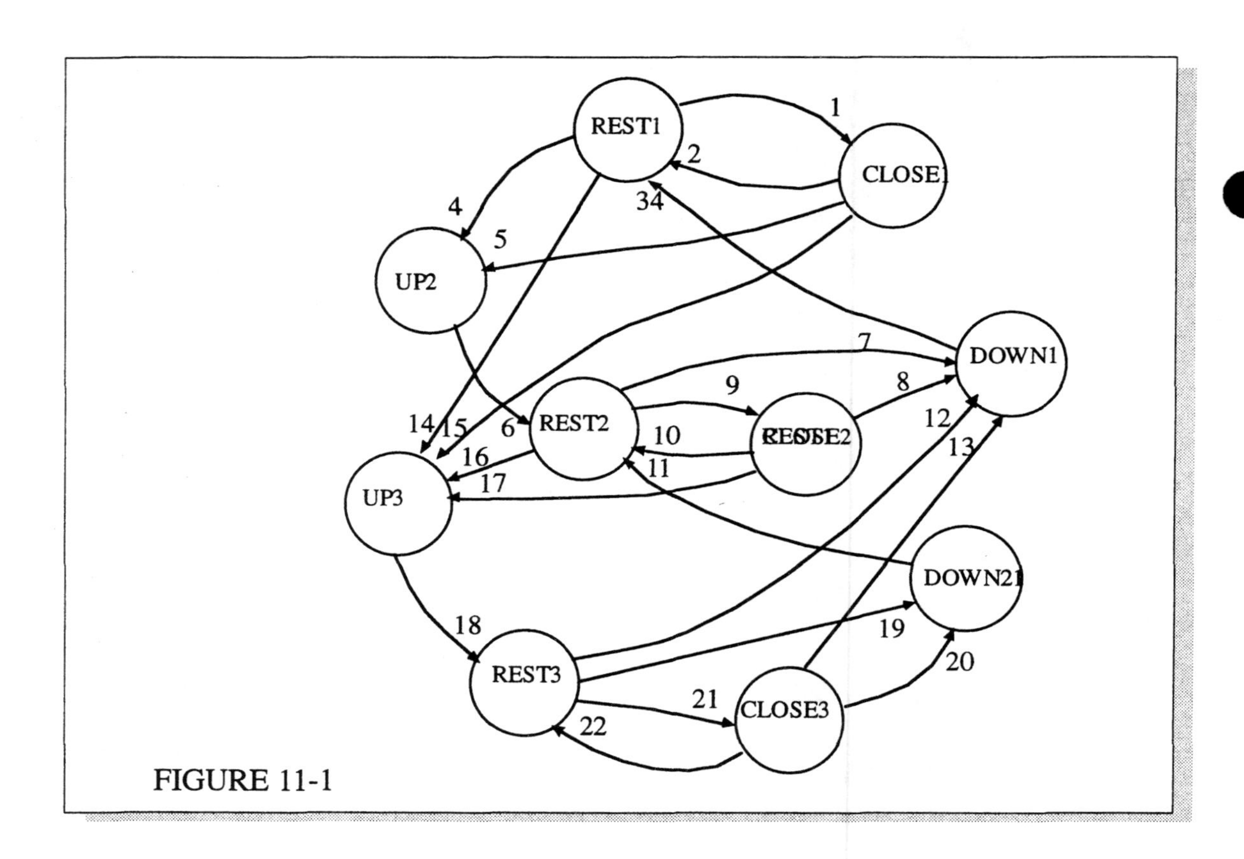

FIGURE 11-1

36. From the state diagram of Problem 35, the ABEL input file is:

```
Module          Control_logic_for_3-story_elevaltor
Title           'Control logic for 3-story elevator in a GAL22V10'
        "Device declaration
                Elevator_cont           DEVICE          'P22V10'
        "Pin declarations
                Clock                           PIN  1,;
                REQ1,REQ2,REQ3                  PIN  2,3,4;
                FLR1,FLR2,FLR3                  PIN 5,6,7;
                OPEN                            PIN 8;
                ARV1,ARV2,ARV3                  PIN 9,10,11;
                L1PB, L1PB_                     PIN  13,14    ISTYPE 'com, buffer';
                L2PB, L2PB_                     PIN  15,16    ISTYPE 'com, buffer';
                L3PB, L3PB_                     PIN  17,18    ISTYPE 'com, buffer';
                DOOR,MOTION,DIR                 PIN 19,20,21  ISTYPE 'reg, buffer';
                RANGE                           PIN 22  ISTYPE 'reg, buffer';

        "State definitions
                CONSTATE        = [DOOR,MOTION,DIR,RANGE];
                REST1           = ^B0000;
                CLOSE1          = ^B1000;
                UP2             = ^B1100;
                UP3             = ^B1101;
                REST2           = ^B0001;
                CLOSE2          = ^B1001;
                DOWN1           = ^B1110;
                DOWN2           = ^B1111;
                REST3           = ^B0011;
                CLOSE3          = ^B1011;
```

```
Equations
        CONSTATE.CLK = Clock;
        L1PB = REQ1 # FLR1 # !L1PB_;
        L1PB_ = !MOTION&!DIR&!RANGE # !L1PB;
        L2PB = REQ2 # FLR2 # !L2PB_;
        L2PB_ = !MOTION&!DIR&RANGE # !L2PB;
        L3PB = REQ3 # FLR3 # !L3PB_;
        L1PB_ = !MOTION&DIR&RANGE # !L3PB;

State_diagram CONSTATE
        State REST1 :   if (L3PB) then UP3 else if (L2PB) then UP2 else CLOSE1;
        State CLOSE1 :  if (L1PB # OPEN) then REST1 else if (L3PB) then UP3 else if (L2PB) then UP2
                        else CLOSE1;
        State UP2:      if (ARV2) then REST2 else UP2;
        State UP3;      if (ARV3) then REST3 else UP3;
        State REST2 :   if (L3PB) then UP3 else if (L1PB) then DOWN1 else CLOSE2;
        State CLOSE2 :  if (L2PB # OPEN) then REST2 else if (L3PB) then UP3 else if (L1PB) then
                        DOWN1 else CLOSE2;
        State DOWN1:    if (ARV1) then REST1 else DOWN1;
        State DOWN2:    if (ARV2) then REST2 else DOWN2;
        State REST3 :   if (L1PB) then DOWN1 else if (L2PB) then DOWN2 else CLOSE3;
        State CLOSE3 :  if (L3PB # OPEN) then REST3 else if (L1PB) then DOWN1 else if (L2PB) then
                        DOWN2 else CLOSE3;

END
```

CHAPTER 12
MEMORY AND STORAGE

1. (a) ROM : no read/write control
(b) RAM

2. They are random access memories because any address can be accessed at any time. You do not have to go through all the preceeding addresses to get to a specific address.

3. ***Address bus*** provides for transfer of address code to memory for accessing any memory location in any order for a read or a write operation.
Data bus provides for transfer of data between the microprocessor and memory or input/output devices.

4. (a) $0A_{16} = 00001010_2 = \mathbf{10_{10}}$
(b) $3F_{16} = 00111111_2 = \mathbf{63_{10}}$
(c) $CD_{16} = 11001101_2 = \mathbf{205_{10}}$

5.

	BIT 0	BIT 1	BIT 2	BIT 3
ROW 0	1	0	0	0
ROW 1	0	0	0	0
ROW 2	0	0	1	0
ROW 3	0	0	0	0

6. See Figure 12-1.

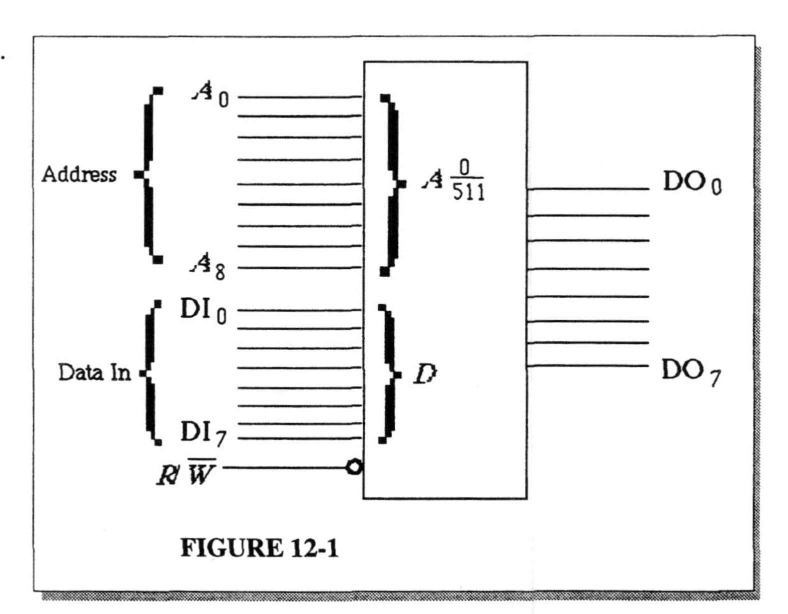

FIGURE 12-1

7. 64k X 8 = 512 X 128 X 8 = **512 rows X 128 8-bit columns**

8. See Figure 12-2.

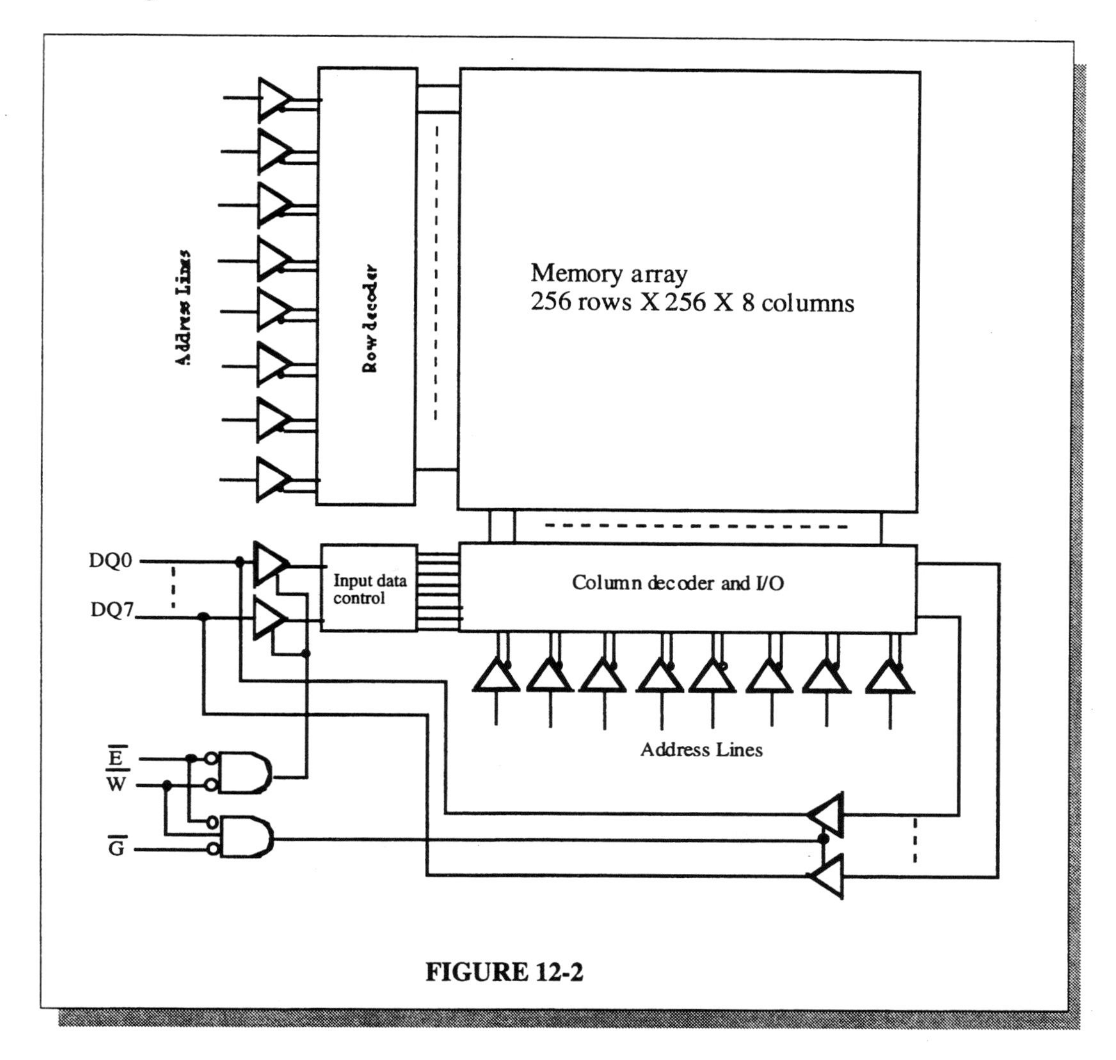

FIGURE 12-2

9. The difference between a SRAM and a DRAM is that data in a SRAM are stored in latches indefinately as long as power is applied while data in a DRAM are stored in capacitors which require periodic refreshing to retain the stored data.

10. The bit capacity of a DRAM with 12 address lines is

$$2^{2\times 12} = 2^{24} = 16{,}777{,}216\,\text{bits} = 16\ \text{Mbits}$$

11.

Inputs		Outputs			
A_1	A_0	O_3	O_2	O_1	O_0
0	0	0	1	0	1
0	1	1	0	0	1
1	0	1	1	1	0
1	1	0	0	1	0

12.

Inputs			Outputs			
A_2	A_1	A_0	O_3	O_2	O_1	O_0
0	0	0	0	1	0	0
0	0	1	1	1	1	1
0	1	0	1	0	1	1
0	1	1	1	0	0	1
1	0	0	1	1	1	0
1	0	1	1	0	0	0
1	1	0	0	0	1	1
1	1	1	0	1	0	1

13.

BCD				Excess-3			
D_3	D_2	D_1	D_0	E_3	E_2	E_1	E_0
0	0	0	0	0	0	1	1
0	0	0	1	0	1	0	0
0	0	1	0	0	1	0	1
0	0	1	1	0	1	1	0
0	1	0	0	0	1	1	1
0	1	0	1	1	0	0	0
0	1	1	0	1	0	0	1
0	1	1	1	1	0	1	0
1	0	0	0	1	0	1	1
1	0	0	1	1	1	0	0

See Figure 12-3.

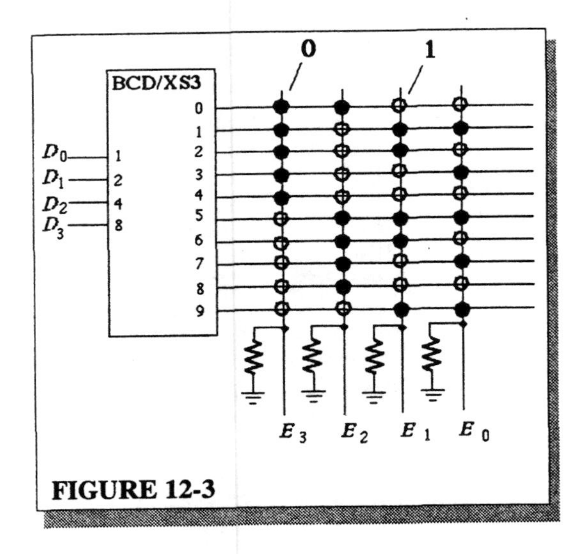

FIGURE 12-3

14. 2^{14} = 16,384 addresses
16,384X8 bits = **131,072 bits**

15. Blown links: 1 through 17, 19 through 23, 25 through 31, 34, 37, 38, and 40 through 47, 53, 55, 58, 61, 62, 63, 65, 67, 69.

	X Input				X Output								
	X_2	X_1	X_0	X^3	2^8	2^7	2^6	2^5	2^4	2^3	2^2	2^1	2^0
0	0	0	0	0	0	0	0	0	0	0	0	0	0
1	0	0	1	1	0	0	0	0	0	0	0	0	1
2	0	1	0	8	0	0	0	0	0	1	0	0	0
3	0	1	1	27	0	0	0	0	1	1	0	1	1
4	1	0	0	64	0	0	1	0	0	0	0	0	0
5	1	0	1	125	0	0	1	1	1	1	1	0	1
6	1	1	0	216	0	1	1	0	1	1	0	0	0
7	1	1	1	343	1	0	1	0	1	0	1	1	1

16.

Address	Contents
A_{13} —— A_0	Q_7 —— Q_0
01001100010011	10101100
11011101011010	00100101
01011010011001	10110011
11010010001110	00101000
01010010100101	10001011
01010000110100	11010101
01001001100001	11001001
11011011100100	01001001
01101110001111	01010010
10111110011010	01001000
10101110011010	11001000

17. 16 k X 4 RAMS can be connected to make a 64 k X 8 RAM as shown in Figure 12-4.

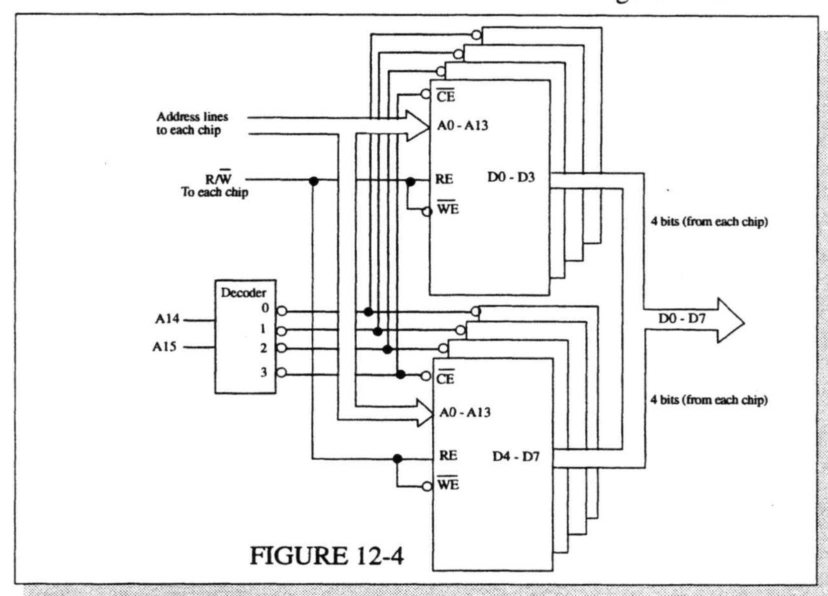

FIGURE 12-4

18. See Figure 12-5.

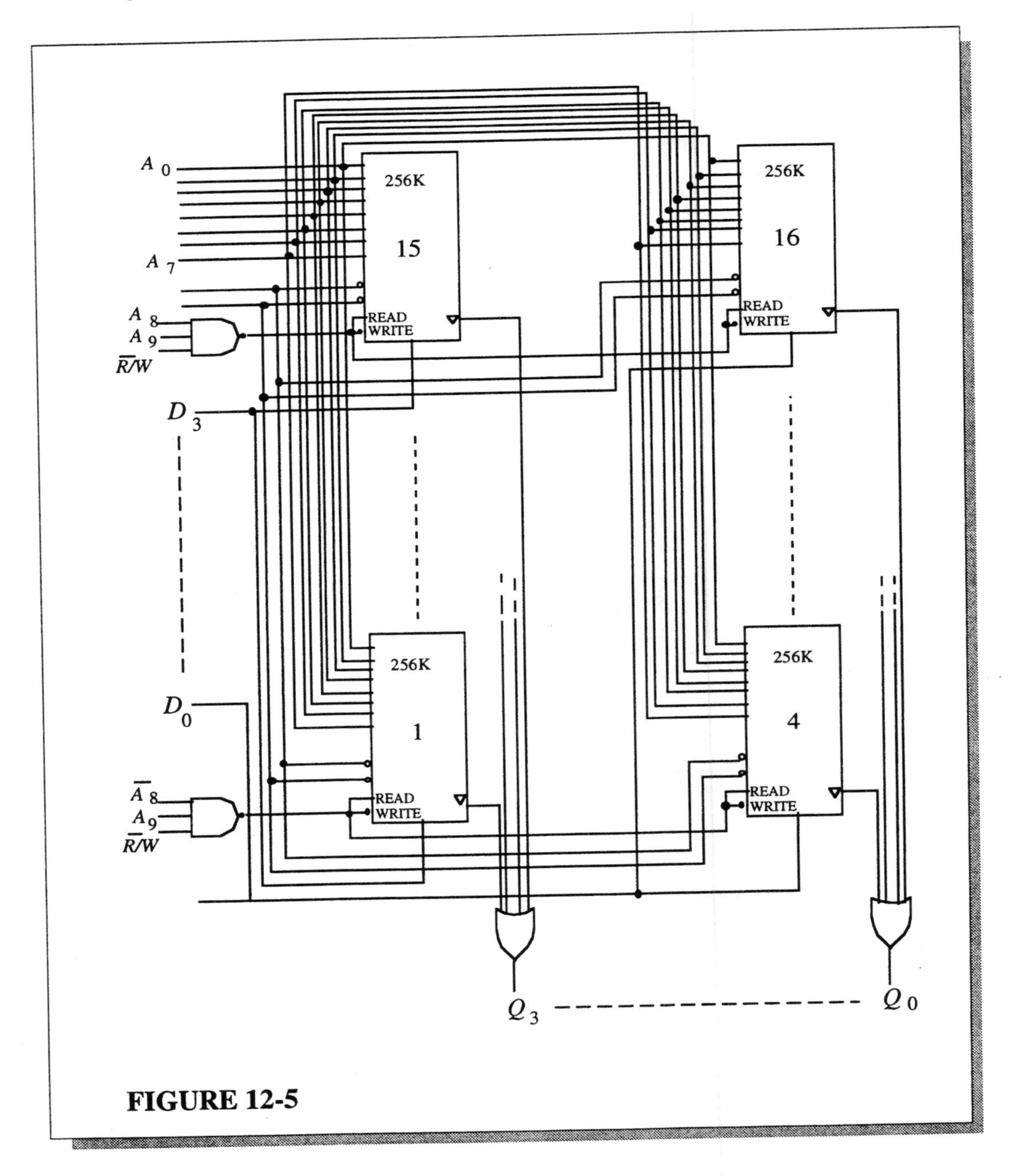

FIGURE 12-5

19. Word length = 8 bits, word capacity = **64 kwords**
Word length = 4 bits, word capacity = **256 kwords**

20. See Figure 12-6.

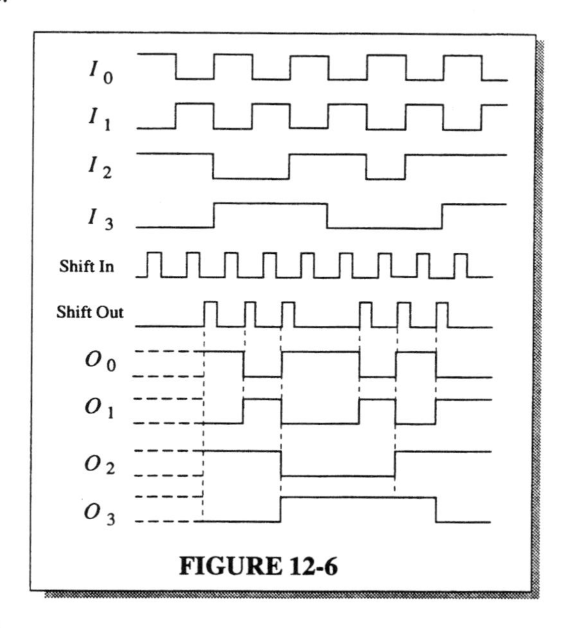

FIGURE 12-6

21. See Figure 12-7.

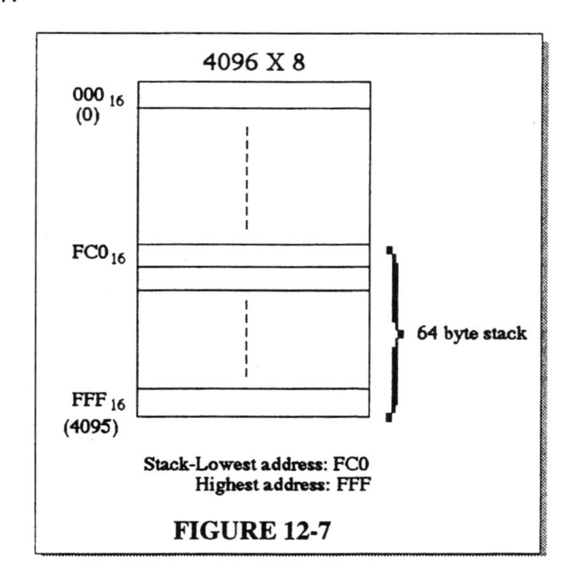

FIGURE 12-7

22. The first byte goes into $\mathbf{FFF_{16}}$.

The last byte (16th) goes into a lower address: $16_{10} = 10_{16}$

$FFF_{16} - 10_{16} = \mathbf{FEF_{16}}$

See Figure 12-8.

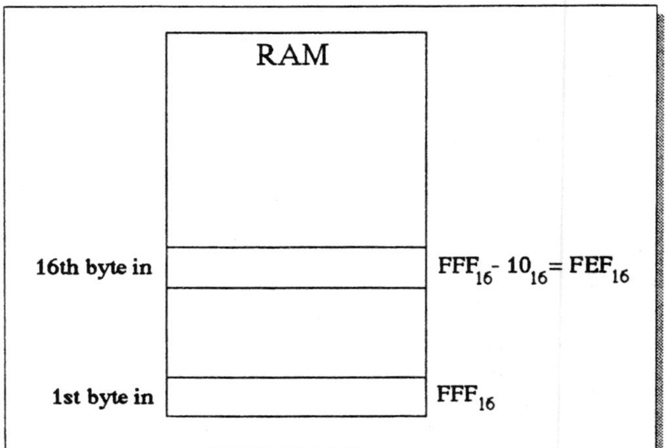

RAM

16th byte in — $FFF_{16} - 10_{16} = FEF_{16}$

1st byte in — FFF_{16}

FIGURE 12-8

23. A hard disk is formatted omtp tracls amd sectors. Each track is divided into a number of sectors with each sector of a track having a physical address. Hard disks typically have from a few hundred to a few thousand.

24. Seek time is the average time required to position the drive head over the track containing the desired data. The latency period is the average time required for the data to move under the drive head.

25. Magnetic tape has a longer access time than disk because data must be accessed sequentially rather than randomly.

26. A magneto-optic disk is a read/write medium using lasers and magnetic fields.
A CD-ROM (compact-disk ROM) is a read-only optical (laser) medium.
A WORM (write-once-read-many) is an optical medium in which data can bewritten once and read many time.

27. The correct checksum is **00100**.
The actual checksum is 01100. The second bit from the left is in error.

28. (a) ROM0: Low address - 00_{16} High address - $1F_{16}$
ROM1: Low address - 20_{16} High address - $3F_{16}$
ROM2: Low address - 40_{16} High address - $5F_{16}$
ROM3: Low address - 60_{16} High address - $7F_{16}$

(b) Same as flow chart in Figure 12-70 in text except that the last data address is specified as $\mathbf{7E_{16}}$ ($7F_{16}$ - 1).

(c) See Figure 12-9.

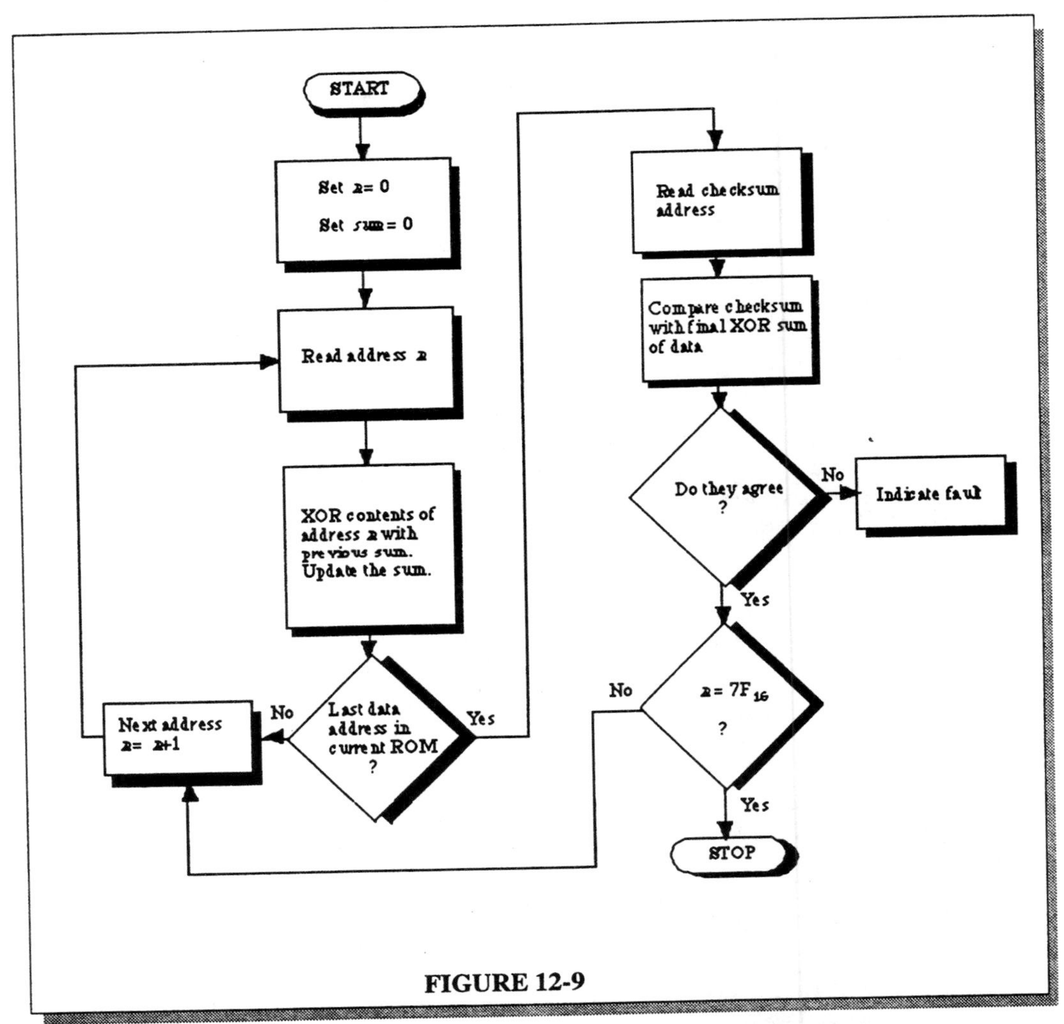

FIGURE 12-9

(d) A single checksum will not isolate the faulty chip. It will only indicate that there is an error in one of the chips.

29.

(a) 40_{16} - $5F_{16}$ is 64 - 95 decimal; ROM 2
(b) 20_{16} - $3F_{16}$ is 32 - 63 decimal; ROM 1
(c) 00_{16} - $7F_{16}$ is 0 - 127 decimal; All ROMs

30. See Figure 12-10.

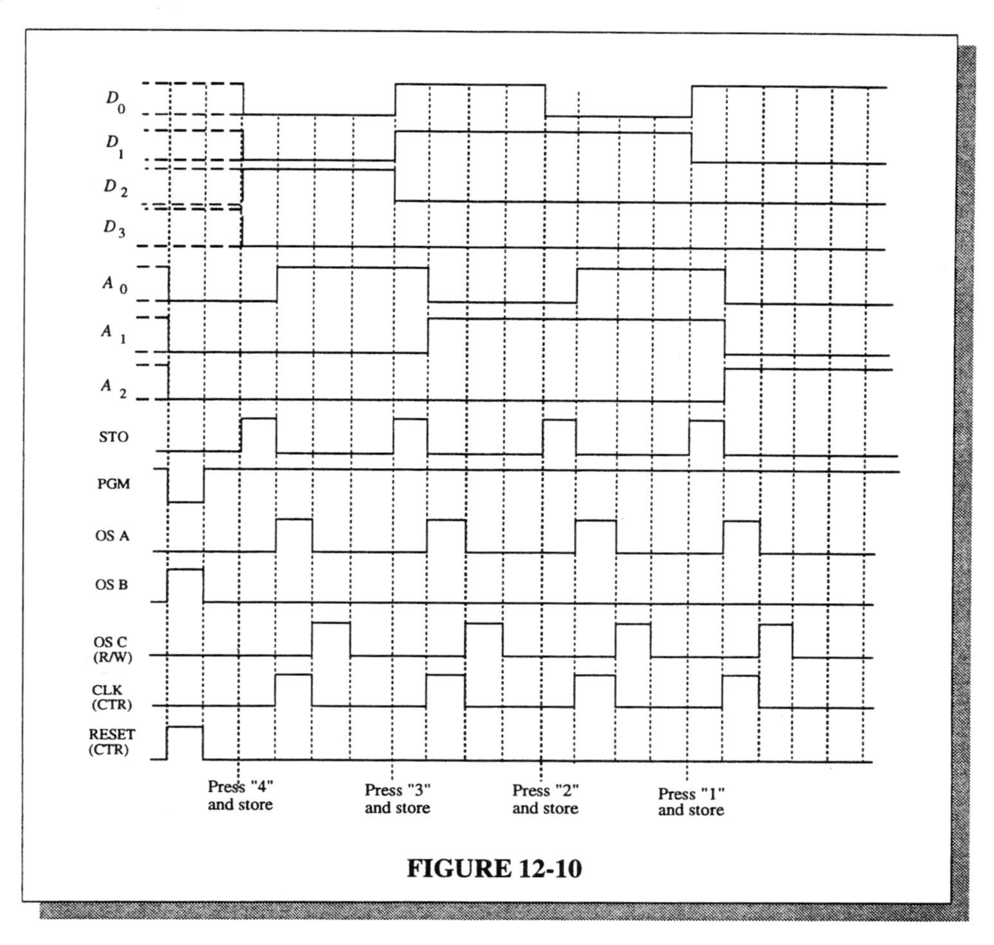

FIGURE 12-10

31. The counter state after pressing the STORE button twice is 0010.

32. The 74LS27 is used to enable the RAM whenever the system is out of the system reset state (any address line is not zero).

33. An advantage of a PROM is that it will retain data when power is off. A disadvantage is that a PROM cannot be readily reprogrammed with a new entry code.

34. See Figure 12-11.

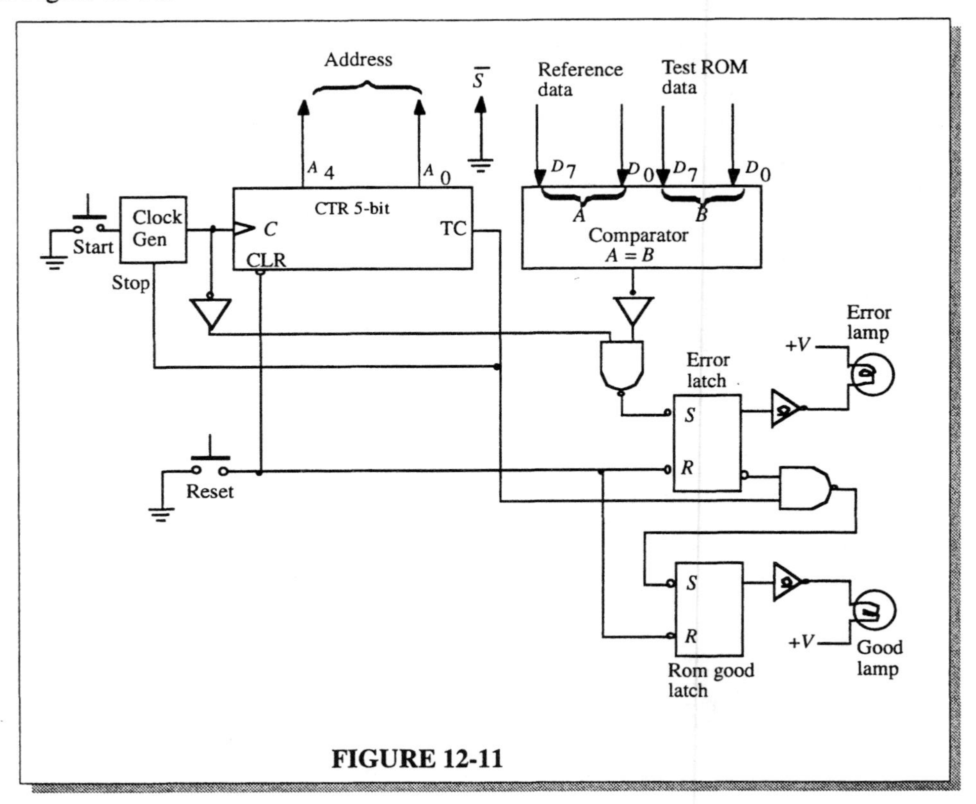

FIGURE 12-11

35. See Figure 12-12.

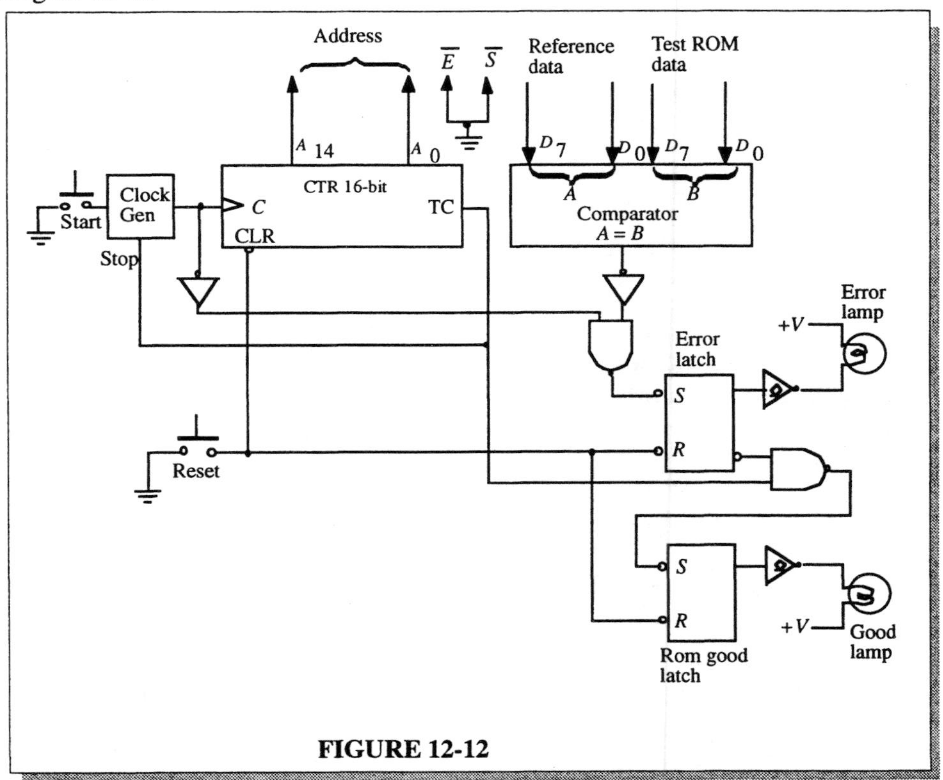

FIGURE 12-12

36. The memory board requires no modification for a 5-bit entry code. The 74189 RAM has a capacity of sixteen 4-bit words and the 3-bit address code can access memory locations 0 through 7.

37. To accomodate a 5-bit entry code, shift register B must be loaded with five 0s instead of four. This is done by connecting pin 14 (input D_3) of the 74LS165 to ground rather that to the HIGH.

CHAPTER 13
INTERFACING

1. See Figure 13-1.

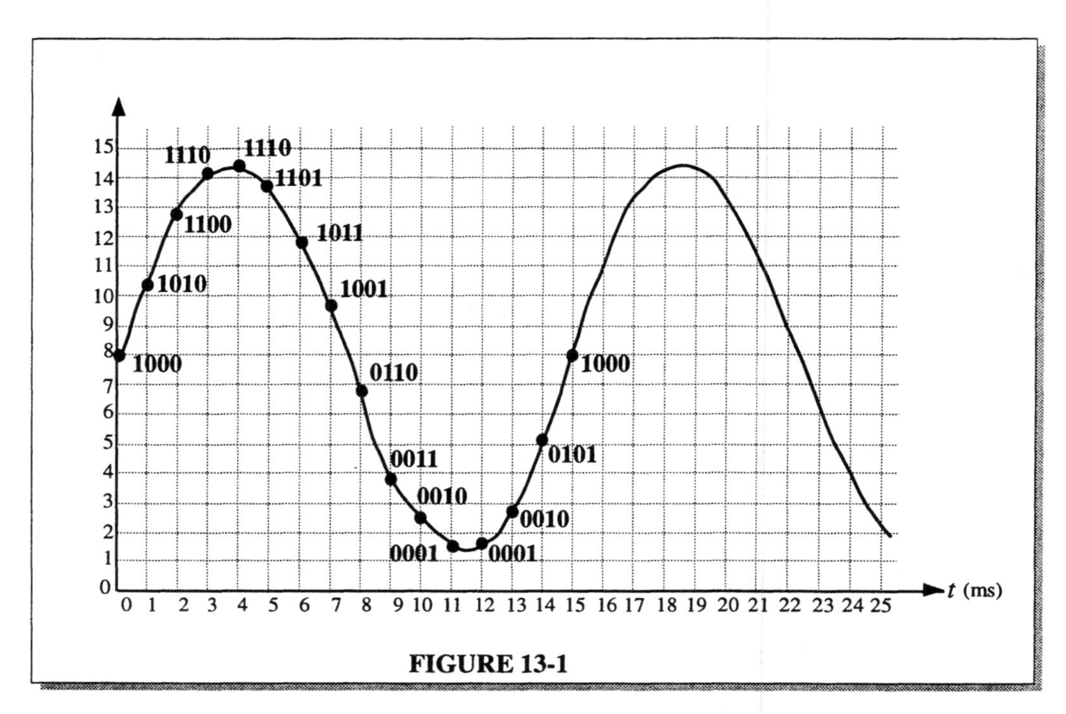

FIGURE 13-1

2. See Figure 13-2.

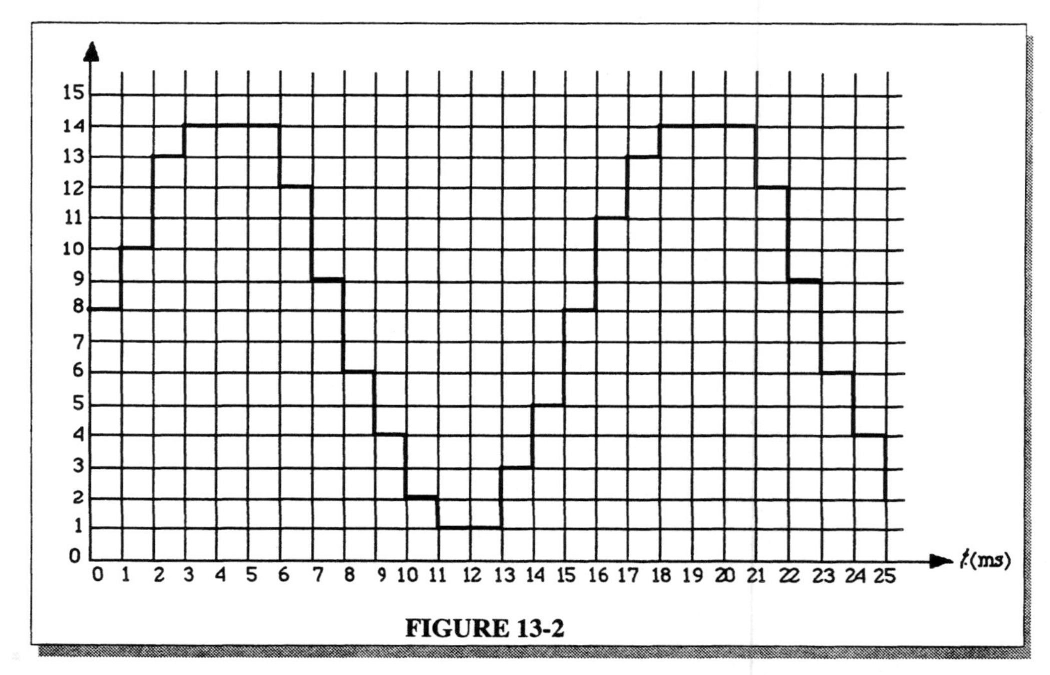

FIGURE 13-2

3. See Figure 13-3.

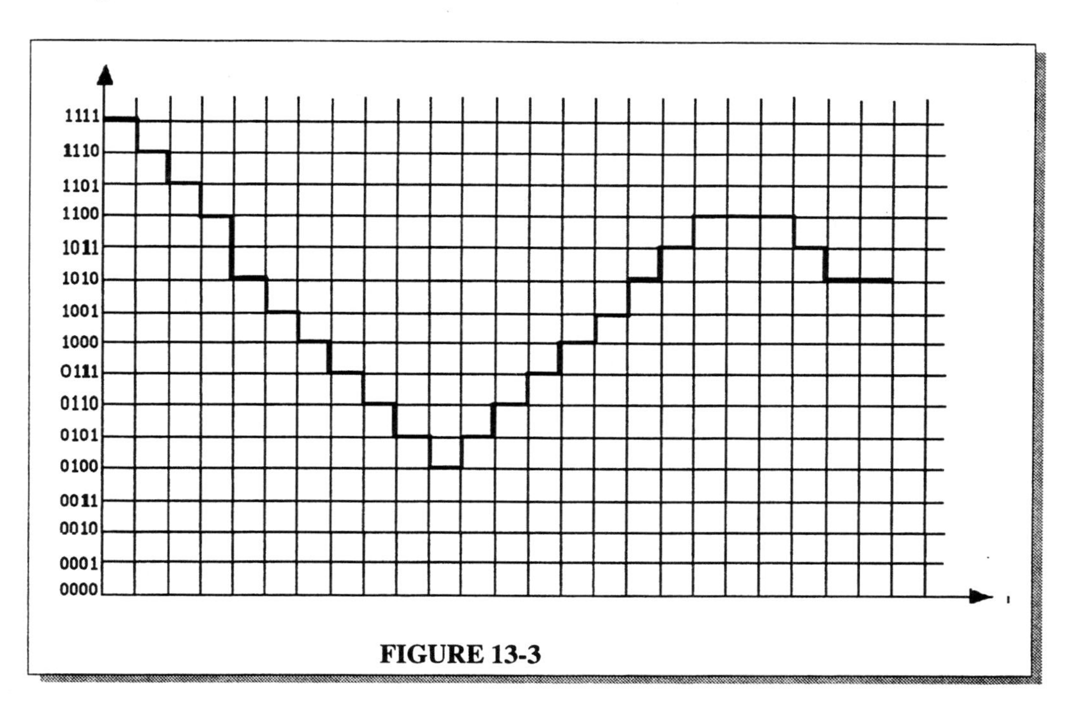

FIGURE 13-3

4. $$\frac{V_{\text{out}}}{V_{\text{in}}} = \frac{2\text{ V}}{10\text{ mV}} = 200$$

5. $$\frac{V_{\text{OUT}}}{V_{\text{IN}}} = \frac{R_{\text{F}}}{R_{\text{IN}}}$$

$$R_{\text{F}} = R_{\text{IN}}\left(\frac{V_{\text{OUT}}}{V_{\text{IN}}}\right) = 1\text{ k}\Omega(330) = 330\text{ k}\Omega$$

6. $$R_0 = 10\text{ k}\Omega$$

$$R_1 = \frac{R_0}{2} = \frac{10\text{ k}\Omega}{2} = 5\text{ k}\Omega$$

$$R_2 = \frac{R_0}{4} = \frac{10\text{ k}\Omega}{4} = 2.5\text{ k}\Omega$$

$$R_3 = \frac{R_0}{8} = \frac{10\text{ k}\Omega}{8} = 1.25\text{ k}\Omega$$

7. See Figure 13-4.

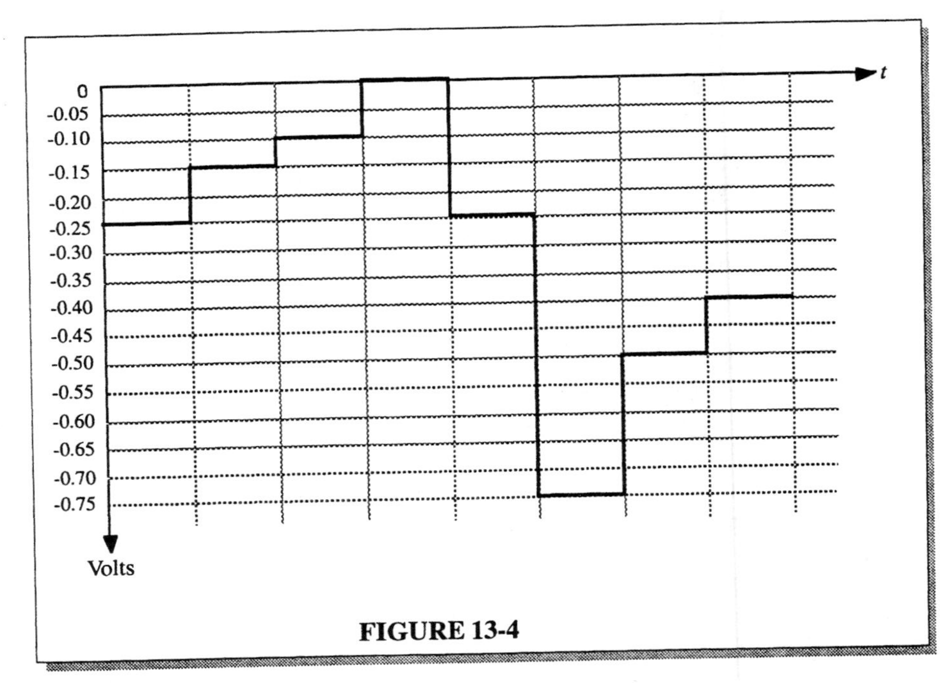

FIGURE 13-4

8. See Figure 13-5.

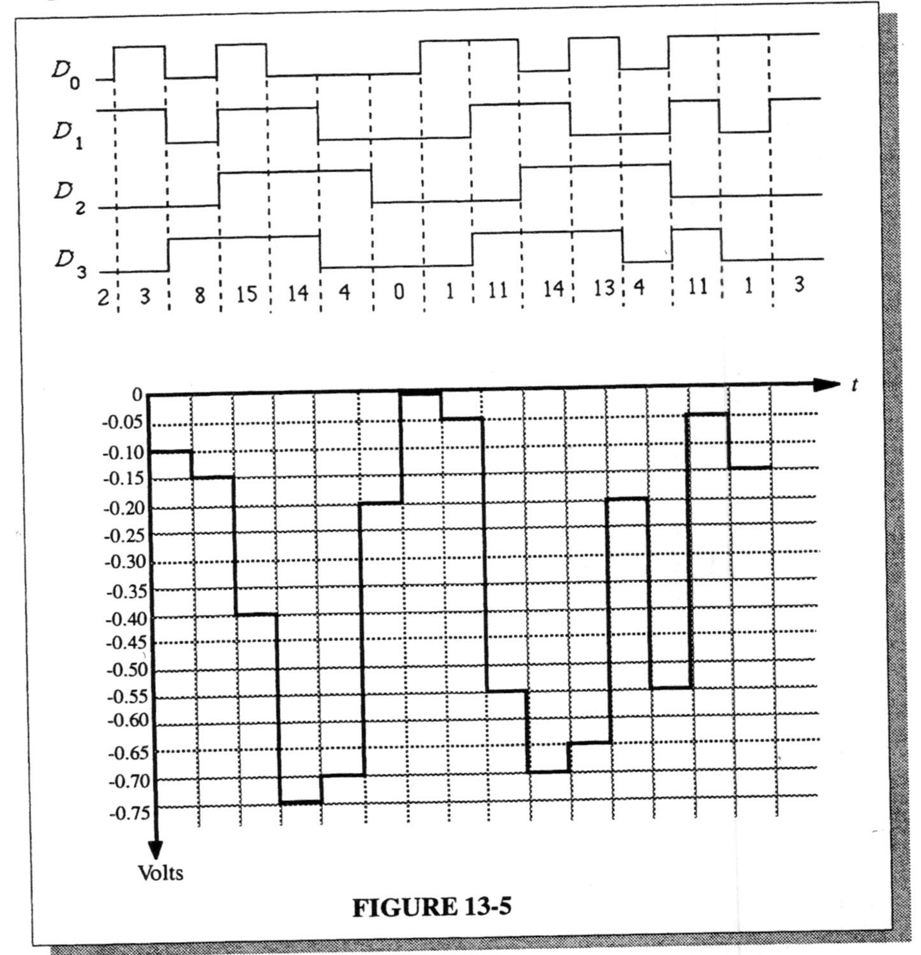

FIGURE 13-5

9. (a) $\left(\frac{1}{(2^3 - 1)}\right)100 = 14.29\ \%$

(b) $\left(\frac{1}{2^{10} - 1}\right)100 = 0.98\ \%$

(c) $\left(\frac{1}{2^{18} - 1}\right)100 = 0.00038\ \%$

10.

001, 010, 011, 101, 110, 111, 111, 111, 111, 110, 101, 101, 110, 110, 110, 101, 100, 011, 010, 001.

See Figure 13-6.

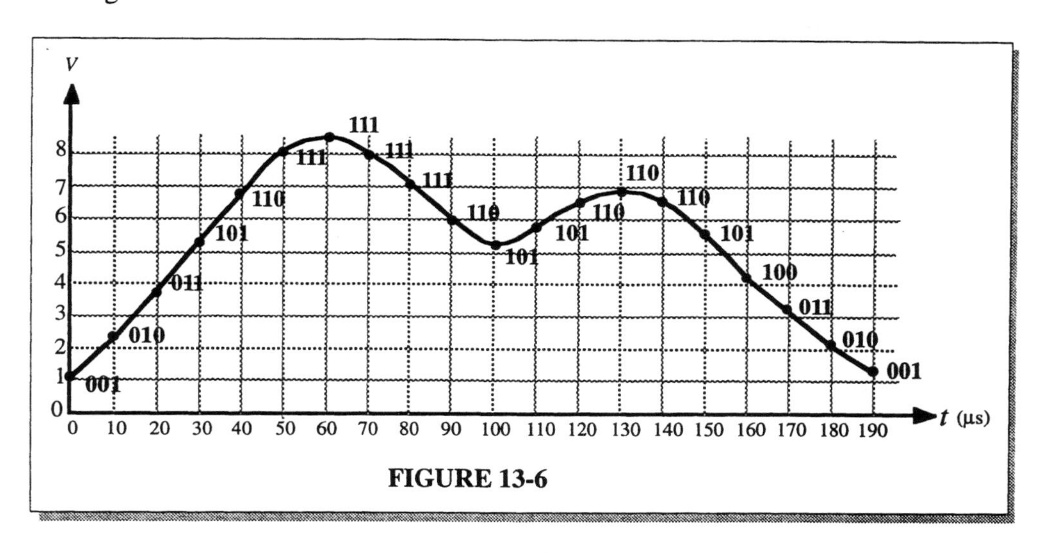

FIGURE 13-6

11.

Output of 3-bit converter: 000, 001, 100, 101, 101, 100, 011, 010, 001, 001, 010, 110, 111, 111, 111, 111, 111, 111, 111, 100.

See Figure 13-7.

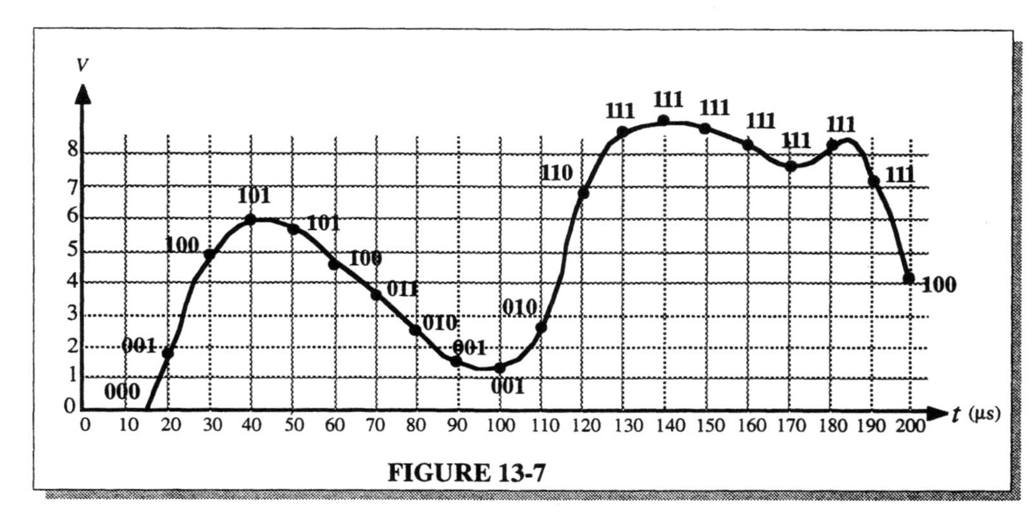

FIGURE 13-7

12. 100, 110, 101, 011, 010, 010,011, 101, 100.

See Figure 13-8.

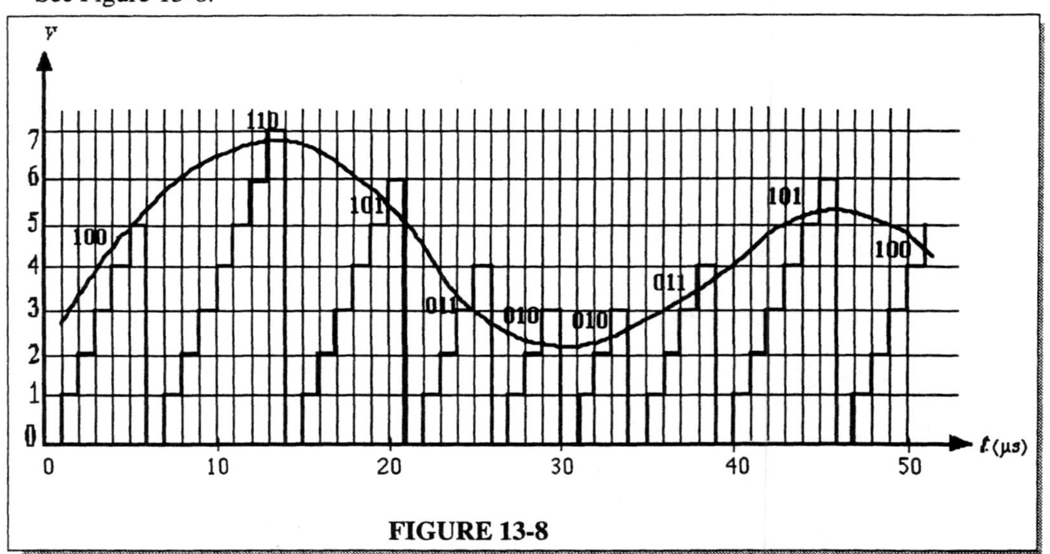

FIGURE 13-8

13. 0000, 0000, 0000, 1110, 1100, 0111, 0110, 0011, 0010, 1100.

See Figure 13-9.

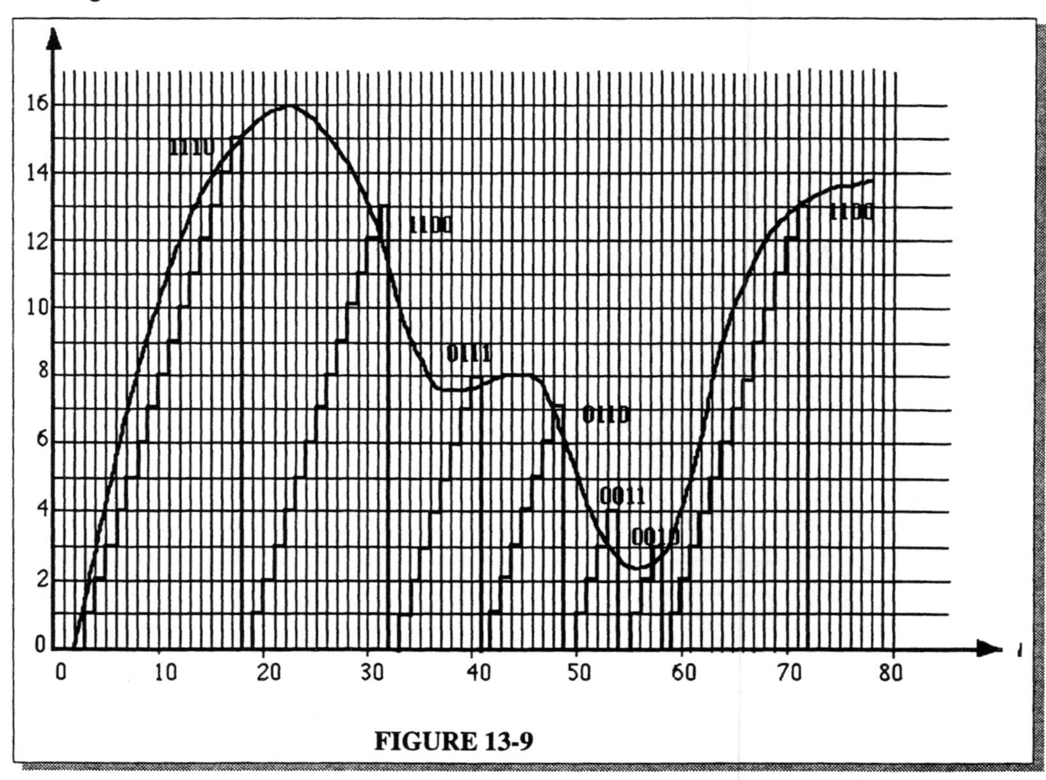

FIGURE 13-9

14. 100, 101, 110, 110, 110, 110, 101, 101, 100, 100, 011, 011, 010, 010, 010, 010, 010, 011, 011, 101, 101, 101.

See Figure 13-10.

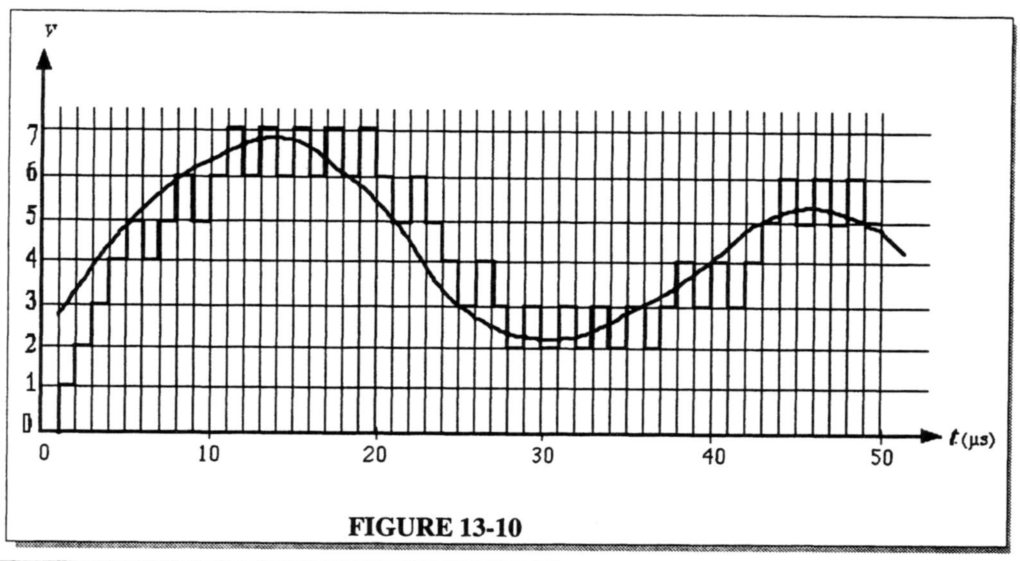

FIGURE 13-10

15.

SAR	Comment
1000	Greater than V_{in}. Reset MSB.
0100	Less than V_{in}. Keep the 1.
0110	Equal to V_{in}. Keep the 1 (final state)

16. See Figure 13-11.

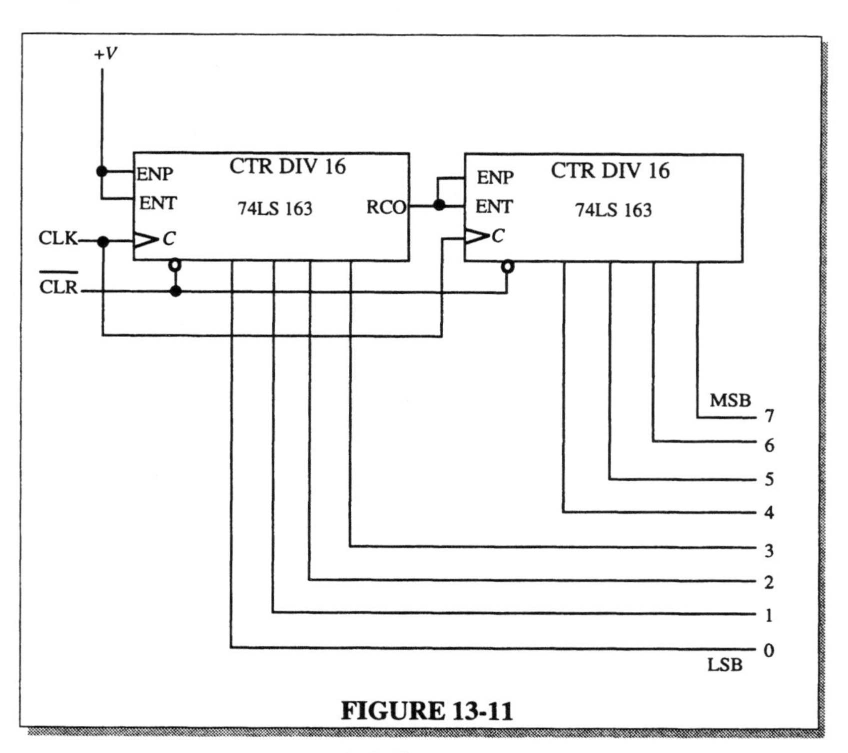

FIGURE 13-11

17. See Figure 13-12.

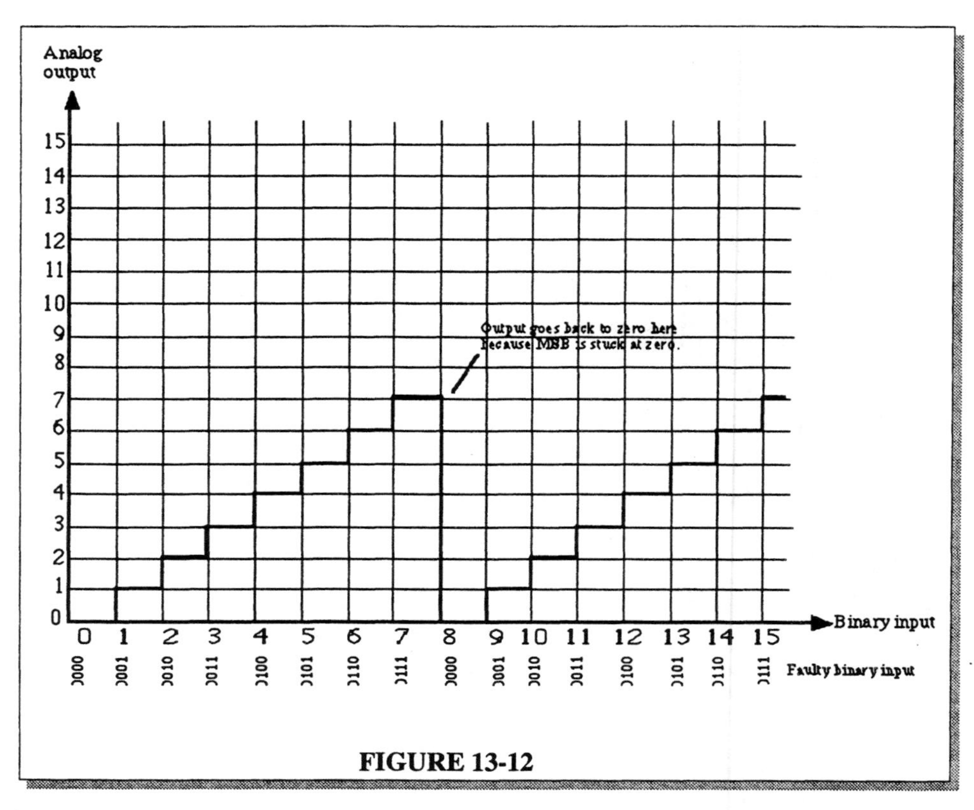

FIGURE 13-12

18. See Figure 13-13.

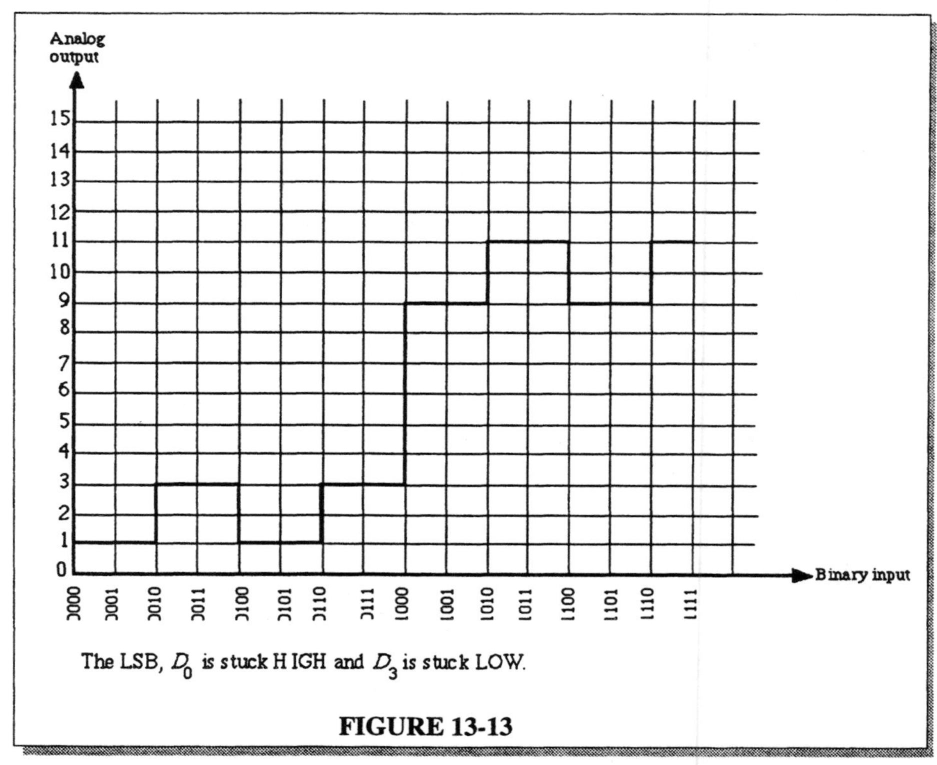

The LSB, D_0 is stuck HIGH and D_3 is stuck LOW.

FIGURE 13-13

19. See Figure 13-14.

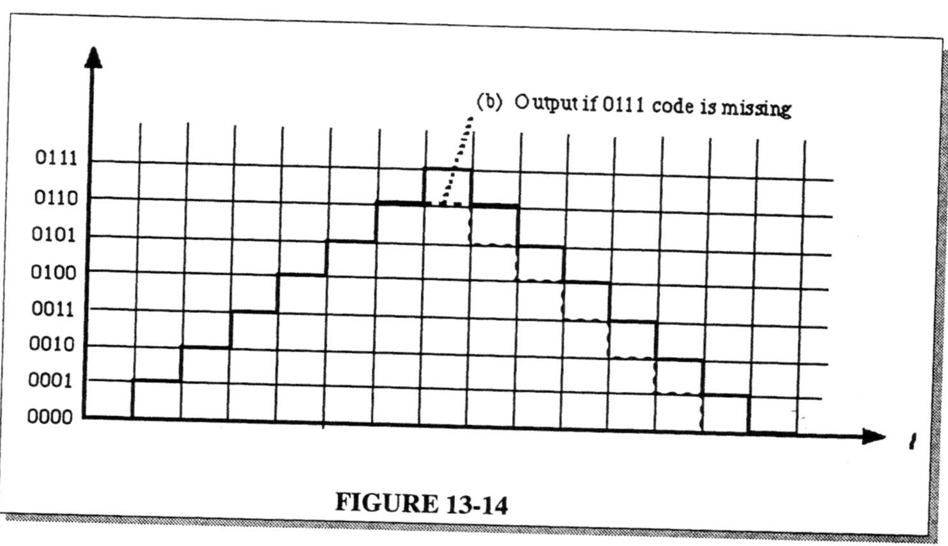

FIGURE 13-14

20. See Figure 13-15.

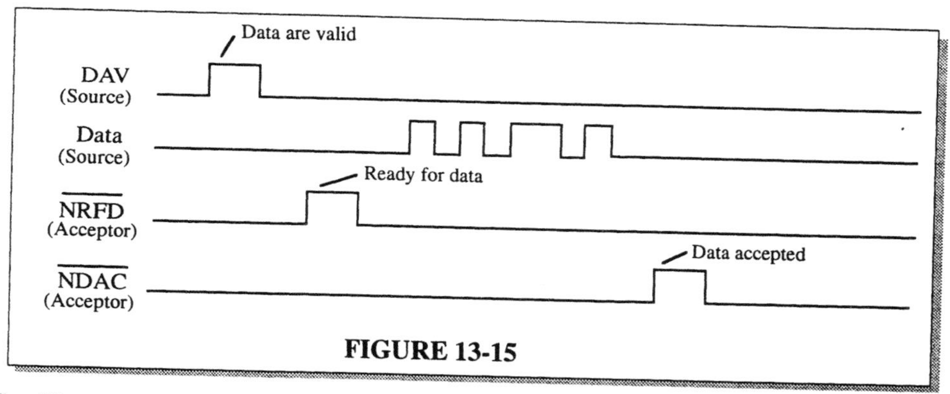

FIGURE 13-15

21. See Figure 13-16.

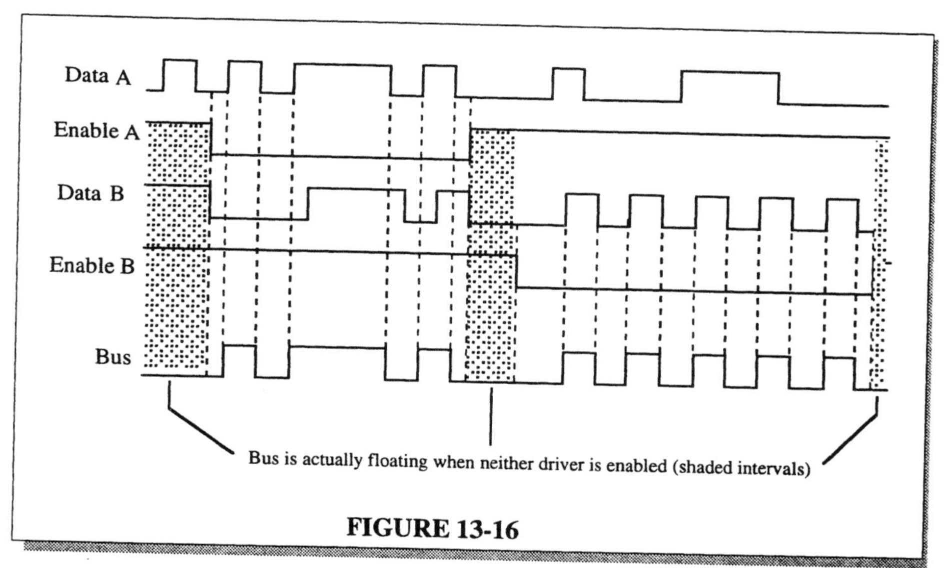

FIGURE 13-16

22. See Figure 13-17.

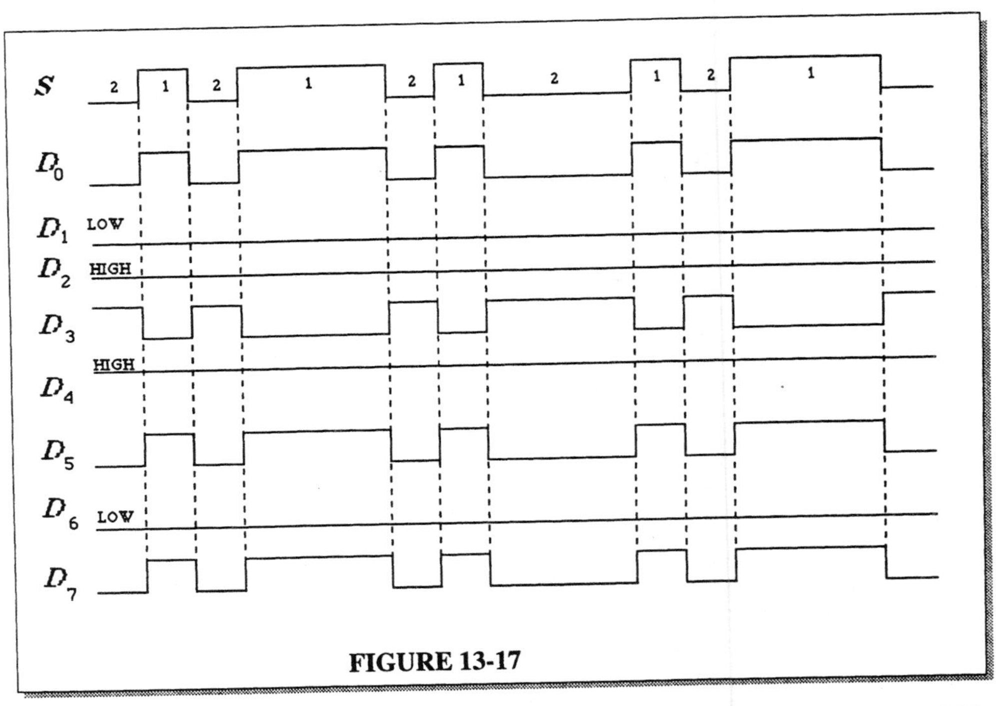

FIGURE 13-17

23. **The local bus is the collection of buses interfacing directly with the processor. The PCI bus is used for expansion devices and is connected to the local bus through a bus controller.**

24. **Plug-and-Play refers to self-configuring hardware that can be installed into and used in a computer system without the need for manual installation of jumpers or setting of switches.**

25. **The PCI bus is a 33 or 66 MHz, 32- or 64-bit, plug-and-play compatibnle expansion bus. ISA is an 8- or 16-bit 8.33 MHz expansion bus. PCI supports 3.3 V supplies while ISA supports 5 V and 12 V supplies.**

26. **A shorter RS-232C cable can support faster communication rates.**

27. **DCE stands for data communications equipment, such as a modem. DTE stands for data terminal equipment, such as a computer. Both acronyms are associated with the RS-232/ EIA-232 standard.**

28. **A USB cable consists of a power line, ground line, and two differerential data lines.**

29. Since the length limit is 15 meters and the minimum separation is 1 meter, **seven** more instruments can be connected with a 1 meter separation.

30. Three data bytes are transferred because the NDAC line goes HIGH three times, each time indicating that a data byte is accepted.

31. A controller is sending data to two listeners. The first two bytes of data (3F and 41) goes to the listener with address 001A. The second two bytes goes to the listener with address 001B. The handshake signals (DAV, NRFD, and NDAC) indicate that the data transfer is successful.. See Figure 13-18.

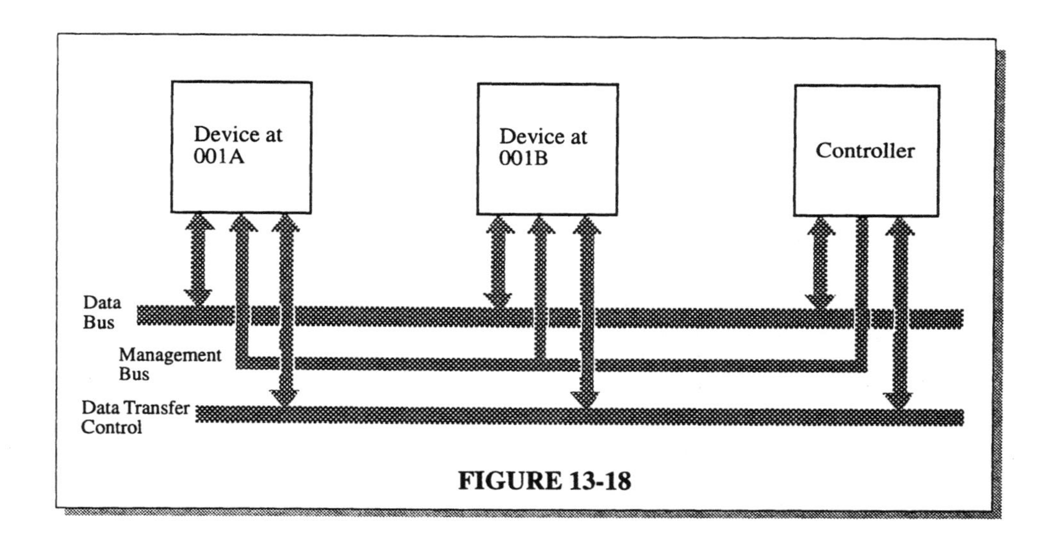

FIGURE 13-18

32. If a talker sends a data byte to a listener on a GPIB system and a DTE sends a data byte to a DCE on an RS-232C system, the RS-232C system will receive the data first. This is because GPIB requires significantly more setup and handshaking than RS-232C.

33. From the specifications of Test Bench 2 activities, the minimum frequency of the select waveform is 100 Hz. The minimum period of the convert waveform is

$$T_{min} = 30\ \mu s + 500\ ns + 100\ ns = 30.6\ \mu s$$

Since there are two conversions per period of the select waveform, the maximum frequency of the select waveform is

$$f_{max} = \frac{1}{2(30.6\ \mu s)} = 16.3\ kHz$$

34. The ADC for the antenna positioning system is 8 bits, so the resolution is

$$\frac{1}{2^8 - 1} = \frac{1}{255} = 0.00392 \text{ or } 0.392\%$$

35. Based on the resolution determined in problem 28, the smallest incremental change in azimuth and elevation is

$$(0.00392)180° = 0.71°$$

36. See Figure 13-19.

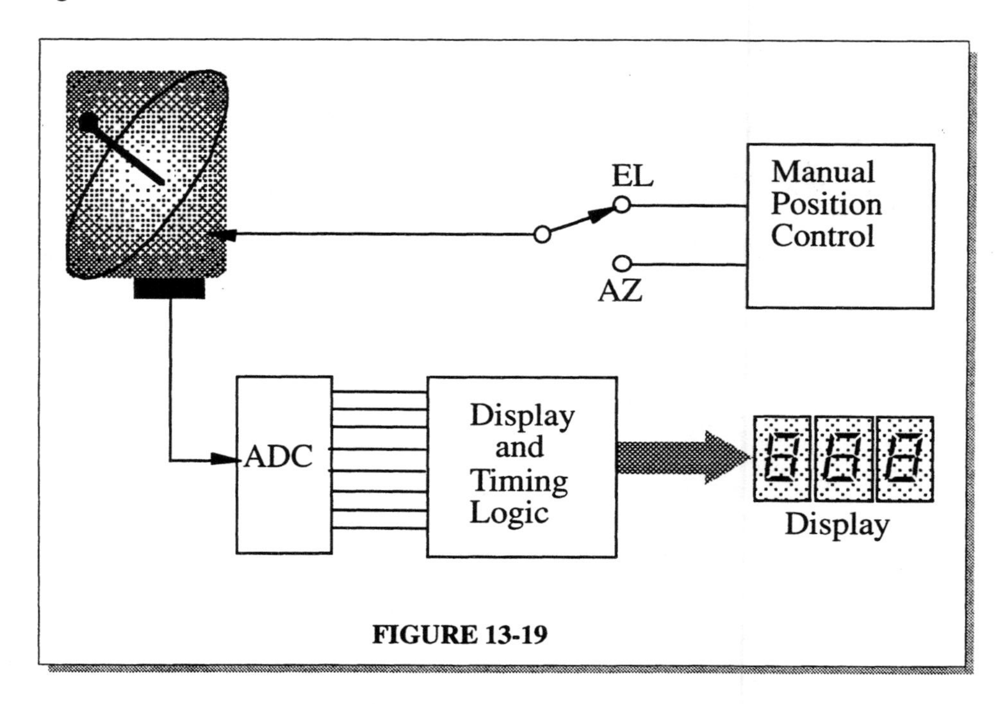

FIGURE 13-19

CHAPTER 14

INTRODUCTION TO MICROPROCESSORS AND COMPUTERS

1 Three basic elements of a microprocessor are *arithmetic logic unit (ALU), register array, and control unit.*

2. (a) A coprrocessor is a microprocessor with a limited instruction set optimized for arithmetic operations.
(b) The coprocessor for the Pentium is on the Pentium chip.

3. The three basic microprocessor buses are:
1. *Address bus*: a one-way bus that carries address information to the memory or to the I/O.
2. *Data bus*: a two-way bus on which data are transferred into the microprocessor or on which the result of an operation is sent out from the microprocessor
3. *Control bus*: a bus used to coordinate the microprocessor's operations and to communicate with external devices.

4. A port is a physical interface on a computer through which data is passed to and from peripherals.

5. (a) The first Intel microprocessor that used a 32-bit address bus was the 80386,
(b) The 80386 could address over 4 Gbytes.

6. The Pentium has a 64-bit data bus.

7. The first Motorola microprocessor that used a 32-bit address bus was the 68020.

8. The major difference between the Motorola MPC601 and the 68000 series is that the 600 series uses RISC architecture whereas the 68000 series uses CISC architecture.

9. A microprocessor repeatedly cycles through *fetch, decode, execute.*

10. Pipelining is the process of fetching and executing at the same time so that more than one instruction can be processed simultaneously.

11. The six segment registers of the 80386 and above are:
CS, DS, SS, ES, FS, GS

12. The code segment (CS) register contains 0F05 and the instruction pointer contains 0100. The physical address is

0F050 + 0100 = 0F150

13. AH and AL are 8-bit registers and represent the high and low part of the 16-bit AX register. The EAX is a 32-bit register which includes the AX register as the lower 16 bits.

14. (a) A *flag* is a bit stored in the flag register that is set or cleared by the processor
(b) A flag indicates a status or a control condition. A *staus* flag is an indicator of a condition after an arithmetic or logic operation. A *control* flag alters processor operations under certain conditions.

15. (a) Instruction pairing allows two instructions to execute at the same time.
(b) Instruction pairing requires two execution units and the 80486 has only one EU.

16. An assembler is a program that translates mnemonics and operands into machine code.

17. The flowchart in Figure 14-1 shows the process for adding numbers from one to ten and saving the results in a memory location named TOTAL.

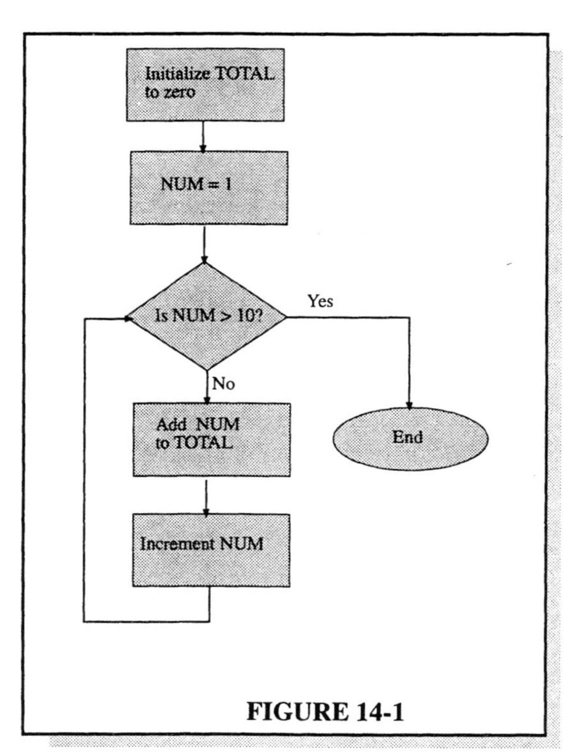

FIGURE 14-1

18. The flowchart in Figuer 14-2 shows how you can count the number of bytes in a string and place the count in a memory location called COUNT. The string starts at a location named START and uses 2H (space) to indicate the end.

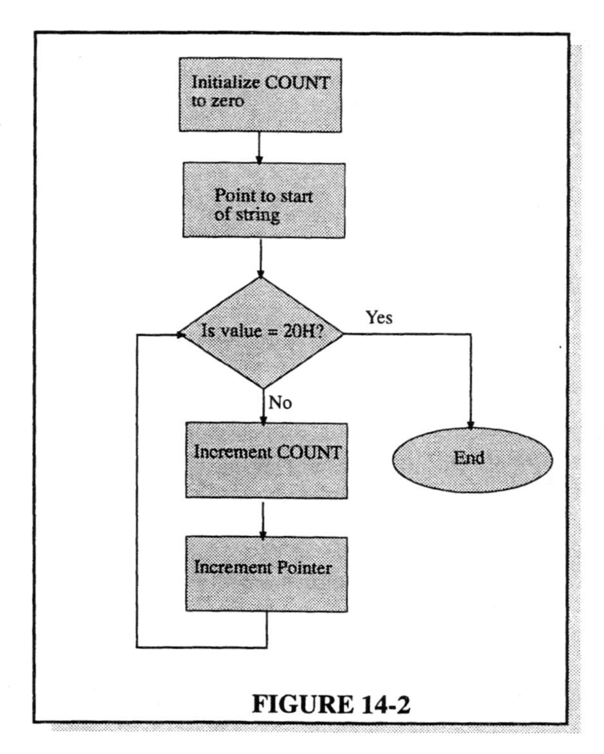

FIGURE 14-2

19. The assembler directive **word ptr** tells the assembler that the location pointed to is 16-bits.

20. When the instruction **mov ax,[bx]** is executed, the word in memory pointed to by the bx register is copied to the ax register.

21. The seven basic types of instruction in Intel processors are:
    1. *Data Transfer*: These instrucrtions are used to copy a byte, a word (16-bits), or a double word (32 bits) betweeen various sources and destinations. A n example is the MOV instruction.
    2. *Arithmetic*: These are instructions for addition, subtraction, multiplication, and division with signed or unsigned numbers. An example is the ADD instruction .
    3. *Bit Manipulation*: These instructions include logical operations, shifts, and rotations. An example is the OR instruction.
    4. *Loops and Jumps*: These instructions are designed to alter the normal sequence of instructions. An example is the JNZ instruction.
    5. *Strings*: A string instruction can copy, load, store, compare, or scan a string either as a byte at a time or a word at a time. An example is the MOVSW instruction.
    6. *Subroutine and Interrupts*: A subroutine is a miniprogram that can be used repeatedly but programmed only once. An example is the CALL instruction.
    7. *Processor Control*: These instructions allow direct control of some of the processor's flags and other miscellaneous tasks. An example is the STC instruction.

22. The $\overline{IORC}$ signal is an I/O read command; the $\overline{MRDC}$ signal is a memory read command.

23. Address lines A0 through A15 are used for I/O read and write operations in the 8086/88.

24. The ALE signal from the bus controller in the 8086/88 tells the address latch that a stable address is present on the address bus.

25. The status bits from the 8086/88 CPU signal the bus controller fro an I/O or read operation.

26. Three of the enhancements to the Pentium that were not in the 8086/88 are:
    1. Piplined architecture
    2. Cache memory for data and instructions
    3. Internal math coprocessor

27. The BIOS work area uses memory locations 400H through 4FFH.

28. Extended memory is the memory locations above the first 1 Mbyte.

29. The status code $\overline{S_1}\overline{S_2}\overline{S_0}$ = 110 indicates a memory write.

30. Dedicated I/O ports are assigned address space in the I/O address space. Memory-mapped I/O ports are assigned within the memory address space. The IN and OUT instructions cannot be used with memory-mapped I/O.

31. 64 k dedicated ports can be connected to a Pentium processor

32. The 8255A IC implements and configures I/O ports.

33. In a polled I/O, the CPU polls each device in turn to see if it needs service; in an interrupt driven system, the peripheral device signals the CPU when it requires service.

34. Vectoring is when the PIC provides a pointer to a service routine.

35. A software interrupt is a program instruction that invokes an interrupt service routine.

36. In a DMA operation, the DMA controller is given control by the CPU and allows data to flow between memory and a peripheral directly, bypassing the CPU.

CHAPTER 15
INTEGRATED CIRCUIT TECHNOLOGIES

1. No, because the $V_{OH(min)}$ is less than the $V_{IH(min)}$. The gate may interpret 2.2 V as a LOW.

2. Yes, they are compatible because the $V_{OL(max)}$ is less than the $V_{IL(max)}$.

3. $V_{NH} = V_{OH(min)} - V_{IH(min)} = 2.4\text{ V} - 2.25\text{ V} = 0.15\text{ V}$
   $V_{NL} = V_{IL(max)} - V_{OL(max)} = 0.65\text{ V} - 0.4\text{ V} = 0.25\text{ V}$

4. The maximum amplitudes equal the noise margins of 0.15 V and 0.25 V.

5. Gate A: $V_{NH} = 2.4\text{ V} - 2\text{ V} = 0.4\text{ V}$
   $V_{NL} = 0.8\text{ V} - 0.4\text{ V} = 0.4\text{ V}$
   Gate B: $V_{NH} = 3.5\text{ V} - 2.5\text{ V} = 1\text{ V}$
   $V_{NL} = 0.6\text{ V} - 0.2\text{ V} = 0.4\text{ V}$
   Gate C: $V_{NH} = 4.2\text{ V} - 3.2\text{ V} = 1\text{ V}$
   $V_{NL} = 0.8\text{ V} - 0.2\text{ V} = 0.6\text{ V}$

   Gate C has the highest noise margins.

6.
$$P_{D(LOW)} = (5\text{ V})(2\text{ mA}) = 10\text{ mW}$$
$$P_{D(HIGH)} = (5\text{ V})(3.5\text{ mA}) = 17.5\text{ mW}$$
$$P_{D(avg)} = \frac{P_{D(LOW)} + P_{D(HIGH)}}{2} = \frac{27.5\text{ mW}}{2} = 13.75\text{ mW}$$

7. The pulse goes through three gates in the shortest path.

   3X4 ns = 12 ns

8.
$$t_{p(avg)} = \frac{t_{pLH} + t_{pHL}}{2} = \frac{2\text{ ns} + 3\text{ ns}}{2} = 2.5\text{ ns}$$

9. Gate A average propagation delay:

$$\frac{t_{pLH} + t_{pHL}}{2} = \frac{1\text{ ns} + 1.2\text{ ns}}{2} = 1.1\text{ ns}$$

   Speed/Power product = (1.1 ns)(15 mW) = 16.5 pJ

   Gate B average propagation delay:

$$\frac{5\text{ ns} + 4\text{ ns}}{2} = 4.5\text{ ns}$$

   Speed/Power product = (4.5 ns)(8 mW) = 36 pJ

Gate C average propagation delay: $\frac{10\text{ ns} + 10\text{ ns}}{2} = 10\text{ ns}$

Speed/Power product = (10 ns)(0.5 mW) = 5 pJ

Gate C has the best speed/power product.

10. Gate A can be operated at the highest frequency because it has the shortest propagation delay.

11. G2 is overloaded because it has 12 unit loads.

12. The network in (a) operate at the highest frequency because the driving gate has fewer loads.

13. (a) ON (b) OFF
(c) OFF (d) ON

14. Unused inputs should be connected as follows:
Negative-OR gate (NAND) to $+V_{CC}$
NAND gate to $+V_{CC}$
NOR gate to ground

15. See Figure 15-1 for another possibleapproach in addition to circuit given in text answers.

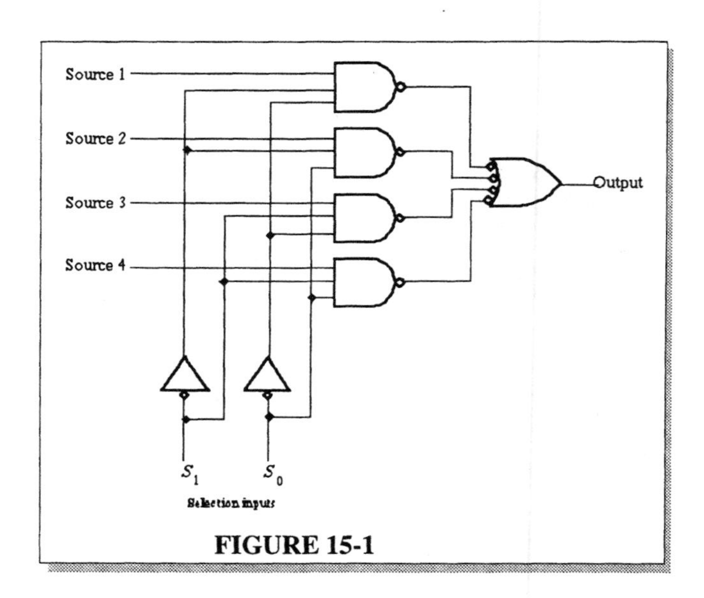

FIGURE 15-1

16.
(a) ON: high voltage on base forward biases the base-emitter junction.
(b) OFF: insufficient voltage on base to forward-bias the base-emitterjunction.
(c) . OFF: emitter is more positive than the base which reverse biases the base-emitter junction.
(d) OFF: base and emitter at same voltage. No forward bias.

17. See Figure 15-2.

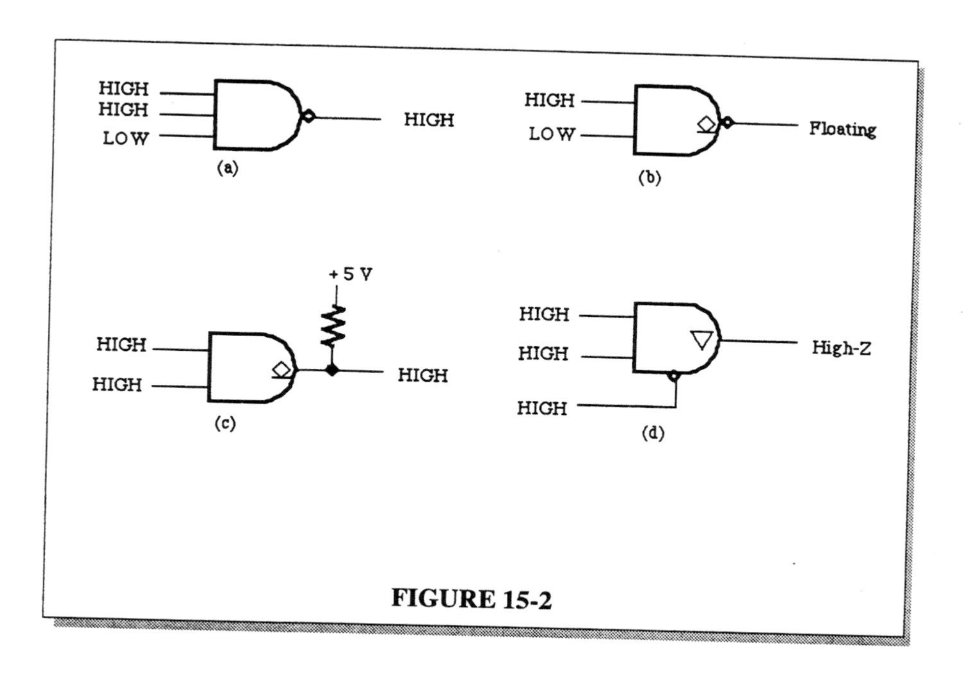

FIGURE 15-2

18. Connect a 1 kΩ pull-up resistor to the unused inputs of the two NAND gates. Connect the unused input of the NOR gate to ground. Connect a pull-up resistor to the open collector of the NOR gate (value depends on load).

19. See Figure 15-3.

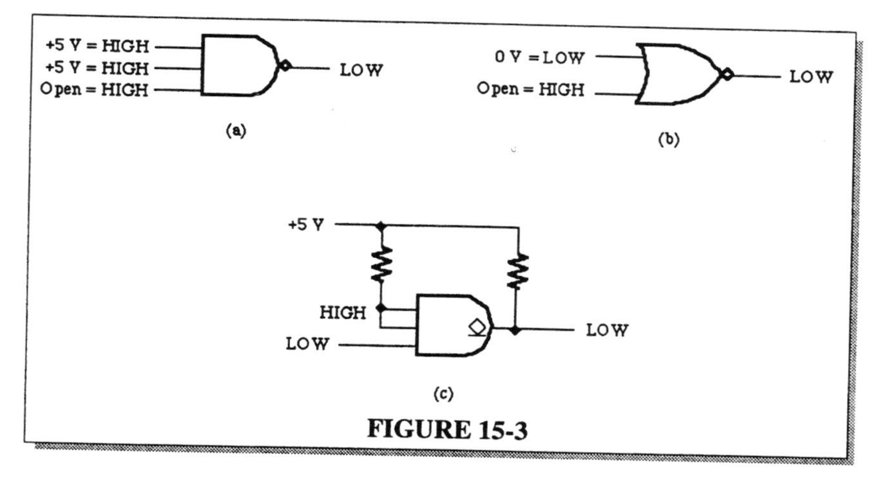

FIGURE 15-3

20. (a) The driving gate output is HIGH, it is sourcing 3 unit loads.
$I_T = 3(40\ \mu A)$ = **120 μA**

(b) The driving gate output is LOW, it is sinking current from 2 unit loads.
$I_T = 2(-1.6\ mA)$ = **-3.2 mA**

(c) G1 output is HIGH, it is sourcing 6 unit loads.
$I_T = 6(40\ \mu A)$ = **240 μA**

G2 output is LOW, it is sinking current from 2 unit loads.
$I_T = 2(-1.6\ mA)$ = **-3.2 mA**

G3 output is HIGH, it is sourcing 2 unit loads.
$I_T = 2(40\ \mu A)$ = **80 μA**

21. See Figure 15-3. Pull-up resistors of second-level inverters are not shown.

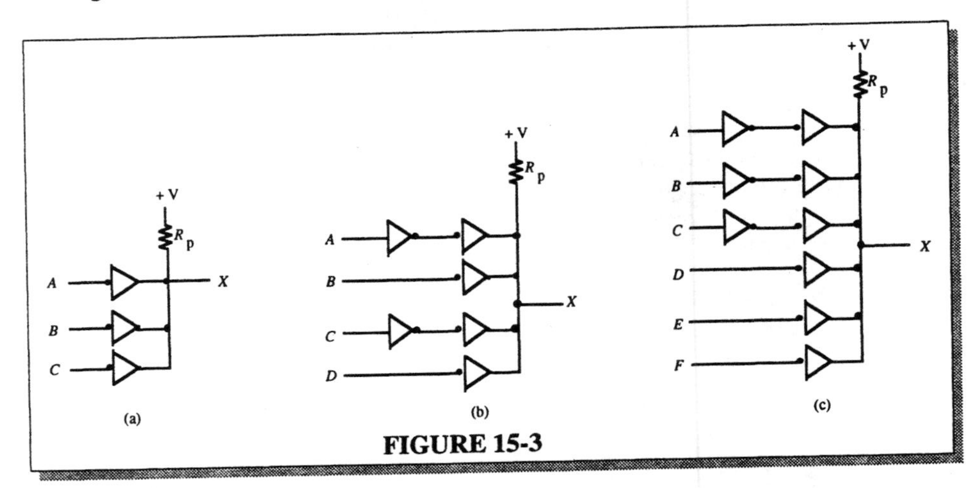

FIGURE 15-3

22. (a) $X = AB\overline{C}\overline{D}$

(b) $X = (\overline{ABC})(\overline{DE})(\overline{FG})$

(c) $X = (\overline{A + B})(\overline{C + D})(\overline{E + F})(\overline{G + H}) = \overline{A}\ \overline{B}\ \overline{C}\ \overline{D}\ \overline{E}\ \overline{F}\ \overline{G}\ \overline{H}$

23. Worst case for determining minimum R_p is when only one gate is sinking all of the current (40 mA maximum).

For 10 UL: $I_L = 10(1.6\text{ mA}) = 16\text{ mA}$
For each gate: $I_{Rp(max)} = I_{OL(max)} - 16\text{ mA} = 40\text{ mA} - 16\text{ mA} = 24\text{ mA}$

$$V_{R_p} = 5\text{ V} - 0.25\text{ V} = 4.75\text{ V}$$

$$R_{p(min)} = \frac{V_{R_p}}{I_{R_p(max)}} = \frac{4.75\text{ V}}{24\text{ mA}} = 198\ \Omega$$

$R_{p(min)}$ for (a), (b), and (c) is the same value

24. See Figure 15-4.

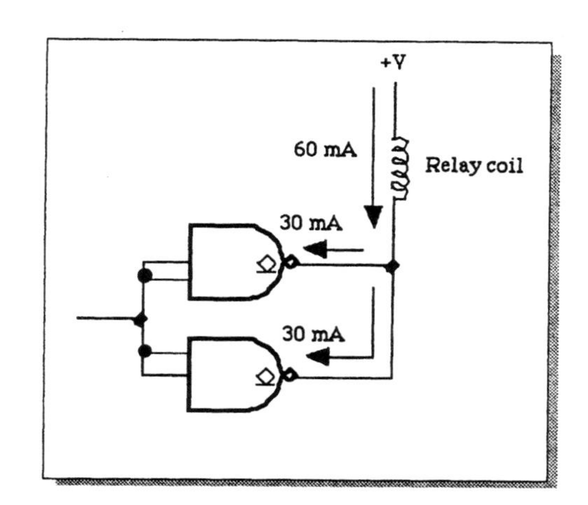

25. **F series**: SPP = 3.3 ns × 6 mW = 19.8 pJ

LS series: SPP = 10 ns × 2.2 mW = 22 pJ

ALS series: SPP = 7 ns × 1.4 mW = 9.8 pJ

ABT series: SPP = 3.2 ns × 17 µW = 0.0544 pJ

HC series: SPP = 7 ns × 2.75 µW = 0.01925 pJ

AC series: SPP = 5 ns × 0.55 µW = 0.00275 pJ

AHC series: SPP = 3.7 ns × 2.75 µW = 0.010175 pJ

LV series: SPP = 9 ns × 1.6 µW = 0.0144 pJ

LVC series: SPP = 4.3 ns × 0.8 µW = 0.00344 pJ

ALVC series: SPP = 3 ns × 0.8 µW = 0.0024 pJ

ALVC has the best (lowest value) speed-power product. It is, however, misleading to compare CMOS and TTL in terms of SPP because the power of CMOS goes up with frequency.

26. (a) ALVC
(b) AHC
(c) AC
(d) ALVC

27. (a) A and B to X: 3(3.3 ns) = 9.9 ns
C and D to X: 2(3.3 ns) = 6.6 ns

(b) A to X1, X2, X3: 2(7 ns) = 14 ms
B to X1: 7 ns
C to X2: 7 ns
D to X3: 7 ns

(c) A, B to X: 3(3.7 ns) = 11.1 ns
C, D to X: 2(3.7 ns) = 7.4 ns

28. (a) HC has an f_{max} = 50 MHz

$$f_{clock} = \frac{1}{50\text{ ns}} = 20\text{ MHz}$$

(b) LS has an f_{max} = 33 MHz

$$f_{clock} = \frac{1}{60\text{ ns}} = 16.7\text{ MHz}$$

(c) AHC has an f_{max} = 170 MHz

$$f_{clock} = \frac{1}{4\text{ ns}} = 250\text{ MHz}$$

Since $f_{clock} > f_{max}$ for the AHC flip-flop, the output will be erratic.

29. ECL operates with nonsaturated BJTs whereas TTL transistors saturate when turned on.

30. (a) Lowest propagation delay - ECL
(b) Lowest power - HCMOS
(c) Lowest speed/power product - HCMOS

PART 2

System Application Solutions

CHAPTER 1
DIGITAL SYSTEM APPLICATION

There are no activities required in this section. The block diagram is given below.

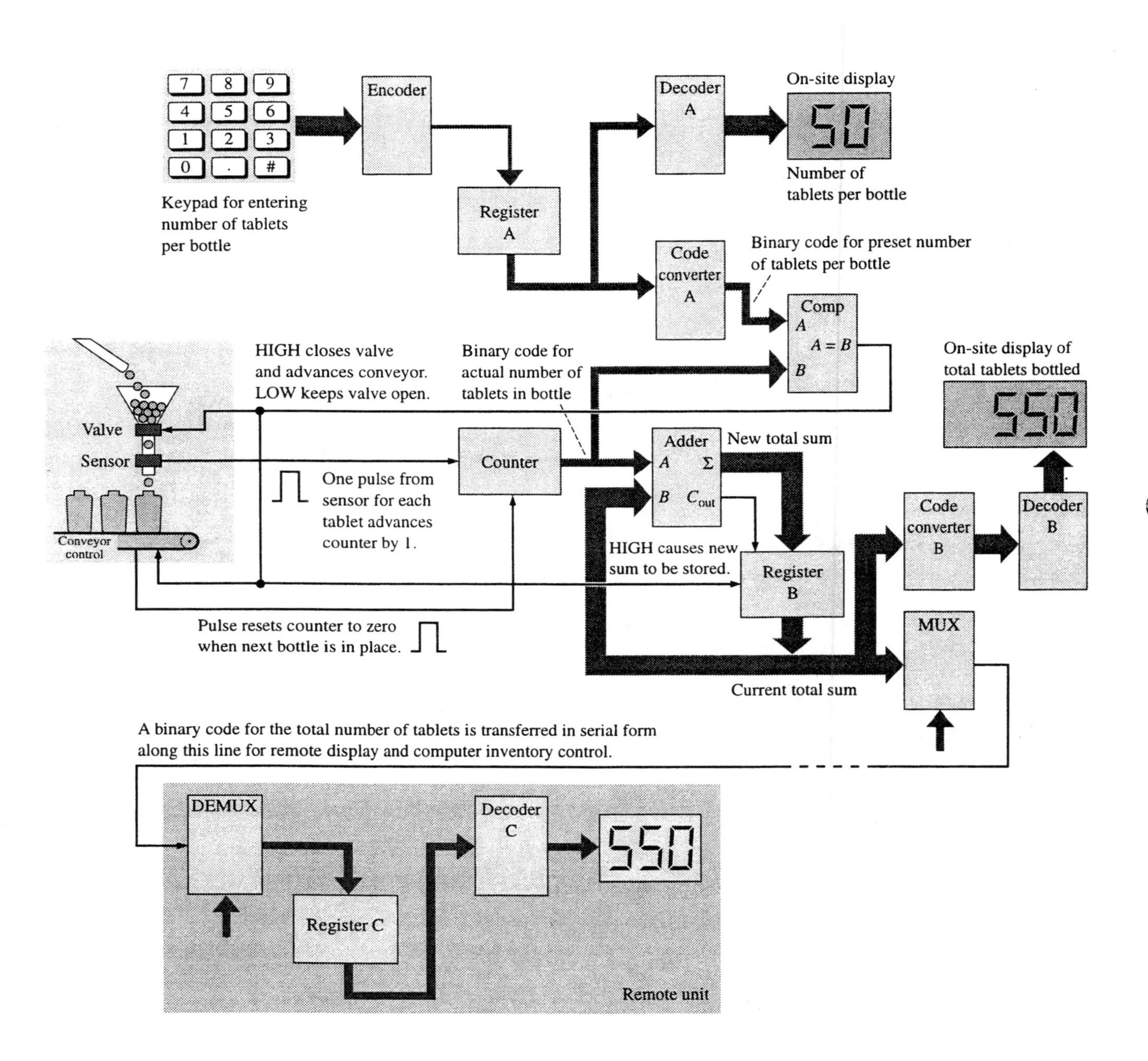

DIGITAL SYSTEM APPLICATION

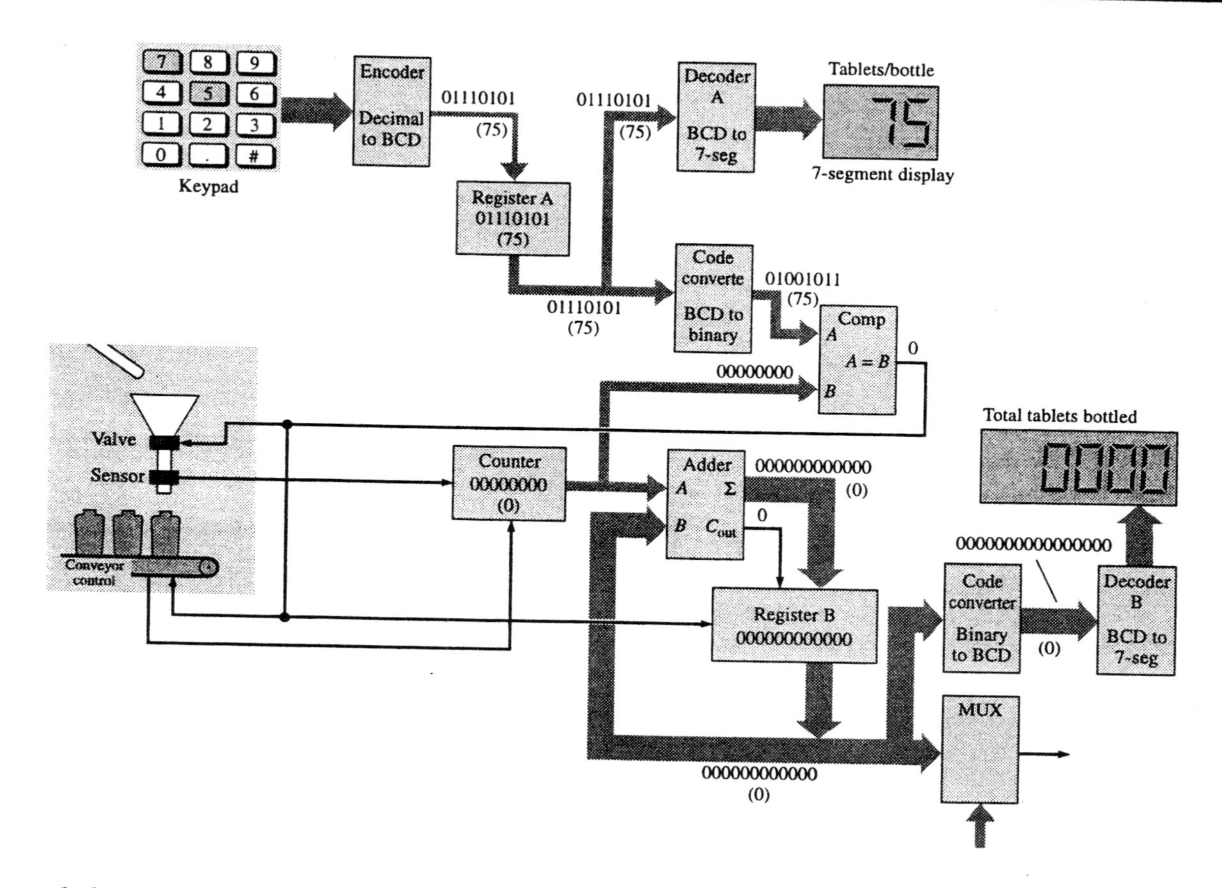

A worksheet master of this diagram that can be copied for a student handout can be found in Part 3.

Activity 1 Initialization

(a) For 60 tablets/bottle, the contents of Register A = **0110000** (BCD).
The A input of the comparator = **00111100** (binary 60_{10}).

(b) The counter's maximum count = **00111100** (binary 60_{10}).

Activity 2 Maximum Count

(a) The highest decimal number that can be represented by 12 bits is 2^{12} - 1 = 4,096.
At 75 tablets/bottle, the total number of bottles is 4,096/75 =54.6. Therefore, the number of *full* bottles is **54**.

(b) After the last bottle (54th) has been filled, the contents of Register B is 75X54 = $4{,}050_{10}$ = **111111111000011** (binary).

(c) When the last bottle is filled, the binary-to-BCD code converter output = **0100000001010000** ($4{,}050_{10}$).

Activity 3 Sequencing

(a) Since Register A always contains the number of tablets/bottle (30 in this case), after any number of bottles the contents of Register A = **0011000** (BCD).
Since the registers as shown are12 bit capacity, the system as shown cannot count 523 bottles. It can only count a maximum of 4095 tablets and is limited to 136 30-tablet bottles. To count 523 30-tablet bottles or more, the registers must be increased to 16 bits. If we assume 16-bit regis ters, the contents of Register B is 30X523 = 15690_{10} = **0011110101001010** (Binary).

(b) Again, assuming 16-bit registers, if the counter is at 00010110 (22_{10}), the output of the adder is 0011110101001010 + 00010110 = **0011110101100000** (22 + 15,690 = 15,712).

(c) **15,712** tablets have been counted.

(d) The number currently on 6-digit display is **15,690** which is the total tablets in 523 bottles. The display is updated when another bottle is filled so the next number will be 15690 + 30 = **15720**.

Activity 4 Hexadecimal

(a) *Figure 2-5:*
Counter: $00000000_2 = \mathbf{00}_{16}$
Comparator input A: $01001011_2 = \mathbf{4B}_{16}$
Register B: $000000000000_2 = \mathbf{000}_{16}$
Adder output: $000000000000_2 = \mathbf{000}_{16}$

(b) *Figure 2-6:*
Counter: $01001010_2 = \mathbf{4A}_{16}$
Comparator input A: $01001011_2 = \mathbf{4B}_{16}$
Register B: $111010100110_2 = \mathbf{EA6}_{16}$
Adder output: $111011110000_2 = \mathbf{EF0}_{16}$

(c) *Figure 2-7:*
Counter: $01001011_2 = \mathbf{4B}_6$
Comparator input A: $01001011_2 = \mathbf{4B}_{16}$
Register B: $111011110001_2 = \mathbf{EF1}_{16}$
Adder output: $111011110001_2 = \mathbf{EF1}_{16}$

(d) If the counter is at $4C_{16}$ and counts nine more tablets, its count is $4C_{16} + 9_{16} = \mathbf{55}_{16}$.

(e) In Figure 2-6, the decimal display 3750_{10} is expressed in hexidecimal as **EA6.**

CHAPTER 3
DIGITAL SYSTEM APPLICATION

System Assignment 1: Analysis and Design

Activity 1

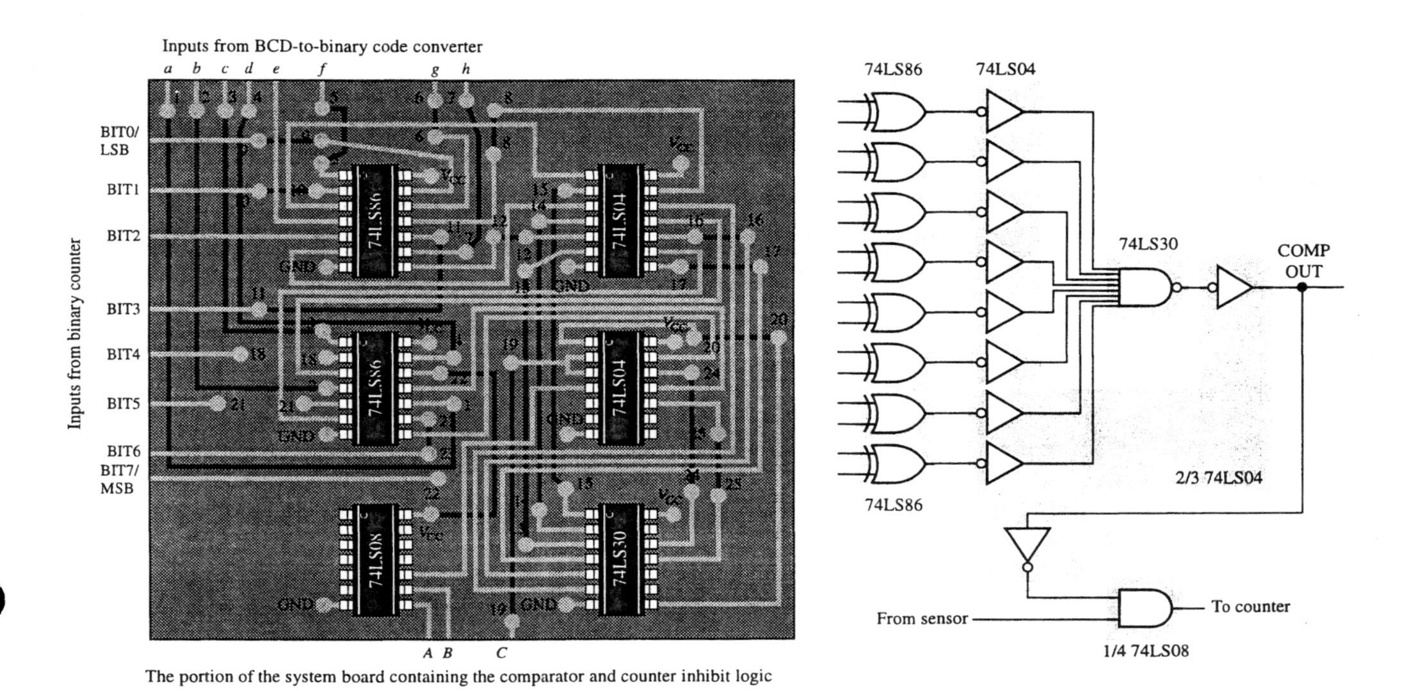

The portion of the system board containing the comparator and counter inhibit logic

Activity 2

From the BIT0 through BIT7 waveforms the binray number for a HIGH comparator output is **00111100 = 60_{10}.** There are 60 tablets going into each bottle.

Activity 3

g = BIT0 = **0**
f = BIT1 = **0**
e = BIT2 = **1**
h = BIT3 = **1**
c = BIT4 = **1**
b = BIT5 = **1**
a = BIT6 = **0**
d = BIT7 = **0**

Activity 4
A- To Counter
B- Sensor
C- Comp Out

Activity 5
For the 74XX series of TTL:
The minimum specified value for V_{CC} is 4.5 V and the maximum is 5.5 V. The reading of 4.85 V is within specifications.
The minimum specified high-level input voltage, $V_{IH(min)}$, is 2 V. The reading of 3.3 V at the BIT5 input exceeds the specified minimum for a HIGH or 1.
The maximum specified low-level input voltage, $V_{IL(max)}$, is 0.8 V. The reading of 0.35 V at the BIT6 input is less than the specified maximum for a LOW or 0.

Activity 6
One approach to reduce the number of ICs is to keep the 74LS86 XOR gates and replace the 74LS04s, the 74LS30, and the 74LS08 (4 ICs) with a 74LS25 dual 4-input NOR and a 7400 quad 2-input NAND (2 ICs) as shown in the figure. Another approach replaces the XOR gates with XNOR gates thus eliminating one inverter IC.

EWB: See EWB Ciruit Solutions (Part 4) for the pc board simulation.

System Application 2: Troubleshooting

Activity 1

The pulser is applying pulses to the input (pin 1) of the 74LS04 inverter. The proble is connected to the output (pin 8) of the 74LS08 AND gate. The OFF condition of the probe does not necessarily indicate at fault. A LOW on any of the inputs to the 74LS30 NAND gate or a LOW on the sensor input of the 74LS08 AND gate would cause the probe to indicate a LOW. To make this a valid test, the inputs to all of the EXOR gates must be such as to produce HIGHs on all inputs to the 74LS30 and the sensor input to the 74LS08 must be HIGH.

Activity 2

Measurement set 1: IC4 inverter (pins 1/2) is faulty. Possible faults are: pin 1 internally open or pin 2 shorted to ground. If an external short is not visible, repair by replacing IC4.

Measurement set 2: A fault is indicated. The trace between IC1 pin6 and IC4 pin 3 is open. Repair the trace.

Measurement set 3:
IC2 XOR gate (pins 8, 9, 10) is faulty. Possible faults are: pin 9 internally open or pin 8 shorted to ground. If an external short is not visible, repair by replacing IC2.

Measurement set 4: The test is inconclusive. A constant low on any of the inputs to the 74LS30 will cause the output to be at a constant HIGH level.

Measurement set 5: IC5 inverter (pins 3/4) is faulty. Possible faults are: pin3 internally open or pin 4 shorted to ground. If an external short is not visible, repair by replacing IC5.

Measurement set 6: The test is inconclusive. A constant low on any of the inputs to the 74LS30 will cause the output to be a constant HIGH level.

Measurement set 7: The test is inconclusive. A LOW on pin 10 of the 74LS08 will keep the output LOW.

CHAPTER 4
DIGITAL SYSTEM APPLICATION

Activity 1

$$b = \overline{D}\,\overline{C}\,\overline{B}\,\overline{A} + \overline{D}\,\overline{C}\,\overline{B}A + \overline{D}\,\overline{C}BA + \overline{D}C\overline{B}\,\overline{A} + \overline{D}CBA + D\overline{C}\,\overline{B}\,\overline{A} + D\overline{C}\,\overline{B}A + \overline{D}\,\overline{C}B\overline{A}$$

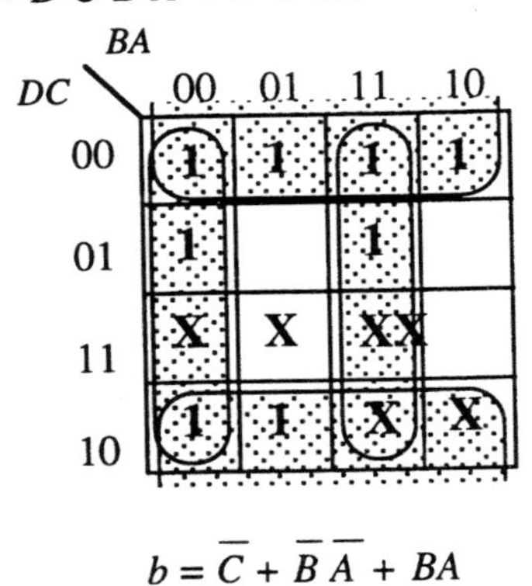

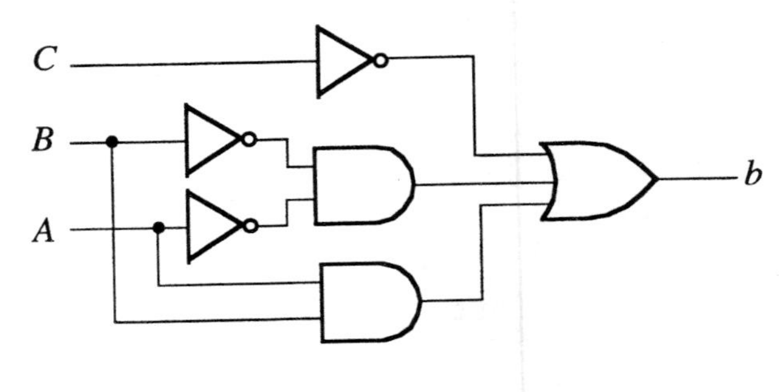

$$b = \overline{C} + \overline{B}\,\overline{A} + BA$$

Activity 2

$$c = \overline{D}\,\overline{C}\,\overline{B}\,\overline{A} + \overline{D}\,\overline{C}\,\overline{B}A + \overline{D}\,\overline{C}BA + \overline{D}C\overline{B}\,\overline{A} + \overline{D}C\overline{B}A + \overline{D}CB\overline{A} + \overline{D}CBA + D\overline{C}\,\overline{B}\,\overline{A} + D\overline{C}\,\overline{B}A$$

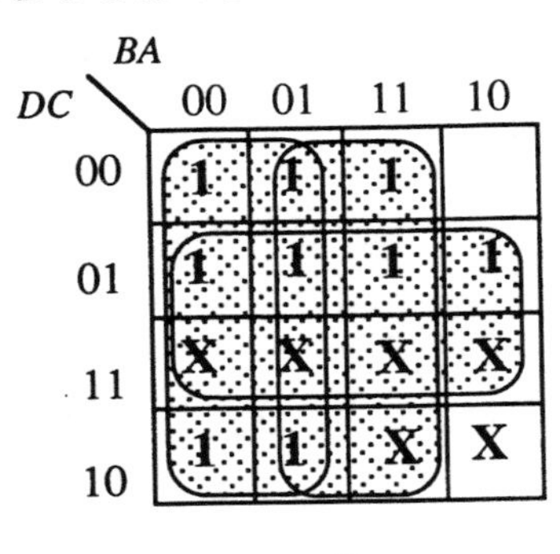

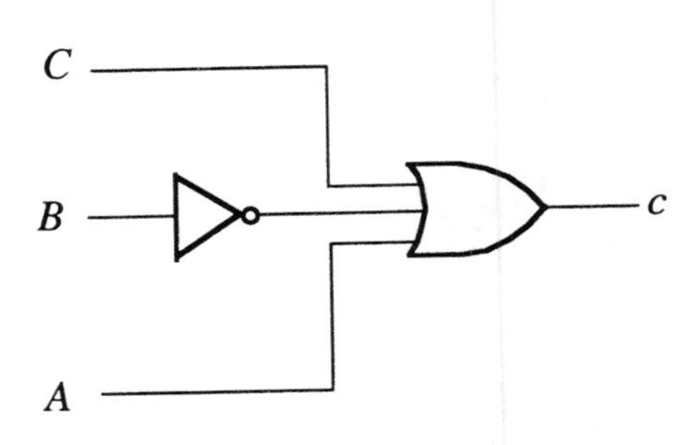

$$c = C + \overline{B} + A$$

Acitvity 3

$$d = \overline{D}\,\overline{C}\,\overline{B}\,\overline{A} + \overline{D}\,\overline{C}B\overline{A} + \overline{D}\,\overline{C}BA + \overline{D}C\overline{B}A + \overline{D}CB\overline{A} + D\overline{C}\,\overline{B}\,\overline{A} + D\overline{C}\,\overline{B}A$$

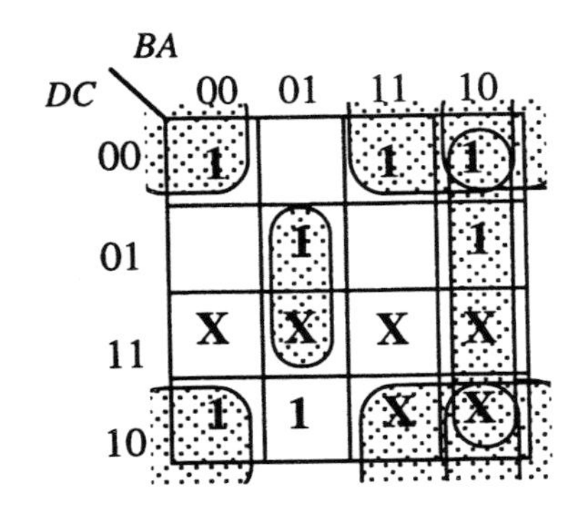

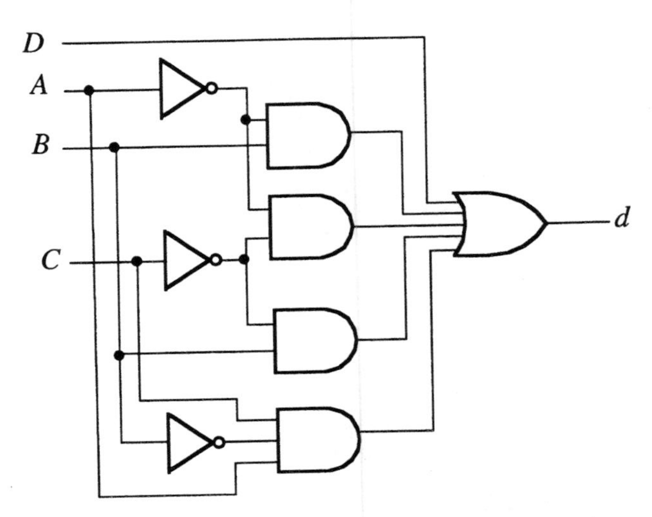

$$d = D + B\overline{A} + \overline{C}\,\overline{A} + \overline{C}B + C\overline{B}A$$

Activity 4

$$e = \overline{D}\,\overline{C}\,\overline{B}\,\overline{A} + \overline{D}\,\overline{C}B\,\overline{A} + \overline{D}CB\,\overline{A} + D\overline{C}\,\overline{B}\,\overline{A}$$

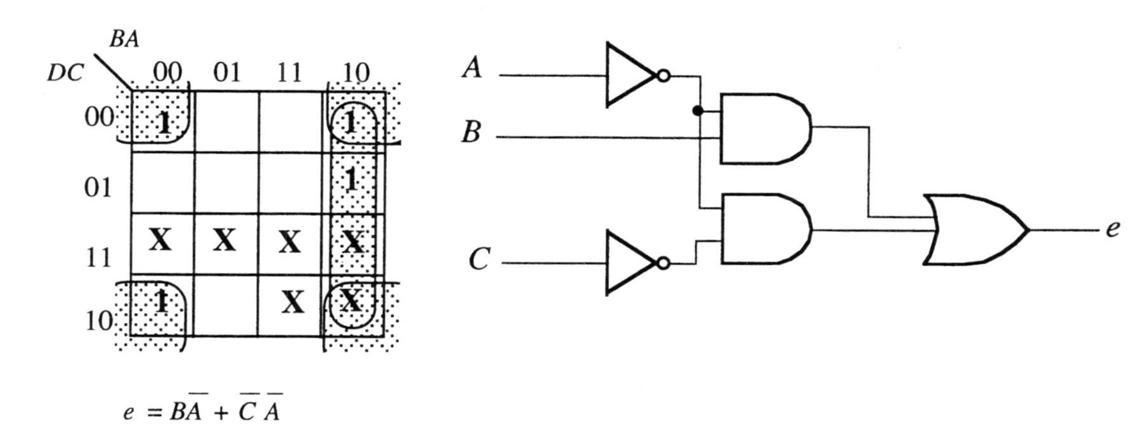

$$e = B\overline{A} + \overline{C}\,\overline{A}$$

Activity 5

$$f = \overline{D}\,\overline{C}\,\overline{B}\,\overline{A} + \overline{D}C\overline{B}\,\overline{A} + \overline{D}C\overline{B}A + \overline{D}CB\overline{A} + D\overline{C}\,\overline{B}\,\overline{A} + D\,\overline{C}\,\overline{B}A$$

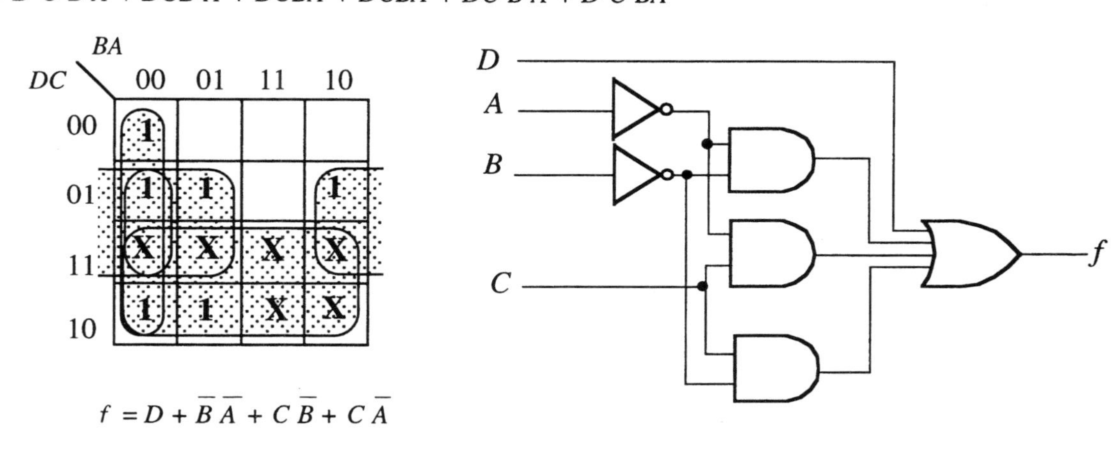

$$f = D + \overline{B}\,\overline{A} + C\,\overline{B} + C\,\overline{A}$$

Activity 6

$$g = \overline{D}\,\overline{C}\,B\,\overline{A} + \overline{D}\,\overline{C}BA + \overline{D}C\overline{B}\,\overline{A} + \overline{D}C\overline{B}A + \overline{D}CB\overline{A} + D\overline{C}\,\overline{B}\,\overline{A} + D\,\overline{C}\,\overline{B}A$$

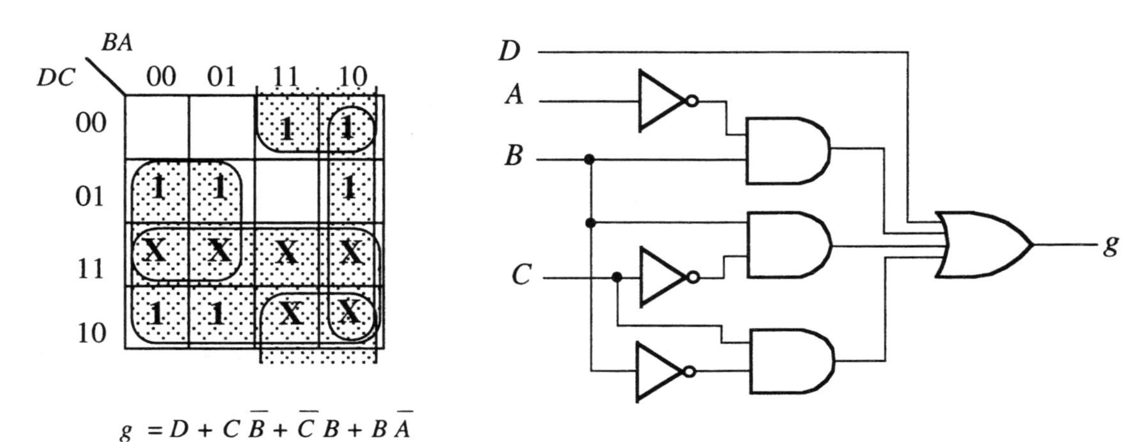

$$g = D + C\,\overline{B} + \overline{C}\,B + B\,\overline{A}$$

Activity 7

The term $\bar{C}\bar{A}$ is used in the logic for segments *a, d*, and *e*
The term $\bar{B}\bar{A}$ is used in the logic for segments *b* and *f*
The term $B\bar{A}$ is used in the logic for segments *d, e*, and *g*
The term $\bar{C}B$ is used in the logic for sements *d* and *g*
The term $C\bar{B}$ is used in the logic for segments *f* and *g*

Activity 8

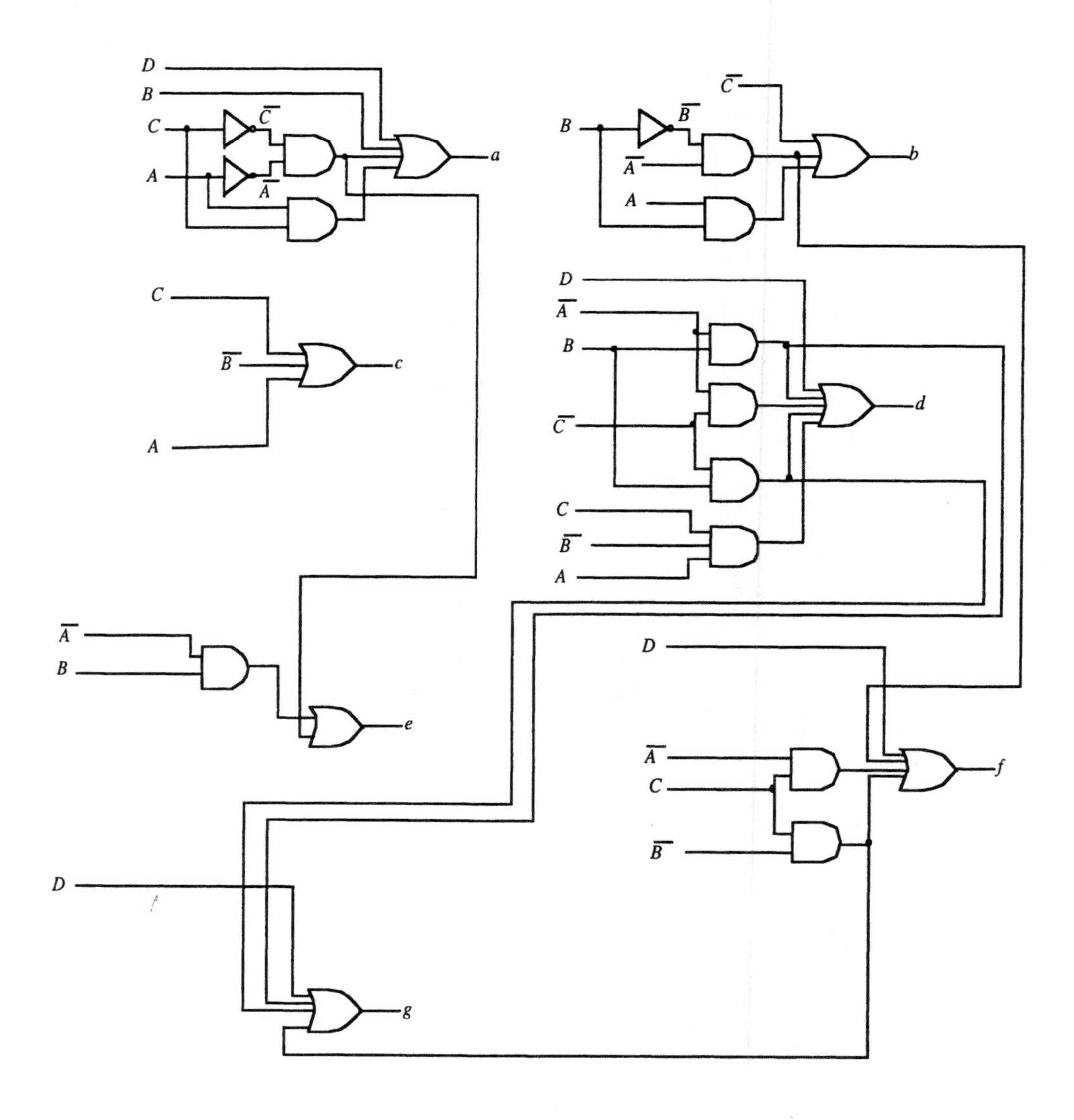

Activity 9

The basic circuit of activity 8 can be implemented with the following ICs:
2 74XX04 Hex Inverter
2 74XX08 Quad 2-input AND gates
1 74XX11 Triple 3-input AND gates
1 74XX27 Triple 3-input NOR gates
2 74XX260 Dual 5-input NOR gates

EWB: See EWB Circuit Solutions (Part 4) for decoding logic simulation.

CHAPTER 5

DIGITAL SYSTEM APPLICATION

System Assignment 1: Design

Activity 1

From Table 5-6 in the text, the conveyor motor logic is developed as follows:

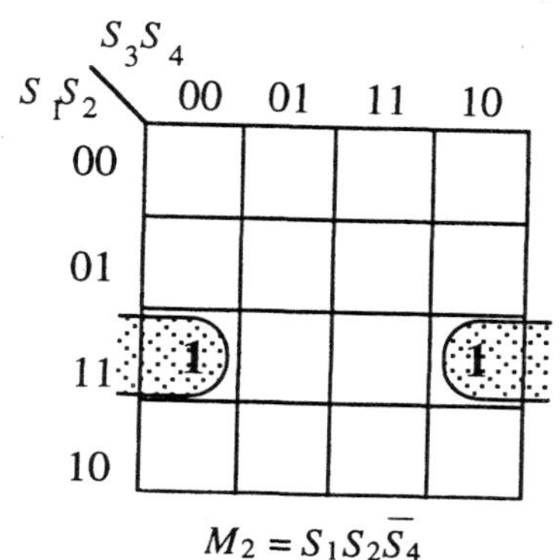

$M_2 = S_1S_2\overline{S_4}$

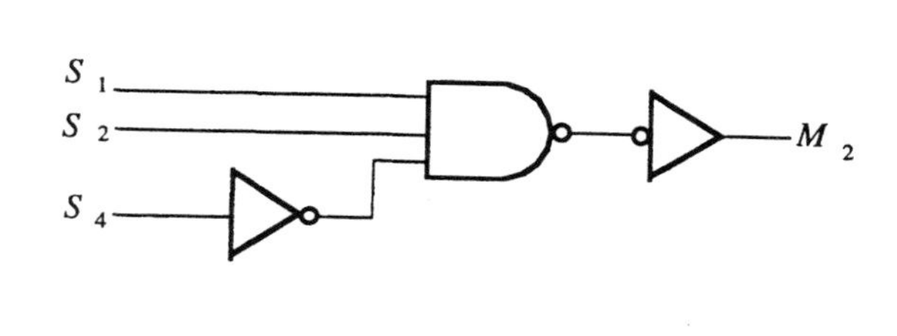

Activity 2

From Tab le 5-6 in the text, the bandsaw motor logic is developed as follows:

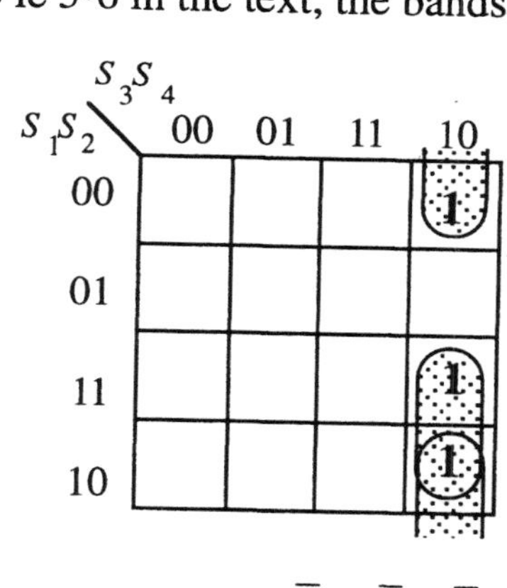

$M_3 = S_1S_3\overline{S_4} + \overline{S_2}S_3\overline{S_4}$

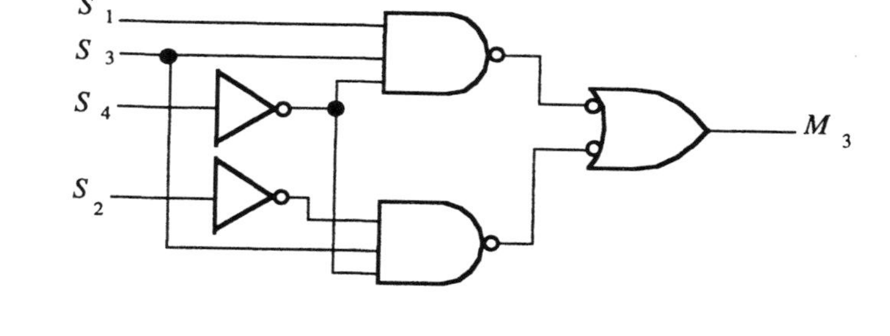

Activity 3

From Table 5-6 in the text, the cross-cut saw motor logic is developed as follows:

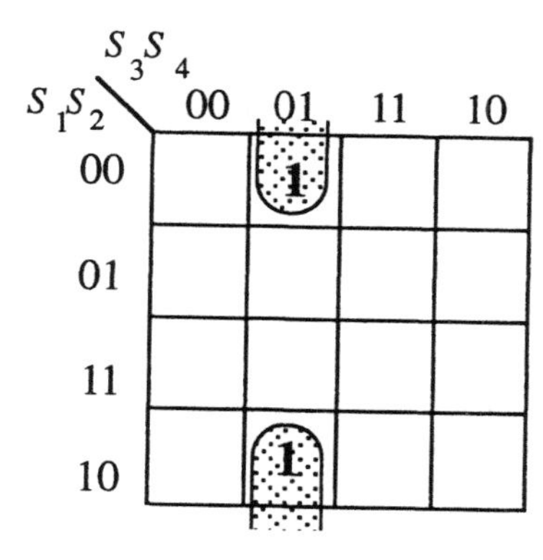

$M_4 = S_2S_3S_4$

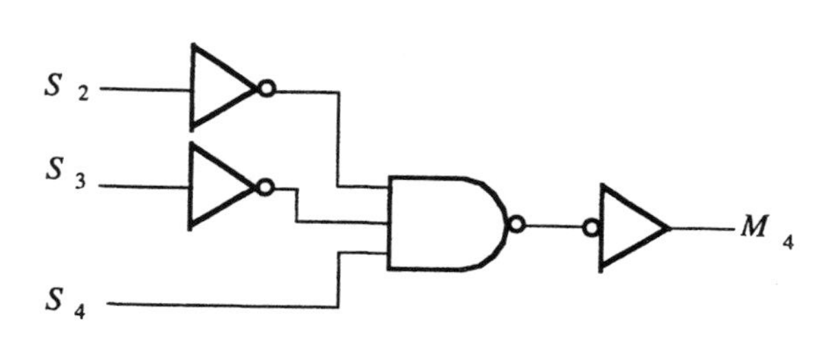

Activity 4
The complete motor control logic diagram is as follows:

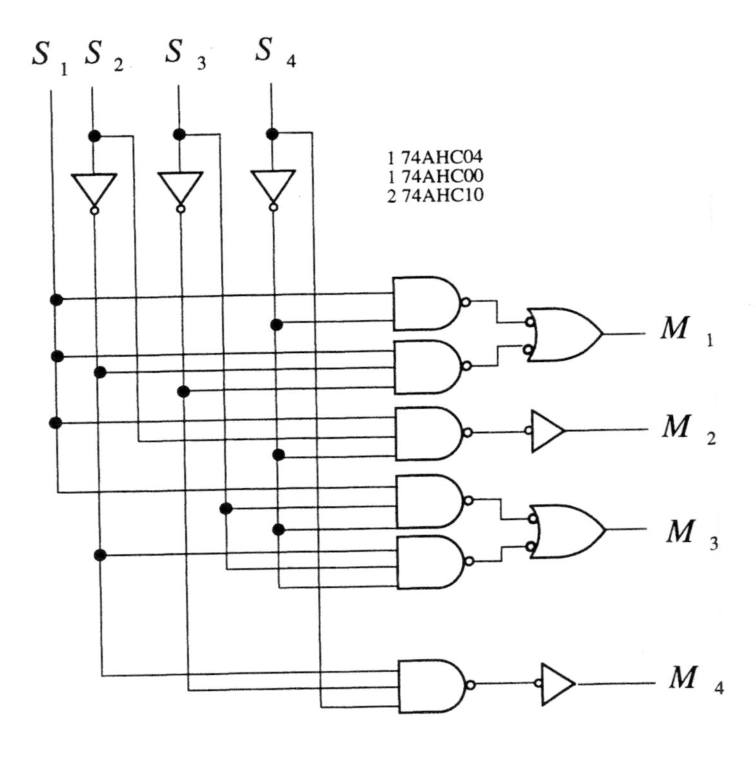

Activity 5
The circuit board is shown below. The logic diagram derived from tracing out the circuit board is shown on the next page. A worksheet master for the circuit board that can be copied for a student handout can be found in Part 3.

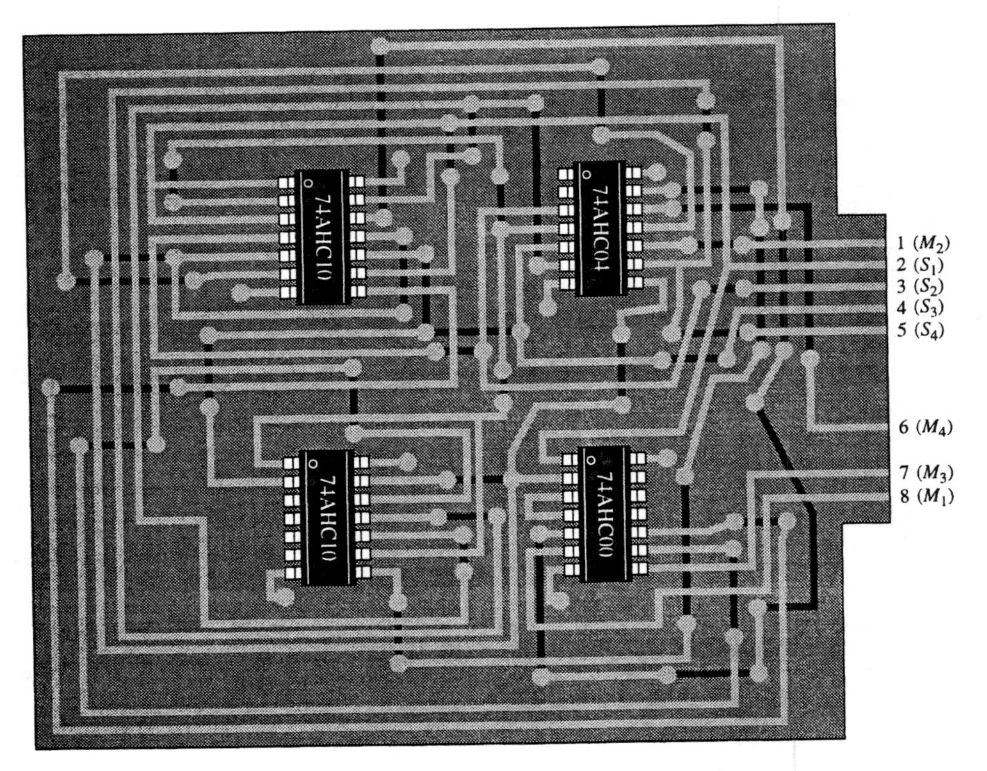

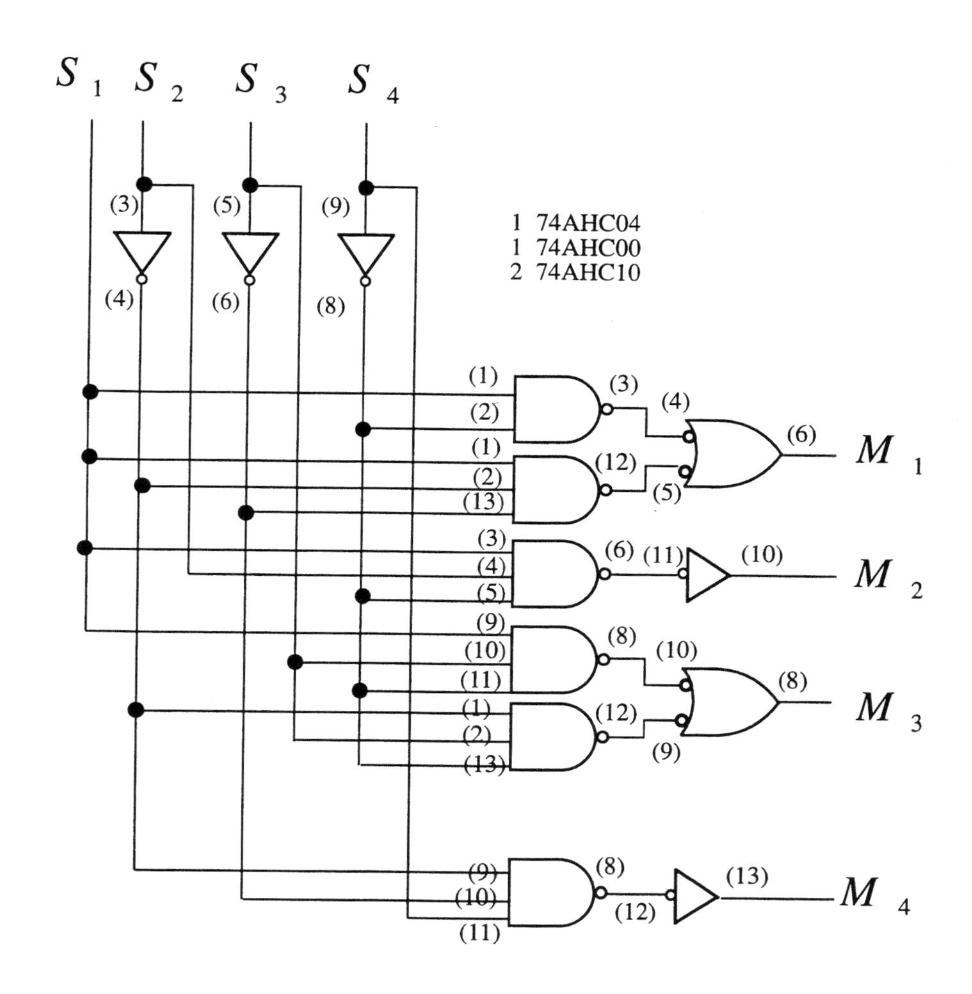

Activity 6
The inputs and outputs are labelled on both the circuit board and on the logic diagram in Activity 5.

System Assignment 2: Testing and Troubleshooting

Activity 1

Display 1: Waveforms are correct.

Display 2: M_1 is LOW for $S_1S_2S_3S_4 = 1000$ when it should be HIGH. The output of gate decoding $S_1\overline{S}_2\overline{S}_3$ or the corresponding input of the following NAND gate is problably open.

Display 3: M_3 is HIGH for $S_1S_2S_3S_4 = 0011$ and $S_1S_2S_3S_4 = 1011$when it should be LOW. The output line of the S_4 inverter or the corresponding input of the following NAND gate is probably open.

Display 4: M_4 is HIGH for $S_1S_2S_3S_4 = 0011$, $S_1S_2S_3S_4 = 0111$, and $S_1S_2S_3S_4 = 1011$ when in should be LOW. Intermittant open outputs of the S_2 and S_3 inverters could be a possibility.

Activiy 2

Hook up switches to the *S* inputs to the board that will connect either ground of +5 V to each input. Hook up LEDs to the *M* outputs. Manually sequence the switches through the sixteen possible states and observe the LED outputs.

EWB: See the EWB circuit solutions (part 4) for the logic simulation.

CHAPTER 6

DIGITAL SYSTEM APPLICATION

System Assignment 1: Design

Activity 1

From Table 6-11 in the text, the logic for the Light Outputs is developed as follows:

$$\overline{MR} = \overline{SO_3} + \overline{SO_4}$$
$$MR = SO_3 SO_4$$
$$\overline{MY} = \overline{SO_2}$$
$$MY = SO_2$$
$$\overline{MG} = \overline{SO_1}$$
$$MG = SO_1$$
$$\overline{SR} = \overline{SO_1} + \overline{SO_2}$$
$$SR = SO_1 SO_2$$
$$\overline{SY} = \overline{SO_4}$$
$$SY = SO_4$$
$$\overline{SG} = \overline{SO_3}$$
$$SG = SO_3$$

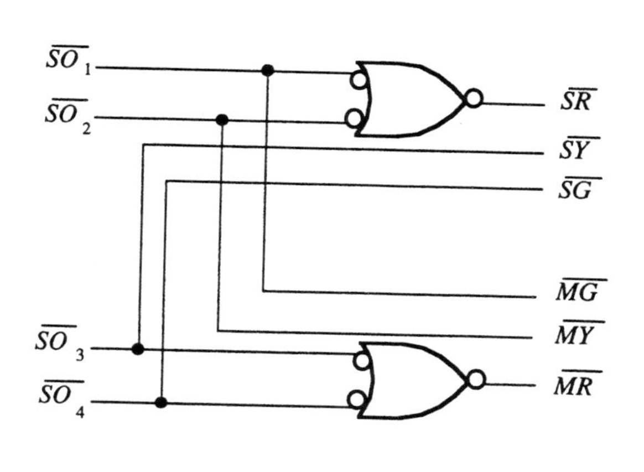

Activity 2

Using the state diagram, the trigger logic is developed as follows:

$$\overline{\text{Long Timer}} = \overline{SO_1} + \overline{SO_3}$$
$$\text{Long Timer} = SO_1 SO_3$$
$$\overline{\text{Short Timer}} = \overline{SO_2} + \overline{SO_4}$$
$$\text{Short Timer} = SO_2 + SO_4$$

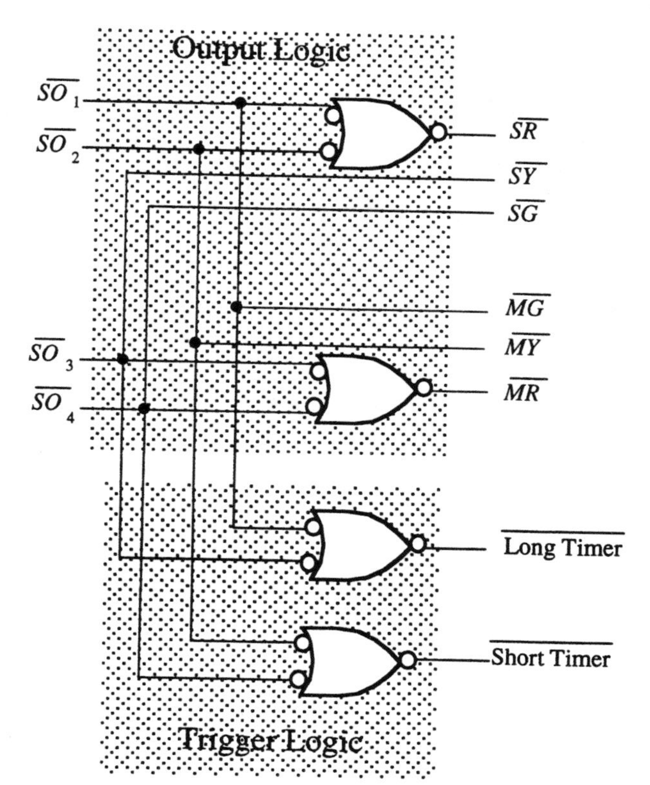

Activity 3

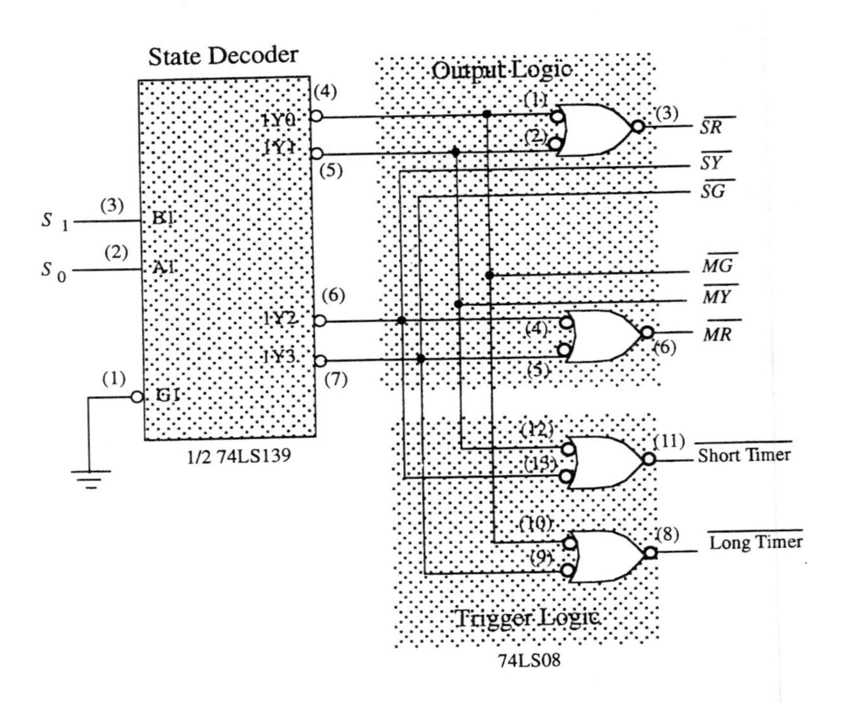

System Assignment 2: Verification and Testing

Activity 1

The logic diagram of the pc board is shown on the next page and is derived from tracing out the interconnections on the board which is shown below. The back side traces are indicated by black lines. A worksheet master of this portion of the board that can be copied for a student handout can be found in Part 3.

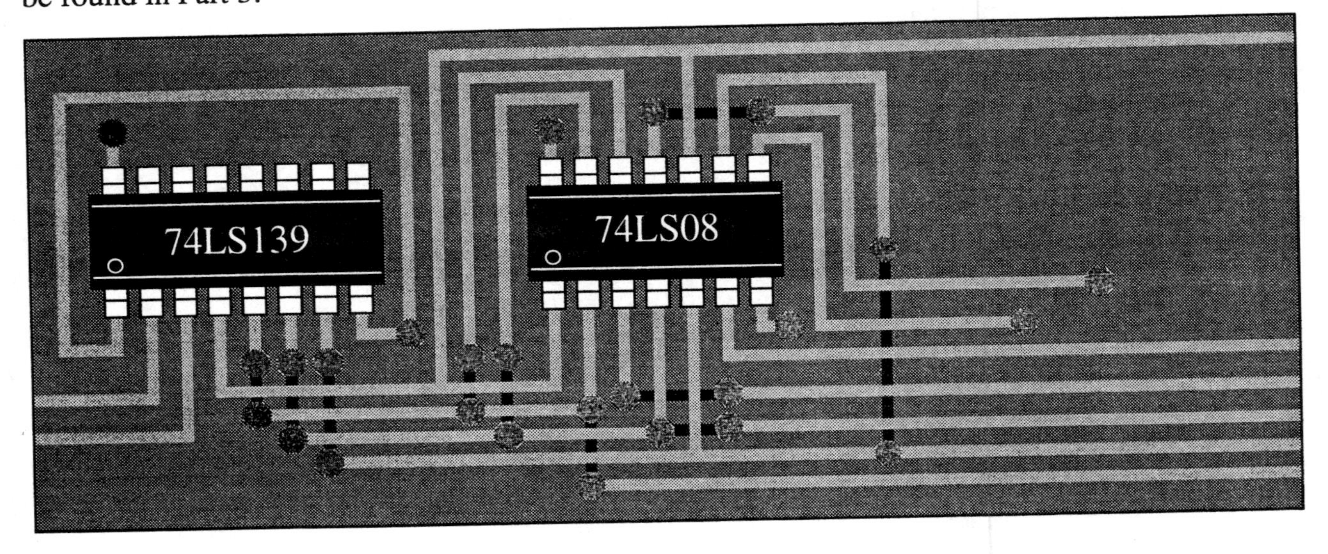

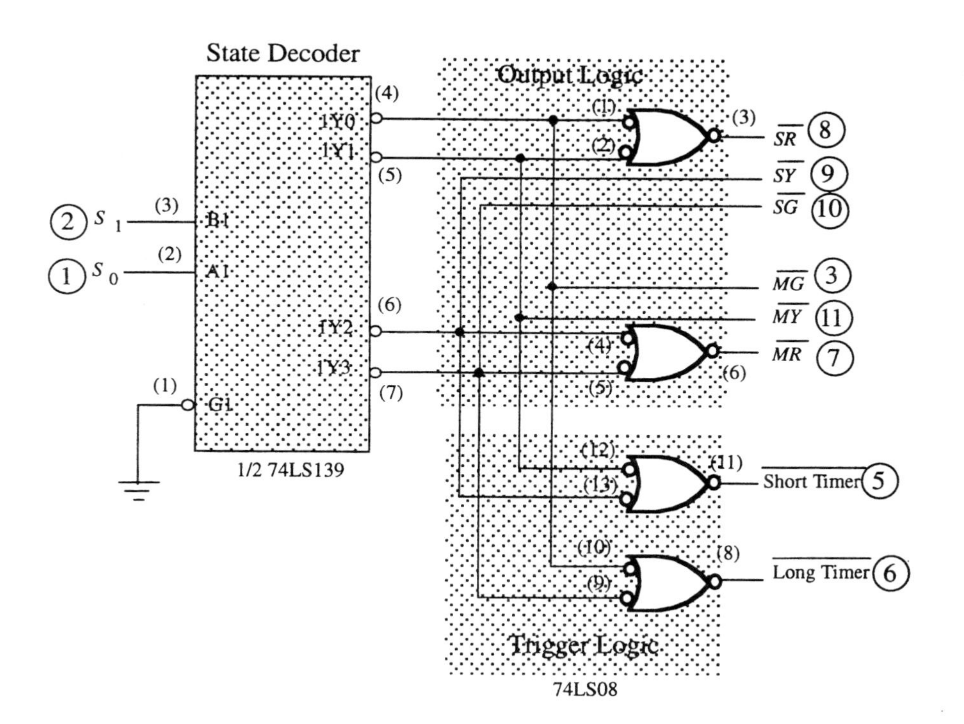

Activity 2
Each of the points on the pc board indicated with circled numbers is shown on the logic diagram in Activity 1 above.

Activity 3
The switches are connected to points 1 and 2 on the pc board as shown in the test bench. For all four possible switch combinations, each of the labeled points are monitored for a HIGH or LOW as indicated in the following table.

S_1	S_0	Gray Code	3	4	5	6	7	8	9	10	11
L	L	00	H	L	L	H	L	H	L	L	L
L	H	01	L	L	H	L	L	H	L	L	H
H	H	11	L	H	L	H	H	L	L	H	L
H	L	10	L	L	H	L	H	L	H	L	L

Activity 4

Case1: Operating properly.
Case 2: Faulty operation because point 8 should be LOW. Pin 1 input of AND gate open.
Case 3: Operating properly.
Case 4: Faulty operation because point 5 should be LOW. Pin 12 input of AND gate open.
Case 5: Operating properly.
Case 6: Faulty operation because points 4, 6, and 7 should be LOW. The 74LS139 has malfunctioned with all of its output HIGH.
Case 7: Operating properly.
Case 8: Faulty operation because point 7 should be LOW. Pin 4 input of AND gate open.

EWB: See the EWB circuit solutions (part 4) for the pc board simulation.

CHAPTER 7
DIGITAL SYSTEM APPLICATION

System Assignment 1: GAL22V10 Programming

Activity 1

The ABEL input file is

```
Module  traffic_light_controller
Title   'Traffic Light State Decoding and Output Logic'

        traffic_light       device              'P22V10';

        S0,S1                       Pin 2,3;
        SHORT,LONG                  Pin 16,17;
        !SG,!SY,!SR                 Pin 18,19,20;
        !MG,!MY,!MR                 Pin 21,.22,23;

Equations
        STATE = [S1,S0];
        MAIN_LIGHTS = [!MR,!MY,!MG];
        SIDE_LIGHTS = [!SR,!SY,!SG];
        TRIGGER = [LONG,SHORT];

Truth_table     (State -> [MAIN_LIGHTS,SIDE_LIGHTS,TRIGGER])
                  0  -> [ 6, 1, 2];
                  1  -> [ 5, 3, 1];
                  2  -> [ 3, 6, 2];
                  3  -> [ 3, 5, 1];
END
```

System Assignment 2: Verification and Testing

Activity 1

For the pc board shown, the circled pin functions are:

1. S0: LSB of the state inputs
2. S1: MSB of the state inputs
3. GND: Power supply ground
4. LONG: Long interval trigger output
5. SHORT: Short interval trigger output
6. $\overline{SG}$: Side street green
7. $\overline{SY}$: Side street yellow
8. $\overline{SR}$: Side strreet red
9. $\overline{MG}$: Main street green
10. $\overline{MY}$: Main street yellow
11. $\overline{MR}$: Main street red
12. VCC: Power supply voltage

Activity 2

To test the circuit, set the switches to the states shown and verify the following logic levels on contacts 4 -11.

S0	S1	4	5	6	7	8	9	10	11
L	L	H	L	H	H	L	L	H	H
L	H	L	H	H	H	L	H	L	H
H	L	H	L	L	H	H	H	H	L
H	H	L	H	H	L	H	H	H	L

System Assignment 1: Circuit Verification and Design

Activity 1

The portion of the Traffic Light Control board containing the state decoder/output logic and the timing circuits appears below. The timing logic diagram in addition to the state decoder/output logic obtained by tracing out the circuit board is shown on the next page. A worksheet master of this portion of the board that can be copied for a student handout can be found in Part 3.

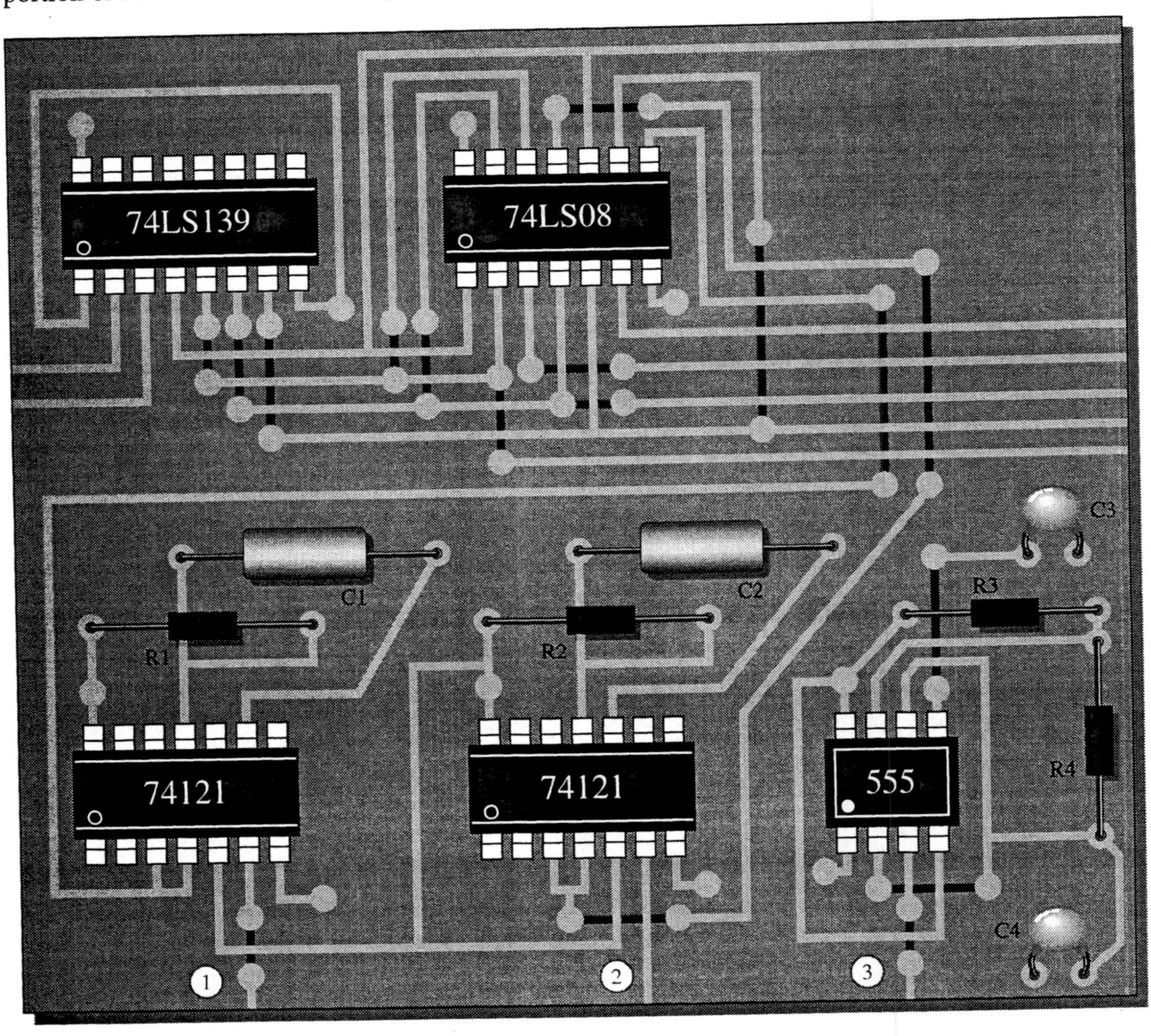

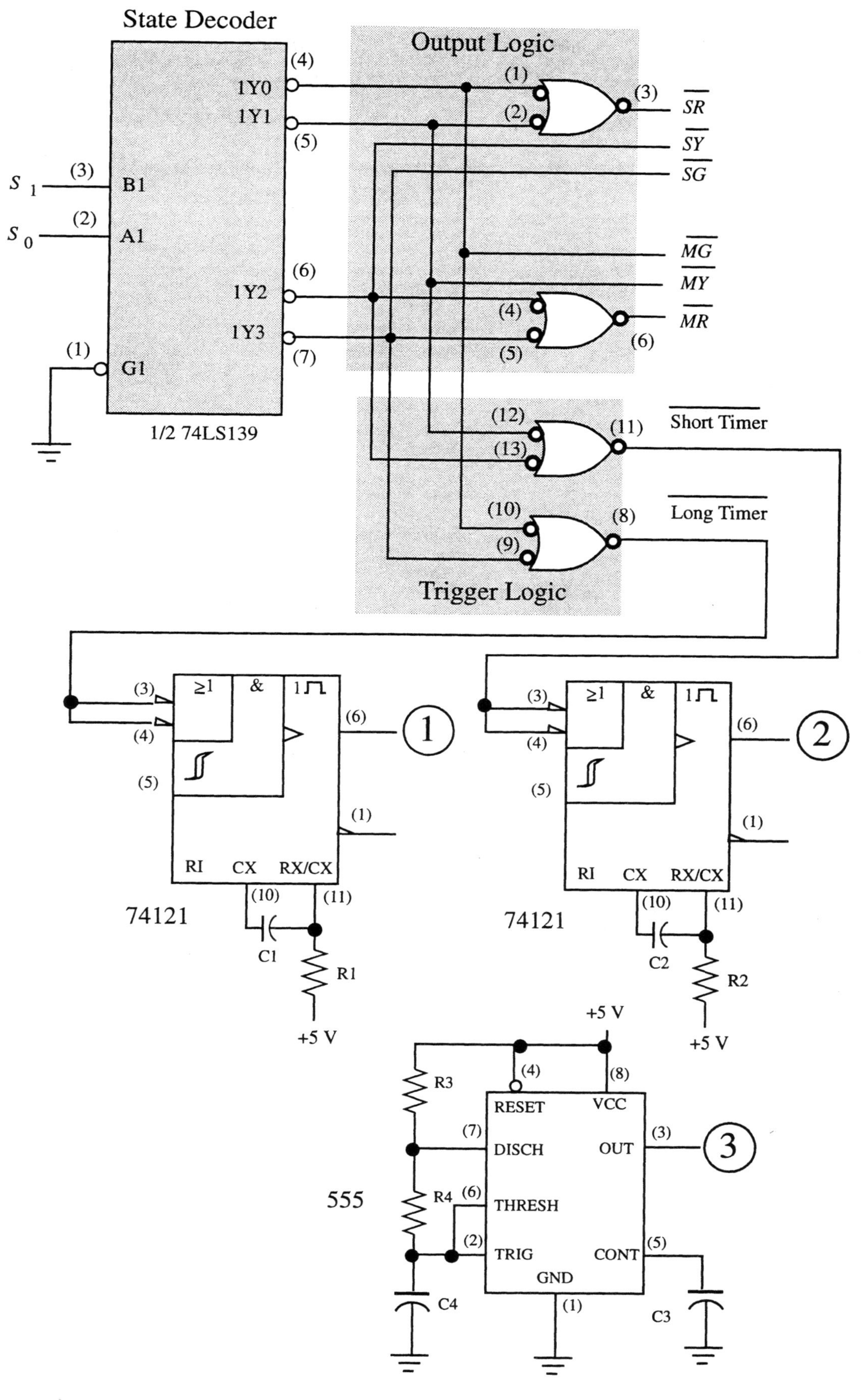

State Decoder
Output Logic
1Y0
1Y1
B1
A1
1Y2
1Y3
G1
1/2 74LS139
SR
SY
SG
MG
MY
MR
Short Timer
Long Timer
Trigger Logic
74121
RI
CX
RX/CX
C1
R1
+5 V
C2
R2
555
RESET
VCC
DISCH
OUT
THRESH
TRIG
CONT
GND
R3
R4
C4
C3

Activity 2

The timing portion of the pc board is correct and properly connected to the state decoder and output logic as determined from the logic diagram derived from tracing out the pc board.

Activity 3

The circled points are indicated on the logic diagram on the preceeding page and their functions are as follows:

1. Long timer output (25 s pulse)
2. Short timer output (4 s pulse)
3. Oscillator output (10 kHz clock)

Activity 4

Short Timer:

Let $C2 = 100\ \mu F$

$4\ s = 0.7R2C2$

$$R2 = \frac{4\ s}{(0.7)(100\ \mu F)} = 57.1\ k\Omega \quad \text{(use 56 k}\Omega\text{)}$$

Long Timer:

Let $C1 = 100\ \mu F$

$4\ s = 0.7R1C1$

$$R1 = \frac{25\ s}{(0.7)(100\ \mu F)} = 357\ k\Omega \quad \text{(use 360 k}\Omega\text{)}$$

Oscillator:

Let $C4 = 0.01\ \mu F$

$$10\ kHz = \frac{1.44}{(R3 + 2R4)C4}$$

$$R3 + 2R4 = \frac{1.44}{(10\ kHz)(0.1\ \mu F)} = 14.4\ k\Omega$$

For $R3 = R4$

$3R3 = 14.4\ k\Omega$

$R3 = R4 = 4.8\ k\Omega$ (use 5.1 kΩ)

System Assignment 2: Testing and Troubleshooting

Activity 1

The switches are set to each of the four possible combinations representing a gray code sequence and the outputs 1, 2, and 3 are observed in accordance with the following table.

S_1	S_0	Gray Code	1	2	3
L	L	00	H (25 s)	L	10 kHz
L	H	01	L	H (4 S)	10 kHz
H	H	11	H (25 s)	L	10 kHz
H	L	10	L	H (4 s)	10 kHz

Activity 2

Case 1: Operating properly.
Case 2: The long and short timer outputs are reversed. This is possibly due to incorrect installation of theexternal resistor and capacitor networks on the one-shots.
Case 3: The short timer (4 s) output is missing. The one-shot is not being triggered due to either a faulty gate in the trigger output logic, an open interconnection, or a malfuction in the one-shot.
Case 4: Nothing seems to be working. The most likely problem is a low of power to the circuit board.

EWB: See the EWB circuit solutions (part 4) for the pc board simulation.

CHAPTER 9
DIGITAL SYSTEM APPLICATION

System Assignment 1: Design

Activity 1

The completed table is as follows:

Present		Next		Input Conditions	FF Inputs	
Q_1	Q_0	Q_1	Q_0		D_1	D_0
0	0	0	0	$T_L + \overline{V}_S$	0	0
0	0	0	1	$\overline{T}_L V_S$	0	1
0	1	0	1	T_S	0	1
0	1	1	1	$\overline{T}_S$	1	1
1	1	1	1	$T_L V_S$	1	1
1	1	1	0	$\overline{T}_L + \overline{V}_S$	1	0
1	0	1	0	T_S	1	0
1	0	0	0	$\overline{T}_S$	0	0

The expression for D_0 is developed as follows:

From the table, the expression for D_0 is

$$D_0 = \overline{Q}_1\overline{Q}_0\overline{T}_L V_S + \overline{Q}_1 Q_0 + Q_1 Q_0 T_L V_S$$

Use the K-map to simplify this expression:

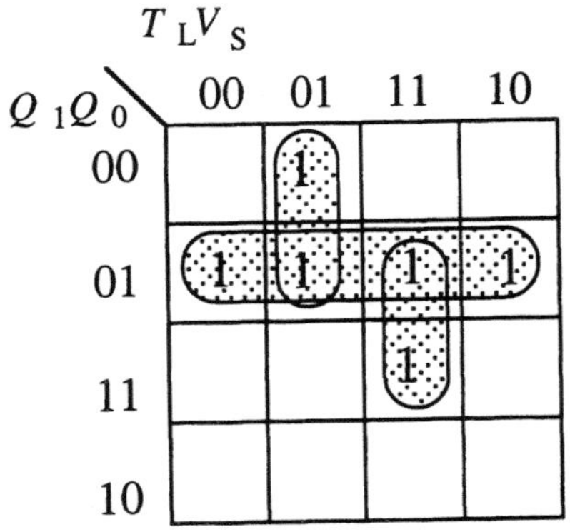

Q_1Q_0 \ T_LV_S	00	01	11	10
00		1		
01	1	1	1	1
11			1	
10				

$$D_0 = \overline{Q}_1\overline{T}_L V_S + \overline{Q}_1 Q_0 + Q_0 T_L V_S$$

Activity 2

The expression for D_1 is developed as follows:

From the table, the expression for D_1 is

$D_1 = \overline{Q}_1\overline{Q}_0\overline{T}_S + Q_1Q_0 + Q_1\overline{Q}_0T_S$

Use the K-map to simplify this expression:

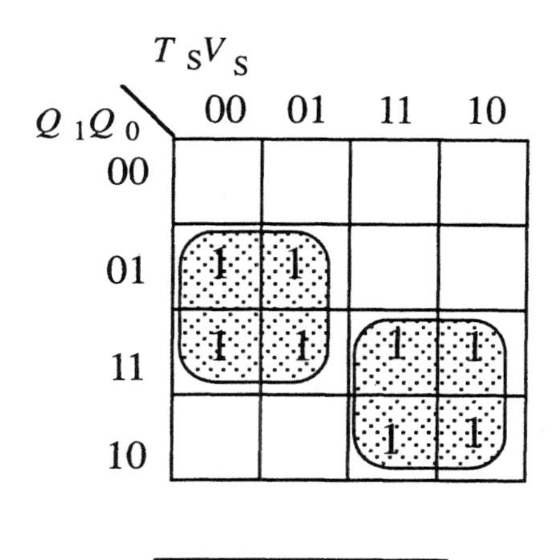

$D_1 = Q_0\overline{T}_S + Q_1T_S$

Activity 3

The sequential logic section of the traffic light control system is implemented as shown below

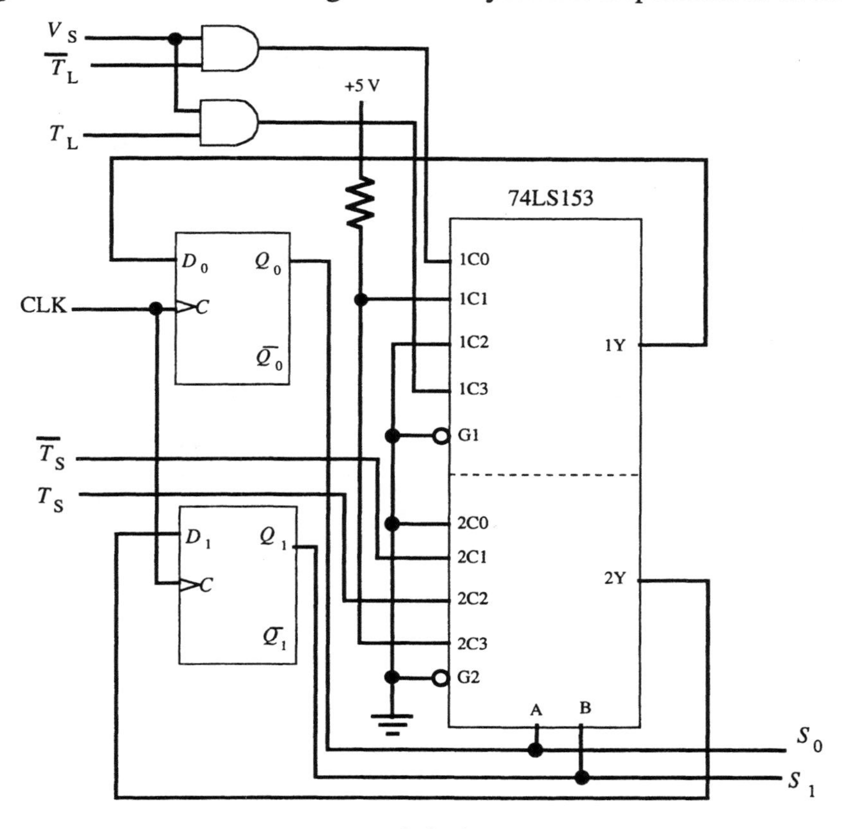

System Assignment 2: Verification

Activity 1

The complete system board is shown below. A worksheet master of the traffic control system board that can be copied for student handouts can be found in Part 3.

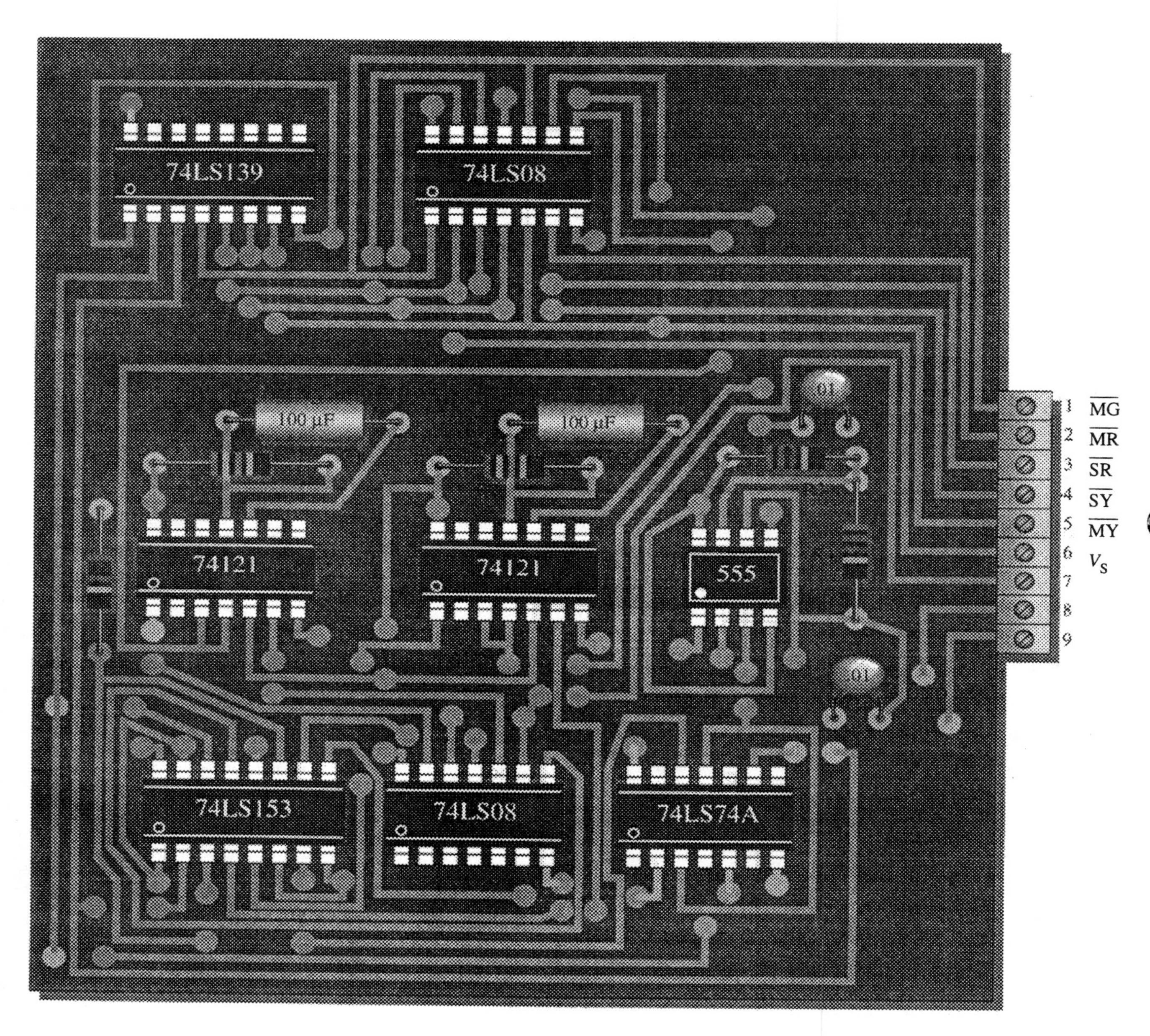

+5 V and GND connections not shown

The diagram of the sequential logic determined by tracing the board is shown below.

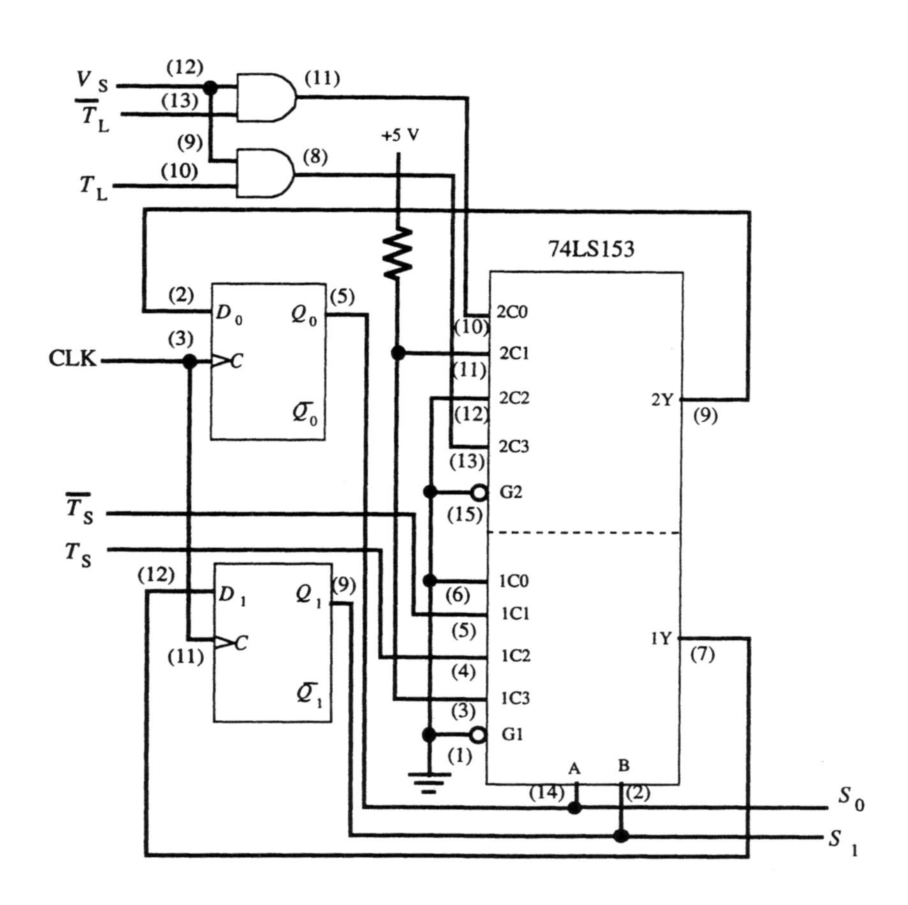

The logic diagram of the entire traffic control board is shown below. The circled numbers represent the input and output pins on the board connector.

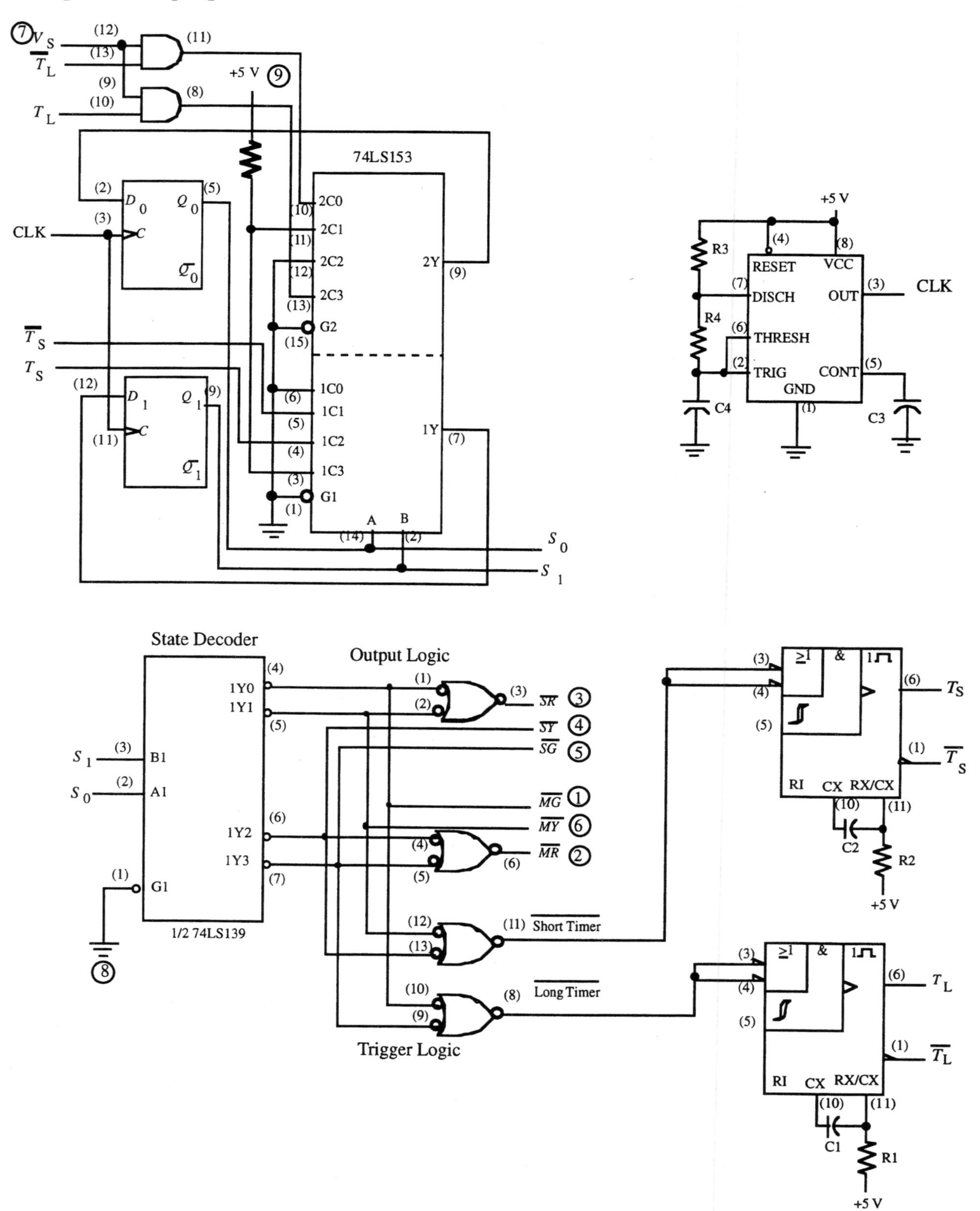

Activity 2
The circuit on the system board is functionally equivalent to the logic diagram developed in Test Bench 1.

Activity 3
The pin numbers on the circuit board connector are indicated on the system logic diagram in Activity 1. The inputs and outputs are as follows:

Pin 1: $\overline{MG}$ (main green)
Pin 2: $\overline{MR}$ (main red)
Pin 3: $\overline{SR}$ (side red)
Pin 4: $\overline{SY}$ (side yellow)
Pin 5: $\overline{SG}$ (side green)
Pin 6: $\overline{MY}$ (main yellow)
Pin 7: V_S (vehicle sensor)
Pin 8: Ground
Pin 9: +5 V

System Assignment 3: Troubleshooting

Activity 1
A test procedure would be to simulate the traffic lights with LEDs and toggle the vehicle sensor switch to simulate cars arriving at the intersection on the side street. If the lights sequence as expected and are on for the correct durations, the circuit is operating properly.

Activity 2

Case 1: Continuous red on side street with the vehicle sensor active. This malfunction could be caused by any one of the following faults: A short to ground on the line from the vehicle sensor input, no clock pulses to the sequential logic (failure of the 555 timer) and the flip-flops stuck in state 00.
Case 2: Continuous red on side street with the vehicle sensor not active is proper operation.
Case 3: The system is missing the yellow caution light indicating a failure in the 4 s one-shot.
Case 4: Each state is lasting only 4 s, indicating a failure of the 25 s one-shot.
Case 5: The states are sequencing with the correct timing but the yellow lights are not coming on. The only possible cause would be that both the MY and SY lines are open perhaps due to faulty connections at the board connector.

EWB: See the EWB circuit solutions (part 4) for the pc board simulation.

CHAPTER 10

DIGITAL SY STEM APPLICATION

System Assignment 1: Analysis

Activity 1: The Decimal-to- BCD Encoder

(a) The decimal -to-BCD encoder converts a keypress to a BCD code.

(b) When a key is pressed, the corresponding encoder input is activated.

(c) When a key is pressed, the BCD code corresponding to the pressed key appears on the encoder outputs.

Activity 2: The Timing Circuits

(a) The gate G_1 produces a trigger pulse to one-shot A when any key is pressed.

(b) As stated in part (a), one-shot A is triggered when any key is pressed and gate G_1 produces a trigger pulse.

(c) One-shot A clocks the BCD code into Shift Registger A and advances the memory address counter. One-shot B generates a delayed trigger to one-shot C. One-shot C advances Shift Register C that disarms the system when the correct code is entered

Activity 3: Shift Registers A and B

(a) Shift Register A stores the most recently entered BCD digit. Shift Register B stores the preprogrammed entry code from the memory.

(b) Shift Register A and Shift Register B each are 4-bit registers.

(c) When a key is pressed, the decimal-to-BCD encoder applies a 4-bit BCD code to the parallel inputs of Shift Register A and one-shot A is triggered via gate G_1 and provides a clock pulse to the shift register to load the 4-bit BCD code.

Activity 4: Comparator

(a) The comparator determines whether each entered code digit is valid or not.

(b) Input bit Set A is the entered code and set B is the programmed code from the memory and stored in Shift Register B.

(c) When the inputs to the comparator are the same, the A=B output goes HIGH.

Activity 5: Shift Register C

(a) Shift Register C controls arming and disarming of the security system.

(b) 1s are initially stored in the left-most positions in the shift register with 0s in the other positions. A 1 is shifted right each time a correct code digit is entered. After the fourth correct digit is entered the 1 is shifted onto the serial output line.

(c) When the serial output is LOW, the system is armed. When the serial output goes HIGH, the system is disarmed.

(d) Each time a correct digit is entered, A 1 in the shift register shifts right one place.

(e) If an incorrect code digit is entered, gate G_2 puts the shift register into the parallel load mode and the pulse from one-shot C clocks the shift register and reinitialize it to 11110000. If a correct code digit is entered, gate G_2 puts the shift register into the serial shift mode so the pulse from one-shot C clocks the shift register and shifts the1 to the right.

System Assignment 2: Design

Activity 1 BCD-to-Decimal Encoder

The 74LS147 can be used for this application.

Activity 2 The Timing Circuits

(a) The 74LS133 13-input NAND gate can be used for gate G_1.

(b) 74121 one-shots can be used for one-shots A, B, and C. The external resistor and capacitor for each one-shot are calculated as follows. Each one-shot has the same values.

$$t_W = 0.7 R_{EXT} C_{EXT}$$

Choose $C_{EXT} = 1\ \mu F$

$$R_{EXT} = \frac{t_W}{0.7 C_{EXT}} = \frac{1\text{ ms}}{(0.7)(1\ \mu F)} = 1.428\text{ k}\Omega \quad (\text{use } 1.5\text{ k}\Omega)$$

Activity 3 Shift Registers A and B

The 74LS195A can be used in this application.

Activity 4 Comparator

The 74LS85 can be used in this application.

Activity 5 Shift Register C

The 74LS165 can be used in this application.

Activity 6 Other Logic

The 74LS08 can be used for the AND and tthe 74LS74A can be used for the flip-flop function.

Activity 7

A logic diagram representing one approach is shown on the next page. Other designs are possible.

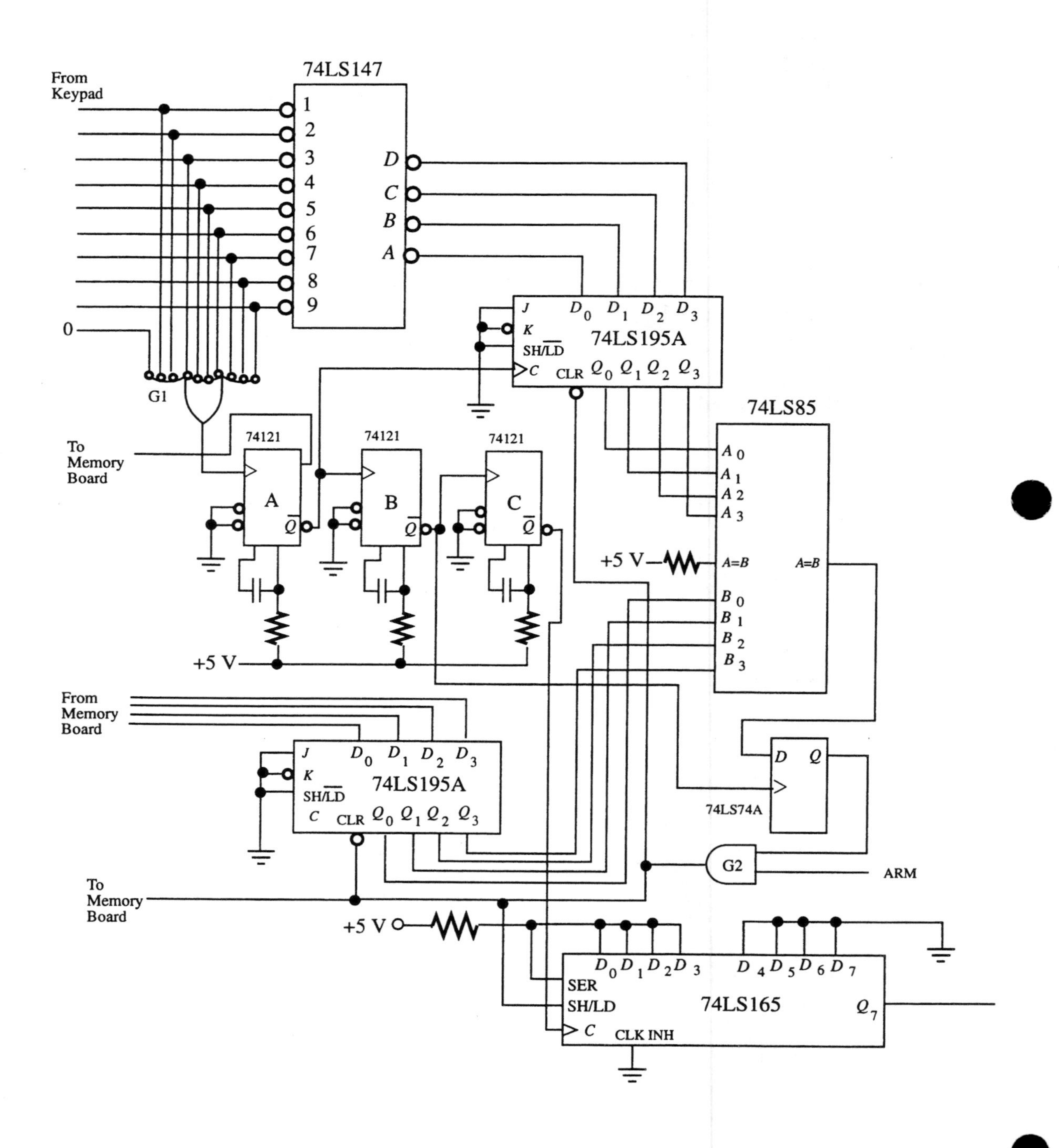

74LS147
From
Keypad
1
2
3
4
5
6
7
8
9
D
C
B
A
0
G1
To
Memory
Board
74121
A
B
C
J
K
SH/LD
C
CLR
D_0 D_1 D_2 D_3
74LS195A
Q_0 Q_1 Q_2 Q_3
74LS85
A_0
A_1
A_2
A_3
+5 V
A=B
B_0
B_1
B_2
B_3
From
Memory
Board
74LS74A
D
Q
G2
ARM
SER
SH/LD
CLK INH
D_4 D_5 D_6 D_7
74LS165
Q_7

System Assignment 3: Verification and Testing

Activity 1

The code entry board is shown below. The logic diagram obtained by tracing the board is shown on the next page. It compares with the logic diagram developed in System Assignment 2. A worksheet master for the code entry board that can be copied for a student handout can be found in Part 3.

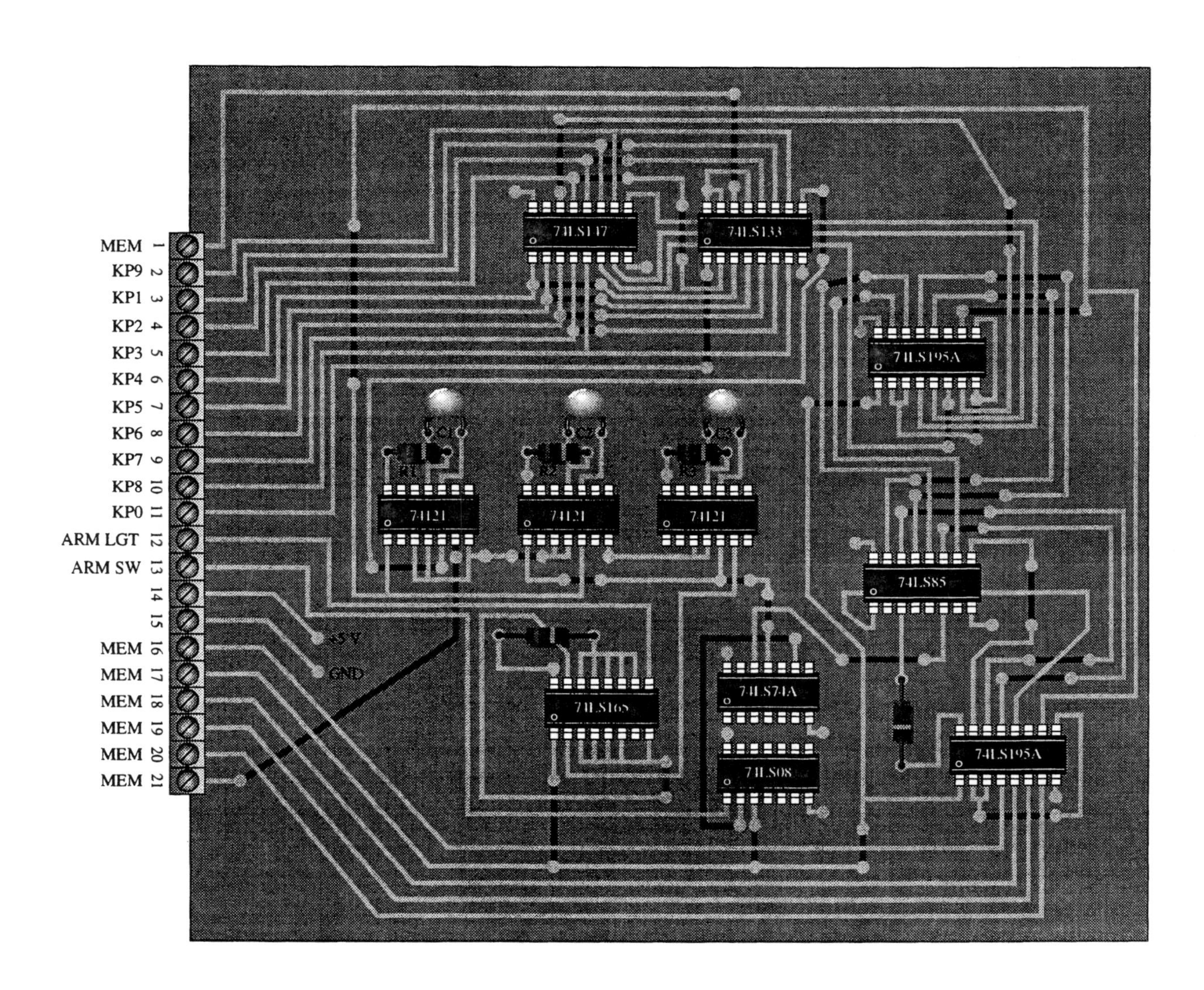

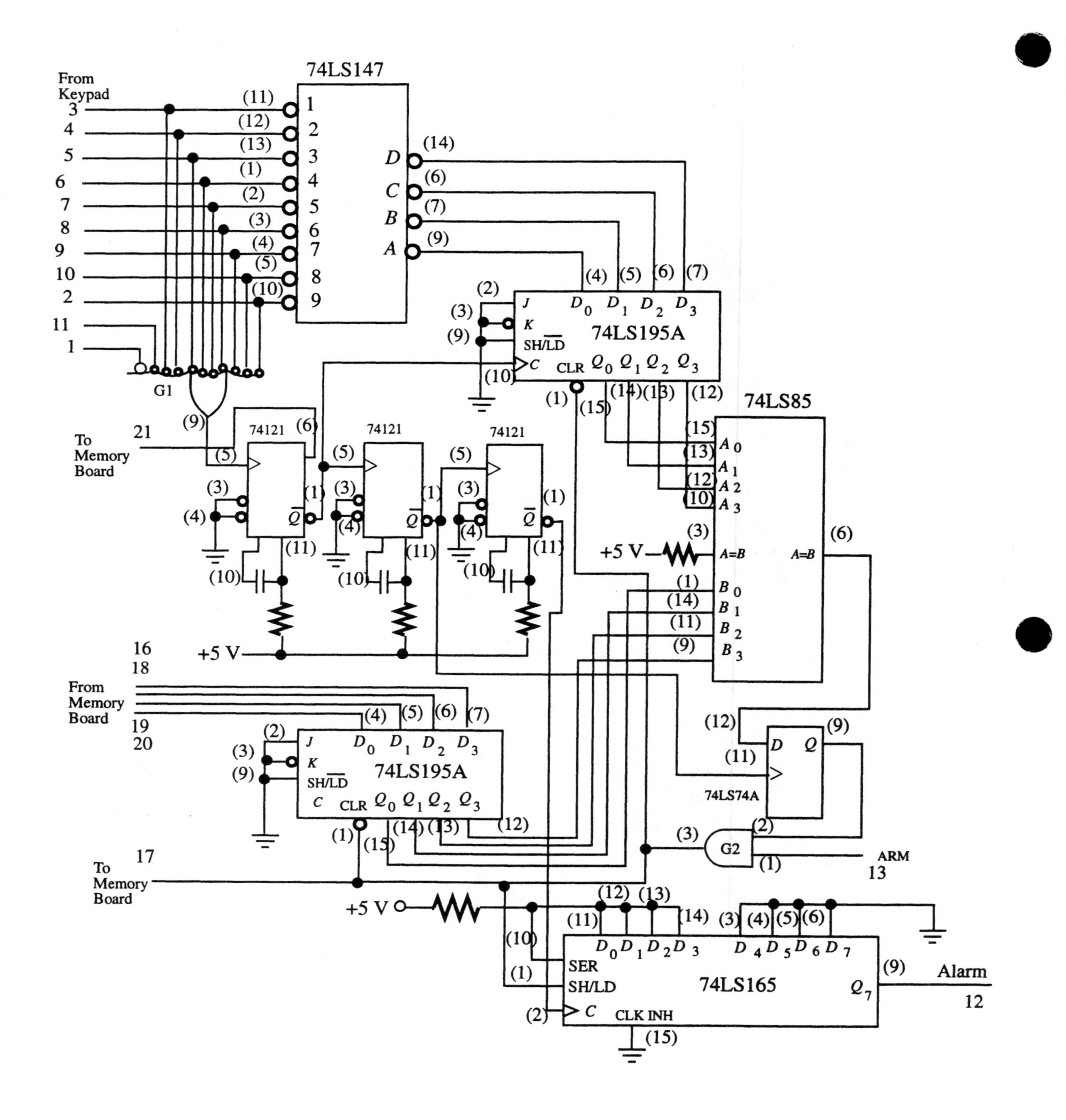
74LS147
From
Keypad
G1
74LS195A
74LS85
74121
74121
74121
To
Memory
Board
+5 V
From
Memory
Board
74LS195A
74LS74A
G2
ARM
To
Memory
Board
74LS165
SER
SH/LD
CLK INH
Alarm

Activity 2

The inputs and outputs are indicated on the logic diagram on the previous page by numbers which correspond to the board connector terminals. The following is a list of the inputs and outputs:

1 "0" key input
2 "9" key input
3 "1" key input
4 "2" key input
5 "3" key input
6 "4" key input
7 "5" key input
8 "6" key input
9 "7" key input
10 "8" key input
11 Program mode inhibit input
12 Sensor/alarm interface and ARMED lamp output
13 ARM switch input
14 + 5 V
15 Ground
16 B0 input from memory
17 Reset output to memory address counter
18 B3 input from memory
19 B2 input from memory
20 B1 input from memory
21 Clock pulse to memory address counter

Activity 3

A test procedure for the board would be to apply inputs to the 74LS147 encoder and trace the signals through the circuit to the one-shots to verify they are trigggering properly. Next the pulse pulses should be traced to the 74LS195A register to verify that the code bits are being cloked through to the 74LS85 comparator. Codes should be applied to the memory inputs to the comparator to verify the comparator is working both when key entry matches the memory code and when it does not. Finally the AND gate, flip-flop, and 74LS165 register should be checked for proper operation by verifying that a 1 is shifted right each time the codes are the same and that reloading occurs each time the codes do not compare.

CHAPTER 11
DIGITAL SY STEM APPLICATION

System Assignment 1: GAL22V10 Programming

Activity 1

From the state diaram of Figure 11-21 and the pin assignment diagram, anABEL input file for the sequential portion of the traffic light contol system is:

```
Module  Traffic_light_controller
Title               'Traffic light controller sequential logic'

"Device declaration
        traffic_seq        DEVICE          'P22V10

"Pin declarations
        CLK                     PIN 1;
        TS,TL,VS                PIN 4,5,6;
        S1_OUT,S0_OUT           PIN 14,15  ISTYPE 'reg,buffer';

        STATE = [S1_OUT,S0_OUT];
        STATE_1 = ^B00;
        STATE_2 = ^B01;
        STATE_3 = ^B11;
        STATE_4 = ^B10;

Equations
        STATE.CLK = Clock;

State_diagram STATE
        STATE_1 : if (!TL&VS) then STATE_2 else STATE_1;
        STATE_2 : if (TS) then STATE_3 else STATE_2;
        STATE_3 : if (!TL# !VS) then STATE_4 else STATE_3;
        STATE_4 : if (!TS) then STATE_1 else STATE_4;
END
```

Activity 2

A complete ABEL file implementing the combinational logic from Chapter 7 and the sequential logic portion from activity 1 is as follows (both truth table and state diagram are shown):

```
Module  Traffic_light_controller
Title               'Traffic light controller logic'

"Device declaration
        traffic_seq         DEVICE          'P22V10

"Pin declarations
        CLK                         PIN 1;
        S0,S1                       PIN 2,3;
        TS,TL,VS                    PIN 4,5,6;
        S1_OUT,S0_OUT               PIN 14,15  ISTYPE 'reg,buffer';
        SHORT,LONG                  PIN 16,17  ISTYPE 'com,buffer';
        !SG,!SY,!SR                 PIN 18,19,20  ISTYPE 'com,buffer';
        !MG,!MY,!MR                 PIN 21,22,23  ISTYPE 'com,buffer';

        STATE = [S1,S0];
        MAIN = [!MR,!MY,!MG];
        SIDE = [!SR,!SY,!SG];
        TRIGGER = [LONG,SHORT];
        STATE_OUT = [S1_OUT,S0_OUT];
        STATE_1 = ^B00;
        STATE_2 = ^B01;
        STATE_3 = ^B11;
        STATE_4 = ^B10;

Equations
        STATE_OUT.CLK = Clock;

Truth_table         (STATE     -> [MAIN,SIDE,TRIGGER])
                     STATE_1->      1,     4,        2;
                     STATE_2->      2,     4,        1;
                     STATE_3->      4,     2,        1;
                     STATE_4->      4,     1,        2;

State_diagram STATE_OUT
        STATE_1 : if (!TL&VS) then STATE_2 else STATE_1;
        STATE_2 : if (TS) then STATE_3 else STATE_2;
        STATE_3 : if (!TL # !VS) then STATE_4 else STATE_3;
        STATE_4 : if (!TS) then STATE_1 else STATE_4;
END
```

System Assignment 2: Verification and Testing

Activity 1

```
Test_vectors
([Clock,TL,TS,VS,STATE] -> [STATE_OUT,MAIN,SIDE,TRIGGER])
[ .c.,  1,   0,  0,   .x.  ] -> [  .x.,  .x.,  .x.,  .x.  ];
[ .c.,  1,   0,  0,   .x.  ] -> [  .x.,  .x.,  .x.,  .x.  ];
[ .c.,  1,   0,  0,   .x.  ] -> [  .x.,  .x.,  .x.,  .x.  ];
[ .c.,  1,  .x., .x.,  0   ] -> [  0 ,   1 ,   4 ,   2   ];
[ .c., .x., .x.,  0,   0   ] -> [  0 ,   1 ,   4 ,   2   ];
[ .c.,  0,  .x.,  1,   1   ] -> [  1 ,   2 ,   4 ,   1   ];
[ .c., .x.,  1,  .x.,  1   ] -> [  1 ,   2 ,   4 ,   1   ];
[ .c., .x.,  0,  .x.,  3   ] -> [  3 ,   4 ,   2 ,   1   ];
[ .c.,  1,  .x.,  1,   3   ] -> [  3 ,   4 ,   2 ,   1   ];
[ .c.,  0,  .x., .x.,  2   ] -> [  2 ,   4 ,   1 ,   2   ];
[ .c., .x.,  1,  .x.,  2   ] -> [  2 ,   4 ,   1 ,   2   ];
[ .c., .x.,  0,  .x.,  0   ] -> [  0 ,   1 ,   4 ,   2   ];
[ .c.,  0 , .x.,  1 ,  1   ] -> [  1 ,   2 ,   4 ,   1   ];
[ .c., .x.,  0,  .x.,  3   ] -> [  3 ,   4 ,   2 ,   1   ];
[ .c., .x., .x.,  0 ,  2   ] -> [  2 ,   4 ,   2 ,   1   ];
[ .c., .x.,  0,  .x.,  0   ] -> [  0 ,   1 ,   4 ,   2   ];
```

Activity 2

The input and output functions are:

1. VCC
2. Main Red
3. Main Yellow
4. Main Green
5. Side Red
6. Side Yellow
7. Side Green
8. Ground
9. Vehicle Sensor

Activity 3

The board logic diagram is shown on the next page.

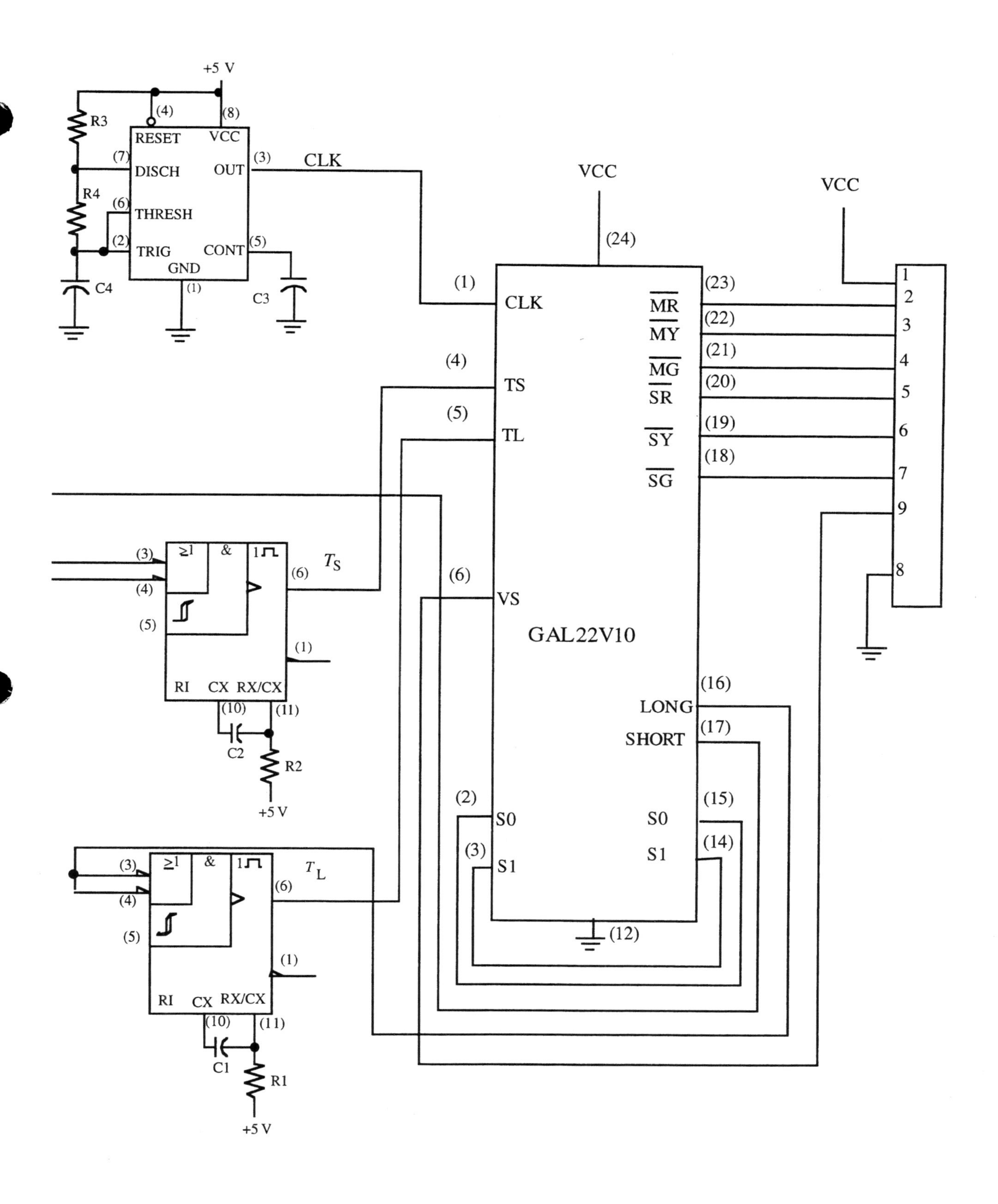

+5 V
R3
(4)
(8)
RESET
VCC
(7)
DISCH
OUT
(3)
CLK
R4
(6)
THRESH
(2)
TRIG
CONT
(5)
GND
C4
(1)
C3
VCC
(24)
VCC
(1)
CLK
(23)
MR
(22)
MY
(21)
MG
(20)
SR
(19)
SY
(18)
SG
1
2
3
4
5
6
7
9
8
(4)
TS
(5)
TL
(3)
≥1
&
1
(6)
T_S
(4)
(5)
(6)
VS
GAL22V10
(1)
RI
CX
RX/CX
(10)
(11)
C2
R2
+5 V
(16)
LONG
(17)
SHORT
(2)
S0
(15)
S0
(3)
S1
(14)
S1
(3)
≥1
&
1
T_L
(6)
(4)
(5)
(12)
(1)
RI
CX
RX/CX
(10)
(11)
C1
R1
+5 V

CHAPTER 12
DIGITAL SYSTEM APPLICATION

System Assignment 1: Design

Activity 1
The 74LS147 decimal-to-BCD encoder can be used for this application.

Activity 2
The 74189 16X4 RAM can be used for this application.

Activity 3
The 74LS161A synchonous 4-bit binary counter can be used for this application.

Activity 4
The 74121 one-shot can be used for this application. The calculations for the external components are as follows:

$t_W = 0.7R_{EXT}C_{EXT}$

Select $C_{EXT} = 0.22\ \mu F$

One-shot A

$$R_{EXT} = \frac{10\ \text{ms}}{(0.7)(0.22\ \mu F)} = 64.94\ \text{k}\Omega\ \ (\text{use } 68\ \text{k}\Omega)$$

One-shots B and C

$$R_{EXT} = \frac{1\ \text{ms}}{(0.7)(0.22\ \mu F)} = 6.49\ \text{k}\Omega\ \ (\text{use } 6.8\ \text{k}\Omega)$$

Activity 5
The 74LS00 quad 2-input NAND gate can be used for this application. An additional requirment for a 3-input NOR gate that is not indicated on the basic block diagram but which will become apparent during the design.

Activity 6
A logic diagram is shown on the next page. This represents one approach. Other designs are possible.

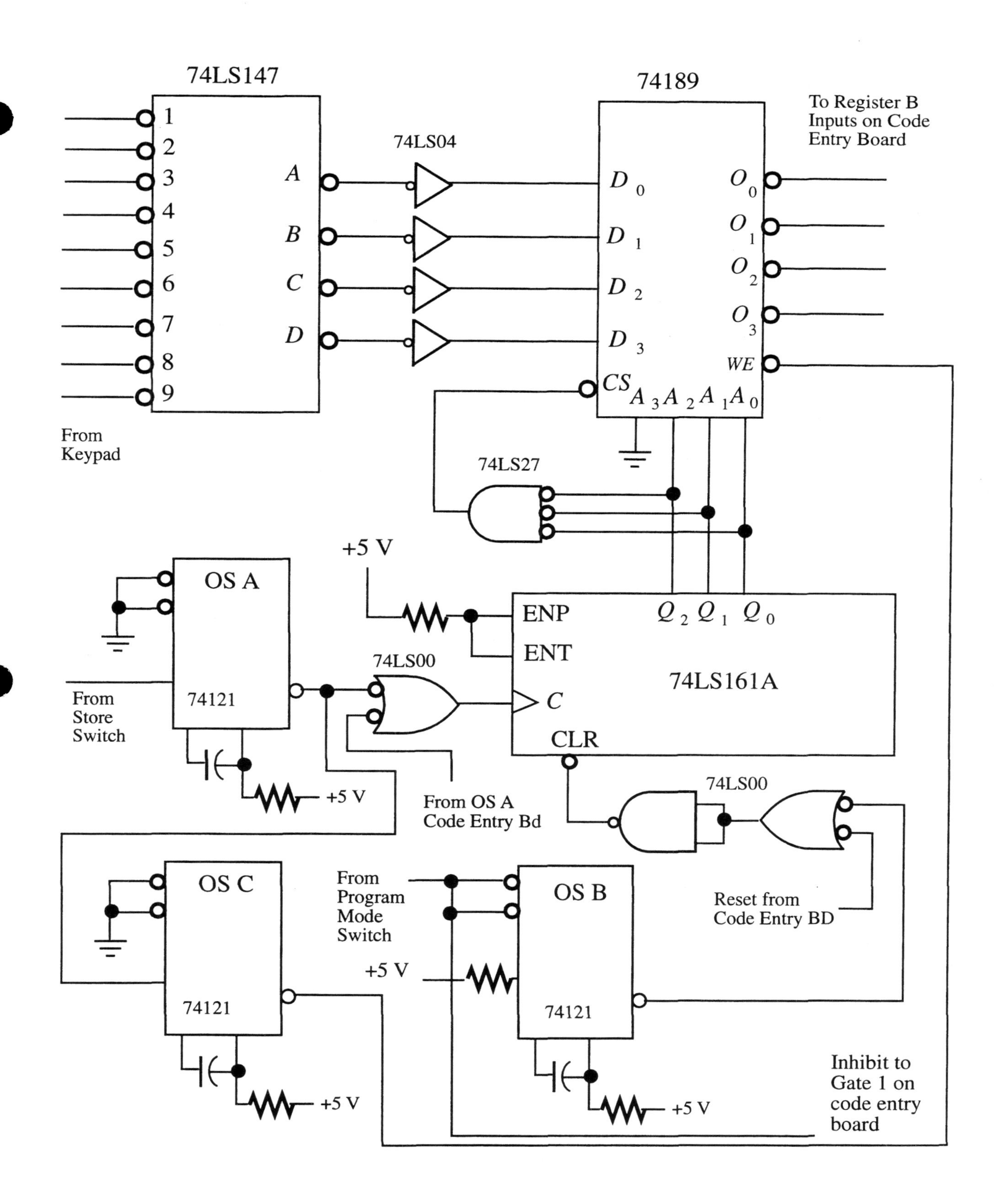
74LS147
74189
To Register B Inputs on Code Entry Board
74LS04
1
2
3
4
5
6
7
8
9
A
B
C
D
D 0
D 1
D 2
D 3
O 0
O 1
O 2
O 3
WE
CS
A 3 A 2 A 1 A 0
From Keypad
74LS27
+5 V
OS A
ENP
ENT
Q 2 Q 1 Q 0
74LS00
74LS161A
From Store Switch
74121
C
CLR
+5 V
From OS A Code Entry Bd
74LS00
OS C
From Program Mode Switch
OS B
Reset from Code Entry BD
+5 V
74121
74121
+5 V
+5 V
Inhibit to Gate 1 on code entry board

Activity 7

The following is a brief description of operation describing the function of each major component. You may wish to instruct the student to add more detail.

The 74LS147 encoder converts each key closure to the complement of the corresponding BCD code. The 74LS04 inverts the BCD bits before they are applied to the 74189 RAM. The 74LS161A counter is initially reset when the PROGRAM MODE switch is closed or by a reset pulse from the code entry board when an incorrect digit is entered. To store a code digit, the PROGRAM MODE switch is closed and the selected key on the key pad is pressed and held while the STORE switch is momentarily closed. One-shot A is triggered by the STORE SWITCH and its output clocks the counter and also triggers one-shot C to put the memory in the WRITE mode. The 74LS27 negative-AND gate decodes the 000 state of the counter and prevents the first memory address which contains all zeros from being used as a code address. Code digits are stored in addresses 1 through 5.

The parts list is as follows:

1 74LS147
1 74189
1 74LS161A
3 74121
1 74LS00
1 74LS27
1 74LS04
2 68 kΩ resistors
4 6.8 kΩ resistor
3 0.22 μF capacitors

System Assignment 2: Verification and Testing

Activity 1

The memory board is shown below and the logic diagram derived from tracing out the board is shown on the next page. The circled numbers indicated connector terminals numbers for inputs and outputs. The diagram compares with the one developed in System Assignment 1. A worksheet master for the memory board that can be copied for a student handout can be found in Part 3.

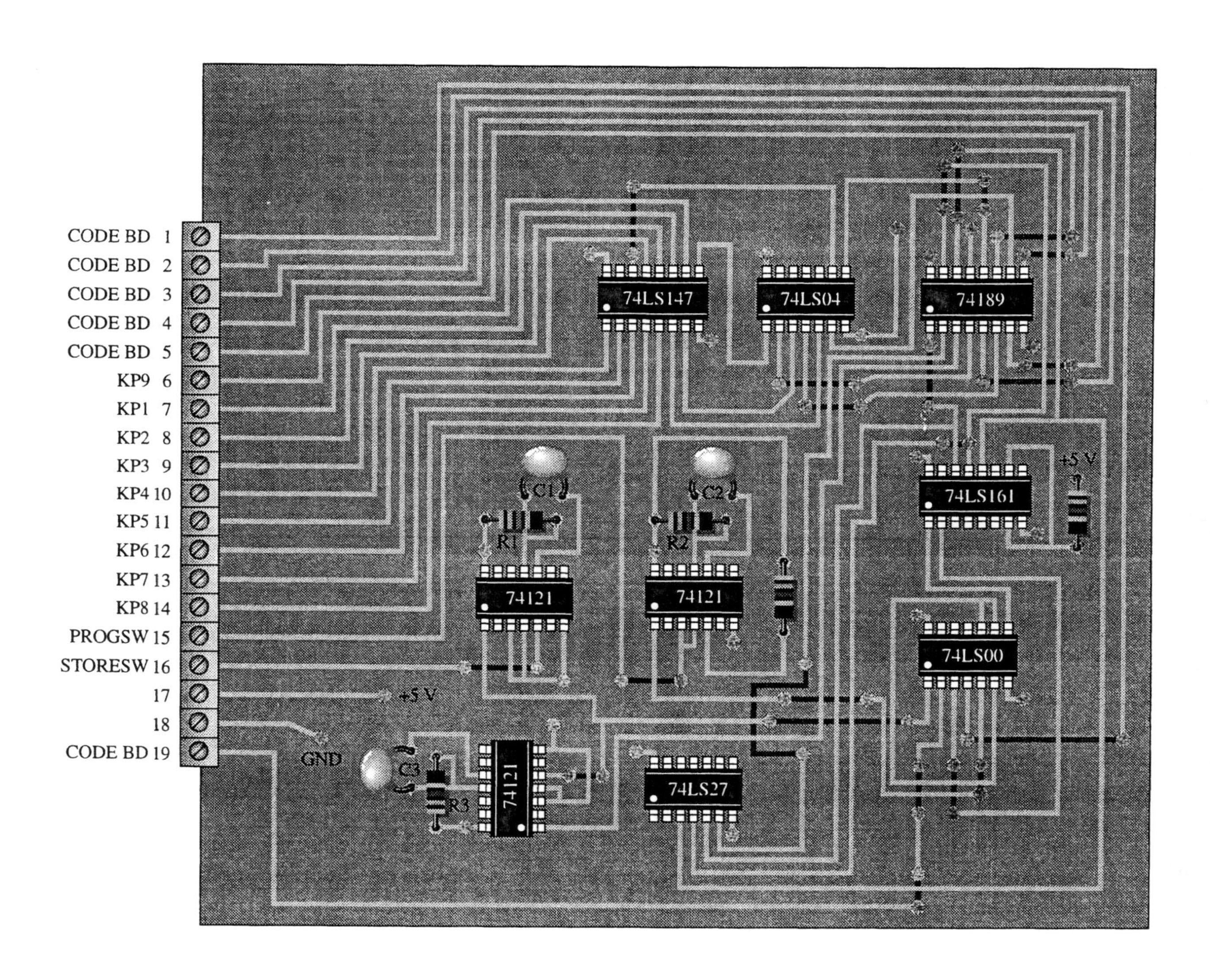

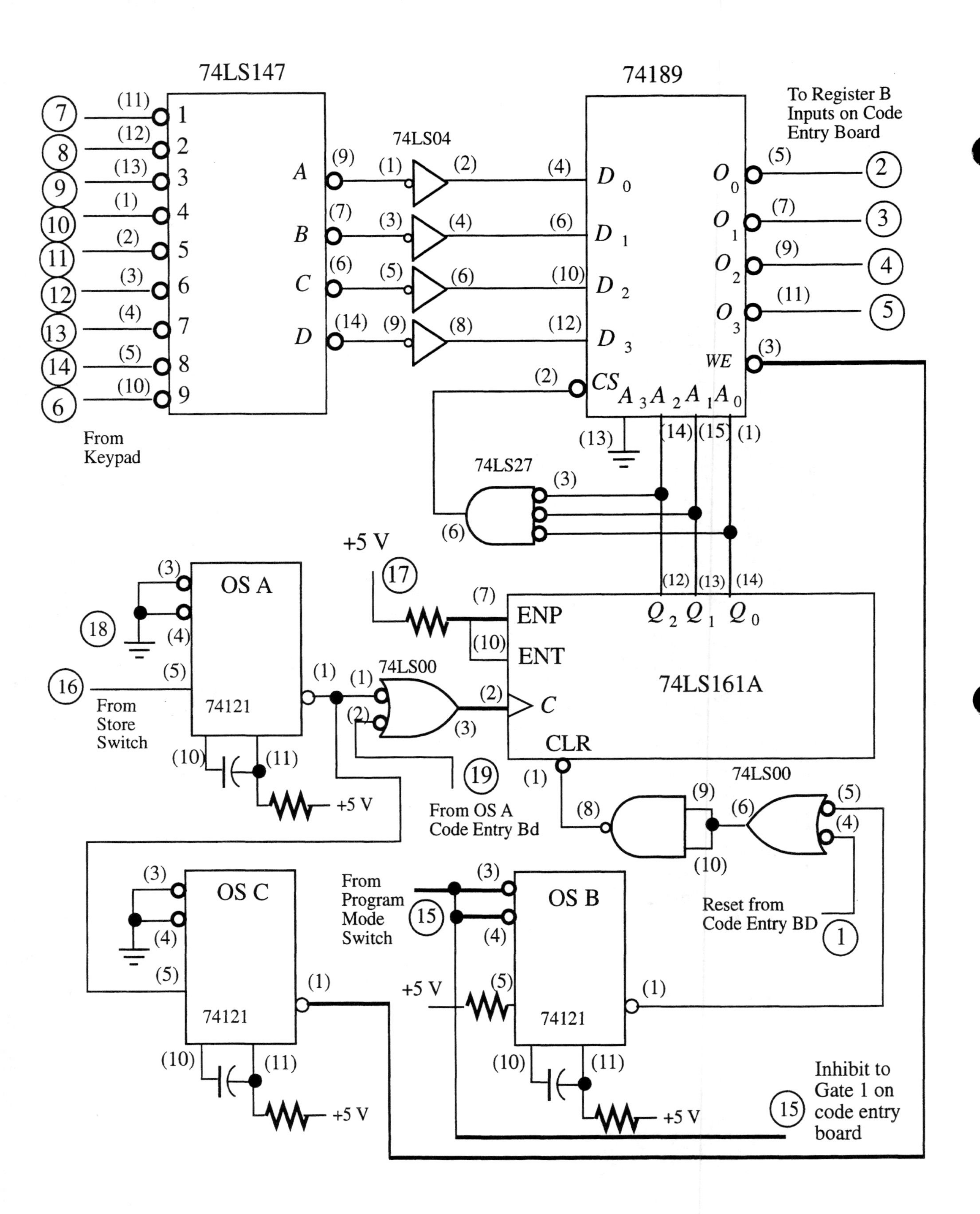

74LS147
74189
74LS04
To Register B Inputs on Code Entry Board
From Keypad
74LS27
+5 V
OS A
74121
From Store Switch
74LS00
74LS161A
ENP
ENT
CLR
From OS A Code Entry Bd
74LS00
Reset from Code Entry BD
OS C
OS B
From Program Mode Switch
Inhibit to Gate 1 on code entry board

Activity 2

The inputs and outputs are shown on the logic diagram on the preceeding page with circled numbers that correspond to connector terminals. The function of each input and output is as follows:

1 Reset from the code entry board.
2 Bit 0 to the code entry board comparator.
3 Bit 1 to the code entry board comparator.
4 Bit 2 to the code entry board comparator.
5 Bit 3 to the code entry board comparator.
6 "9" key input.
7 "1" key input.
8 "2" key input.
9 "3" key input.
10 "4" key input.
11 "5" key input.
12 "6" key input.
13 "7" key input.
14 "8" key input.
15 Program mode switch input. Also goes to inhibit input on entry code board.
16 Store switch input.
17 +5 V
18 Ground
19 Clock pulse from code entry board.

Activity 3

Using an actual key pad or switches to simulate the keys, program various 4-digit codes into the memory using a program mode switch and a store switch. Then apply a series of four pulses to connector terminal 1 and monitor the four outputs of the 74189 RAM for the proper sequence of digits.

System Assignment 3: The Complete System

Activity 1

The point-to-point list for connecting all of the components in the system is as follows:

Code Entry Board	Memory Board	Other Components
1		keypad "0"
2	6	keypad "9"
3	7	keypad "1"
4	8	keypad "2"
5	9	keypad "3"
6	10	keypad "4"
7	11	keypad "5"
8	12	keypad "6"
9	13	keypad "7"
10	14	keypad "8"
11	15	Tog Switch 1 (Prog Mode)
12		Buffer/Driver Bd 2
13		PB Switch B 1 (ARM)
14	17	+5 V Power supply
15	18	Ground
16	2	
17	1	
18	5	
19	4	
20	3	
21	19	
	16	PB Switch A 1 (Store)
		Tog Switch 2 to ground
		PB Swtich A 2 to +5 V
		PB Switch B 2 to +5V
		Buffer/Driver Bd 3 to LED2
		LED 1 to +5 V

Activity 2

The basic test procedure should include steps to program security codes into the memory and verify that proper code entry into an armed system will disarm and incorrect code entry will reset the sytem and not allow it to be disarmed. Students can be required to make a detailed list of steps.

Activity 3

The technical report can be as detailed as you require but should include at least the following elements:

1. How to store a security code.
2. How to gain access with the code.
3. How to arm and disarm the system.
4. How to know when the system is armed or disarmed.
5. System operation on power-up, backup power, etc.

Activity 4

(a) LED remains off when ARM switch is pressed. Possible faults are defective LED, defective buffer/driver, no power, defective shift register on the code entry board, or parallel input (pin 6) of the shift register is open.

(b) ARMED light remains on when correct code is entered. Possible faults are defective (bouncy) keypad, defective buffer driver, defective encoder, defective RAM, SHIFT/LOAD input of register shorted to ground, faulty comparator.

(c) ARMED light turns off after first digit is entered. Possible faults are input D_7 (pin 6) of shift register open or defective buffer/driver

CHAPTER 13
DIGITAL SYSTEM APPLICATION

System Assignment 1: Design of Display Logic

Activity 1
One approach is to use 74185A binary-to-BCD converters. Four of these devices is required to convert 8 binary bits to a 3-digit BCD code. Another approach is to use programmable logic devices.

Activity 2
The 74XX95 4-bit parallel acces shift register can be used for the azimuth and elevation registers. Three 7495s are required for each register.

Activity 3
The 74XX47 BCD-to-7 segment decoder/driver can be used in this application. Six of these devices is required. The MAN-1 7-segment display with 330 W limiting resistors can be used. Six of these devices is required.

Activity 4
A diagram of the display logic is shown on the following page. Separate clock pulses to the azimuth and the elevation registers are required to alternately store the azimuth position code and the elevation position code.

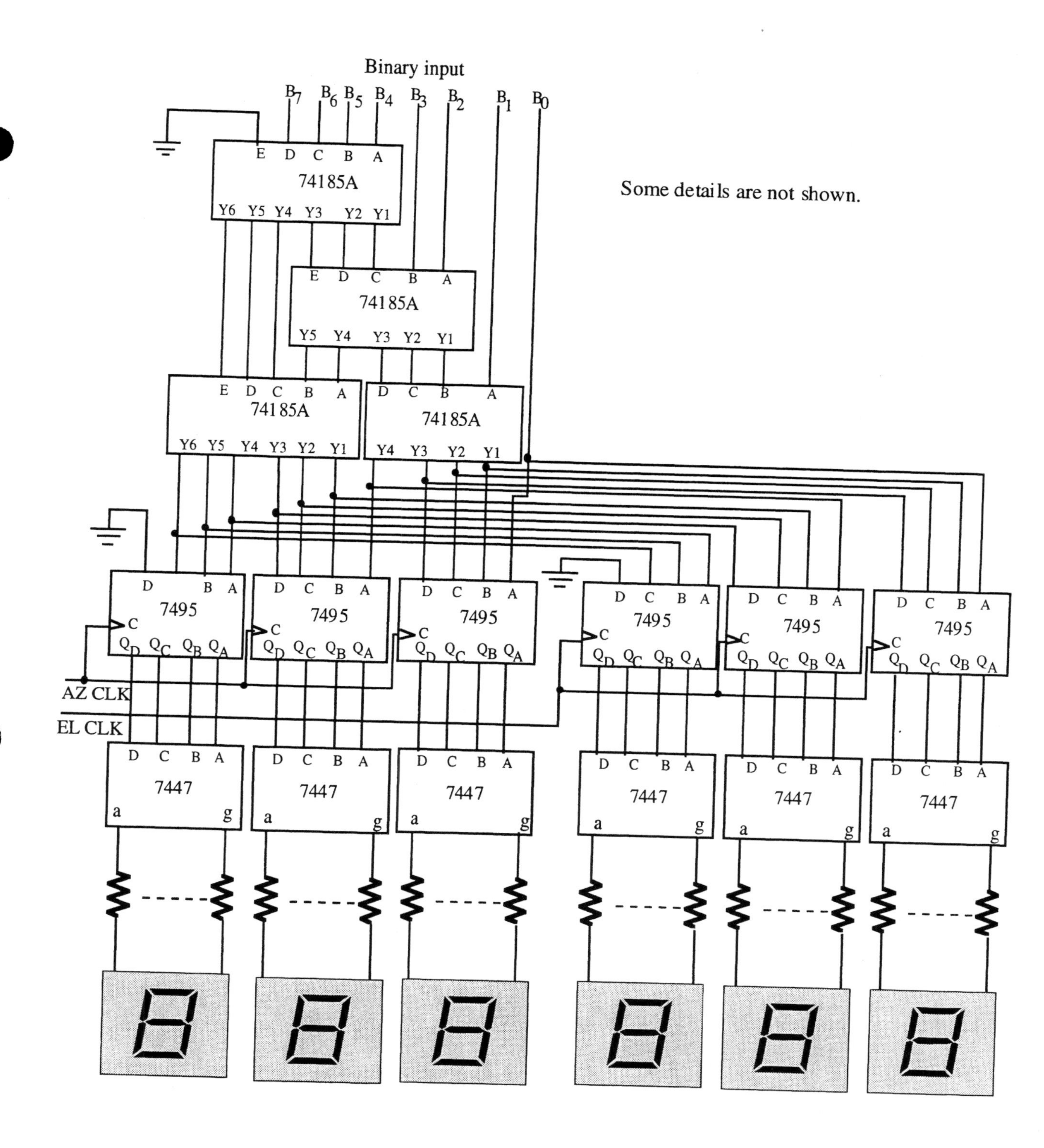
Binary input
B7 B6 B5 B4 B3 B2 B1 B0
Some details are not shown.
E D C B A
74185A
Y6 Y5 Y4 Y3 Y2 Y1
E D C B A
74185A
Y5 Y4 Y3 Y2 Y1
E D C B A
74185A
Y6 Y5 Y4 Y3 Y2 Y1
D C B A
74185A
Y4 Y3 Y2 Y1
7495
7495
7495
7495
7495
7495
AZ CLK
EL CLK
7447
7447
7447
7447
7447
7447

System Assignment 2: Design of Timing Logic

Activity 1 Channel Select Waveform

The minimum frequency is 100 Hz and it can be any value above that as long as the timing criteria specified in the following acitvities are met. A 555 timer configured as an astable multivibrator with a duty cycle of approximately 50% and a frequnecy of approximately 1 kHz is shown below.

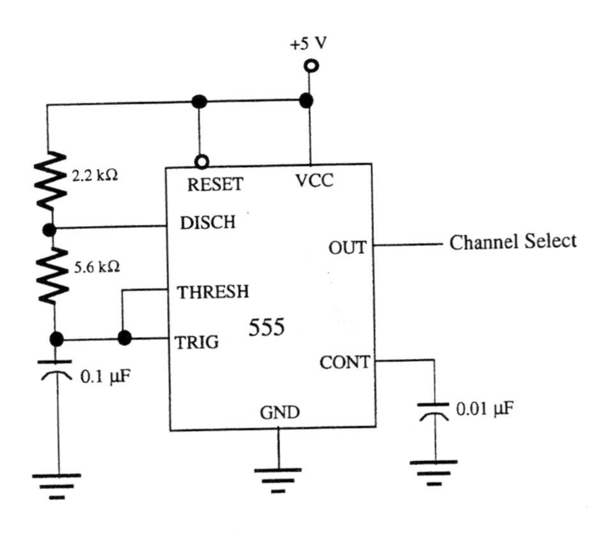

Activity 2 Convert Waveform

OS A triggers on the positive transitions and OS B triggers on the negative transitions of the Channel Select waveform to provide delays of approximately 150 ns . OS A and OS B pulses are ORed to trigger OS C which produces a Convert pulse with a width of approximately 700 ns. The logic diagram is shown be-

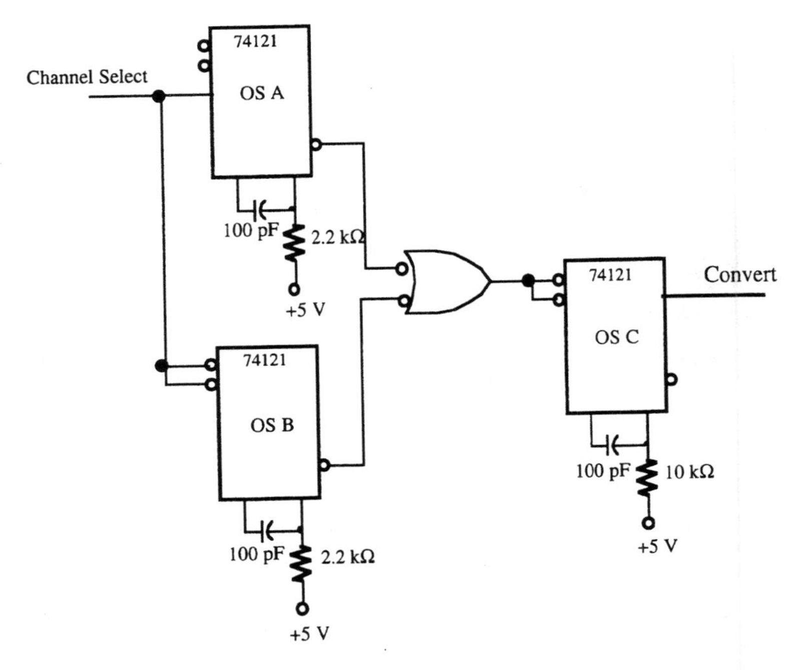

Activity 3 Data Enable Waveform

The $\overline{\text{Date Ready}}$ from the ADC triggers the one shot to produce a negative -going pulse of approximately 700 ns for the $\overline{\text{Data Enable}}$ as shown below.

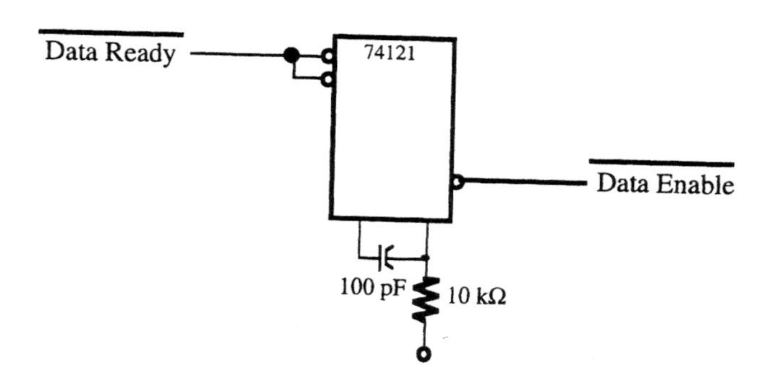

Activity 4 Azimuth and Elevation Register Clocks

Two separate clocks are required to clock the azimth and elevation data into the appropriate register. A logic diagram and the corresponding timing diagram illustrates a method of achieving this. The 7495 registers clock on the negative-going edges of the clock pulses.

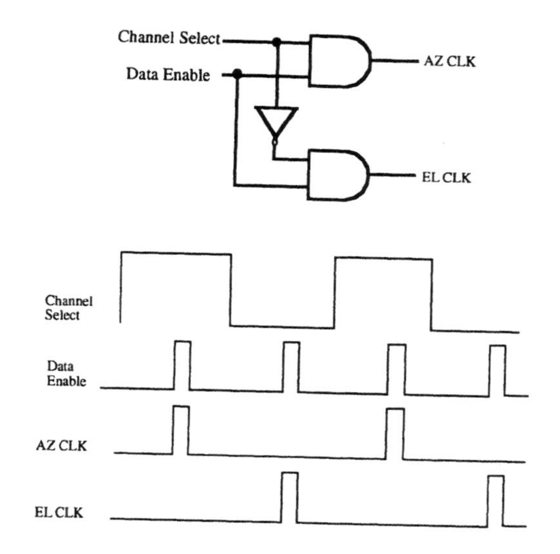

Activity 5

The full timing logic diagram is shown on the next page. The only connections to the display logic are the AZ CLK and EL CLK.

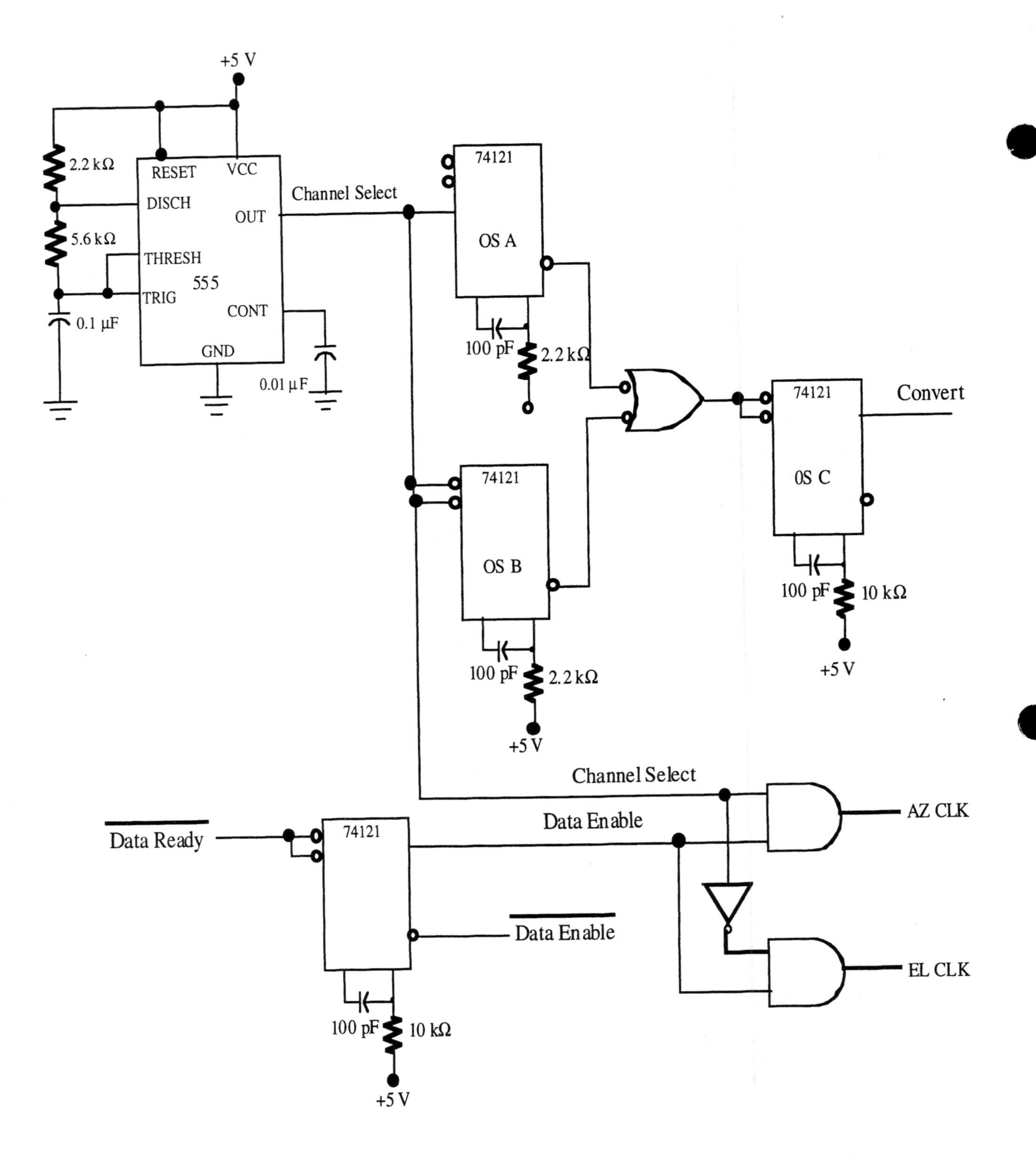
+5 V
2.2 kΩ
RESET
VCC
DISCH
OUT
5.6 kΩ
THRESH
555
TRIG
CONT
0.1 μF
GND
0.01 μF
Channel Select
74121
OS A
100 pF
2.2 kΩ
74121
Convert
0S C
100 pF
10 kΩ
+5 V
74121
OS B
100 pF
2.2 kΩ
+5 V
Channel Select
AZ CLK
Data Ready
74121
Data Enable
Data Enable
EL CLK
100 pF
10 kΩ
+5 V

PART 3

System Applications

Worksheet Masters

Chapter 2

Worksheet Master:

Tablet Counting/Control System Block Diagram

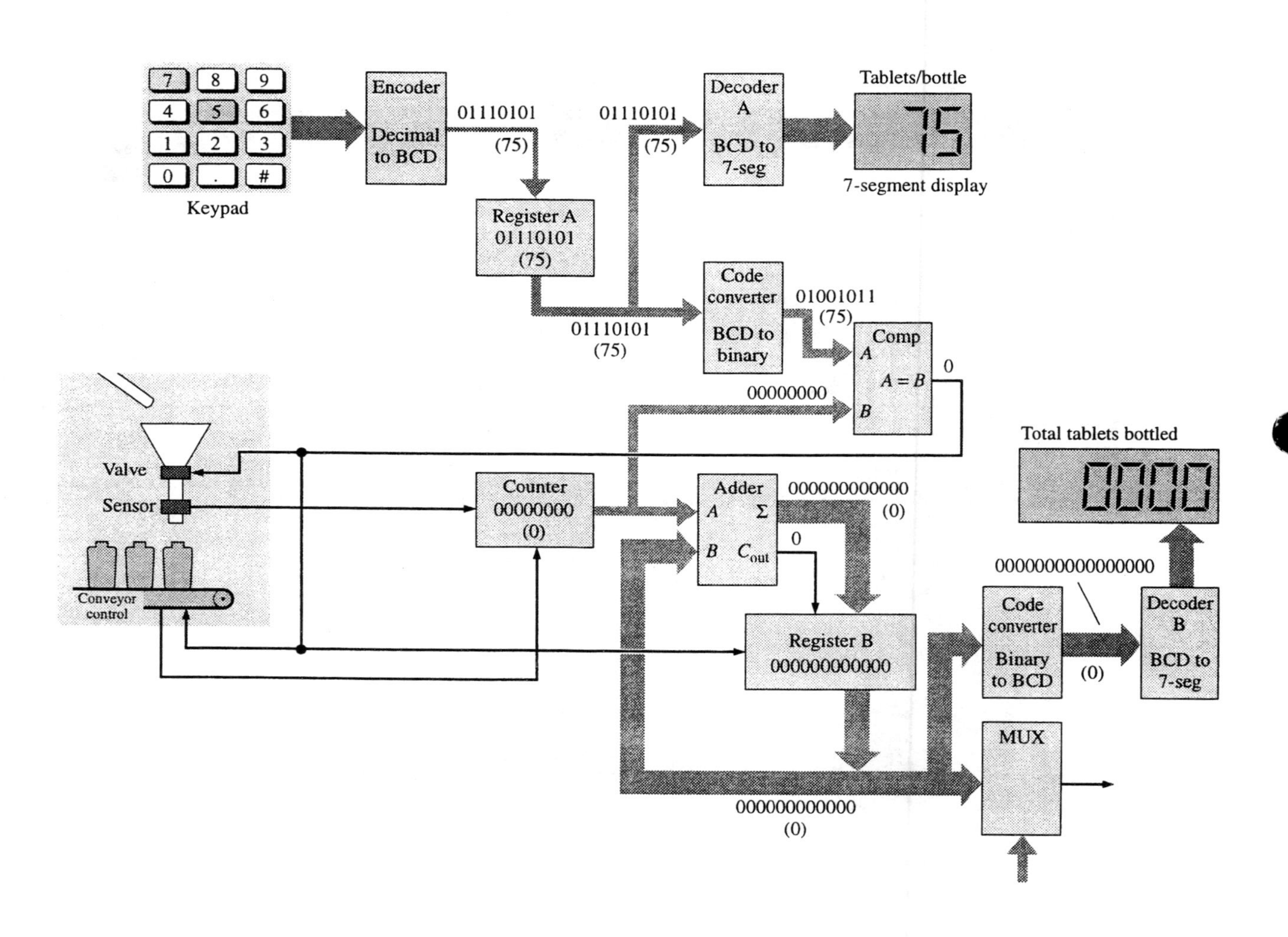

Chapter 3

Worksheet Master:

Comparator Portion of System Board

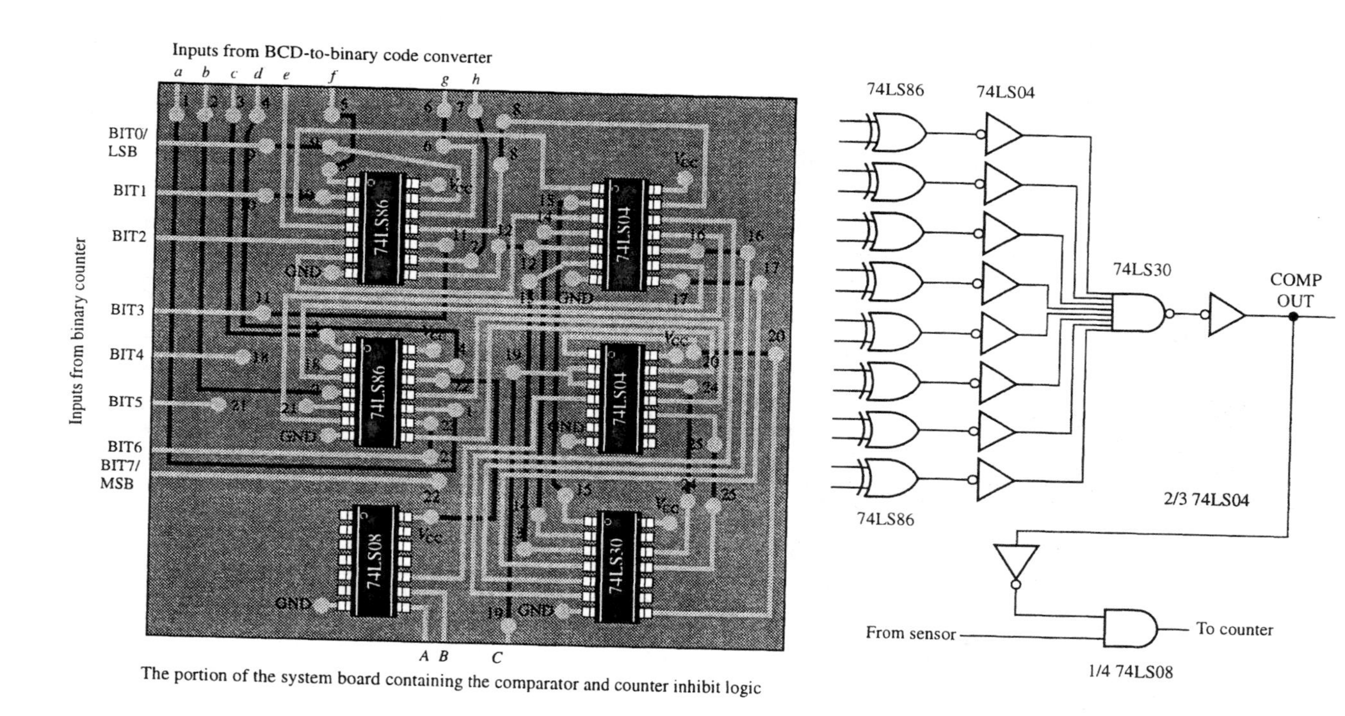

The portion of the system board containing the comparator and counter inhibit logic

Chapter 5

Worksheet Master:

Lumber Processing Logic Board

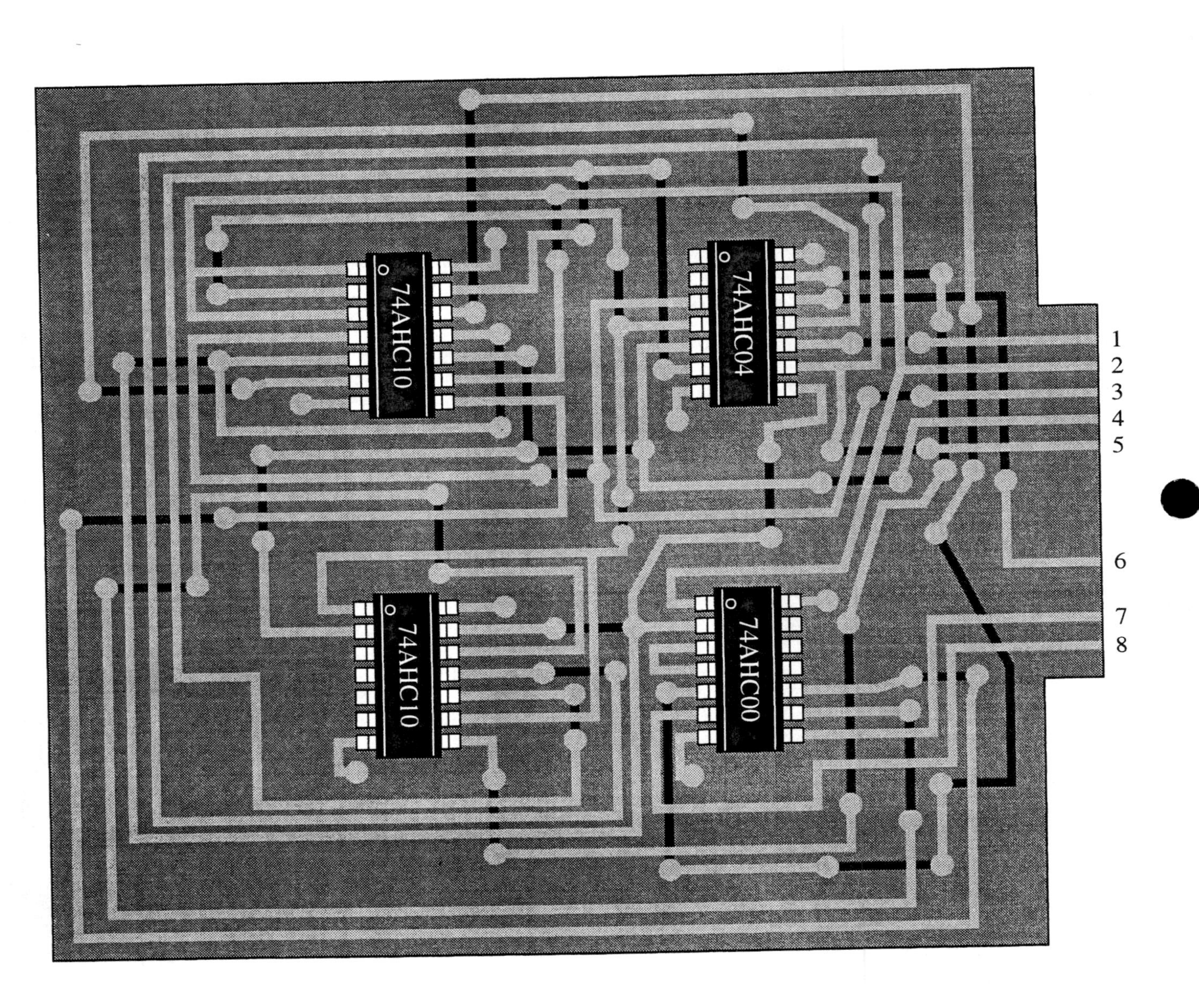

Chapter 6

Worksheet Master:

State Decoder/Output Logic on the Traffic Light Control board.

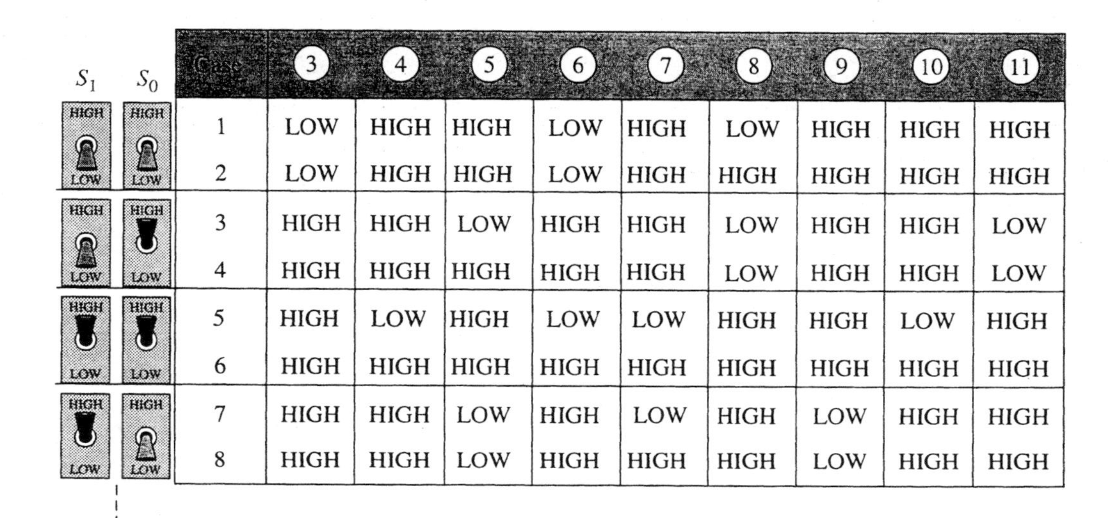

Case	3	4	5	6	7	8	9	10	11
1	LOW	HIGH	HIGH	LOW	HIGH	LOW	HIGH	HIGH	HIGH
2	LOW	HIGH	HIGH	LOW	HIGH	HIGH	HIGH	HIGH	HIGH
3	HIGH	HIGH	LOW	HIGH	HIGH	LOW	HIGH	HIGH	LOW
4	HIGH	HIGH	HIGH	HIGH	HIGH	LOW	HIGH	HIGH	LOW
5	HIGH	LOW	HIGH	LOW	LOW	HIGH	HIGH	LOW	HIGH
6	HIGH	HIGH	HIGH	HIGH	HIGH	HIGH	HIGH	HIGH	HIGH
7	HIGH	HIGH	LOW	HIGH	LOW	HIGH	LOW	HIGH	HIGH
8	HIGH	HIGH	LOW	HIGH	HIGH	HIGH	LOW	HIGH	HIGH

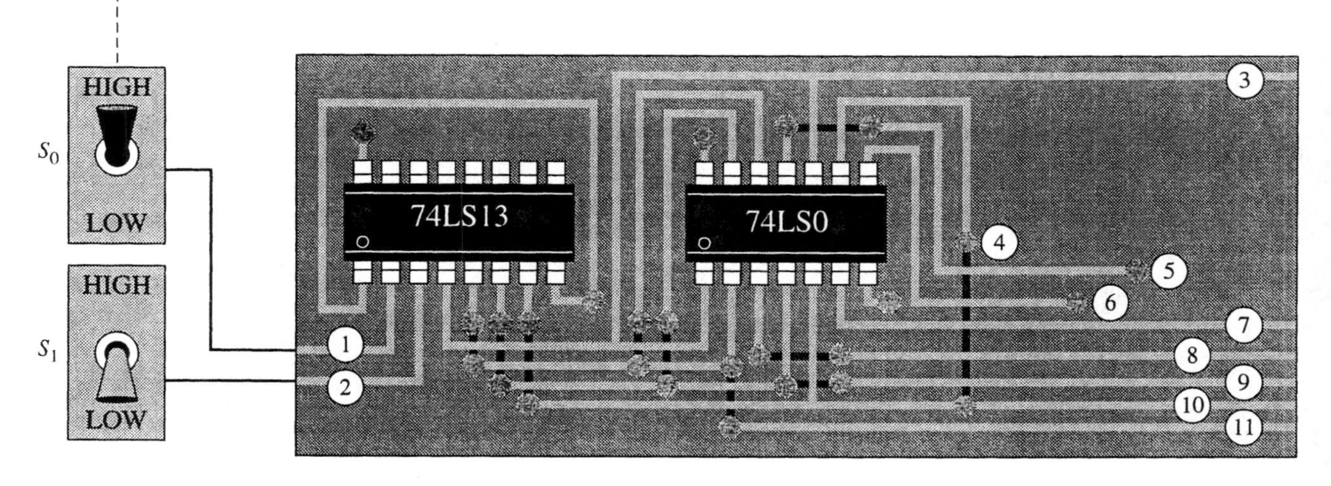

Chapter 8

Worksheet Master:

State Decoder/Output Logic/Timing Logic on the Traffic Light Control Board

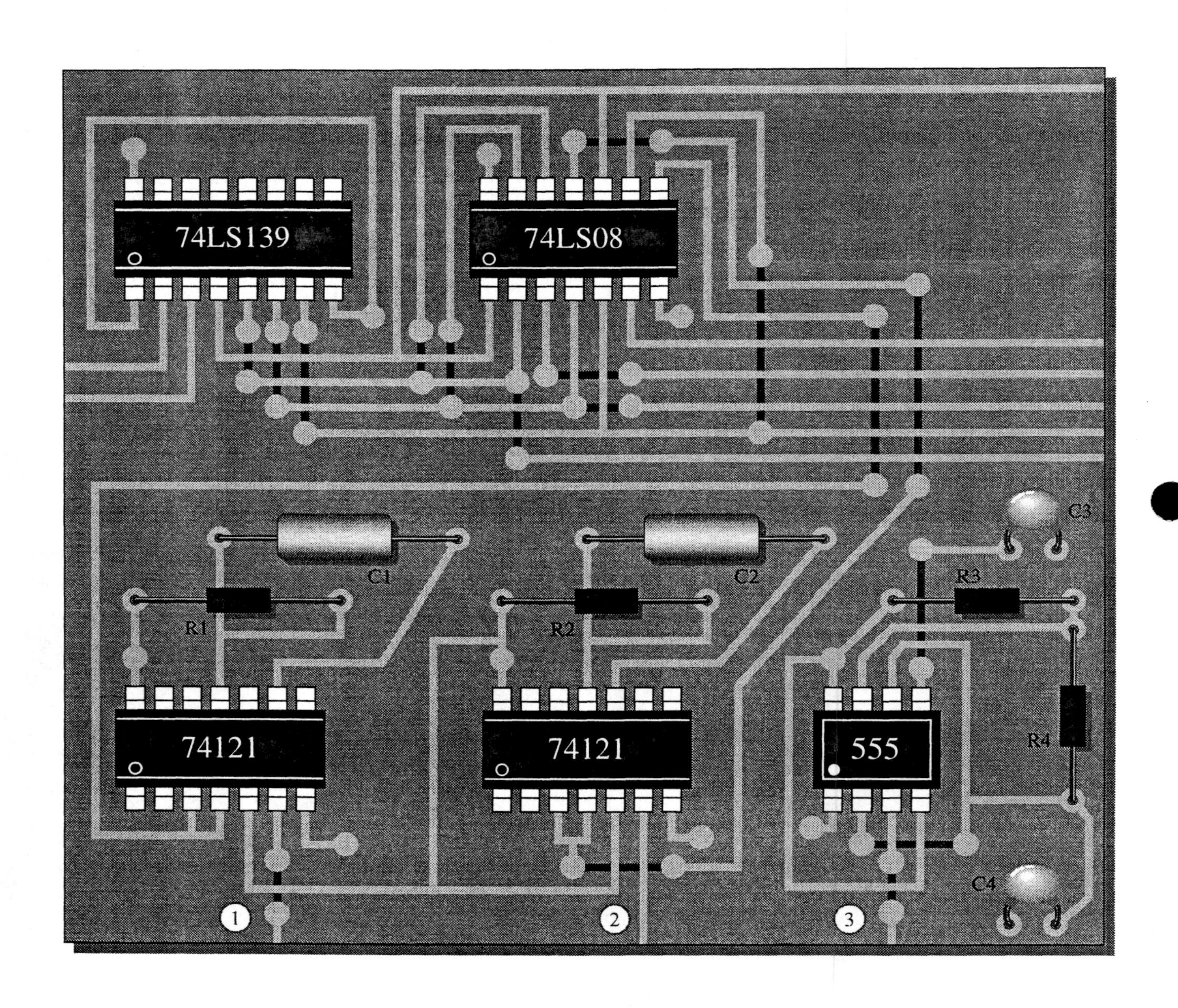

Chapter 9

Worksheet Master:

Complete Traffic Light Control board

Chapter 10

Worksheet Master:

Security Code Entry Logic Board

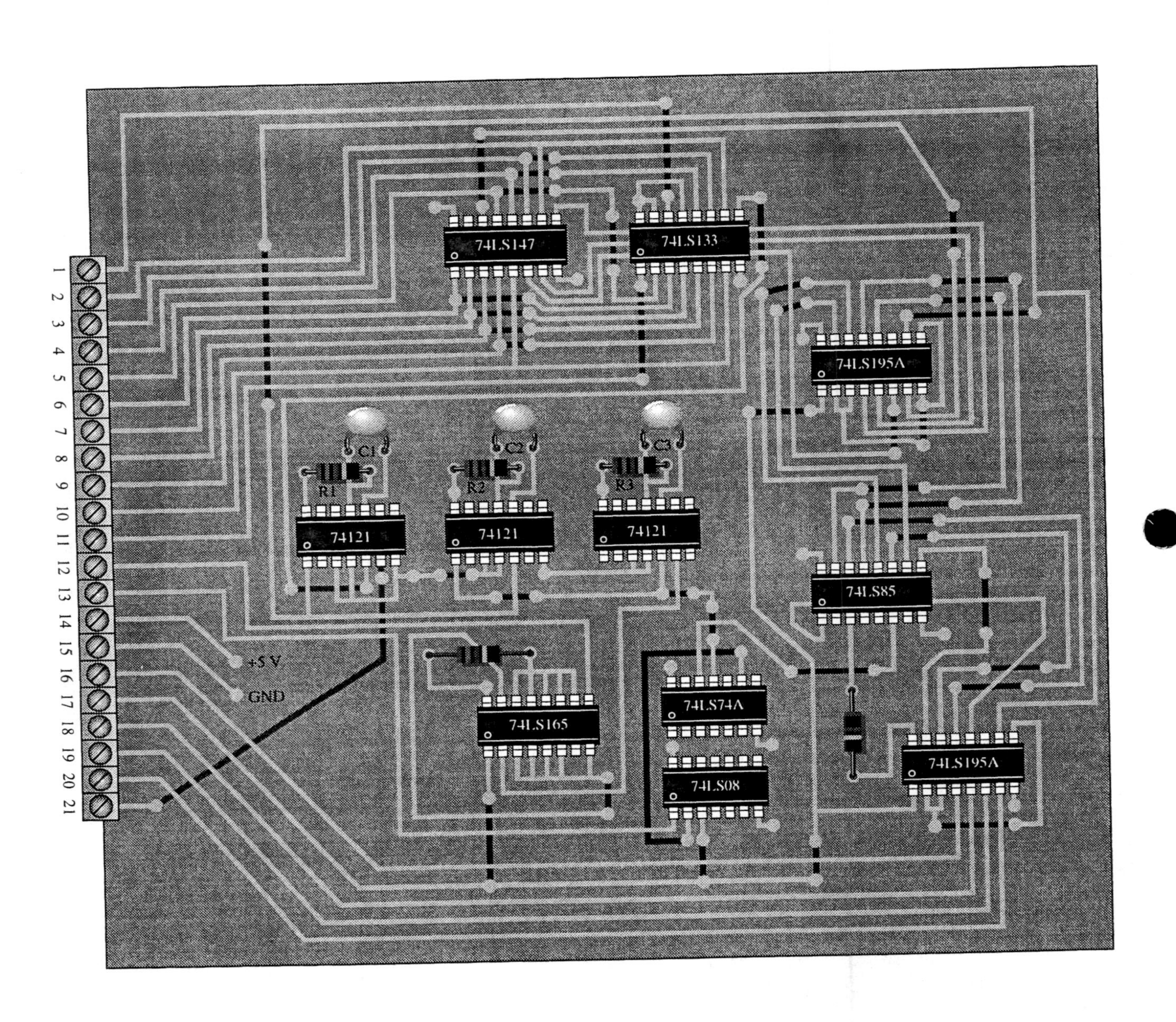

Chapter 12

Worksheet Master:

Memory Logic Board

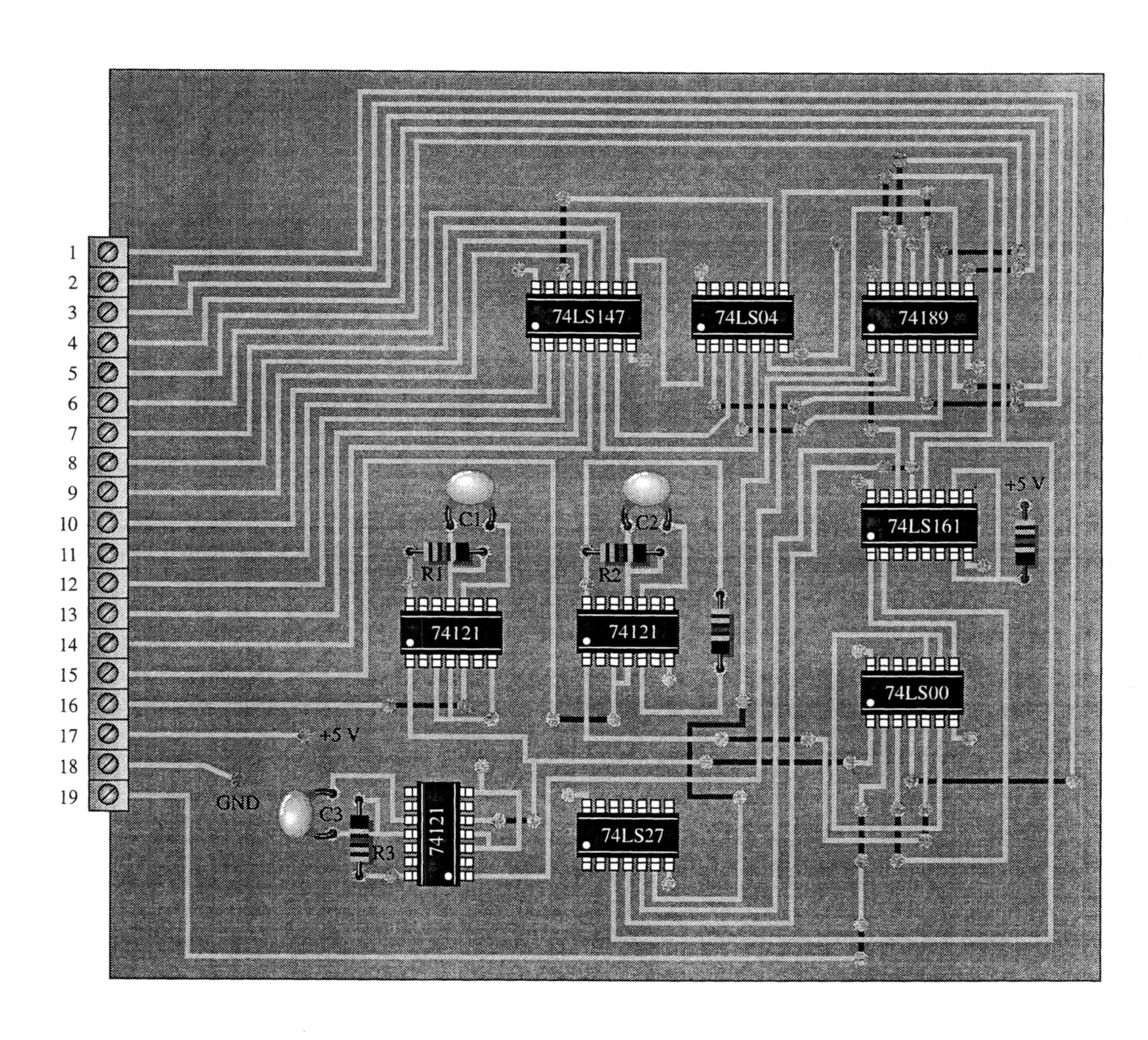

PART 4

EWB Circuit Simulation Results

Prepared by Gary Snyder

System Applications

Chapter 1:

There is no EWB system application for this chapter.

Chapter 2:

There is no EWB system application for this chapter.

Chapter 3:

The 8-bit comparator circuit EWB test setup is as shown:

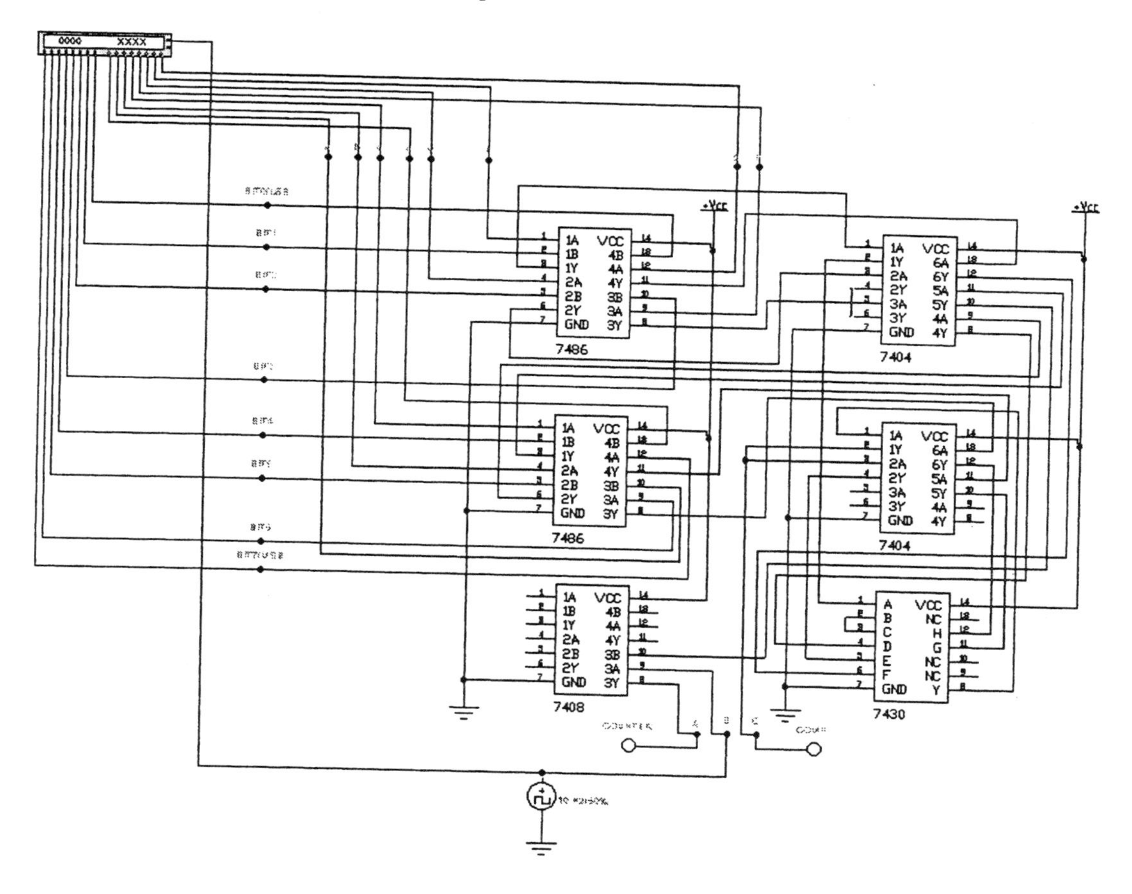

Chapter 4:

The 7-segment decoding logic EWB test setup is as shown. When the circuit is activated, the 7-segment display should repeatedly cycle from 0 to 9.

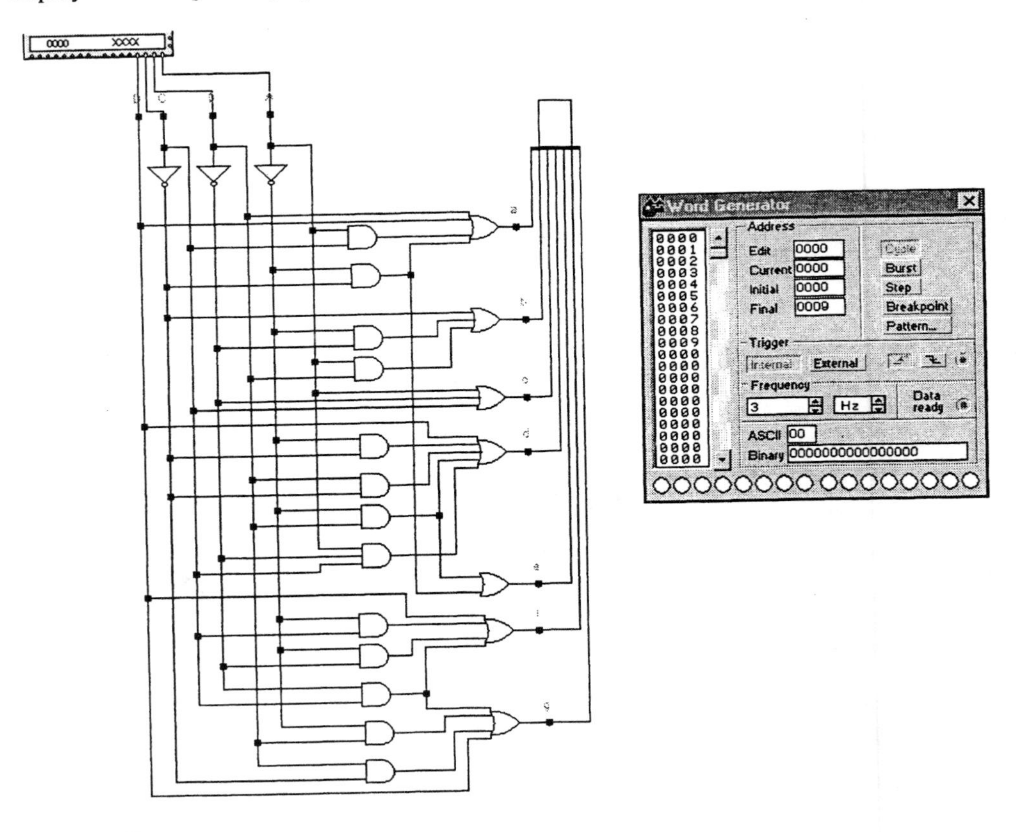

Chapter 5:

The sawmill control EWB test setup is as shown:

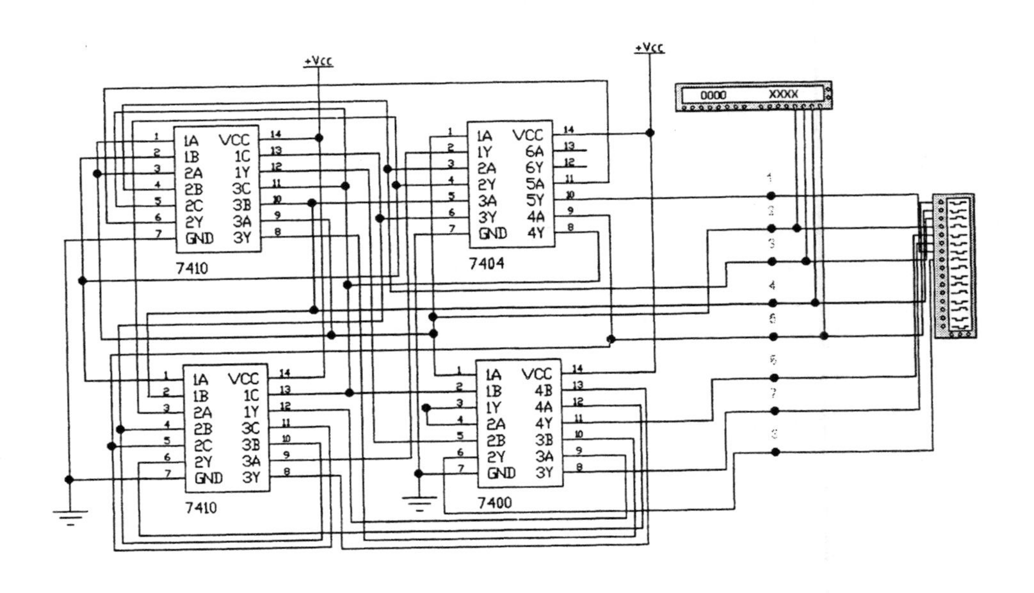

Chapter 6:

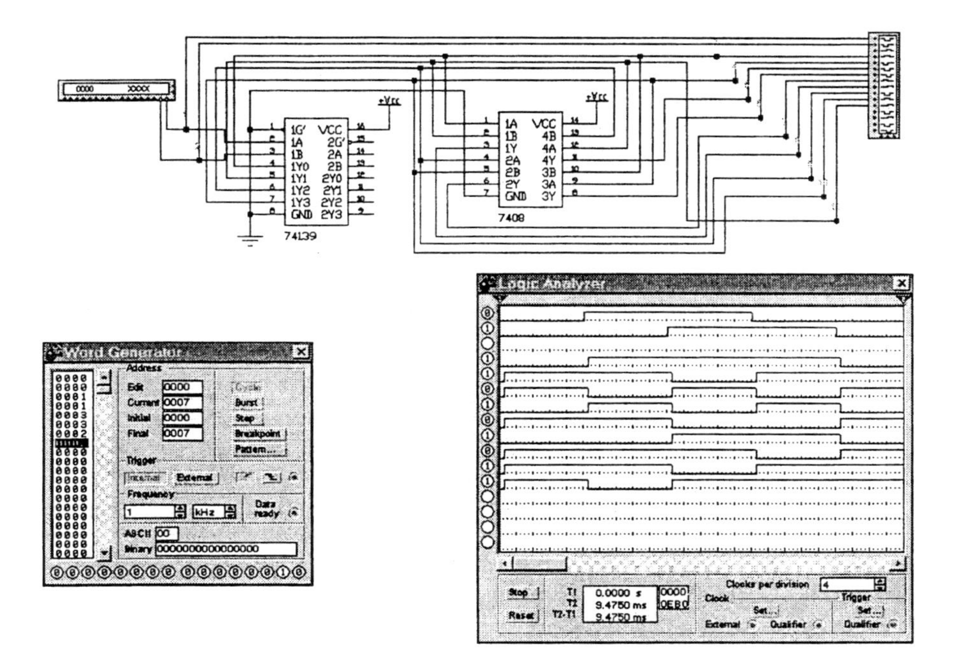

The system application setup and results are as shown.

Chapter 7:

There is no EWB system application for this chapter.

Chapter 8:

The system application setup is as shown:

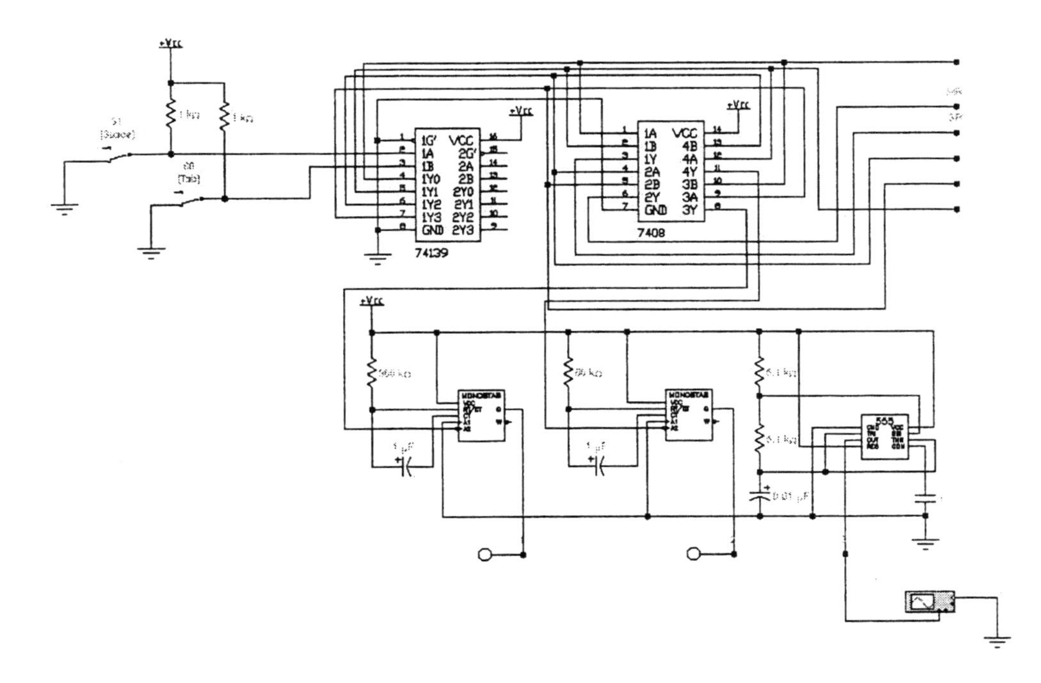

Chapter 9:

The system application setup is as shown on p. 277. The LEDs are arranged to simulate the lights of actual stoplights, and a switch used to indicate the presence of cars on the side street.

Chapter 10:

The system application setup is as shown on p. 278. The pattern to simulate a successful keypad entry is shown on the word generator below.

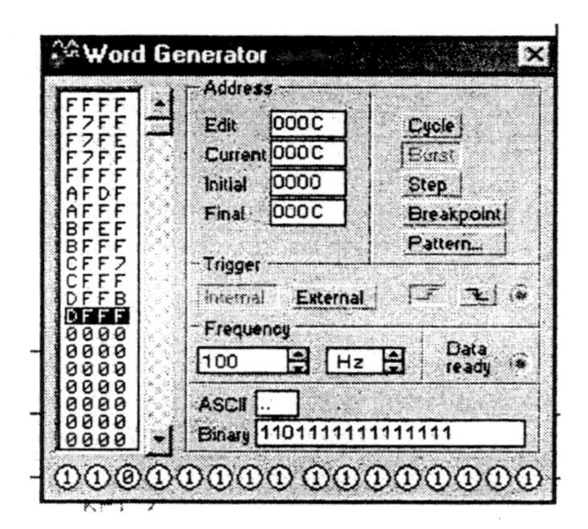

Chapter 11:

There is no EWB system application for this chapter.

Chapter 12:

There is no EWB system application for this chapter.

Chapter 13:

There is no EWB system application for this chapter.

Chapter 14:

There is no EWB system application for this chapter.

Chapter 15:

There is no EWB system application for this chapter.

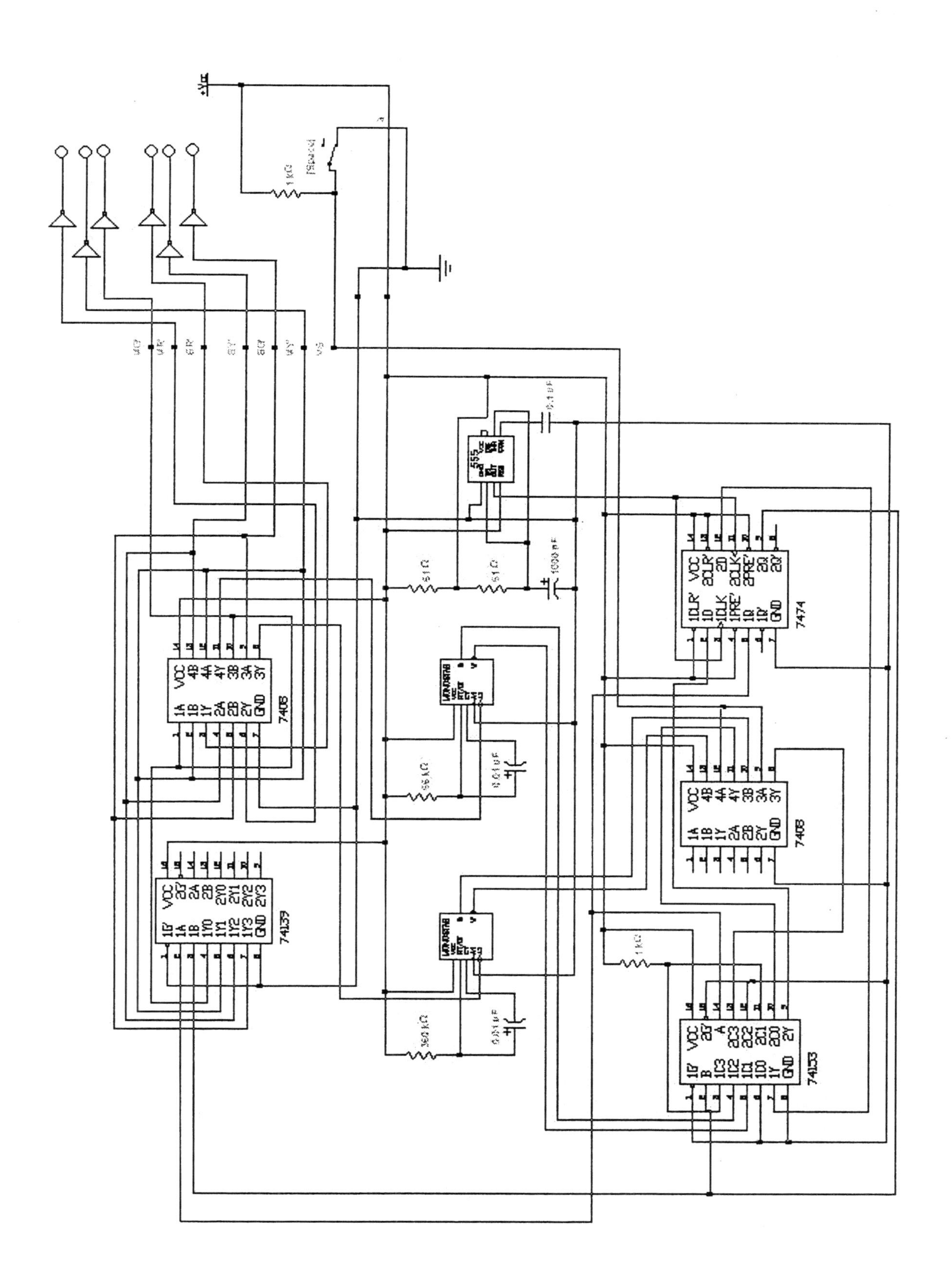
7408
74139
7474
7408
74153
555

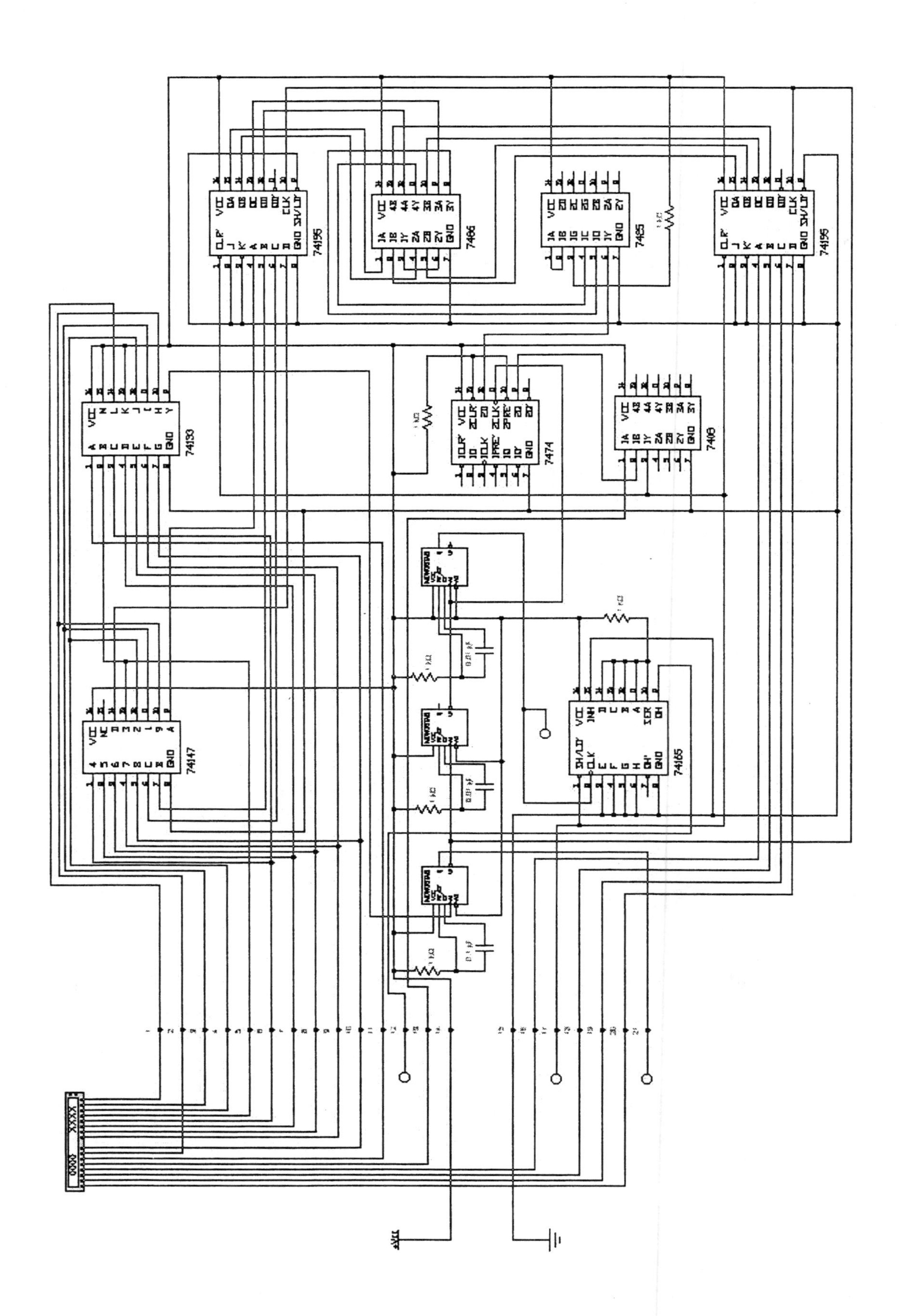

Chapter Troubleshooting Problems:

Note: The circuit restrictions password for the EWB troubleshooting circuits on the DF7 CD-ROM is **book** (all lower-case). If this does not work, try entering **BOOK** (all upper-case).

Chapter 1:

There are no EWB troubleshooting problems for this chapter.

Chapter 2:

There are no EWB troubleshooting problems for this chapter.

Chapter 3:

46. The truth table for the AND gate is as shown. From the table the output follows A so that A is probably shorted through to the output.

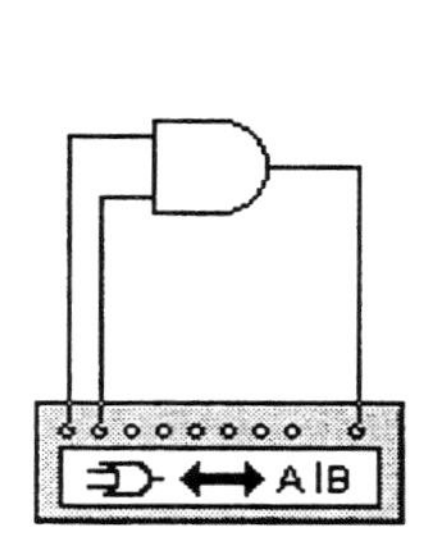

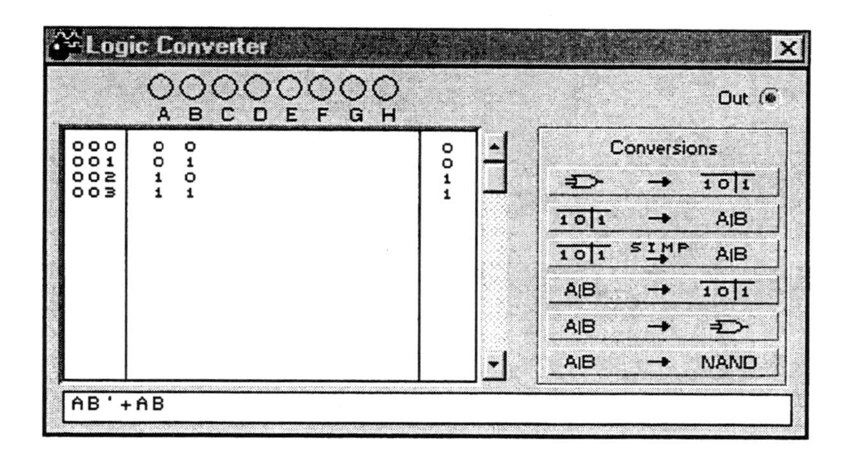

47. The truth table for the NAND gate is as shown. From the table the output is LOW only when both inputs are HIGH, and HIGH when either input is LOW, so that the inputs appear to be shorted together.

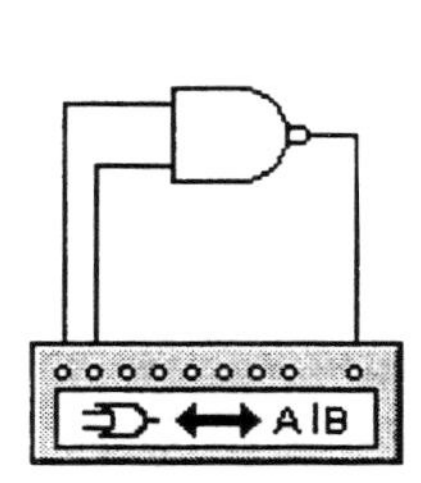

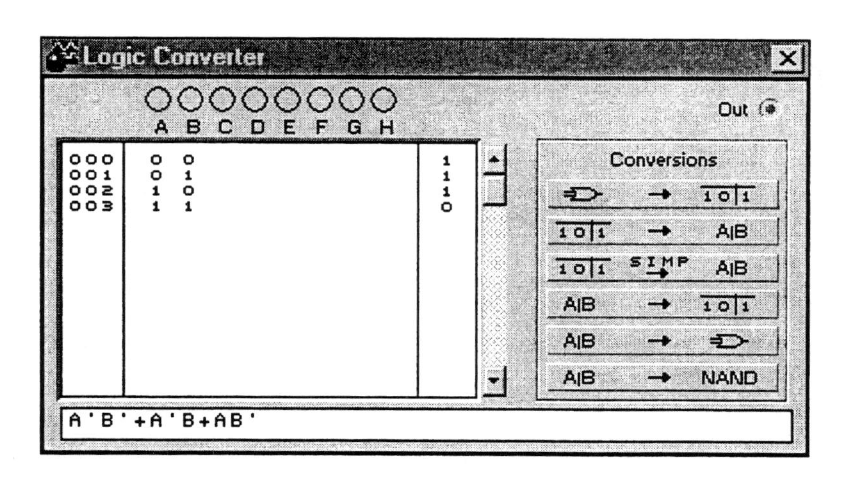

48. The truth table for the NOR gate is as shown. From the table the output is that of a NOR function, so that there does not appear to be any detectable fault.

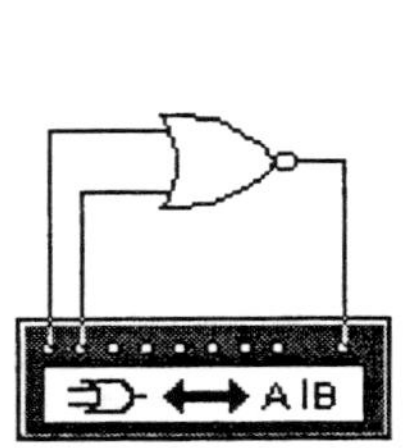

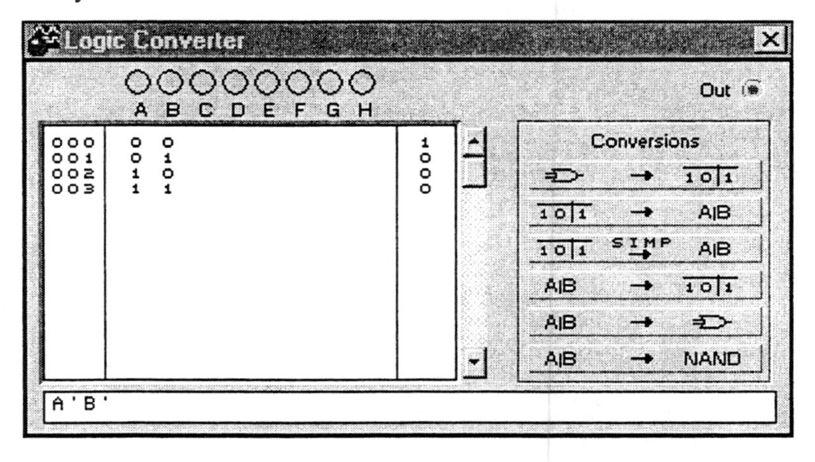

49. The truth table for the XOR gate is as shown. From the table, the output is LOW for any combination of inputs, so that the output is probably open.

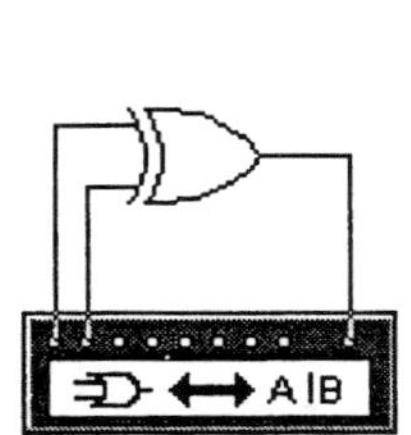

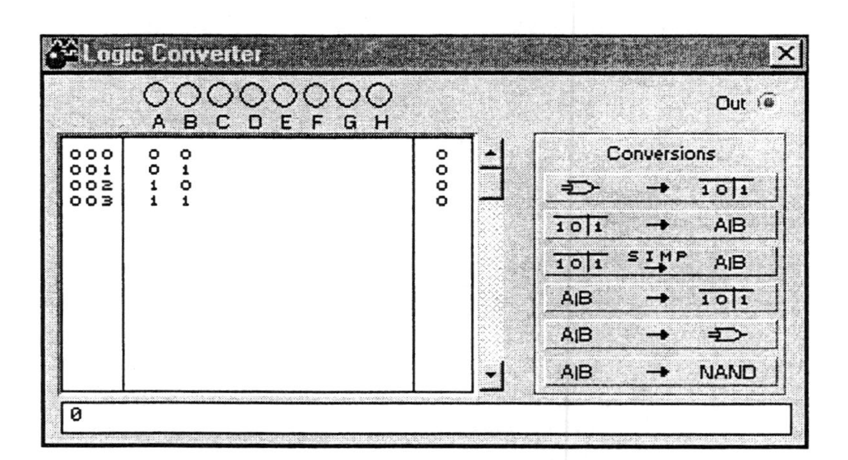

Chapter 4:

61. When the circuit is activated the 7-segment display shows that the binary inputs 0000, 0010, 0100, 0110, and 1000 produce errors. The one thing common to these inputs is having the LSB (Bit A) LOW. The most likely culprit is that the A' inverter input or output is open, preventing proper recognition of these binary patterns.

62. When the circuit is activated the 7-segment display shows an error only for the input 0110, when the e-segment is blank. The most likely fault is that the bottom input of the e segment OR-gate is open. The output connected to this input cannot be at fault, as the a-segment to which it also connects operates properly.

63. When the circuit is activated the 7-segment display shows that the b-segment never comes on for any of the inputs that should activate it. The most likely fault is that the output of the b-segment OR-gate is open.

Chapter 5:

46. When the circuit is activated the input and output waveforms are as shown. From the diagram, the output pattern for the input value A'BC is incorrect, isolating the fault to G1 or the OR gate input connected to the output of G1. Further examination would be able to show that the output of G1 does not go high for B=1 and C=1, but it is not possible with EWB to track the cause further to determine whether the output or one of the two inputs is at fault. This is because an open on an AND gate input acts like a 0 in EWB, which forces the output to 0 regardless of the states of either input. At the same time, an open output also acts as a 0 regardless of the state of either input. This is an example of a fault that cannot be fully isolated.

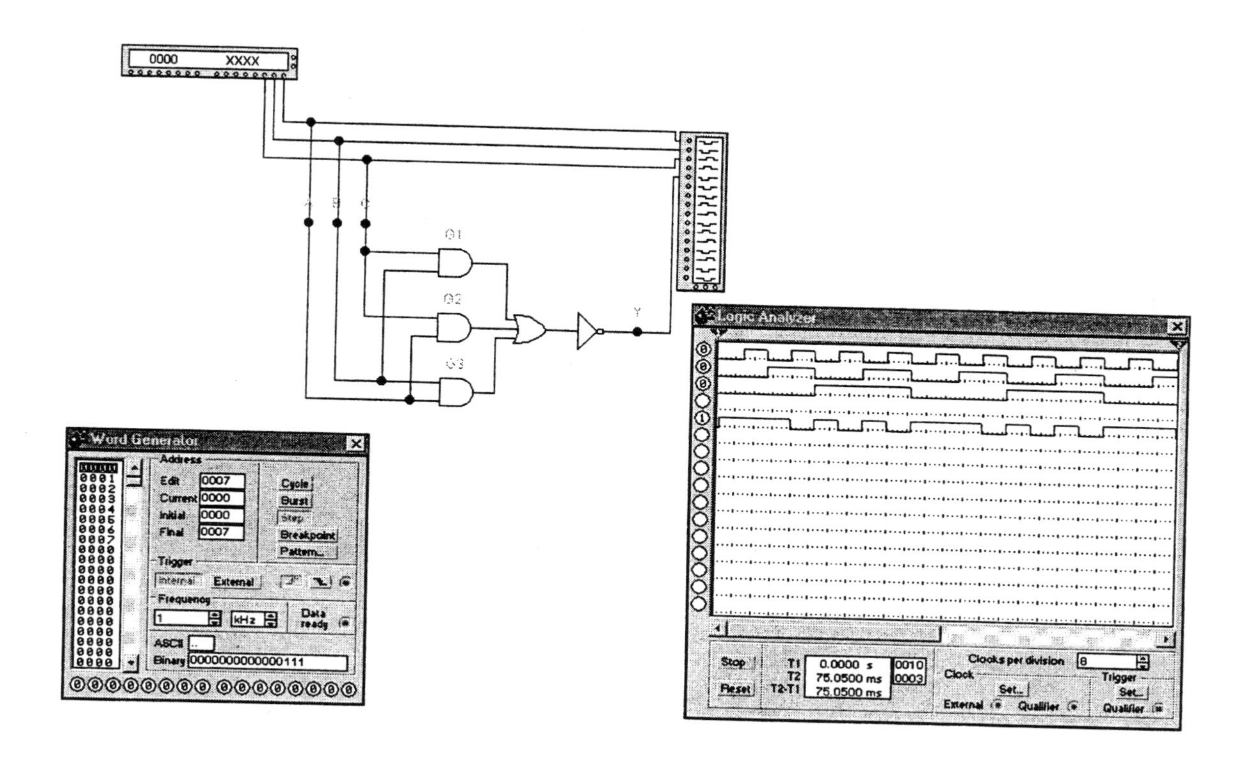

47. When the circuit is activated the input and output waveforms are as shown. From the diagram, the output corresponding to the term ABC' is incorrect, indicating that either the AND gate decoding the ABC' input is faulty or the OR gate input connected to the AND gate is open. The probes attached to the circuit will reveal that the AND gate is operating correctly, and that the input to the OR gate must be open.

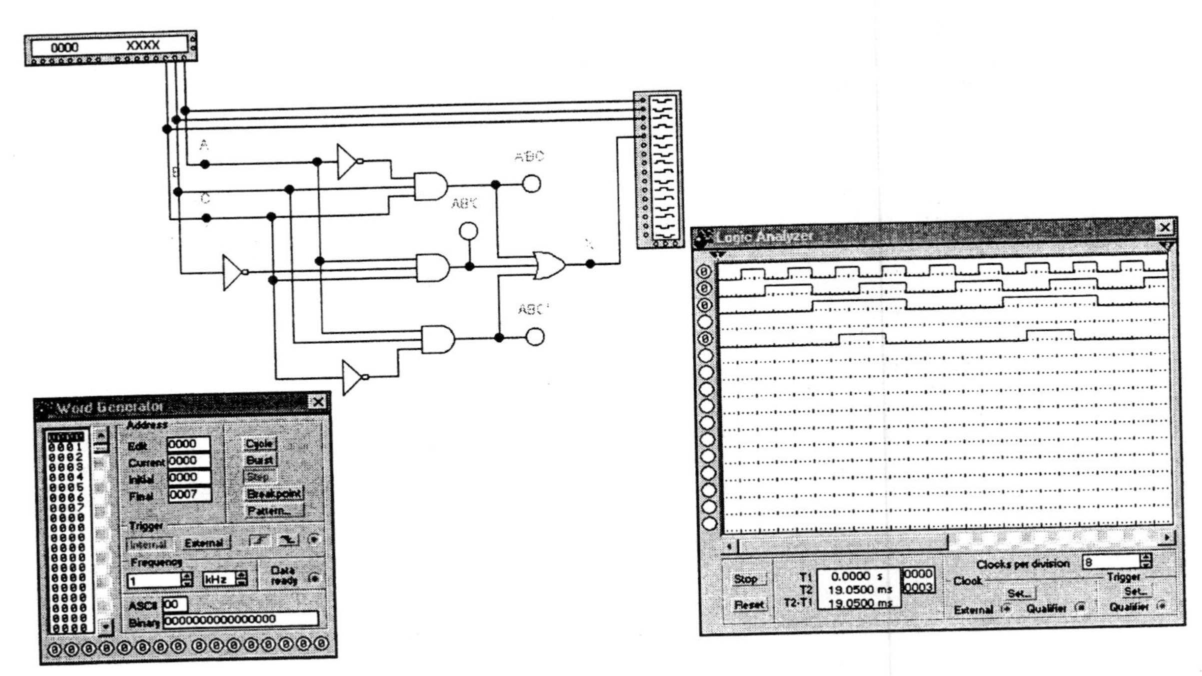

48. When the circuit is activated the input and output waveforms are as shown. From the diagram it appears that the output of Y2 is always high, indicating that, since the input C is changing states, the input to the inverter is open.

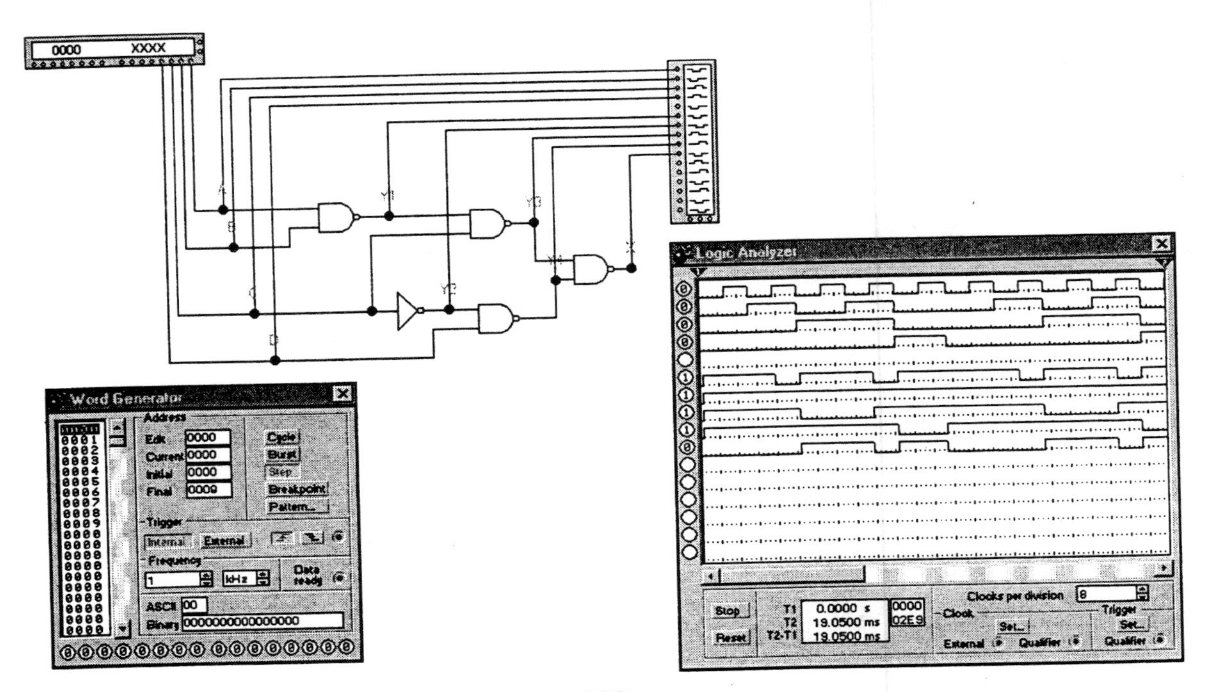

49. When the circuit is activated the input and output waveforms are as shown. In this case the circuit is operating properly.

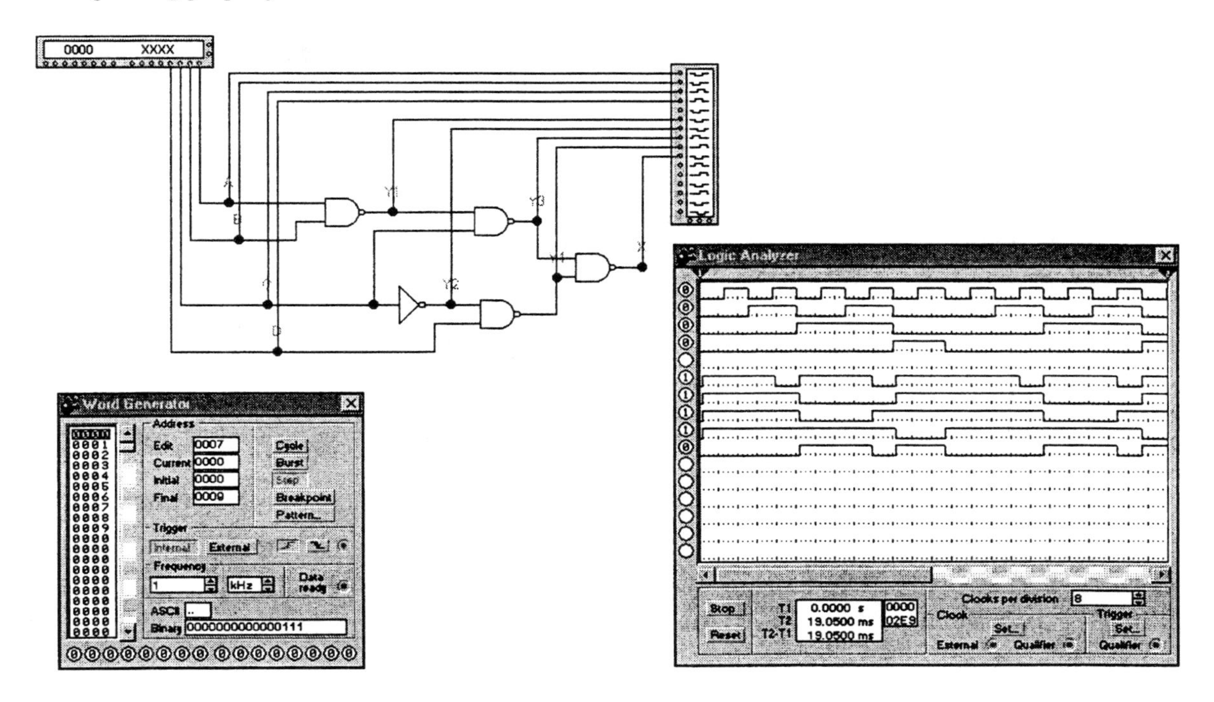

Chapter 6:

49. From the logic analyzer waveforms, it appears that C_O1 or C_O2 is open, as $A_1 = 1$ and $B_1 = 1$ do not result in a carry into the next adder.

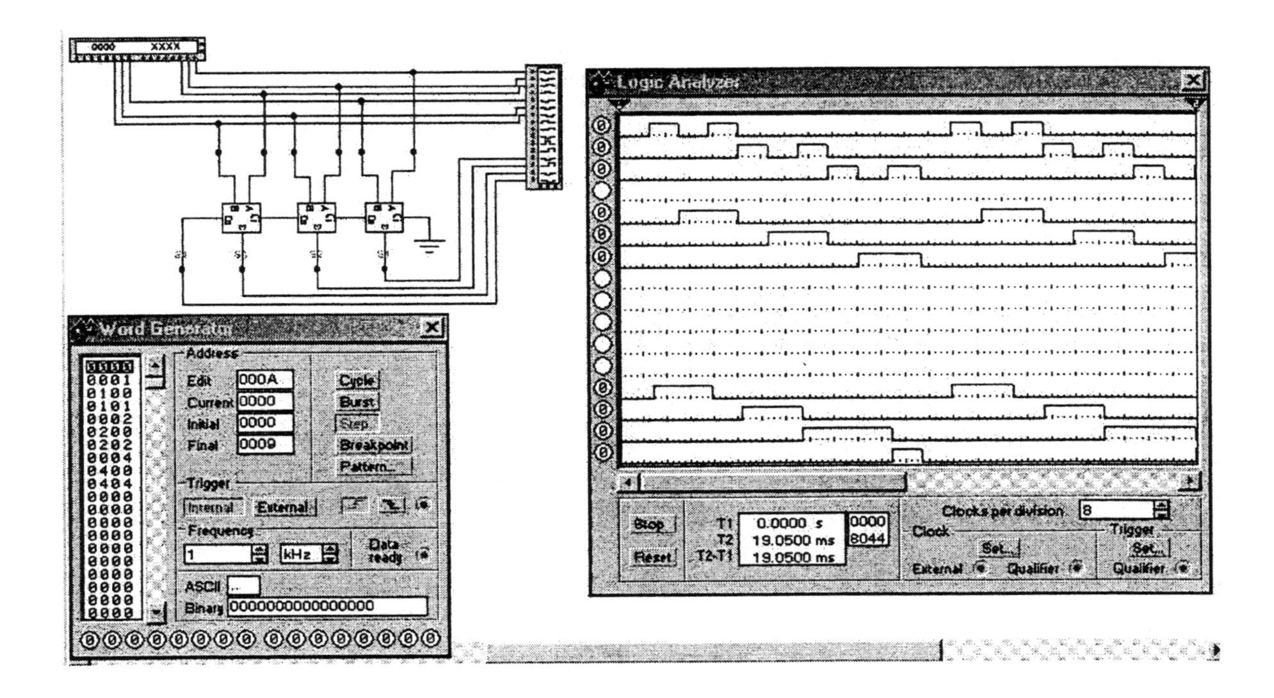

50. From the logic analyzer waveforms, it appears that outputs 3 and 4 are shorted together.

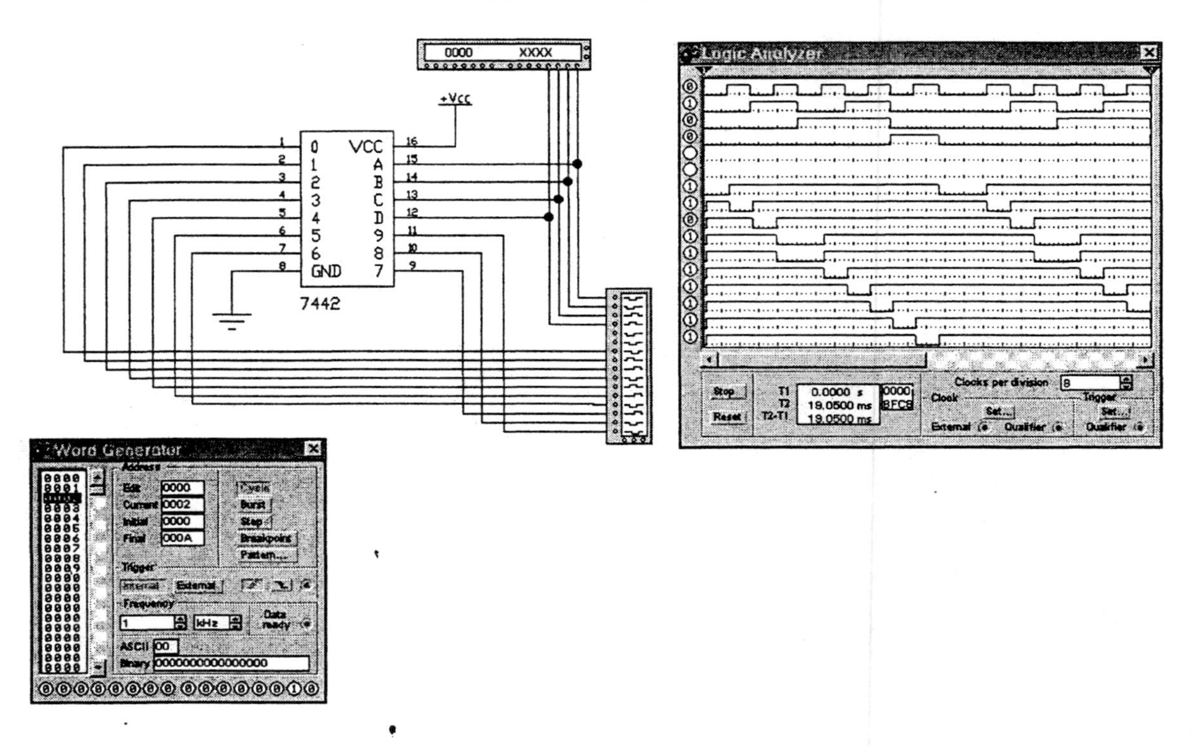

51. From the logic analyzer waveforms, it appears that input 2 of the LSB decoder is open or stuck high.

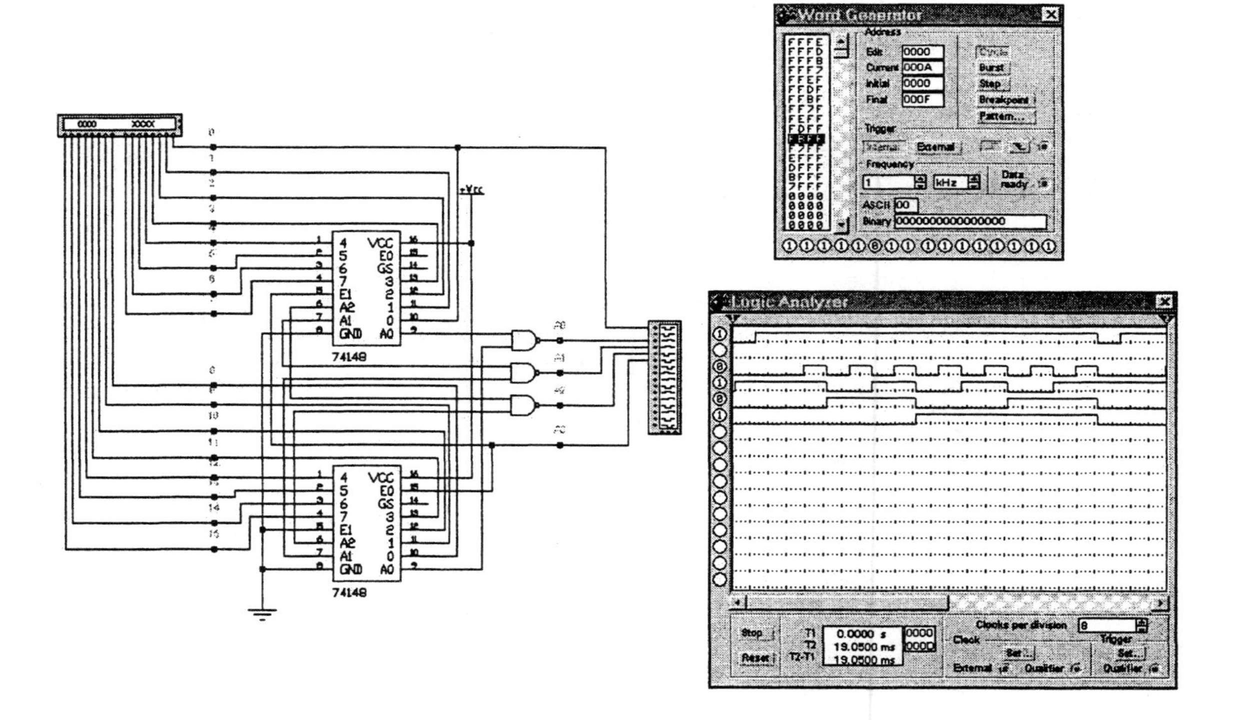

52. From the logic analyzer waveforms, it appears that input 1 of the top multiplexer is open or shorted to ground.

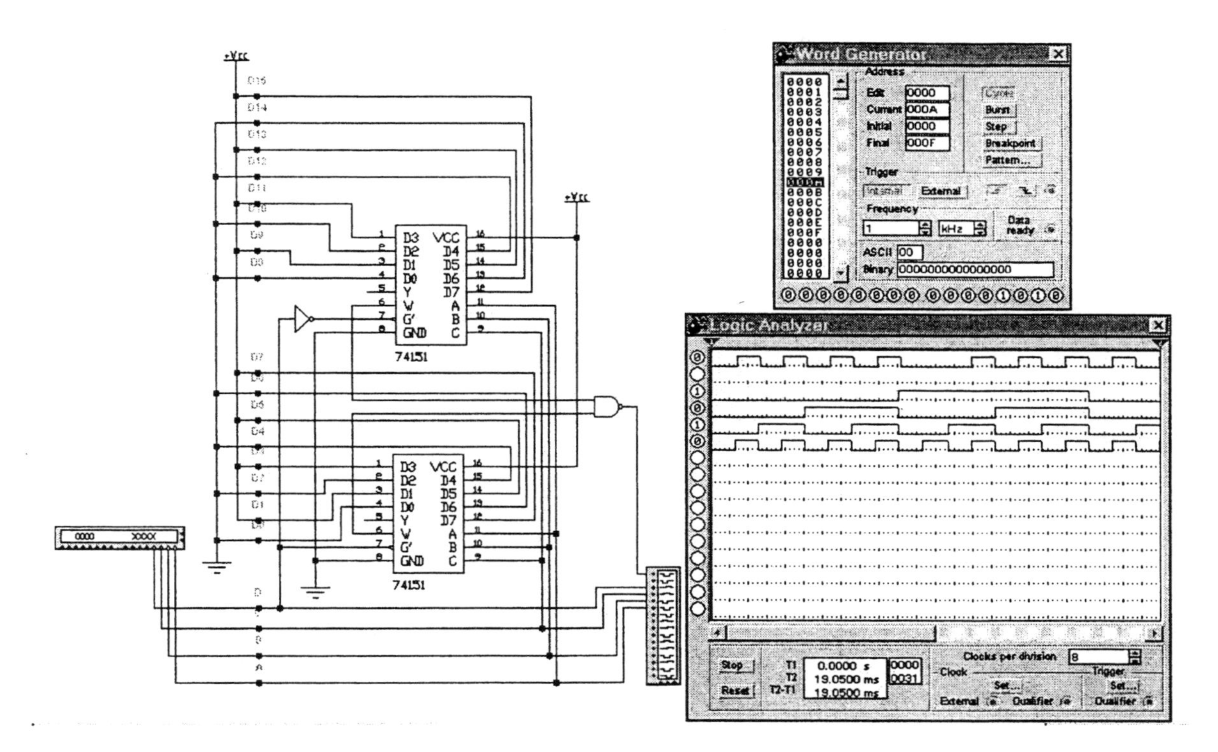

Chapter 7:

There are no EWB troubleshooting problems for this chapter.

Chapter 8:

48. From the logic analyzer waveforms, it appears that the /Q output of U1 is open or possible shorted to ground.

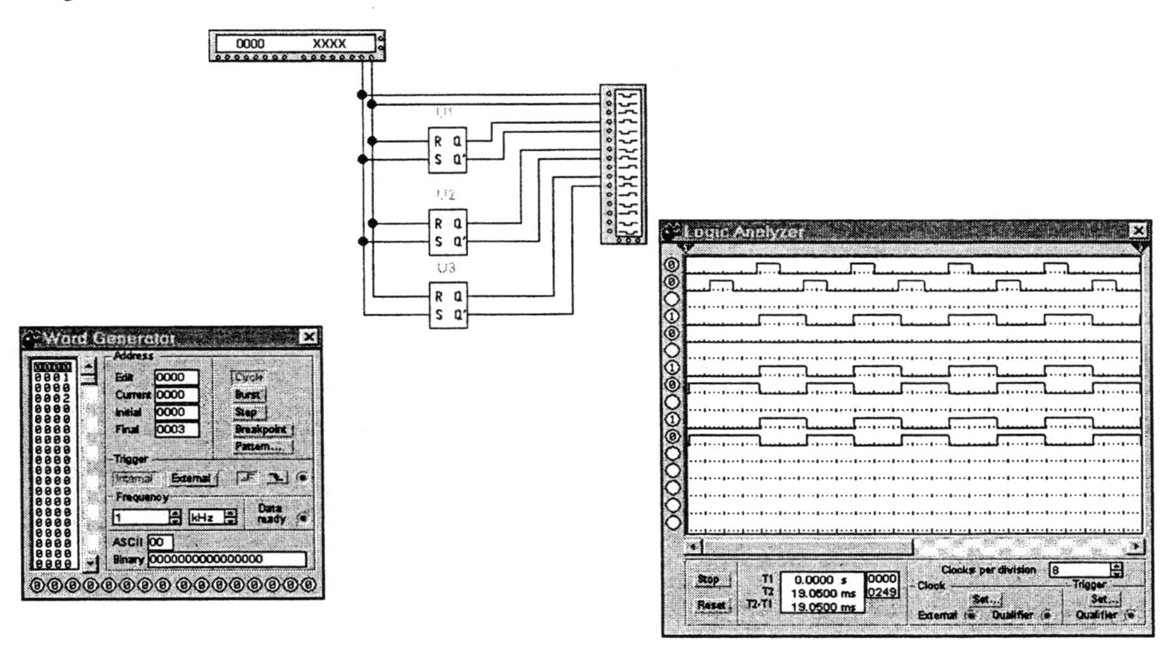

49. From the logic analyzer waveforms, it appears that the K input of Q2 is open.

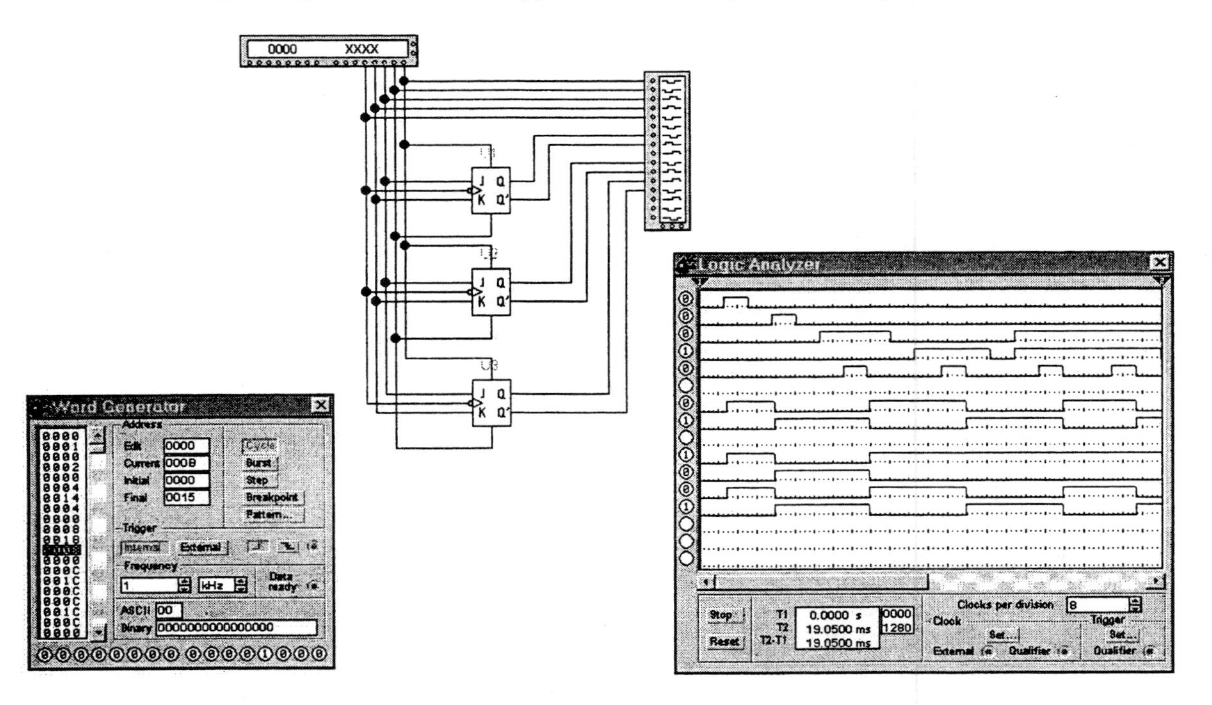

50. From the logic analyzer waveforms, it appears that the /PRE input of U1 is open.

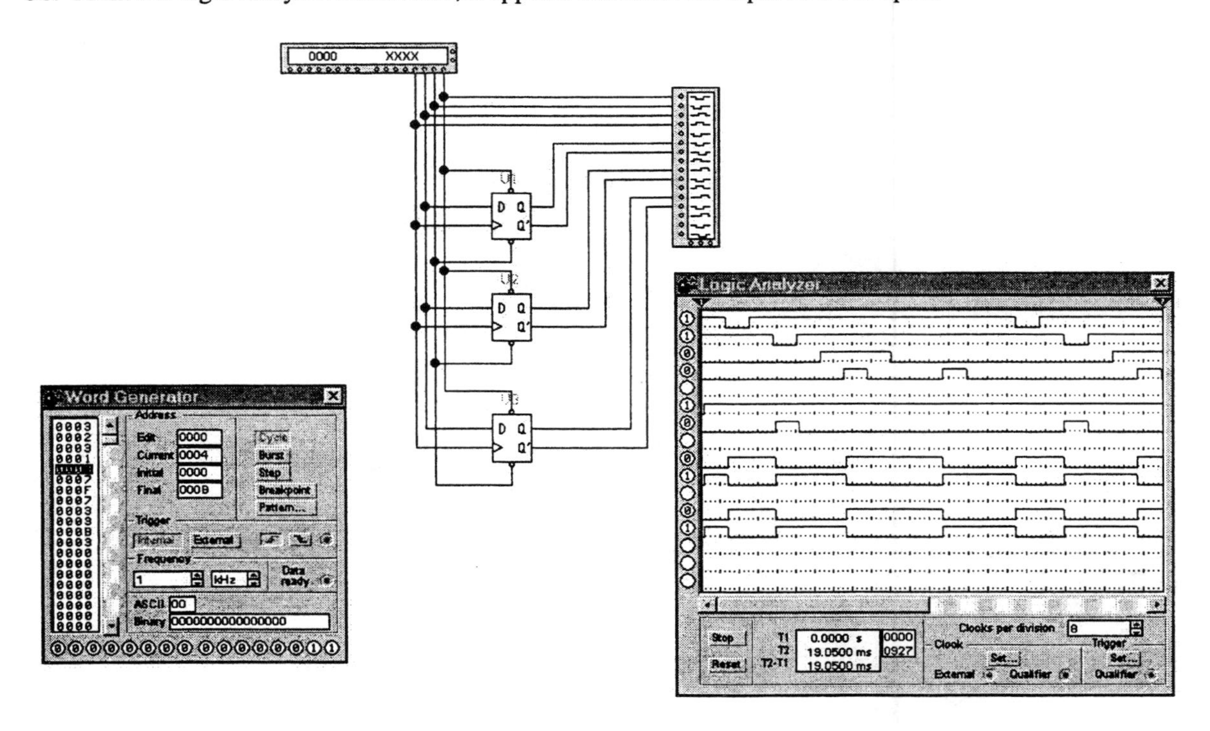

51. The middle one-shot circuit is faulty. Measurement shows that the RC time constant is 1/10th what it should be, so R or C is 1/10th its nominal value. In this case, R = 100 Ω rather than 1 kΩ.

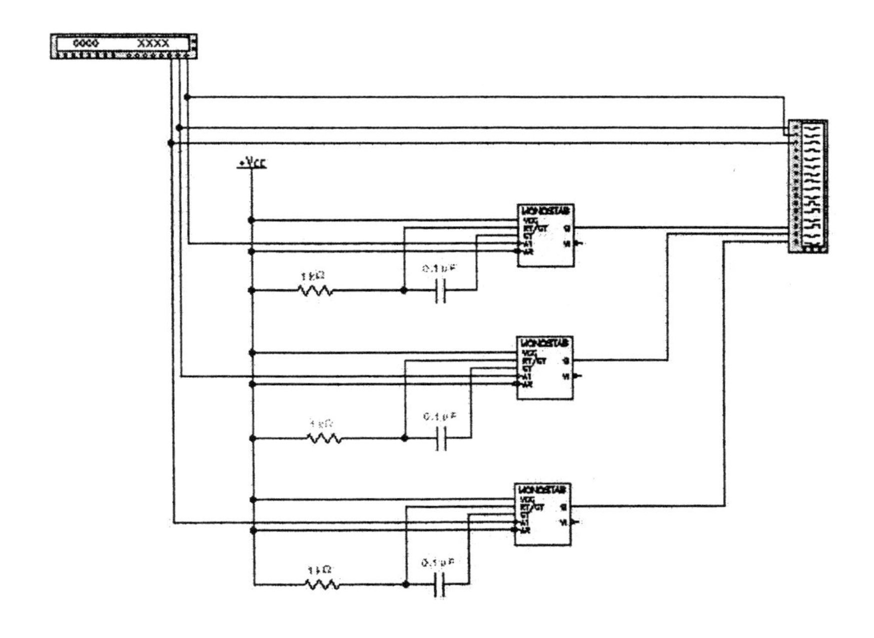

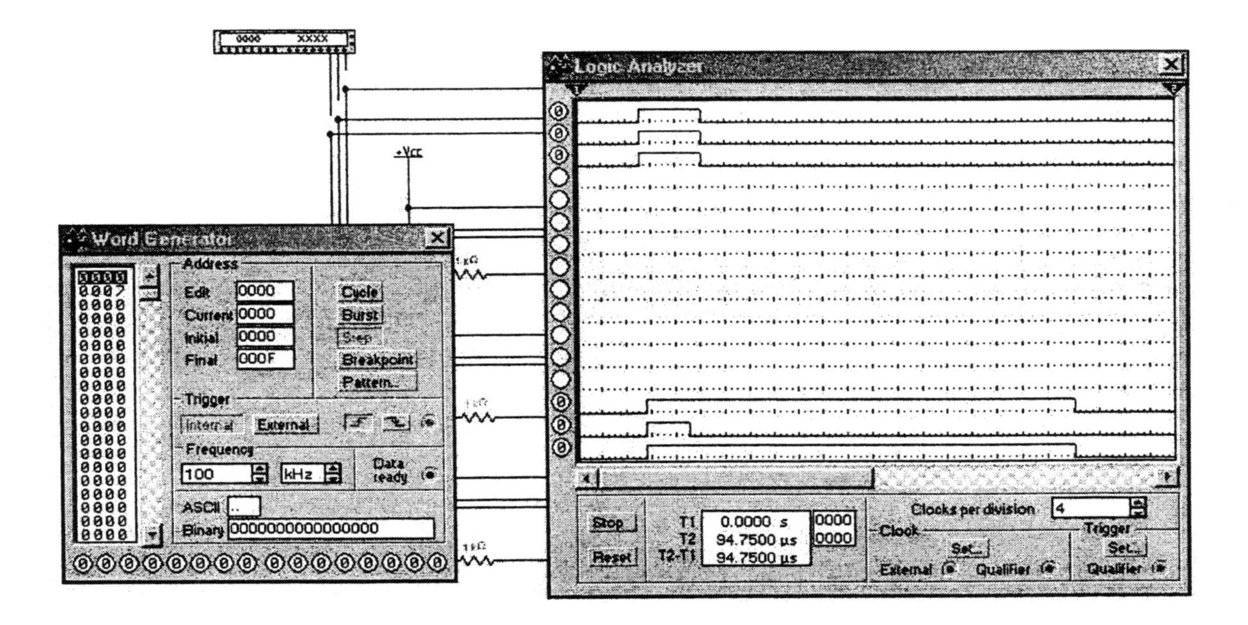

52. From the logic analyzer waveforms, it appears that the clock input of U2 is open. It is possible that the /PRE input of U2 is shorted to ground but an open between Q of U1 and CLK of U2 is more likely.

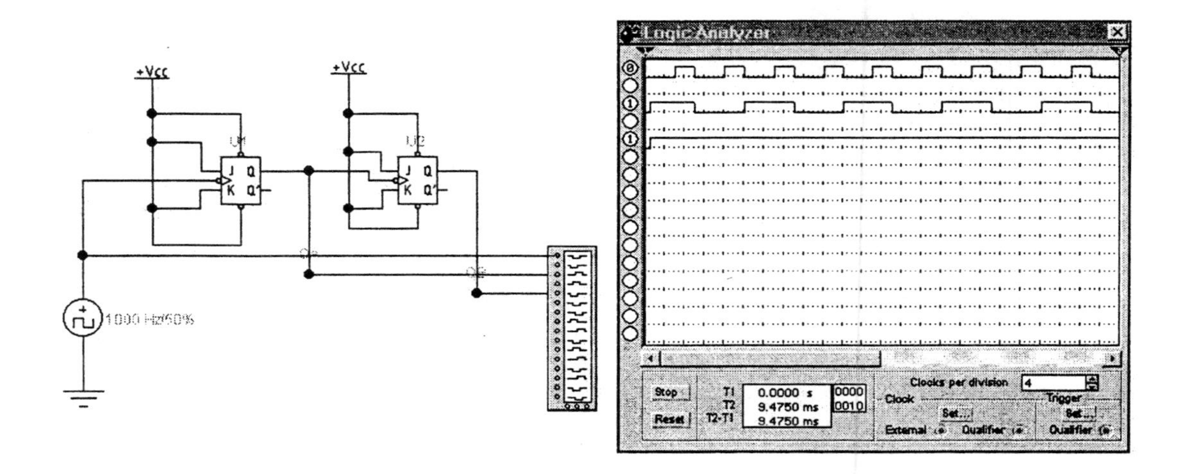

Chapter 9:

55. From the logic analyzer waveforms, it appears that the Q_2 output from U3 is open.

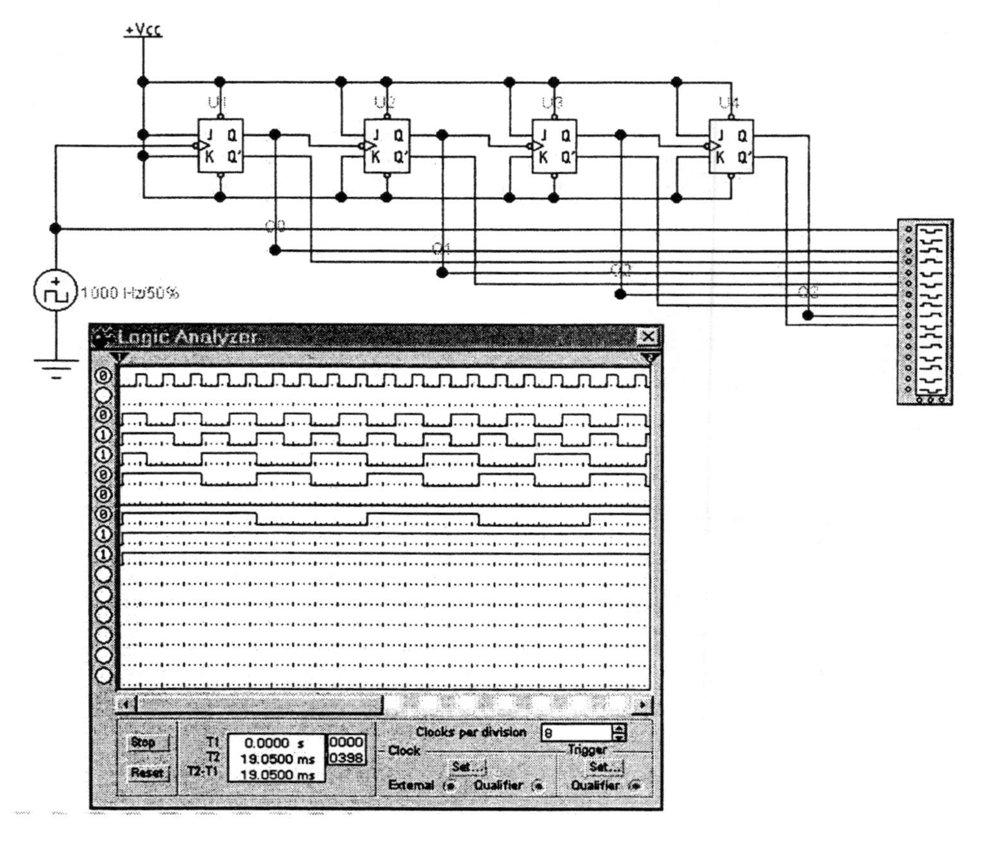

56. The /PRE input of U1 is open and appears to be HIGH.

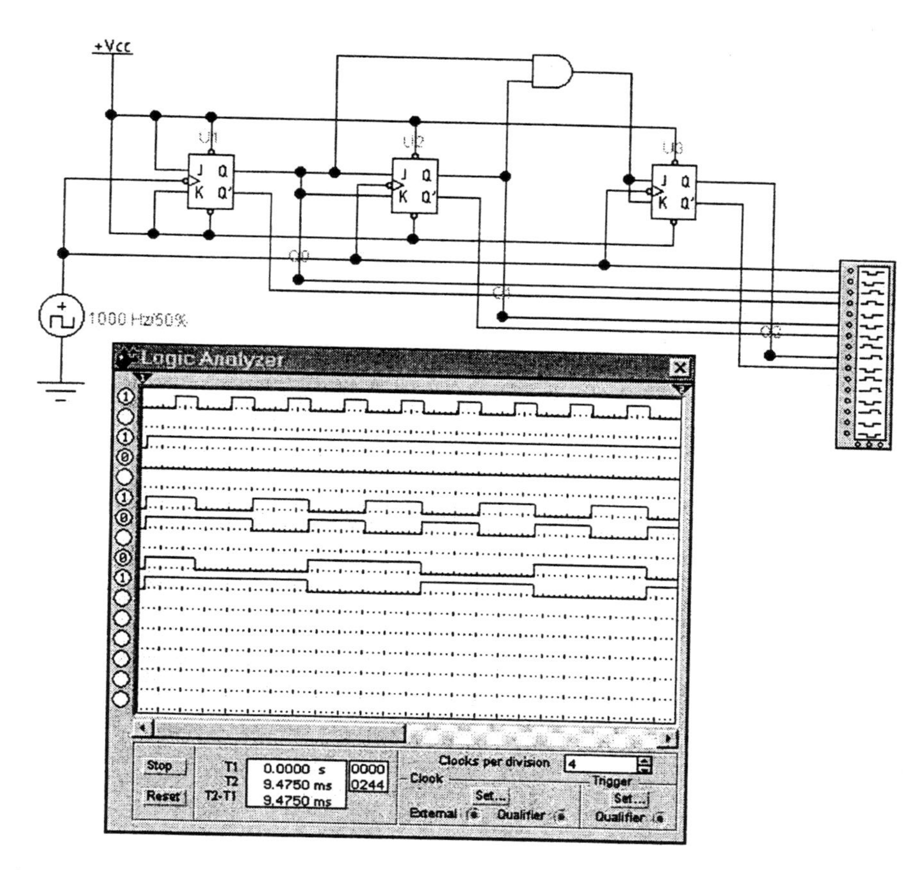

57. From the logic analyzer waveforms, it appears that the Q_3 input to G3 is open.

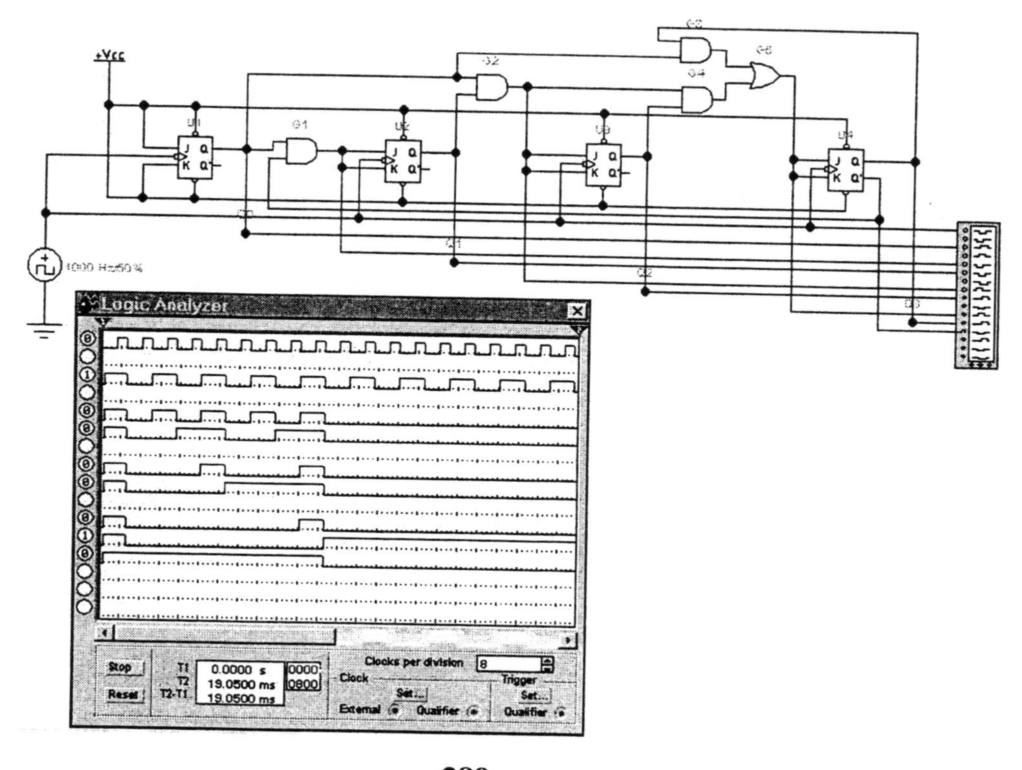

58. From the logic analyzer waveforms the counter is functioning normally. There is no fault.

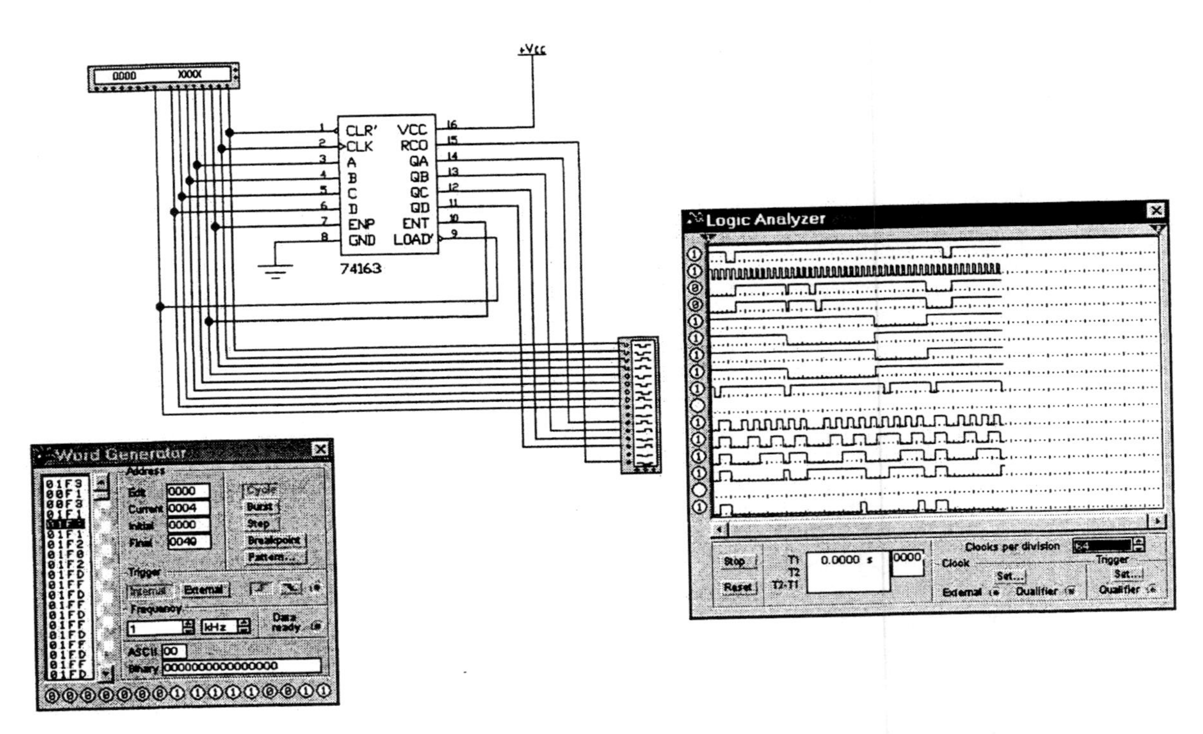

59. From the logic analyzer waveforms, it appears that the D input of the 74190 is open.

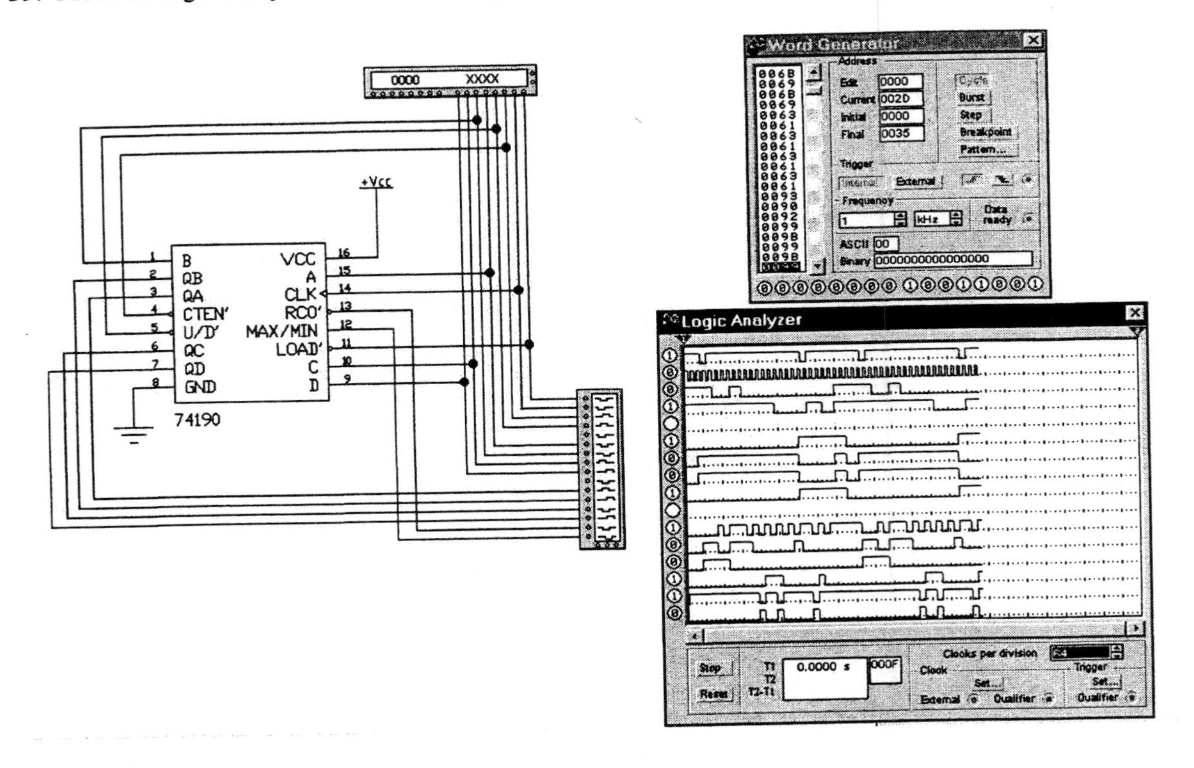

Chapter 10:

47. From the logic analyzer waveforms, it appears that the CLOCK input of U3 is open.

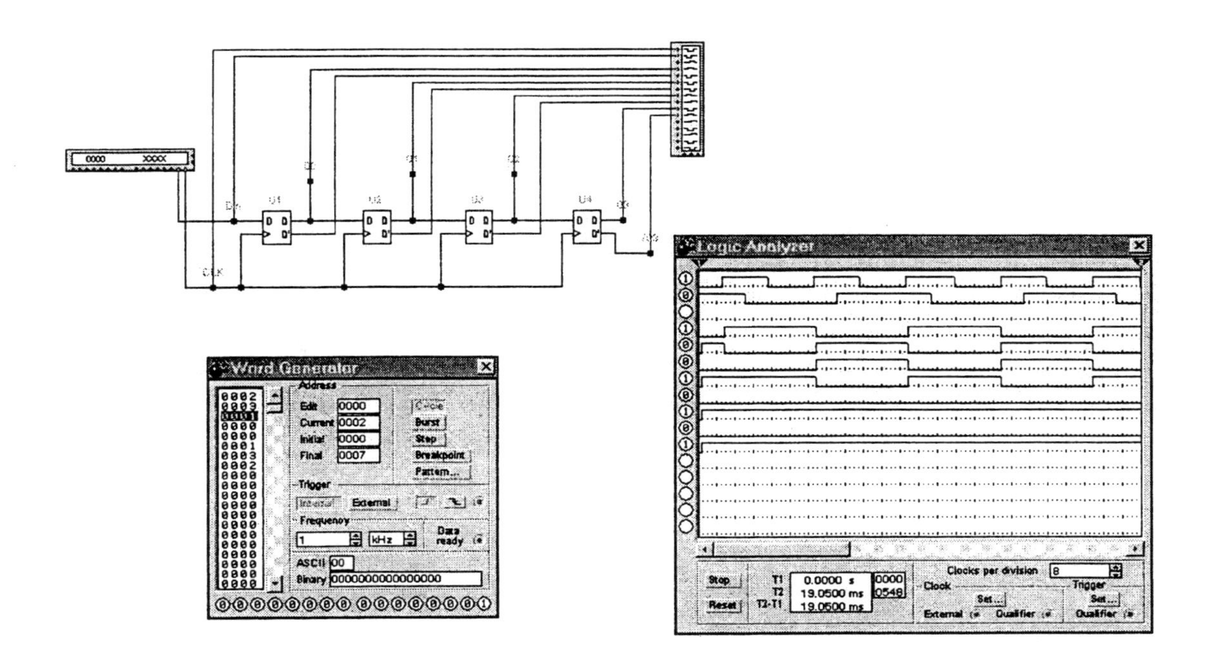

48. From the logic analyzer waveforms, it appears that there is no fault in the circuit.

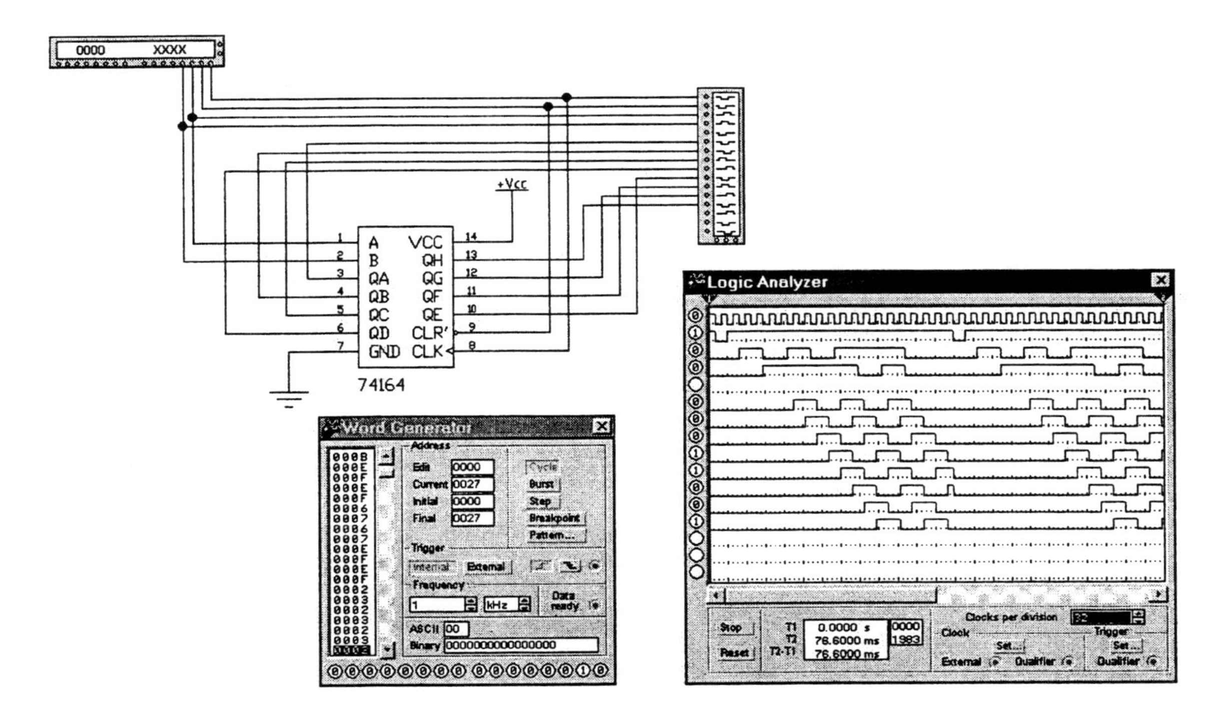

49. From the logic analyzer waveforms, it appears that the D input (Pin 14) is open and appears LOW.

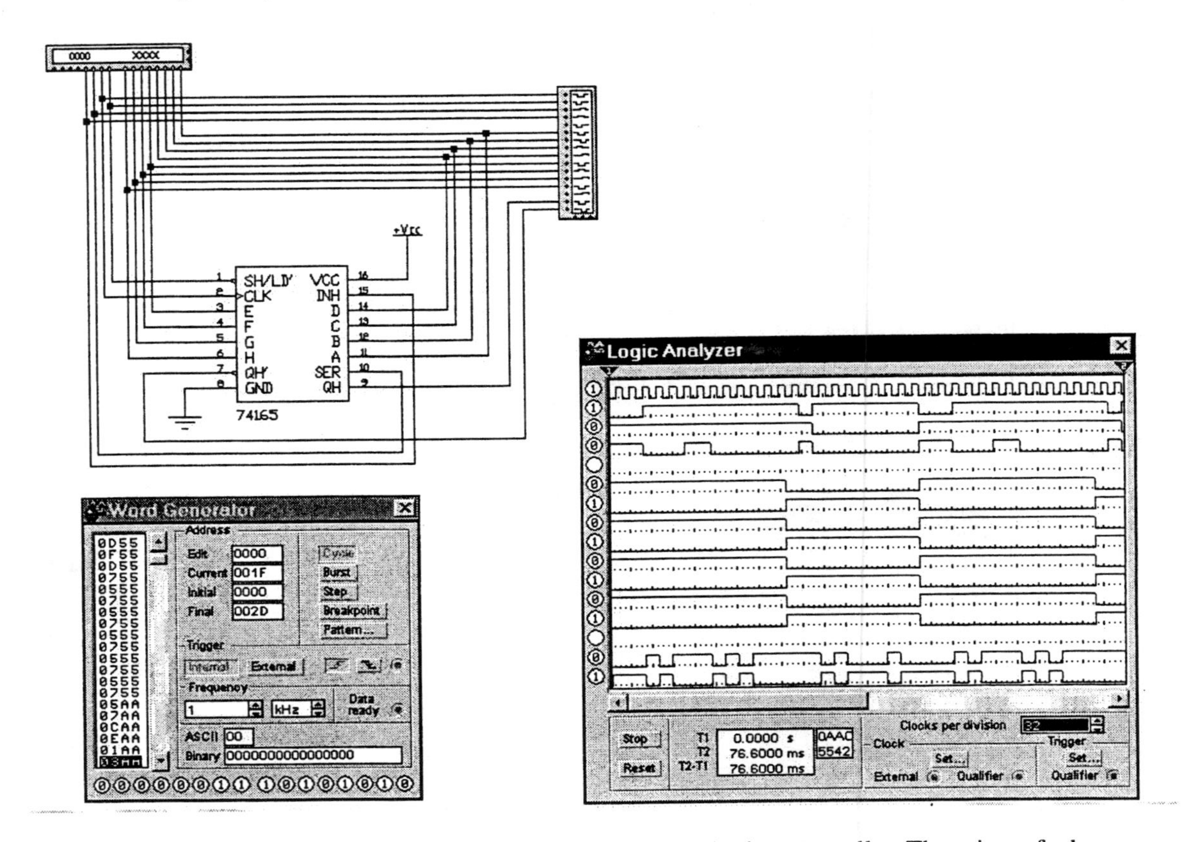

50. From the logic analyzer waveforms the shift register is functioning normally. There is no fault indicated.

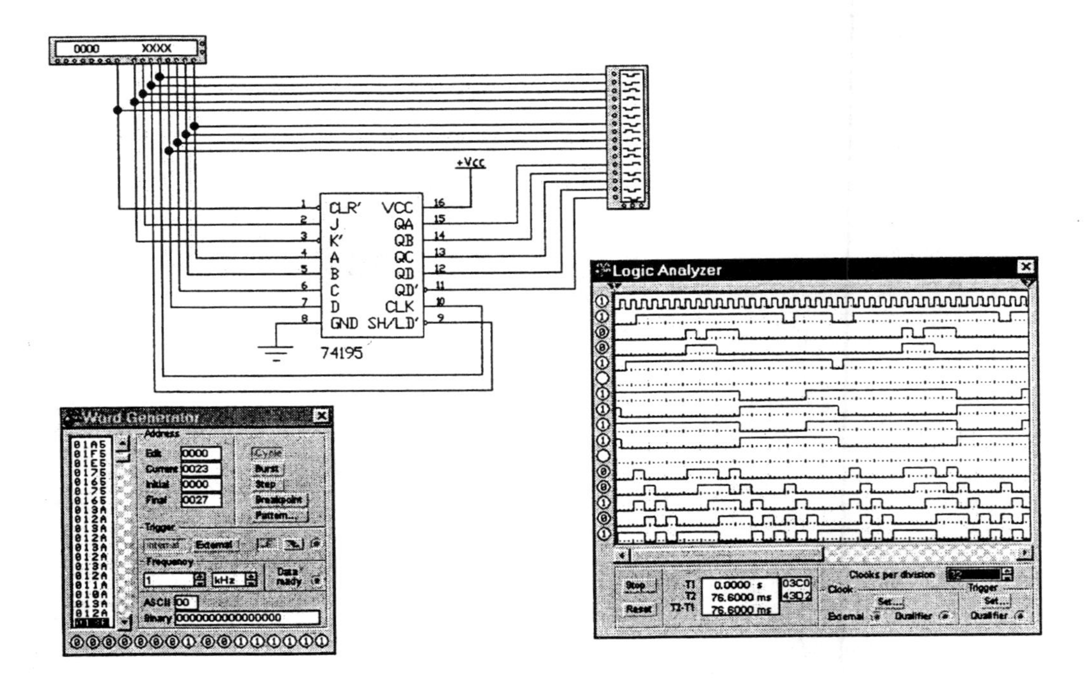

51. From the logic analyzer waveforms, it appears that the CLOCK or data input of U6 is open.

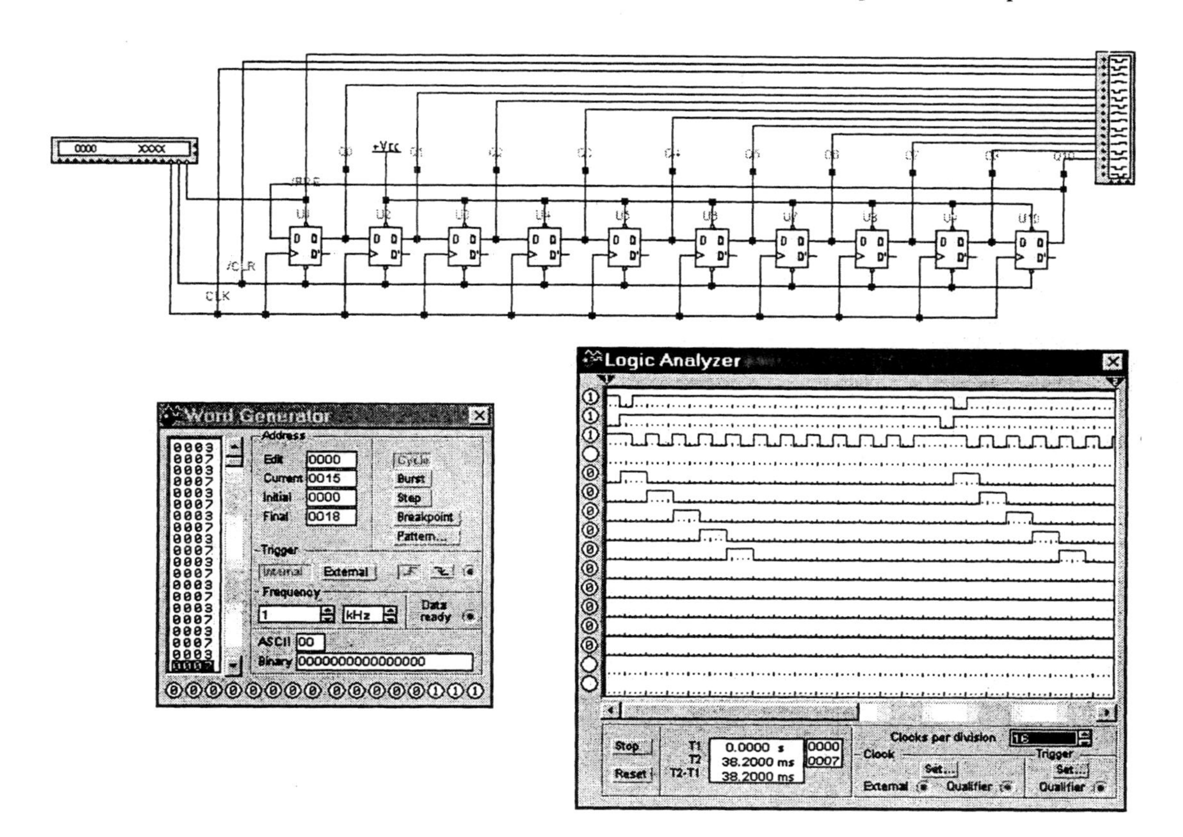

Chapter 11:

37. From the logic analyzer waveforms there is no fault with this circuit.

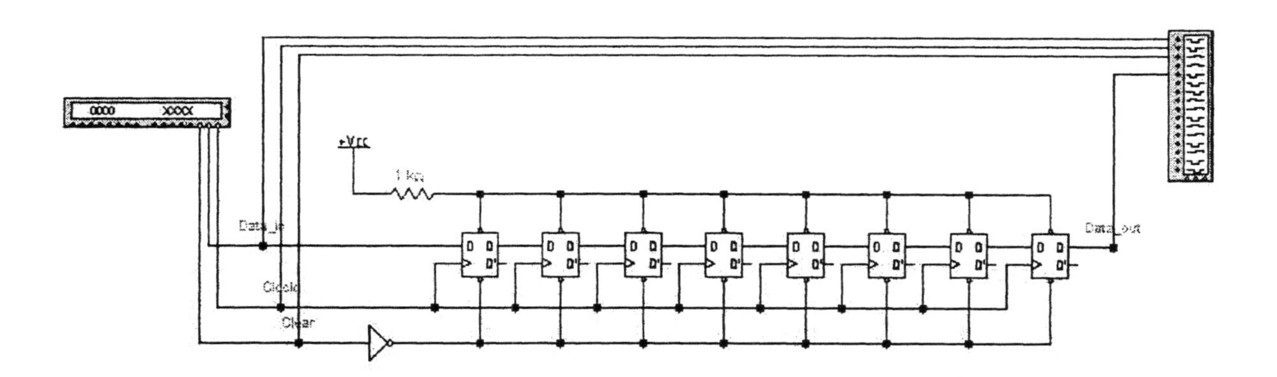

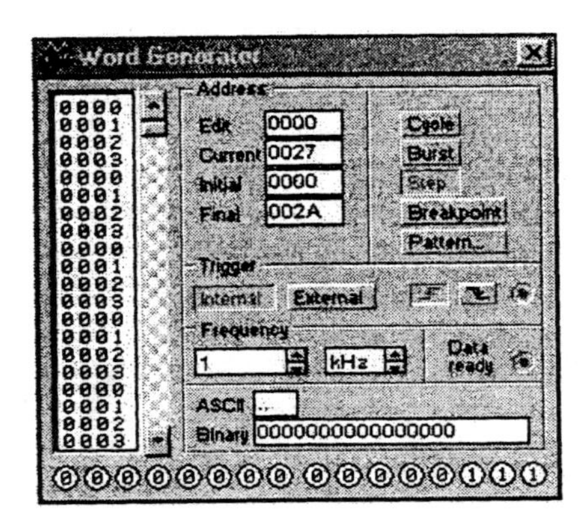

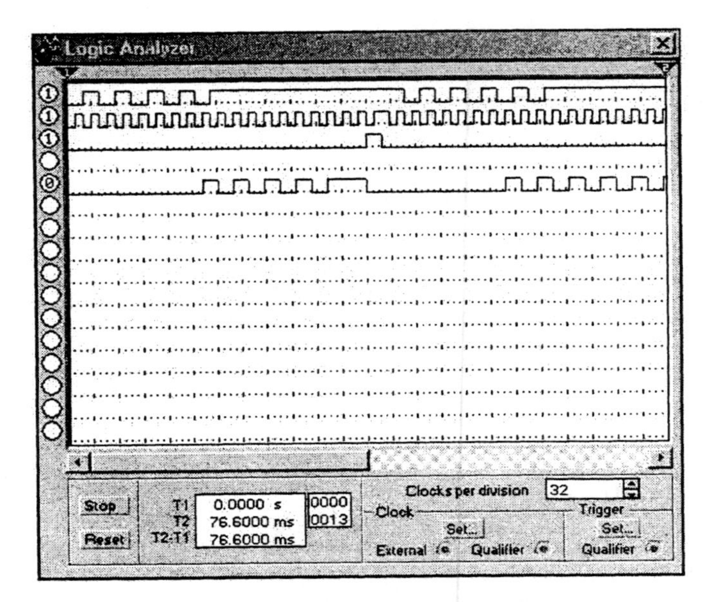

38. From the logic analyzer waveforms, it appears that the /CLEAR input is not working.

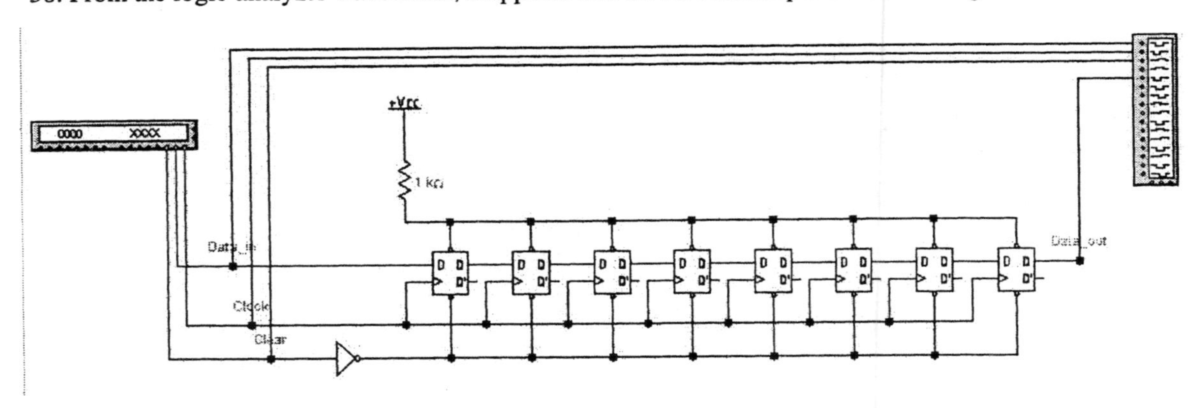

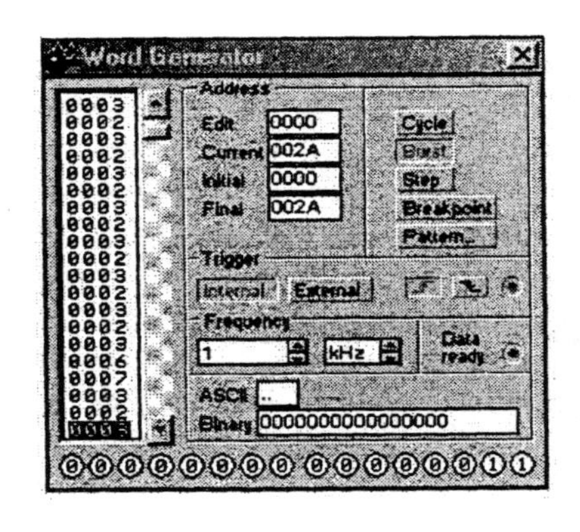

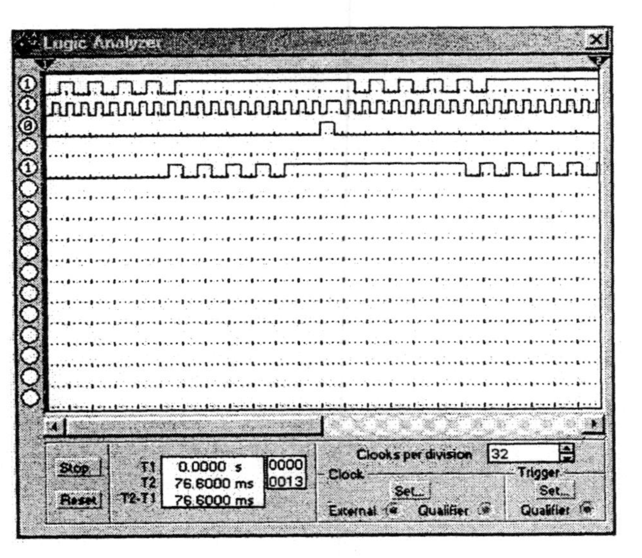

39. From the logic analyzer waveforms, it appears that there is no fault in the circuit.

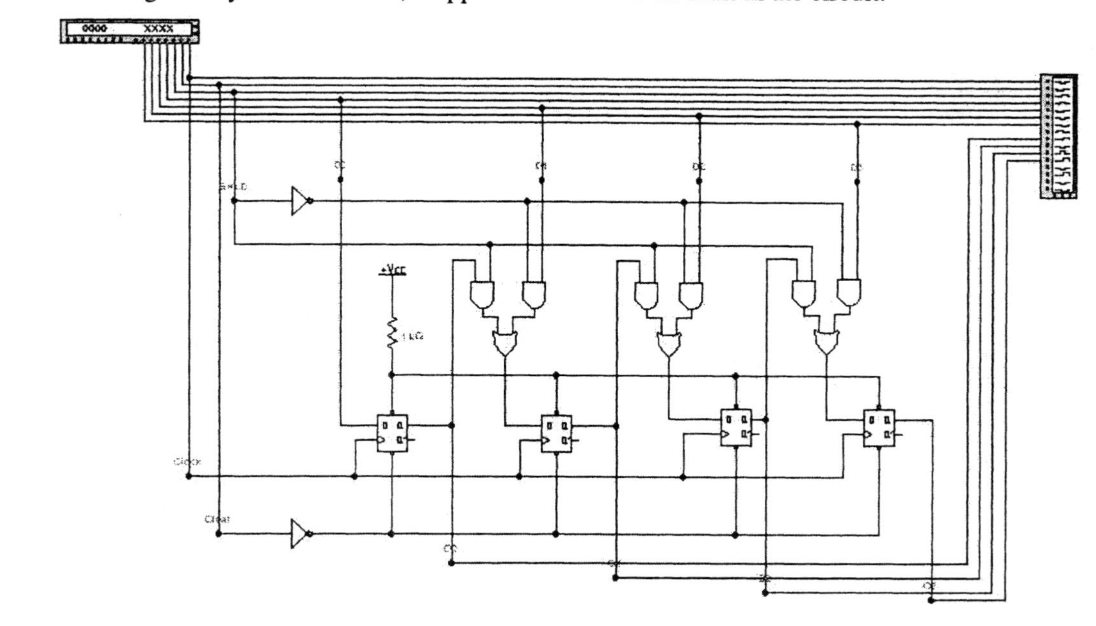

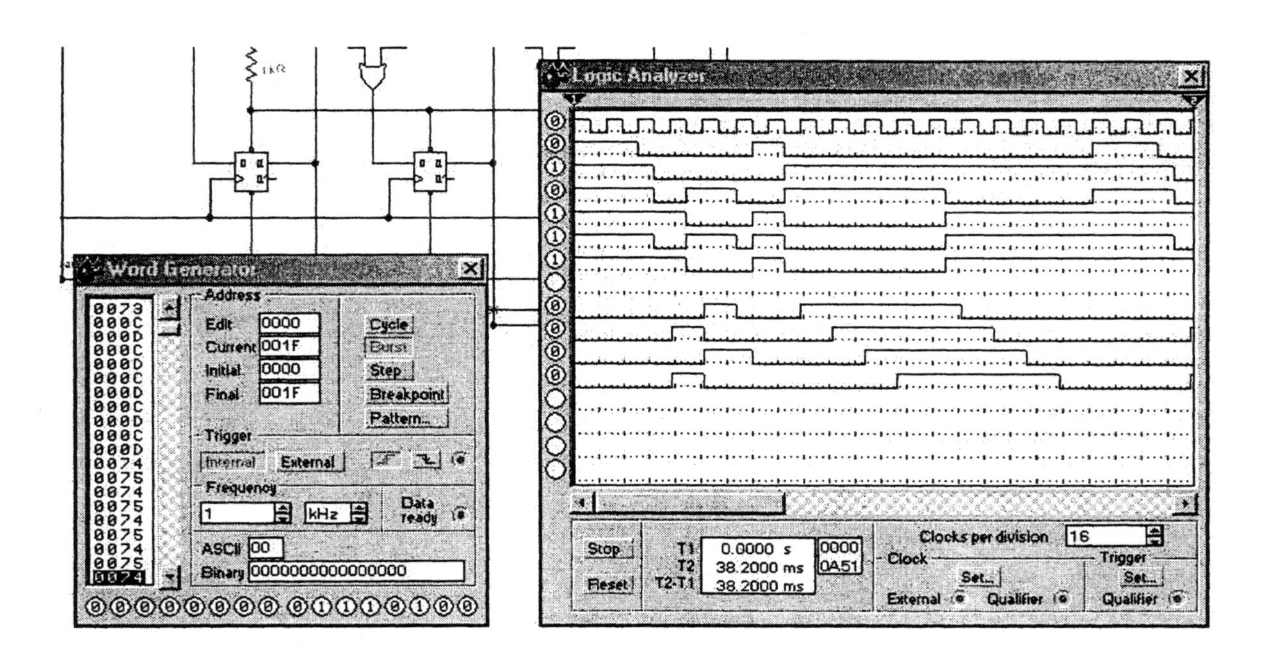

40. From the logic analyzer waveforms it appears that the SHIFT input of FF2 or the gate to it is open.

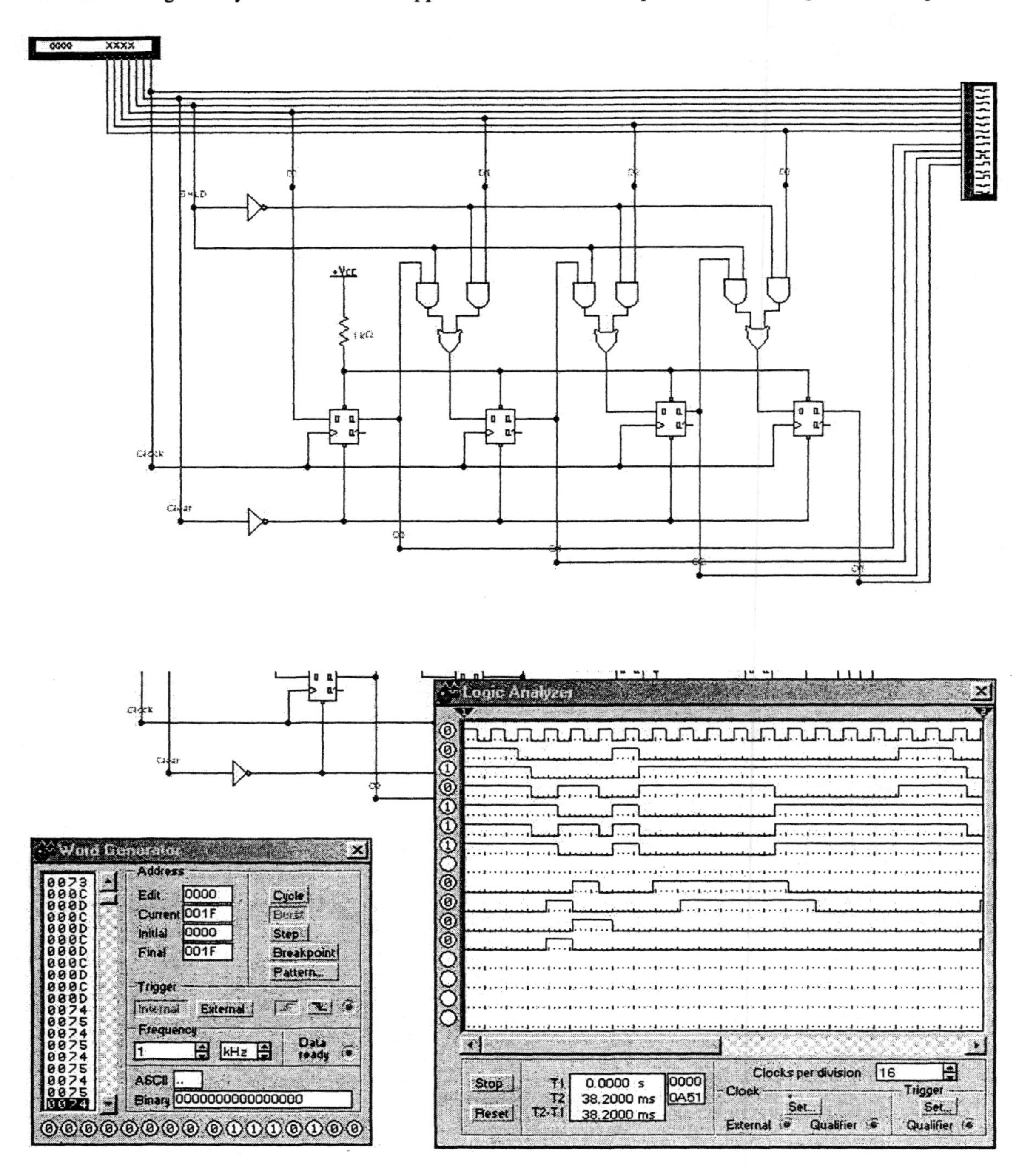

Chapter 12:

There are no EWB problems for this chapter.

Chapter 13:

37. There is actually no problem with the circuit, but the sampling frequency is too low, causing what is known as an *aliasing error* on the oscilloscope. The sampling frequency must be at least twice the highest frequency contained in the waveform to be sampled (called the *Nyquist criterion*) to prevent aliasing.

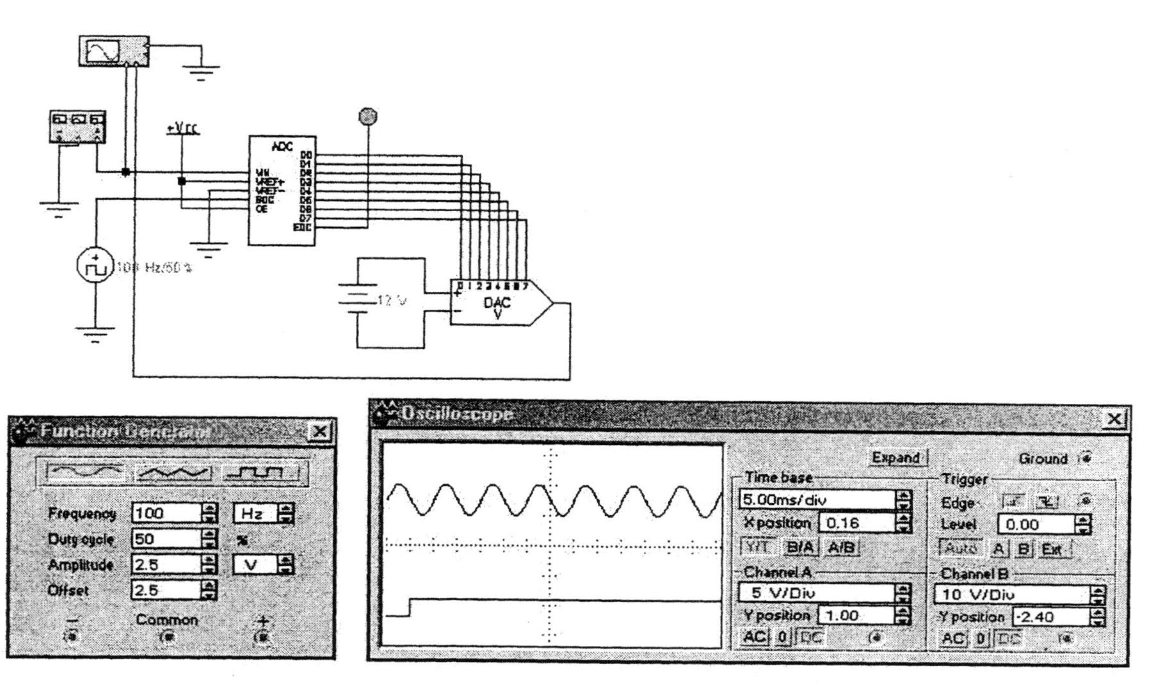

38. From the oscilloscope and logic analyzer waveforms, it appears that the D5 output is open or stuck LOW.

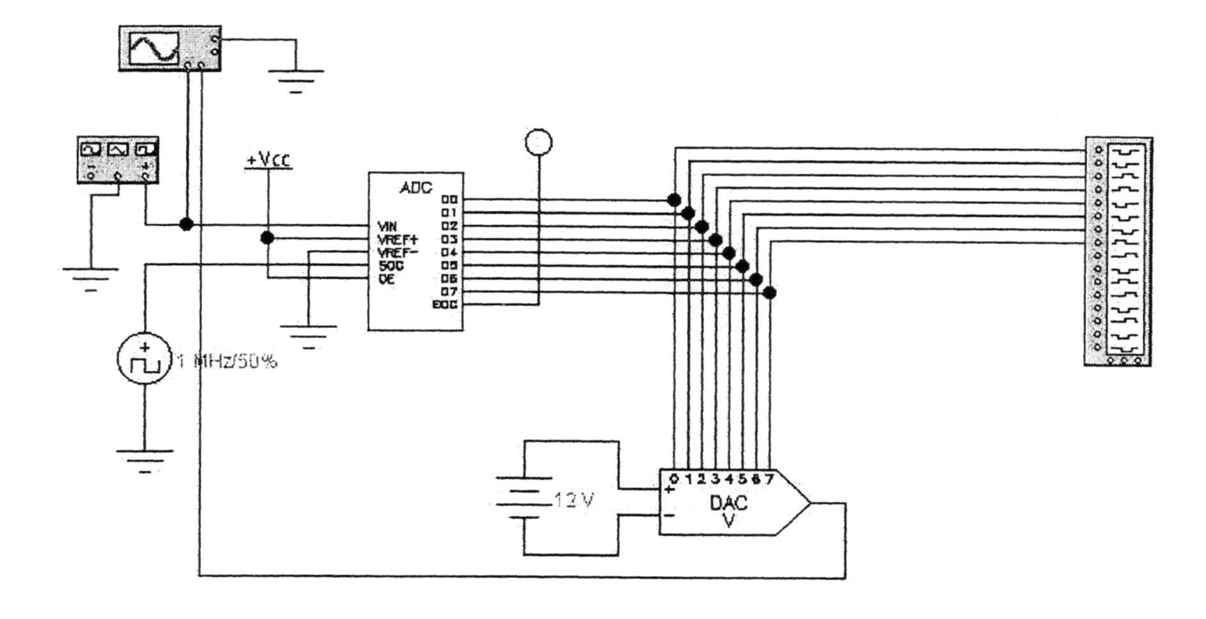

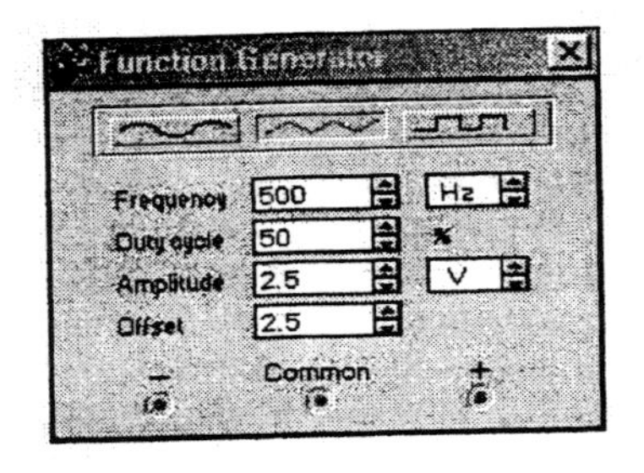

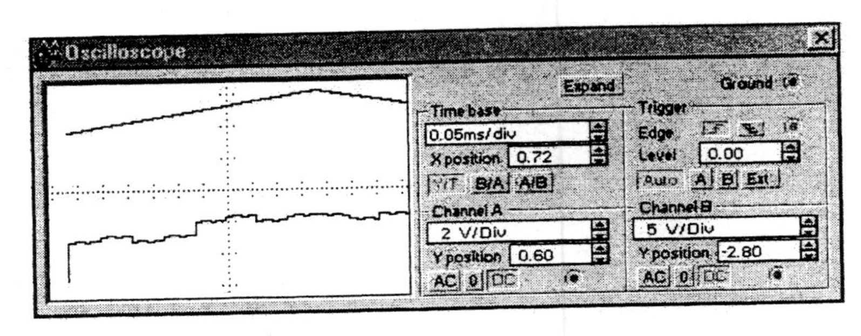

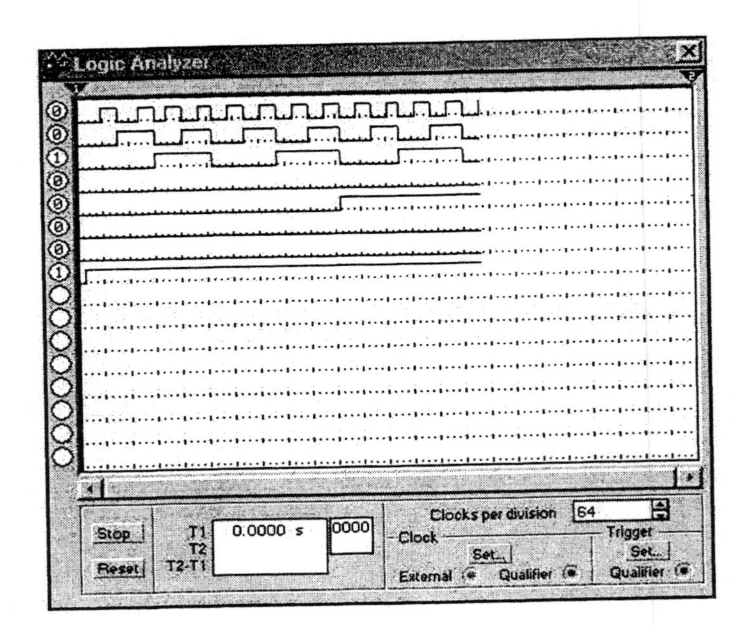

39. From the oscilloscope and logic analyzer waveforms shown, there appears to be no fault in the circuit.

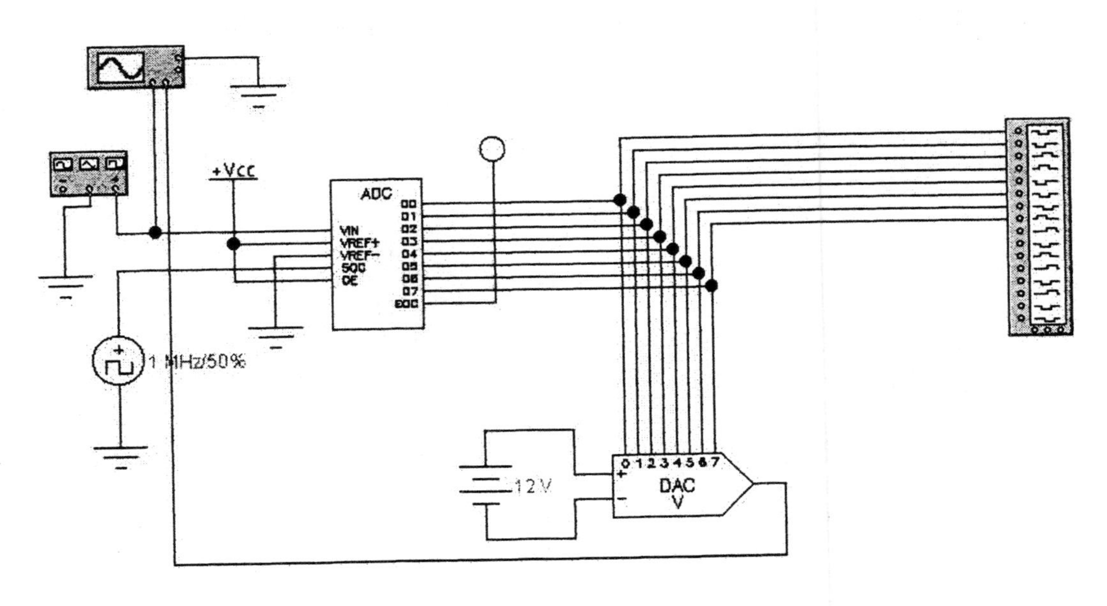

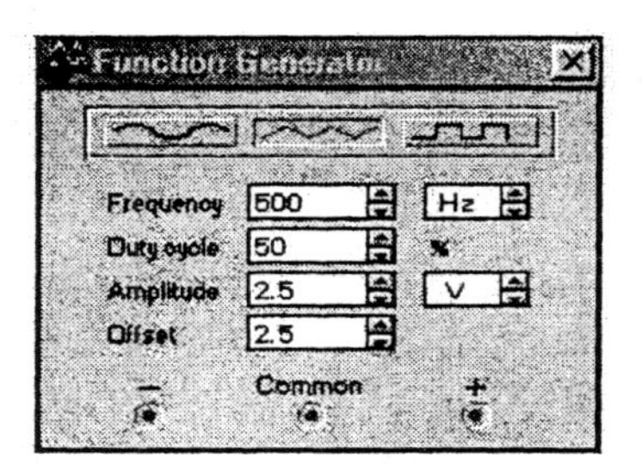

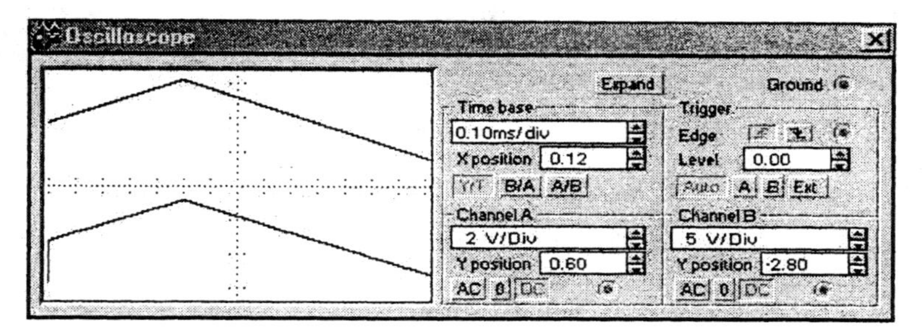

40. From the oscilloscope and logic analyzer waveforms shown, it appears that the D1 and D2 signals are shorted together.

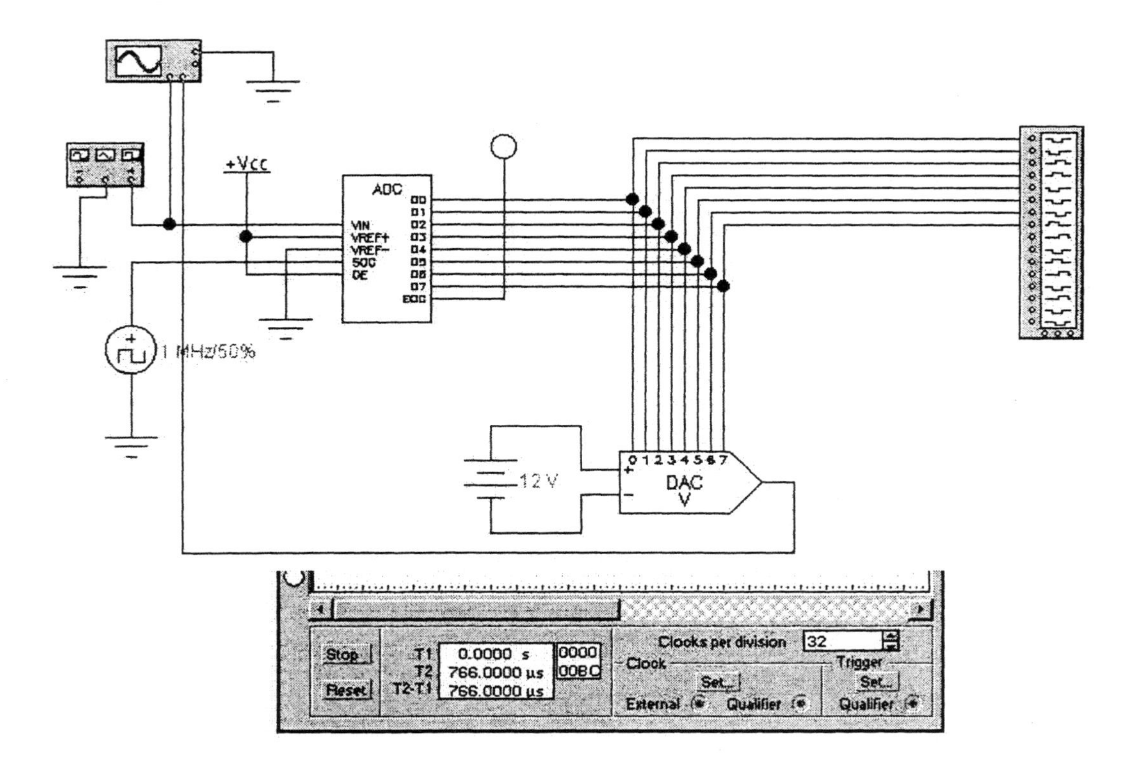

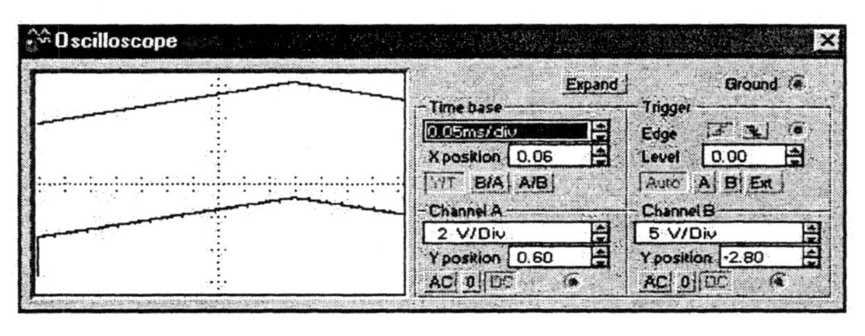

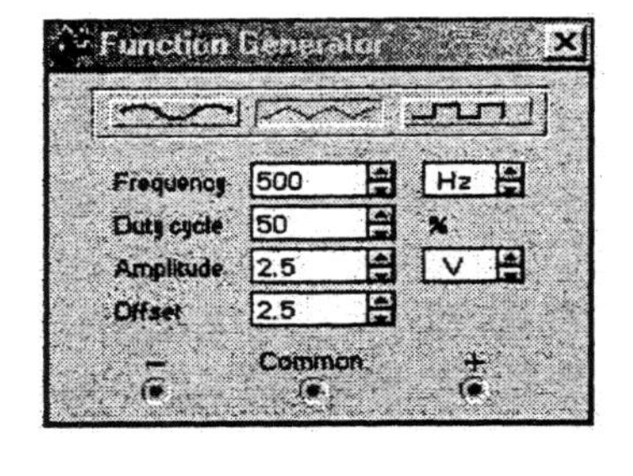

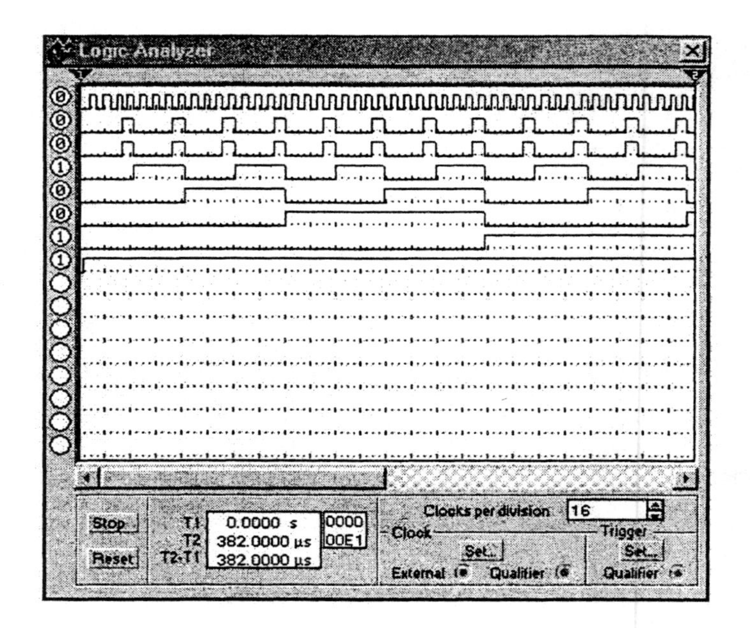

Chapter 14:

There are no EWB troubleshooting problems for this chapter.

Chapter 15:

There are no EWB troubleshooting problems for this chapter.

PART 5

Overview of

IEEE Standard 91-1984

Explanation of Logic Symbols

Courtesy of Texas Instruments

1.0 INTRODUCTION

The International Electrotechnical Commission (IEC) has been developing a very powerful symbolic language that can show the relationship of each input of a digital logic circuit to each output without showing explicitly the internal logic. At the heart of the system is dependency notation, which will be explained in Section 4.

The system was introduced in the USA in a rudimentary form in IEEE/ANSI Standard Y32.14-1973. Lacking at that time a complete development of dependency notation, it offered little more than a substitution of rectangular shapes for the familiar distinctive shapes for representing the basic functions of AND, OR, negation, etc. This is no longer the case.

Internationally, Working Group 2 of IEC Technical Committee TC 3 has prepared a new document (Publication 617-12) that consolidates the original work started in the mid 1960's and published in 1972 (Publication 117-15) and the amendments and supplements that have followed. Similarly for the USA, IEEE Committee SCC 11.9 has revised the publication IEEE Std 91/ANSI Y32.14. Now numbered simply IEEE Std 91-1984, the IEEE standard contains all of the IEC work that has been approved, and also a small amount of material still under international consideration. Texas Instruments is participating in the work of both organizations and this document introduces new logic symbols in accordance with the new standards. When changes are made as the standards develop, future editions will take those changes into account.

The following explanation of the new symbolic language is necessarily brief and greatly condensed from what the standards publications will contain. This is not intended to be sufficient for those people who will be developing symbols for new devices. It is primarily intended to make possible the understanding of the symbols used in various data books and the comparison of the symbols with logic diagrams, functional block diagrams, and/or function tables will further help that understanding.

2.0 SYMBOL COMPOSITION

A symbol comprises an outline or a combination of outlines together with one or more qualifying symbols. The shape of the symbols is not significant. As shown in Figure 1, general qualifying symbols are used to tell exactly what logical operation is performed by the elements. Table I shows general qualifying symbols defined in the new standards. Input lines are placed on the left and output lines are placed on the right. When an exception is made to that convention, the direction of signal flow is indicated by an arrow as shown in Figure 11.

All outputs of a single, unsubdivided element always have identical internal logic states determined by the function of the element except when otherwise indicated by an associated qualifying symbol or label inside the element.

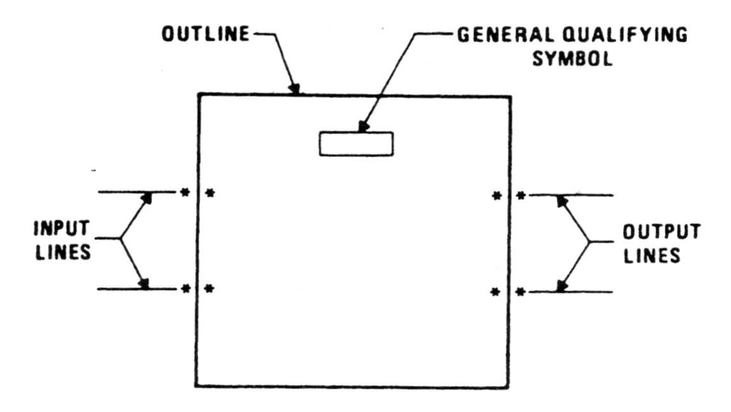

Figure 1. Symbol Composition

The outlines of elements may be abutted or embedded in which case the following conventions apply. There is no logic connection between the elements when the line common to their outlines is in the direction of signal flow. There is at least one logic connection between the elements when the line common to their outlines is perpendicular to the direction of signal flow. The number of logic connections between elements will be clarified by the use of qualifying symbols and this is discussed further under that topic. If no indications are shown on either side of the common line, it is assumed there is only one connection.

When a circuit has one or more inputs that are common to more than one element of the circuit, the common-control block may be used. This is the only distinctively shaped outline used in the IEC system. Figure 2 shows that unless otherwise qualified by dependency notation, an input to the common-control block is an input to each of the elements below the common-control block.

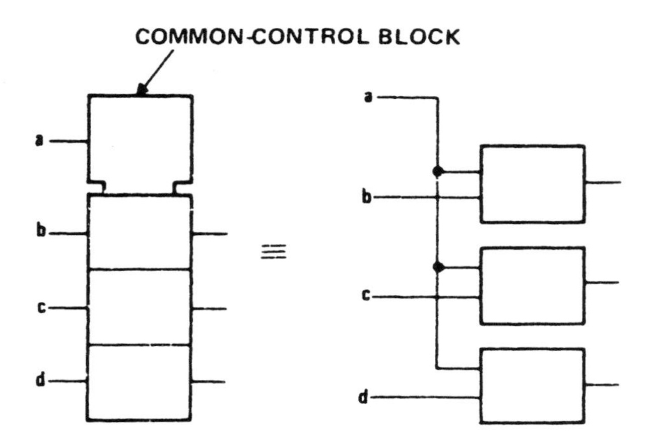

Figure 2. Common-Control Block

A common output depending on all elements of the array can be shown as the output of a common-output element. Its distinctive visual feature is the double line at its top. In addition the common-output element may have other inputs as shown in Figure 3. The function of the common-output element must be shown by use of a general qualifying symbol.

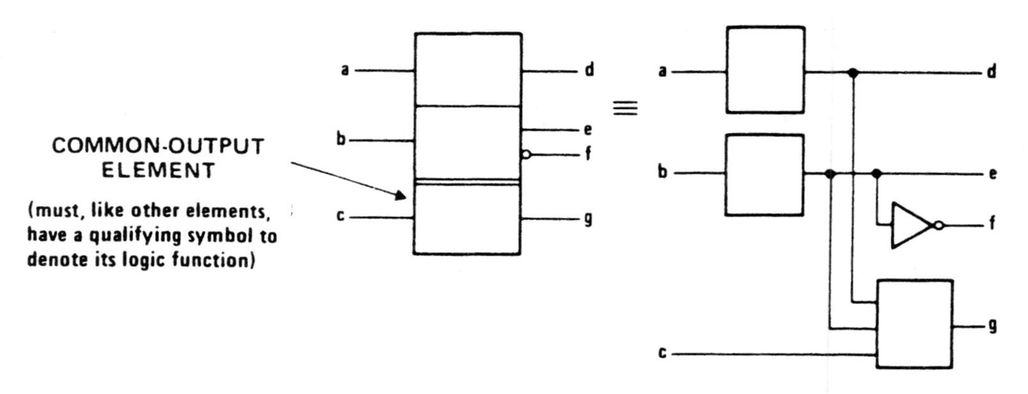

Figure 3. Common-Output Element

3.0 QUALIFYING SYMBOLS

3.1 General Qualifying Symbols

Table I shows general qualifying symbols defined by IEEE Standard 91. These characters are placed near the top center or the geometric center of a symbol or symbol element to define the basic function of the device represented by the symbol or of the element.

3.2 Qualifying Symbols for Inputs and Outputs

Qualifying symbols for inputs and outputs are shown in Table II and will be familiar to most users with the possible exception of the logic polarity and analog signal indicators. The older logic negation indicator means that the external 0 state produces the internal 1 state. The internal 1 state means the active state. Logic negation may be used in pure logic diagrams; in order to tie the external 1 and 0 logic states to the levels H (high) and L (low), a statement of whether positive logic (1 = H, 0 = L) or negative logic (1 = L, 0 = H) is being used is required or must be assumed. Logic polarity indicators eliminate the need for calling out the logic convention and are used in various data books in the symbology for actual devices. The presence of the triangular polarity indicator indicates that the L logic level will produce the internal 1 state (the active state) or that, in the case of an output, the internal 1 state will produce the external L level. Note how the active direction of transition for a dynamic input is indicated in positive logic, negative logic, and with polarity indication.

The internal connections between logic elements abutted together in a symbol may be indicated by the symbols shown in Table II. Each logic connection may be shown by the presence of qualifying symbols at one or both sides of the common line and if confusion can arise about the numbers of connections, use can be made of one of the internal connection symbols.

Table I. General Qualifying Symbols

SYMBOL	DESCRIPTION	CMOS EXAMPLE	TTL EXAMPLE
&	AND gate or function.	'HC00	SN7400
≥ 1	OR gate or function. The symbol was chosen to indicate that at least one active input is needed to activate the output.	'HC02	SN7402
$= 1$	Exclusive OR. One and only one input must be active to activate the output.	'HC86	SN7486
=	Logic identity. All inputs must stand at the same state.	'HC86	SN74180
2k	An even number of inputs must be active.	'HC280	SN74180
2k + 1	An odd number of inputs must be active.	'HC86	SN74ALS86
1	The one input must be active.	'HC04	SN7404
▷ or ◁	A buffer or element with more than usual output capability (symbol is oriented in the direction of signal flow).	'HC240	SN74S436
⌷	Schmitt trigger; element with hysteresis.	'HC132	SN74LS18
X/Y	Coder, code converter (DEC/BCD, BIN/OUT, BIN/7-SEG, etc.).	'HC42	SN74LS347
MUX	Multiplexer/data selector.	'HC151	SN74150
DMUX or DX	Demultiplexer.	'HC138	SN74138
Σ	Adder.	'HC283	SN74LS385
P−Q	Subtracter.	*	SN74LS385
CPG	Look-ahead carry generator	'HC182	SN74182
π	Multiplier.	*	SN74LS384
COMP	Magnitude comparator.	'HC85	SN74LS682
ALU	Arithmetic logic unit.	'HC181	SN74LS381
⎍	Retriggerable monostable.	'HC123	SN74LS422
1⎍	Nonretriggerable monostable (one-shot)	'HC221	SN74121
G ⎍⎍	Astable element. Showing waveform is optional.	*	SN74LS320
!G ⎍⎍	Synchronously starting astable.	*	SN74LS624
G! ⎍⎍	Astable element that stops with a completed pulse.	*	*
SRGm	Shift register. m = number of bits.	'HC164	SN74LS595
CTRm	Counter. m = number of bits; cycle length = 2^m.	'HC590	SN54LS590
CTR DIVm	Counter with cycle length = m.	'HC160	SN74LS668
RCTRm	Asynchronous (ripple-carry) counter; cycle length = 2^m.	'HC4020	*
ROM	Read-only memory.	*	SN74187
RAM	Random-access read/write memory.	'HC189	SN74170
FIFO	First-in, first-out memory.	*	SN74LS222
I=0	Element powers up cleared to 0 state.	*	SN74AS877
I=1	Element powers up set to 1 state.	'HC7022	SN74AS877
Φ	Highly complex function; "gray box" symbol with limited detail shown under special rules.	*	SN74LS608

*Not all of the general qualifying symbols have been used in TI's CMOS and TTL data books, but they are included here for the sake of completeness.

Table II. Qualifying Symbols for Inputs and Outputs

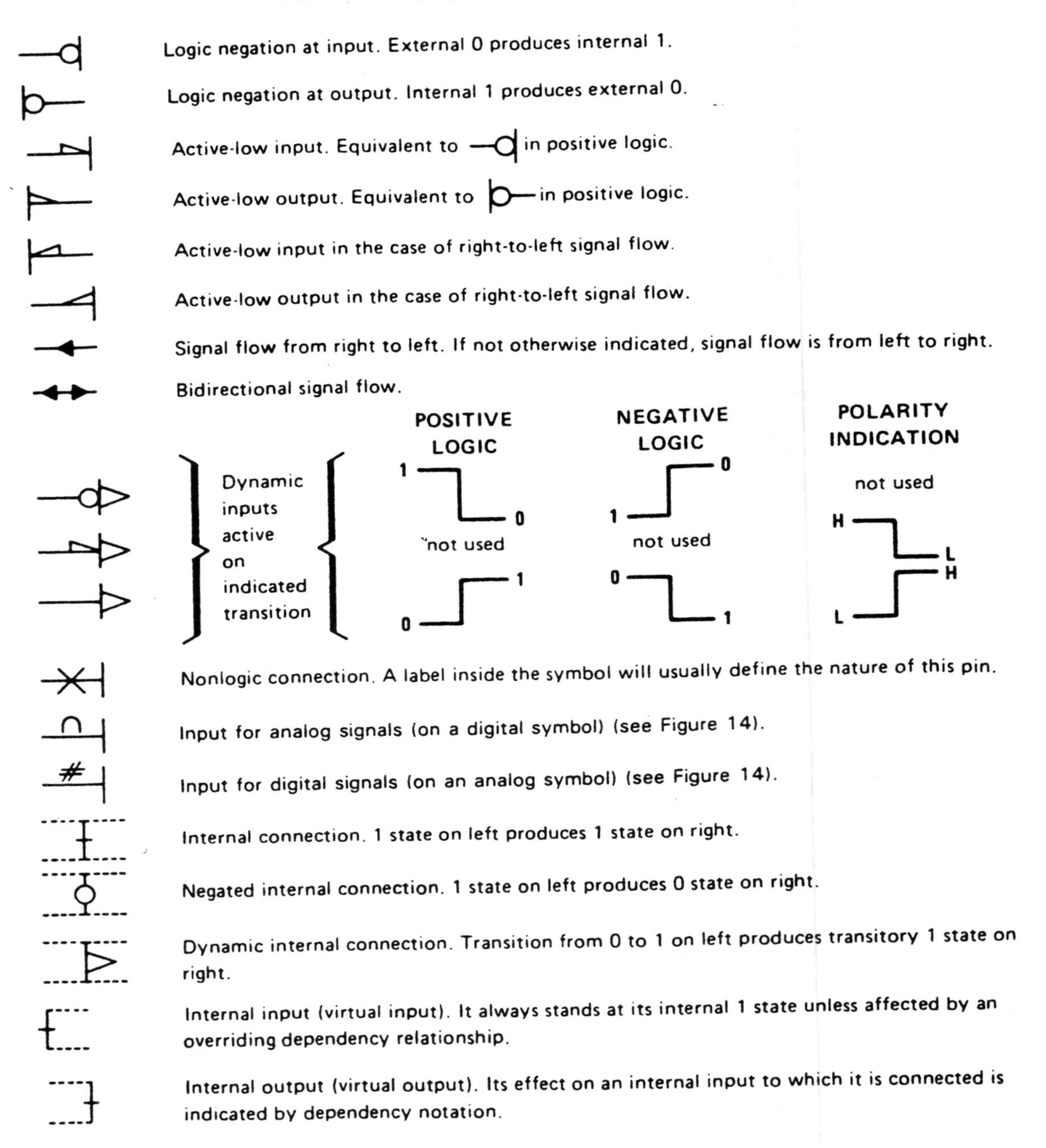

The internal (virtual) input is an input originating somewhere else in the circuit and is not connected directly to a terminal. The internal (virtual) output is likewise not connected directly to a terminal. The application of internal inputs and outputs requires an understanding of dependency notation, which is explained in Section 4.

Table III. Symbols Inside the Outline

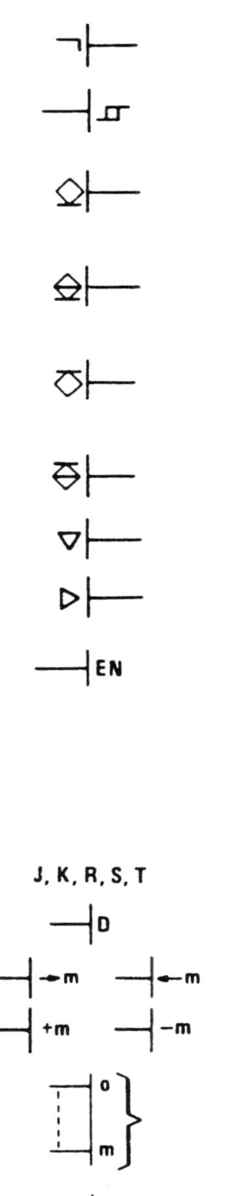

Postponed output (of a pulse-triggered flip-flop). The output changes when input initiating change (e.g., a C input) returns to its initial external state or level. See § 5.

Bi-threshold input (input with hysteresis).

N-P-N open-collector or similar output that can supply a relatively low-impedance L level when not turned off. Requires external pull-up. Capable of positive-logic wired-AND connection.

Passive-pull-up output is similar to N-P-N open-collector output but is supplemented with a built-in passive pull-up.

N-P-N open-emitter or similar output that can supply a relatively low-impedance H level when not turned off. Requires external pull-down. Capable of positive-logic wired-OR connection.

Passive-pull-down output is similar to N-P-N open-emitter output but is supplemented with a built-in passive pull-down.

3-state output.

Output with more than usual output capability (symbol is oriented in the direction of signal flow).

Enable input
When at its internal 1-state, all outputs are enabled.
When at its internal 0-state, open-collector and open-emitter outputs are off, three-state outputs are at normally defined internal logic states and at external high-impedance state, and all other outputs (e.g., totem-poles) are at the internal 0-state.

Usual meanings associated with flip-flops (e.g., R = reset, T = toggle)

Data input to a storage element equivalent to: S R

Shift right (left) inputs, m = 1, 2, 3, etc. If m = 1, it is usually not shown.

Counting up (down) inputs, m = 1, 2, 3, etc. If m = 1, it is usually not shown.

Binary grouping. m is highest power of 2.

The contents-setting input, when active, causes the content of a register to take on the indicated value.

The content output is active if the content of the register is as indicated.

Input line grouping . . . indicates two or more terminals used to implement a single logic input.

e.g., The paired expander inputs of SN7450. X X̄ E

Fixed-state output always stands at its internal 1 state. For example, see SN74185.

In an array of elements, if the same general qualifying symbol and the same qualifying symbols associated with inputs and outputs would appear inside each of the elements of the array, these qualifying symbols are usually shown only in the first element. This is done to reduce clutter and to save time in recognition. Similarly, large identical elements that are subdivided into smaller elements may each be represented by an unsubdivided outline. The SN54HC242 or SN54LS440 symbol illustrates this principle.

3.3 Symbols Inside the Outline

Table III shows some symbols used inside the outline. Note particularly that open-collector (open-drain), open-emitter (open-source), and three-state outputs have distinctive symbols. Also note that an EN input affects all of the outputs of the circuit and has no effect on inputs. When an enable input affects only certain outputs and/or affects one or more inputs, a form of dependency notation will indicate this (see 4.9). The effects of the EN input on the various types of outputs are shown.

It is particularly important to note that a D input is always the data input of a storage element. At its internal 1 state, the D input sets the storage element to its 1 state, and at its internal 0 state it resets the storage element to its 0 state.

The binary grouping symbol will be explained more fully in Section 8. Binary-weighted inputs are arranged in order and the binary weights of the least-significant and the most-significant lines are indicated by numbers. In this document weights of input and output lines will be represented by powers of two usually only when the binary grouping symbol is used, otherwise decimal numbers will be used. The grouped inputs generate an internal number on which a mathematical function can be performed or that can be an identifying number for dependency notation (Figure 28). A frequent use is in addresses for memories.

Reversed in direction, the binary grouping symbol can be used with outputs. The concept is analogous to that for the inputs and the weighted outputs will indicate the internal number assumed to be developed within the circuit.

Other symbols are used inside the outlines in accordance with the IEC/IEEE standards but are not shown here. Generally these are associated with arithmetic operations and are self-explanatory.

When nonstandardized information is shown inside an outline, it is usually enclosed in square brackets [like these].

4.0 DEPENDENCY NOTATION

4.1 General Explanation

Dependency notation is the powerful tool that sets the IEC symbols apart from previous systems and makes compact, meaningful, symbols possible. It provides the means of denoting the relationship between inputs, outputs, or inputs and outputs without actually showing all the elements and interconnections involved. The information provided by dependency notation supplements that provided by the qualifying symbols for an element's function.

In the convention for the dependency notation, use will be made of the terms ''affecting'' and ''affected.'' In cases where it is not evident which inputs must be considered as being the affecting or the affected ones (e.g., if they stand in an AND relationship), the choice may be made in any convenient way.

So far, eleven types of dependency have been defined and all of these are used in various TI data books. X dependency is used mainly with CMOS circuits. They are listed below in the order in which they are presented and are summarized in Table IV following 4.12.

Section	Dependency Type or Other Subject
4.2	G, AND
4.3	General Rules for Dependency Notation
4.4	V, OR
4.5	N, Negate (Exclusive-OR)
4.6	Z, Interconnection
4.7	X, Transmission
4.8	C, Control
4.9	S, Set and R, Reset
4.10	EN, Enable
4.11	M, Mode
4.12	A, Address

4.2 G (AND) Dependency

A common relationship between two signals is to have them ANDed together. This has traditionally been shown by explicitly drawing an AND gate with the signals connected to the inputs of the gate. The 1972 IEC publication and the 1973 IEEE/ANSI standard showed several ways to show this AND relationship using dependency notation. While ten other forms of dependency have since been defined, the ways to invoke AND dependency are now reduced to one.

In Figure 4 input **b** is ANDed with input **a** and the complement of **b** is ANDed with **c**. The letter G has been chosen to indicate AND relationships and is placed at input **b**, inside the symbol. A number considered appropriate by the symbol designer (1 has been used here) is placed after the letter G and also at each affected input. Note the bar over the 1 at input **c**.

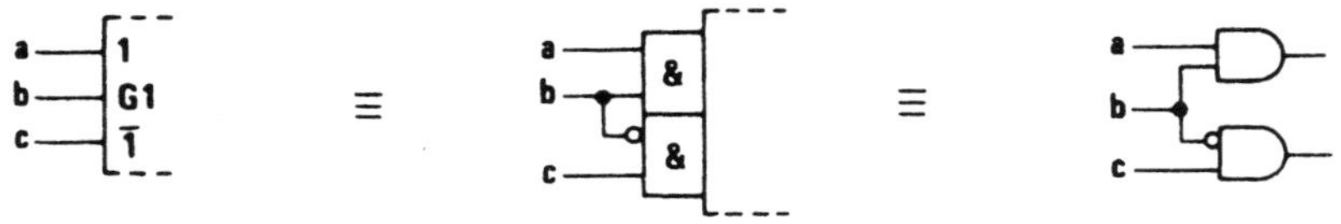

a — 1
b — G1
c — $\overline{1}$
≡
a, b — &
b, c — &
≡
a, b, c (AND gates)

Figure 4. G Dependency Between Inputs

In Figure 5, output **b** affects input **a** with an AND relationship. The lower example shows that it is the internal logic state of **b**, unaffected by the negation sign, that is ANDed. Figure 6 shows input **a** to be ANDed with a dynamic input **b**.

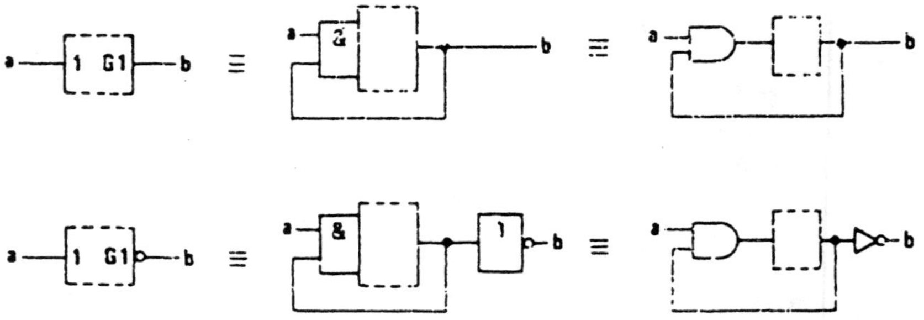

Figure 5. G Dependency Between Outputs and INputs

Figure 6. G Dependency with a Dynamic Input

The rules for G dependency can be summarized thus:

> When a G*m* input or output (*m* is a number) stands at its internal 1 state, all inputs and outputs affected by G*m* stand at their normally defined internal logic states. When the G*m* input or output stands at its 0 state, all inputs and outputs affected by G*m* stand at their internal 0 states.

4.3 Conventions for the Application of Dependency Notation in General

The rules for applying dependency relationships in general follow the same pattern as was illustrated for G dependency.

Application of dependency notation is accomplished by:

1) labeling the input (or output) *affecting* other inputs or outputs with the letter symbol indicating the relationship involved (e.g., G for AND) followed by an identifying number, appropriately chosen, and
2) labeling each input or output *affected* by that affecting input (or output) with that same number.

If it is the complement of the internal logic state of the affecting input or output that does the affecting, then a bar is placed over the identifying numbers at the affected inputs or outputs (Figure 4).

If two affecting inputs or outputs have the same letter and same identifying number, they stand in an OR relationship to each other (Figure 7).

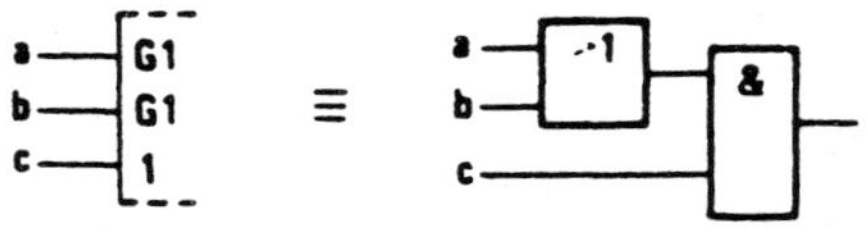

Figure 7. ORed Affecting Inputs

If the affected input or output requires a label to denote its function (e.g., "D"), this label will be *prefixed* by the identifying number of the affecting input (Figure 15).

If an input or output is affected by more than one affecting input, the identifying numbers of each of the affecting inputs will appear in the label of the affected one, separated by commas. The normal reading order of these numbers is the same as the sequence of the affecting relationships (Figure 15).

If the labels denoting the functions of affected inputs or outputs must be numbers (e.g., outputs of a coder), the identifying numbers to be associated with both affecting inputs and affected inputs or outputs will be replaced by another character selected to avoid ambiguity, e.g., Greek letters (Figure 8).

Figure 8. Substitution for Numbers

4.4 V (OR) Dependency

The symbol denoting OR dependency is the letter V (Figure 9).

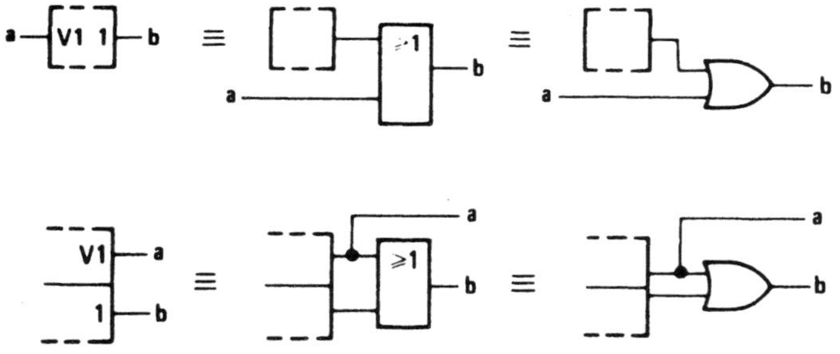

Figure 9. V (OR) Dependency

When a V*m* input or output stands at its internal 1 state, all inputs and outputs affected by V*m* stand at their internal 1 states. When the V*m* input or output stands at its internal 0 state, all inputs and outputs affected by V*m* stand at their normally defined internal logic states.

4.5 N (Negate) (Exclusive-OR) Dependency

The symbol denoting negate dependency is the letter N (Figure 10). Each input or output affected by an N*m* input or output stands in an Exclusive-OR relationship with the N*m* input or output.

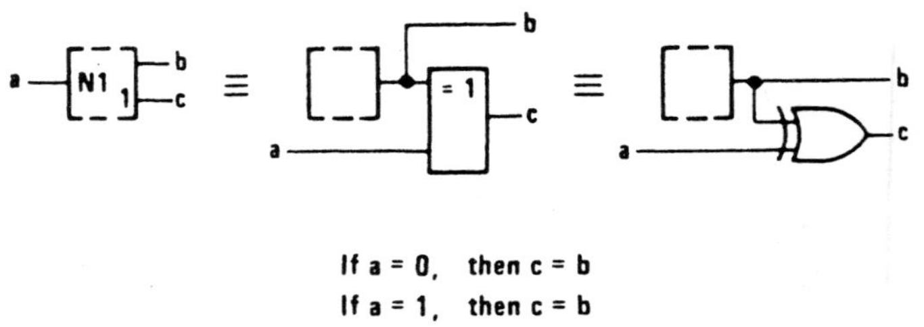

N1
=1
a
b
c
If a = 0, then c = b
If a = 1, then c = b

Figure 10. N (Negate) (Exclusive-OR) Dependency

When an N*m* input or output stands at its internal 1 state, the internal logic state of each input and each output affected by N*m* is the complement of what it would otherwise be. When an N*m* input or output stands at its internal 0 state, all inputs and outputs affected by N*m* stand at their normally defined internal logic states.

4.6 Z (Interconnection) Dependency

The symbol denoting interconnection dependency is the letter Z.

Interconnection dependency is used to indicate the existence of internal logic connections between inputs, outputs, internal inputs, and/or internal outputs.

The internal logic state of an input or output affected by a Z*m* input or output will be the same as the internal logic state of the Z*m* input or output, unless modified by additional dependency notation (Figure 11).

Z1
where
where
G1
1Z2
&

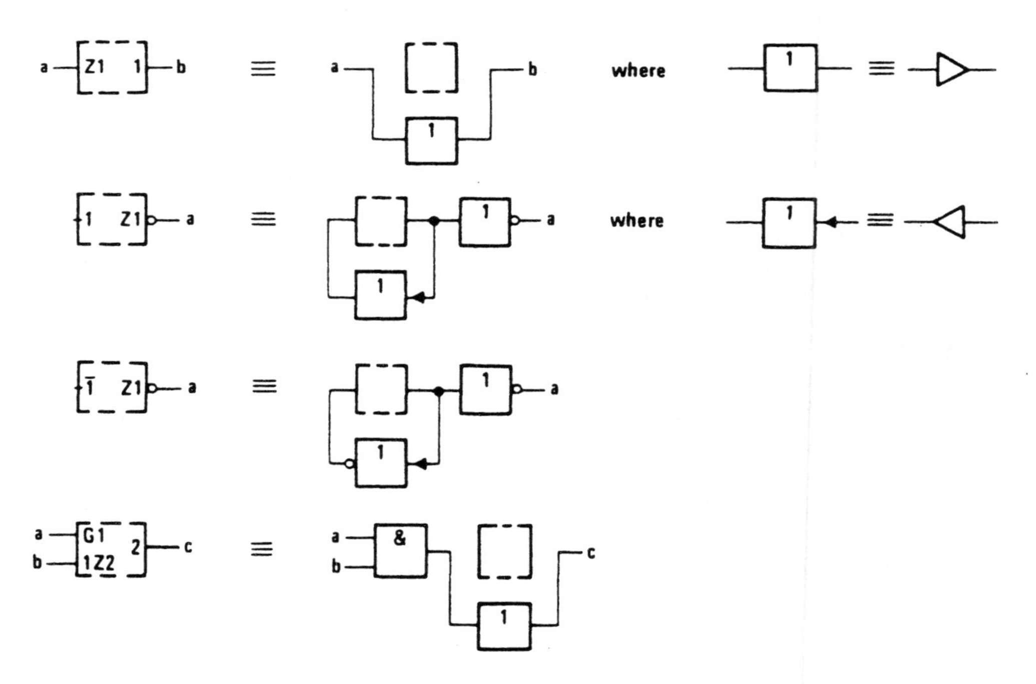

Figure 11. Z (Interconnection) Dependency

4.7 X (Transmission) Dependency

The symbol denoting transmission dependency is the letter X.

Transmission dependency is used to indicate controlled bidirectional connections between affected input/output ports (Figure 12).

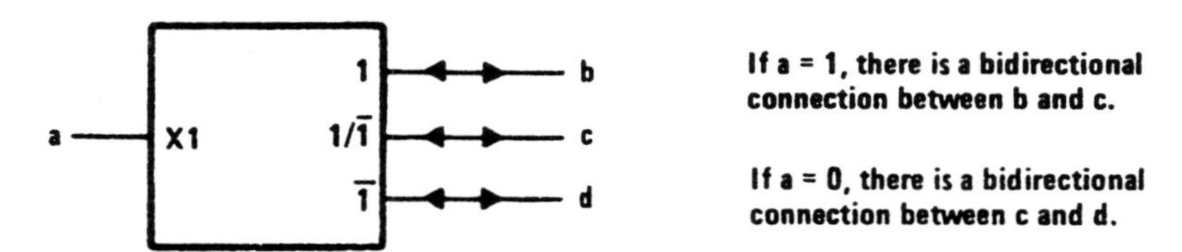

Figure 12. X (Transmission) Dependency

When an X*m* input or output stands at its internal 1 state, all input-output ports affected by this X*m* input or output are bidirectionally connected together and stand at the same internal logic state or analog signal level. When an X*m* input or output stands at its internal 0 state, the connection associated with this set of dependency notation does not exist.

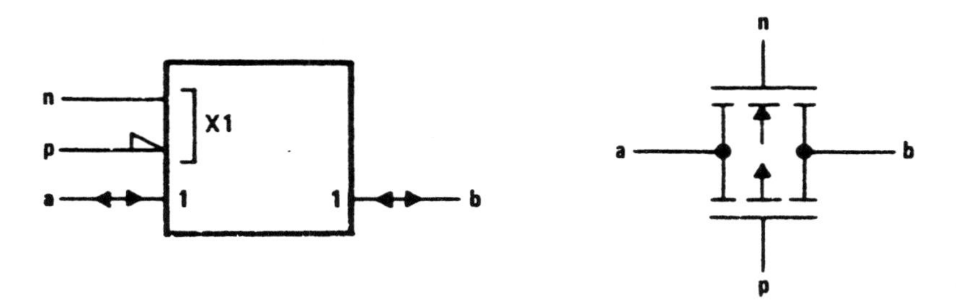

Figure 13. CMOS Transmission Gate Symbol and Schematic

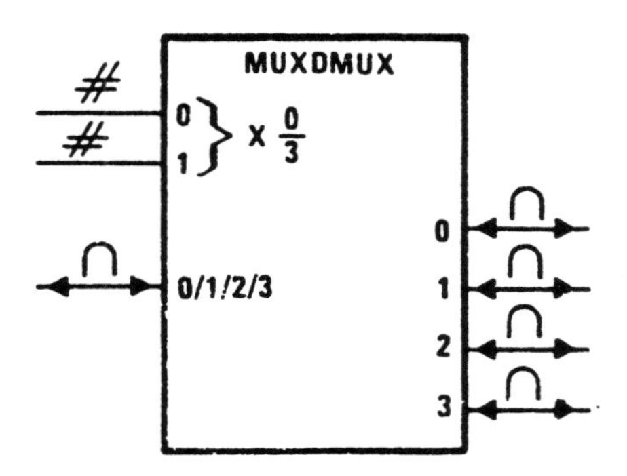

Figure 14. Analog Data Selector (Multiplexer/Demultiplexer)

Although the transmission paths represented by X dependency are inherently bidirectinal, use is not always made of this property. This is analogous to a piece of wire, which may be constrained to carry current in only one direction. If this is the case in a particular application, then the directional arrows shown in Figures 12, 13, and 14 would be omitted.

4.8 C (Control) Dependency

The symbol denoting control dependency is the letter C.

Control inputs are usually used to enable or disable the data (D, J, K, R, or S) inputs of storage elements. They may take on their internal 1 states (be active) either statically or dynamically. In the latter case the dynamic input symbol is used as shown in the third example of Figure 15.

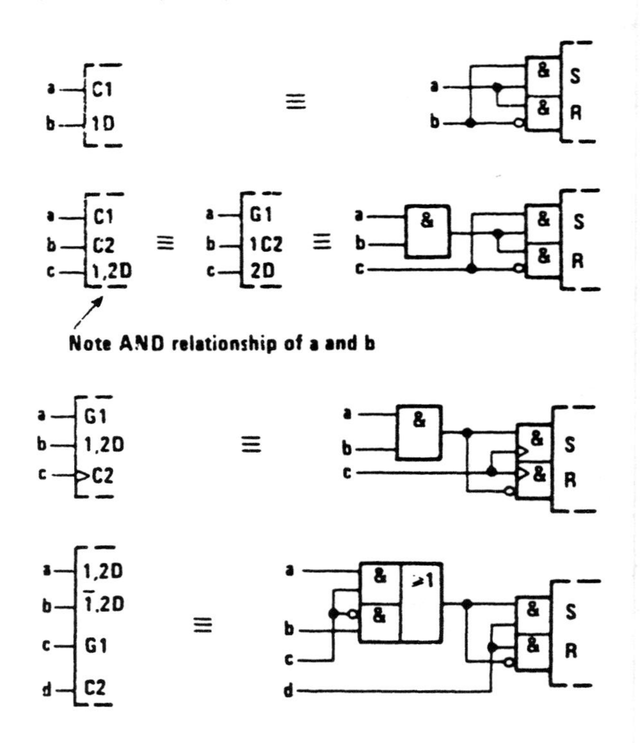

Input c selects which of a or b is stored when d goes low.

Figure 15. C (Control) Dependency

When a C*m* input or output stands at its internal 1 state, the inputs affected by C*m* have their normally defined effect on the function of the element, i.e., these inputs are enabled. When a C*m* input or output stands at its internal 0 state, the inputs affected by C*m* are disabled and have no effect on the function of the element.

4.9 S (Set) and R (Reset) Dependencies

The symbol denoting set dependency is the letter S. The symbol denoting reset dependency is the letter R.

Set and reset dependencies are used if it is necessary to specify the effect of the combination R = S = 1 on a bistable element. Case 1 in Figure 16 does not use S or R dependency.

When an S*m* input is at its internal 1 state, outputs affected by the S*m* input will react, regardless of the state of an R input, as they normally would react to the combination S = 1, R = 0. See cases 2, 4, and 5 in Figure 16.

When an R*m* input is at its internal 1 state, outputs affected by the R*m* input will react, regardless of the state of an S input, as they normally would react to the combination S = 0, R = 1. See cases 3, 4, and 5 in Figure 16.

When an S*m* or R*m* input is at its internal 0 state, it has no effect.

Note that the noncomplementary output patterns in cases 4 and 5 are only pseudo stable. The simultaneous return of the inputs to S = R = 0 produces an unforeseeable stable and complementary output pattern.

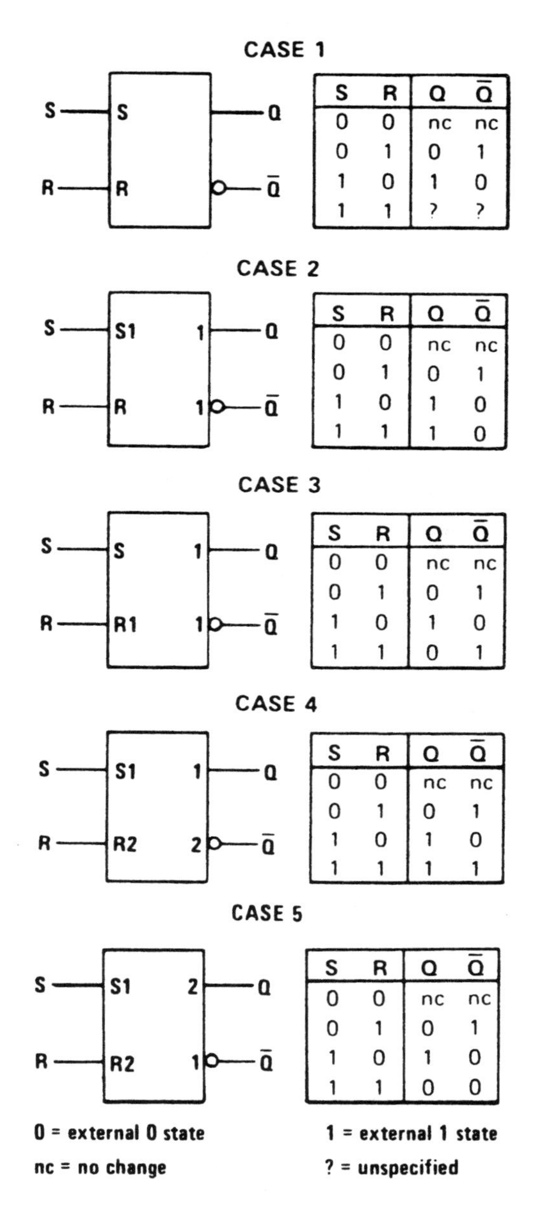

CASE 1

S	R	Q	$\bar{Q}$
0	0	nc	nc
0	1	0	1
1	0	1	0
1	1	?	?

CASE 2

S	R	Q	$\bar{Q}$
0	0	nc	nc
0	1	0	1
1	0	1	0
1	1	1	0

CASE 3

S	R	Q	$\bar{Q}$
0	0	nc	nc
0	1	0	1
1	0	1	0
1	1	0	1

CASE 4

S	R	Q	$\bar{Q}$
0	0	nc	nc
0	1	0	1
1	0	1	0
1	1	1	1

CASE 5

S	R	Q	$\bar{Q}$
0	0	nc	nc
0	1	0	1
1	0	1	0
1	1	0	0

Figure 16. S (Set) and R (Reset) Dependencies

4.10 EN (Enable) Dependency

The symbol denoting enable dependency is the combination of letters EN.

An EN*m* input has the same effect on outputs as an EN input, see 3.1, but it affects only those outputs labeled with the identifying number *m*. It also affects those inputs labeled with the identifying number *m*. By contrast, an EN input affects all outputs and no inputs. The effect of an EN*m* input on an affected input is identical to that of a C*m* input (Figure 17).

When an EN*m* input stands at its internal 1 state, the inputs affected by EN*m* have their normally defined effect on the function of the element and the outputs affected by this input stand at their normally defined internal logic states, i.e., these inputs and outputs are enabled.

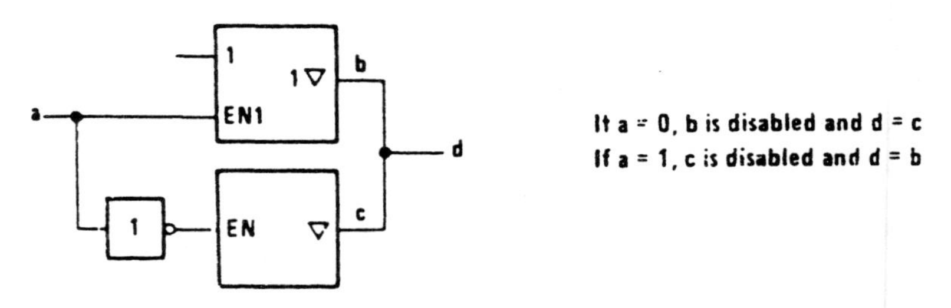

Figure 17. EN (Enable) Dependency

When an EN*m* input stands at its internal 0 state, the inputs affected by EN*m* are disabled and have no effect on the function of the element, and the outputs affected by EN*m* are also disabled. Open-collector outputs are turned off, three-state outputs stand at their normally defined internal logic states but externally exhibit high impedance, and all other outputs (e.g., totem-pole outputs) stand at their internal 0 states.

4.11 M (MODE) Dependency

The symbol denoting mode dependency is the letter M.

Mode dependency is used to indicate that the effects of particular inputs and outputs of an element depend on the mode in which the element is operating.

If an input or output has the same effect in different modes of operation, the identifying numbers of the relevant affecting M*m* inputs will appear in the label of that affected input or output between parentheses and separated by solidi (Figure 22).

4.11.1 M Dependency Affecting Inputs

M dependency affects inputs the same as C dependency. When an M*m* input or M*m* output stands at its internal 1 state, the inputs affected by this M*m* input or M*m* output have their normally defined effect on the function of the element, i.e., the inputs are enabled.

When an M*m* input or M*m* output stands at its internal 0 state, the inputs affected by this M*m* input or M*m* output have no effect on the function of the element. When an affected input has several sets of labels separated by solidi (e.g., C4/2—/3+), any set in which the identifying number of the M*m* input or M*m* output appears has no effect and is to be ignored. This represents disabling of some of the functions of a multifunction input.

When an EN*m* input stands at its internal 1 state, the inputs affected by EN*m* have their normally defined effect on the function of the element and the outputs affected by this input stand at their normally defined internal logic states, i.e., these inputs and outputs are enabled.

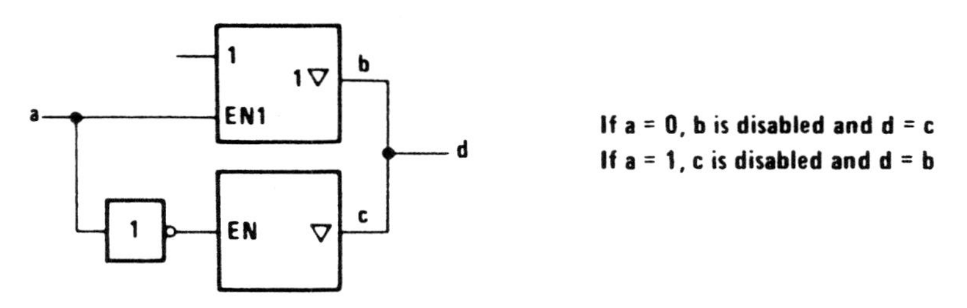

Figure 17. EN (Enable) Dependency

When an EN*m* input stands at its internal 0 state, the inputs affected by EN*m* are disabled and have no effect on the function of the element, and the outputs affected by EN*m* are also disabled. Open-collector outputs are turned off, three-state outputs stand at their normally defined internal logic states but externally exhibit high impedance, and all other outputs (e.g., totem-pole outputs) stand at their internal 0 states.

4.11 M (MODE) Dependency

The symbol denoting mode dependency is the letter M.

Mode dependency is used to indicate that the effects of particular inputs and outputs of an element depend on the mode in which the element is operating.

If an input or output has the same effect in different modes of operation, the identifying numbers of the relevant affecting M*m* inputs will appear in the label of that affected input or output between parentheses and separated by solidi (Figure 22).

4.11.1 M Dependency Affecting Inputs

M dependency affects inputs the same as C dependency. When an M*m* input or M*m* output stands at its internal 1 state, the inputs affected by this M*m* input or M*m* output have their normally defined effect on the function of the element, i.e., the inputs are enabled.

When an M*m* input or M*m* output stands at its internal 0 state, the inputs affected by this M*m* input or M*m* output have no effect on the function of the element. When an affected input has several sets of labels separated by solidi (e.g., C4/2→/3+), any set in which the identifying number of the M*m* input or M*m* output appears has no effect and is to be ignored. This represents disabling of some of the functions of a multifunction input.

The circuit in Figure 18 has two inputs, **b** and **c**, that control which one of four modes (0, 1, 2, or 3) will exist at any time. Inputs **d**, **e**, and **f** are D inputs subject to dynamic control (clocking) by the **a** input. The numbers 1 and 2 are in the series chosen to indicate the modes so inputs **e** and **f** are only enabled in mode 1 (for parallel loading) and input **d** is only enabled in mode 2 (for serial loading). Note that input **a** has three functions. It is the clock for entering data. In mode 2, it causes right shifting of data, which means a shift away from the control block. In mode 3, it causes the contents of the register to be incremented by one count.

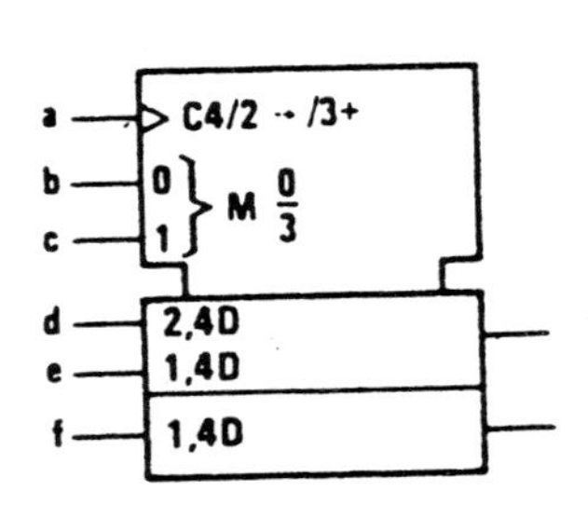

Note that all operations are synchronous.

In MODE 0 (b = 0, c = 0), the outputs remain at their existing states as none of the inputs has an effect.

In MODE 1 (b = 1, c = 0), parallel loading takes place thru inputs e and f.

In MODE 2 (b = 0, c = 1), shifting down and serial loading thru input d take place.

In MODE 3 (b = c = 1), counting up by increment of 1 per clock pulse takes place.

Figure 18. M (Mode) Dependency Affecting Inputs

4.11.2 M Dependency Affecting Outputs

When an M*m* input or M*m* output stands at its internal 1 state, the affected outputs stand at their normally defined internal logic states, i.e., the outputs are enabled.

When an M*m* input or M*m* output stands at its internal 0 state, at each affected output any set of labels containing the identifying number of that M*m* input or M*m* output has no effect and is to be ignored. When an output has several different sets of labels separated by solidi (e.g., 2,4/3,5), only those sets in which the identifying number of this M*m* input or M*m* output appears are to be ignored.

Figure 19 shows a symbol for a device whose output can behave like either a 3-state output or an open-collector output depending on the signal applied to input **a**. Mode 1 exists when input **a** stands at its internal 1 state and, in that case, the three-state symbol applies and the open-element symbol has no effect. When a = 0, mode 1 does not exist so the three-state symbol has no effect and the open-element symbol applies.

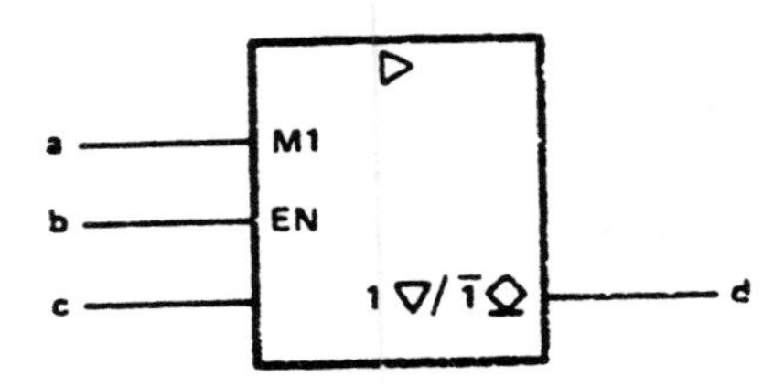

Figure 19. Type of Output Determined by Mode

In Figure 20, if input **a** stands at its internal 1 state establishing mode 1, output **b** will stand at its internal 1 state only when the content of the register equals 9. Since output **b** is located in the common-control block with no defined function outside of mode 1, the state of this output outside of mode 1 is not defined by the symbol.

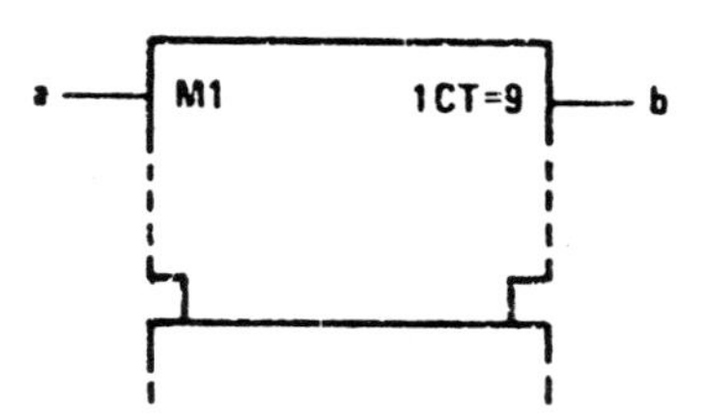

Figure 20. An Output of the Common-Control Block

In Figure 21, if input **a** stands at its internal 1 state establishing mode 1, output **b** will stand at its internal 1 state only when the content of the register equals 15. If input **a** stands at its internal 0 state, output **b** will stand at its internal 1 state only when the content of the register equals 0.

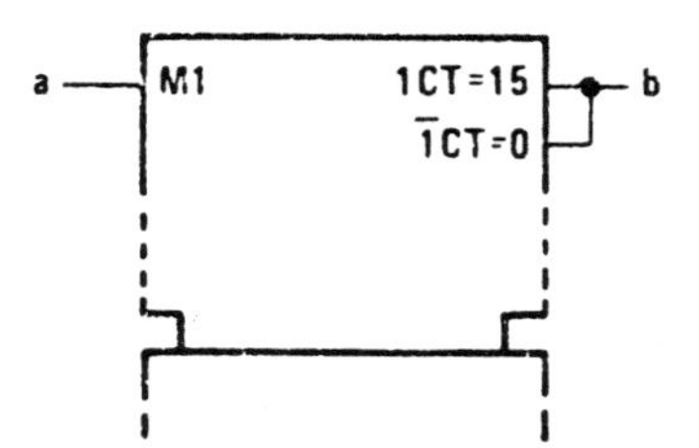

Figure 21. Determining and Output's Function

In Figure 22 inputs **a** and **b** are binary weighted to generate the numbers 0, 1, 2, or 3. This determines which one of the four modes exists.

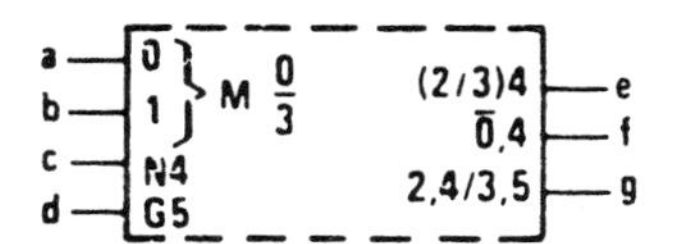

Figure 22. Dependent Relationships Affected by Mode

At output **e** the label set causing negation (if **c** = 1) is effective only in modes 2 and 3. In modes 0 and 1 this output stands at its normally defined state as if it had no labels. At output **f** the label set has effect when the mode is not 0 so output **e** is negated (if **c** = 1) in modes 1, 2, and 3. In mode 0 the label set has no effect so the output stands at its normally defined state. In this example $\overline{0}$,4 is equivalent to (1/2/3)4. At output **g** there are two label sets. The first set, causing negation (if **c** = 1), is effective only in mode 2. The second set, subjecting **g** to AND dependency on **d**, has effect only in mode 3.

Note that in mode 0 none of the dependency relationships has any effect on the outputs, so **e**, **f**, and **g** will all stand at the same state.

4.12 A (Address) Dependency

The symbol denoting address dependency is the letter A.

Address dependency provides a clear representation of those elements, particularly memories, that use address control inputs to select specified sections of a multildimensional arrays. Such a section of a memory array is usually called a word. The purpose of address dependency is to allow a symbolic presentation of the entire array. An input of the array shown at a particular element of this general section is common to the corresponding elements of all selected sections of the array. An output of the array shown at a particular element of this general section is the result of the OR function of the outputs of the corresponding elements of selected sections.

Inputs that are not affected by any affecting address input have their normally defined effect on all sections of the array, whereas inputs affected by an address input have their normally defined effect only on the section selected by that address input.

An affecting address input is labeled with the letter A followed by an identifying number that corresponds with the address of the particular section of the array selected by this input. Within the general section presented by the symbol, inputs and outputs affected by an A*m* input are labeled with the letter A, which stands for the identifying numbers, i.e., the addresses, of the particular sections.

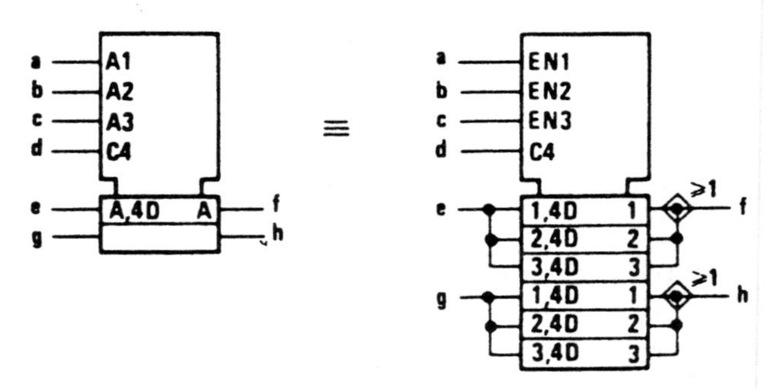

Figure 23. A (Address) Dependency

Figure 23 shows a 3-word by 2-bit memory having a separate address line for each word and uses EN dependency to explain the operation. To select word 1, input **a** is taken to its 1 state, which establishes mode 1. Data can now be clocked into the inputs marked "1,4D." Unless words 2 and 3 are also selected, data cannot be clocked in at the inputs marked "2,4D" and "3,4D." The outputs will be the OR functions of the selected outputs, i.e., only those enabled by the active EN functions.

The identifying numbers of affecting address inputs correspond with the addresses of the sections selected by these inputs. They need not necessarily differ from those of other affecting dependency-inputs (e.g., G, V, N, . . .), because in the general section presented by the symbol they are replaced by the letter A.

If there are several sets of affecting A*m* inputs for the purpose of independent and possibly simultaneous access to sections of the array, then the letter A is modified to 1A, 2A, ... Because they have access to the same sections of the array, these sets of A inputs may have the same identifying numbers. The symbols for 'HC170 or SN74LS170 make use of this.

Figure 24 is another illustration of the concept.

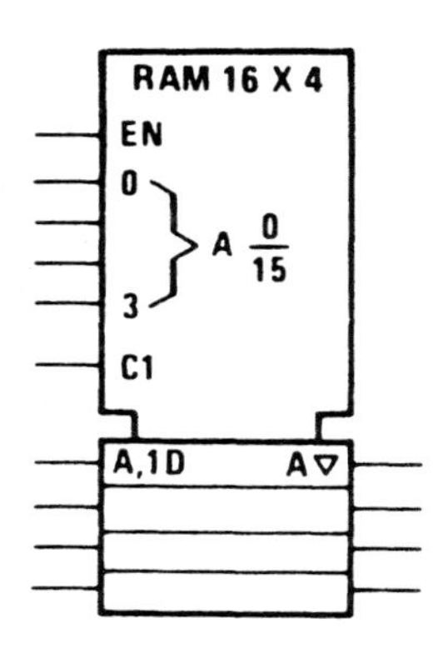

Figure 24. Array of 16 Sections of Four Transparent Latches with 3-State Outputs Comprising a 16-Word X 4-Bit Random-Access Memory

Table IV. Summary of Dependency Notation

TYPE OF DEPENDENCY	LETTER SYMBOL*	AFFECTING INPUT AT ITS 1-STATE	AFFECTING INPUT AT ITS 0-STATE
Address	A	Permits action (address selected)	Prevents action (address not selected)
Control	C	Permits action	Prevents action
Enable	EN	Permits action	Prevents action of inputs ◇outputs off ▽outputs at external high impedance, no change in internal logic state Other outputs at internal 0 state
AND	G	Permits action	Imposes 0 state
Mode	M	Permits action (mode selected)	Prevents action (mode not selected)
Negate (Ex-OR)	N	Complements state	No effect
Reset	R	Affected output reacts as it would to S = 0, R = 1	No effect
Set	S	Affected output reacts as it would to S = 1, R = 0	No effect
OR	V	Imposes 1 state	Permits action
Transmission	X	Bidirectional connection exists	Bidirectional connection does not exist
Interconnection	Z	Imposes 1 state	Imposes 0 state

*These letter symbols appear at the AFFECTING input (or output) and are followed by a number. Each input (or output) AFFECTED by that input is labeled with that same number. When the labels EN, R, and S appear at inputs without the following numbers, the descriptions above do not apply. The action of these inputs is described under "Symbols Inside the Outline," see 3.3.

5.0 BISTABLE ELEMENTS

The dynamic input symbol, the postponed output symbol, and dependency notation provide the tools to differentiate four main types of bistable elements and make synchronous and asynchronous inputs easily recognizable (Figure 25). The first column shows the essential distinguishing features; the other columns show examples.

Transparent latches have a level-operated control input. The D input is active as long as the C input is at its internal 1 state. The outputs respond immediately. Edge-triggered elements accept data from D, J, K, R, or S inputs on the active transition of C. Pulse-triggered elements require the setup of data before the start of the control pulse; the C input is considered static since the data must be maintained as long as C is at its 1 state. The output is postponed until C returns to its 0 state. The data-lock-out element is similar to the pulse-triggered version except that the C input is considered dynamic in that shortly after C goes through its active transition, the data inputs are disabled and data does not have to be held. However, the output is still postponed until the C input returns to its initial external level.

Notice that synchronous inputs can be readily recognized by their dependency labels (1D, 1J, 1K, 1S, 1R) compared to the asynchronous inputs (S, R), which are not dependent on the C inputs.

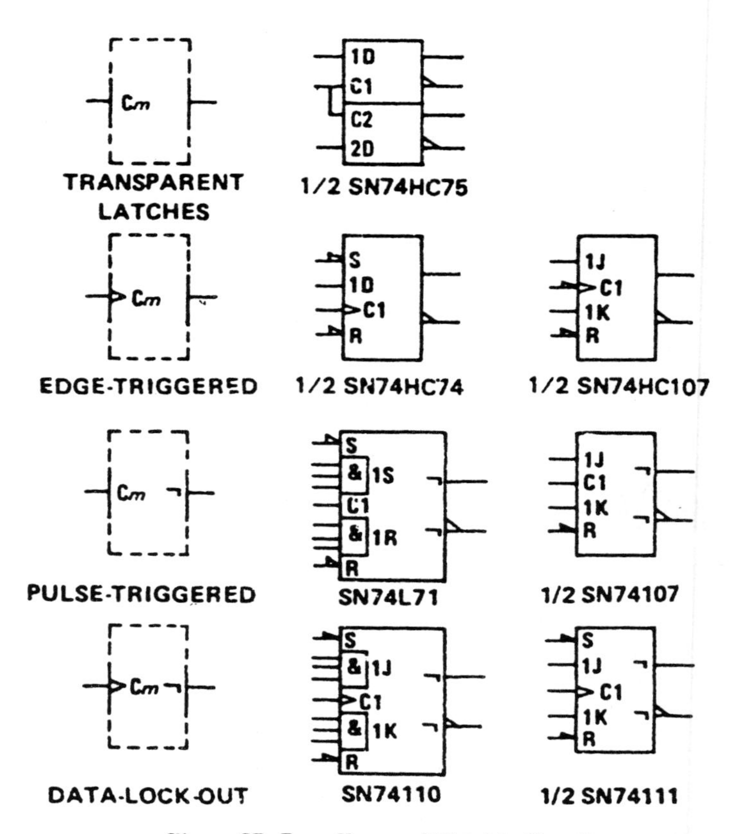

Figure 25. Four Types of Bistable Circuits

6.0 CODERS

The general symbol for a coder or code converter is shown in Figure 26. X and Y may be replaced by appropriate indications of the code used to represent the information at the inputs and at the outputs, respectively.

Figure 26. Coder General Symbol

Indication of code conversion is based on the following rule:

> Depending on the input code, the internal logic states of the inputs determine an internal value. This value is reproduced by the internal logic states of the outputs, depending on the output code.

The indication of the relationships between the internal logic states of the inputs and the internal value is accomplished by:

1) labeling the inputs with numbers. In this case the internal value equals the sum of the weights associated with those inputs that stand at their internal 1-state, or by
2) replacing X by an appropriate indication of the input code and labeling the inputs with characters that refer to this code.

The relationships between the internal value and the internal logic states of the outputs are indicated by:

1) labeling each output with a list of numbers representing those internal values that lead to the internal 1-state of that output. These numbers shall be separated by solidi as in Figure 27. This labeling may also be applied when Y is replaced by a letter denoting a type of dependency (see Section 7). If a continuous range of internal values produces the internal 1 state of an output, this can be indicated by two numbers that are inclusively the beginning and the end of the range, with these two numbers separated by three dots (e.g., 4 . . . 9 = 4/5/6/7/8/9) or by
2) replacing Y by an appropriate indiction of the output code and labeling the outputs with characters that refer to this code as in Figure 28.

Alternatively, the general symbol may be used together with an appropriate reference to a table in which the relationship between the inputs and outputs is indicated. This is a recommended way to symbolize a PROM after it has been programmed.

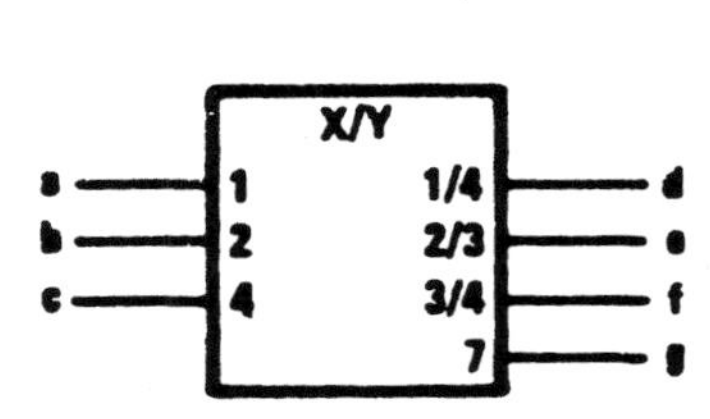

FUNCTION TABLE

INPUTS			OUTPUTS			
c	b	a	g	f	e	d
0	0	0	0	0	0	0
0	0	1	0	0	0	1
0	1	0	0	0	1	0
0	1	1	0	1	1	0
1	0	0	0	1	0	1
1	0	1	0	0	0	0
1	1	0	0	0	0	0
1	1	1	1	0	0	0

Figure 27. An X/Y Code Converter

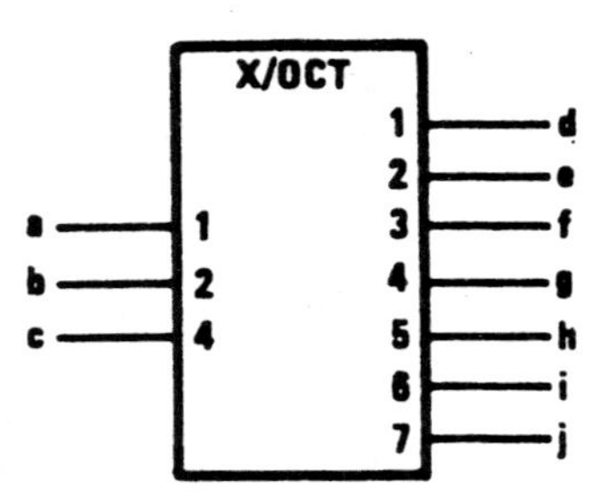

INPUTS			OUTPUTS						
c	b	a	j	i	h	g	f	e	d
0	0	0	0	0	0	0	0	0	0
0	0	1	0	0	0	0	0	0	1
0	1	0	0	0	0	0	0	1	0
0	1	1	0	0	0	0	1	0	0
1	0	0	0	0	0	1	0	0	0
1	0	1	0	0	1	0	0	0	0
1	1	0	0	1	0	0	0	0	0
1	1	1	1	0	0	0	0	0	0

Figure 28. An X/Octal Code Converter

7.0 USE OF A CODER TO PRODUCE AFFECTING INPUTS

It often occurs that a set of affecting inputs for dependency notation is produced by decoding the signals on certain inputs to an element. In such a case use can be made of the symbol for a coder as an embedded symbol (Figure 29).

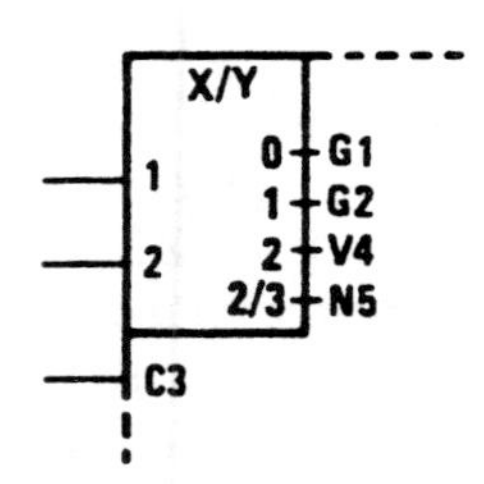

Figure 29. Producing Various Types of Dependencies

If all affecting inputs produced by a coder are of the same type and their identifying numbers shown at the outputs of the coder, Y (in the qualifying symbol X/Y) may be replaced by the letter denoting the type of dependency. The indications of the affecting inputs should then be omitted (Figure 30).

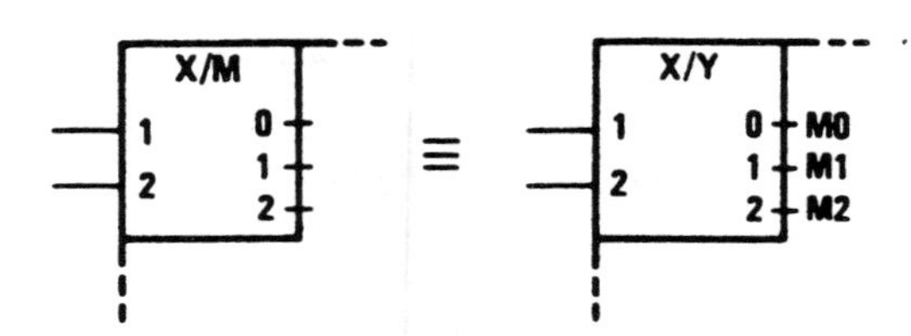

Figure 30. Producing One Type of Dependency

8.0 USE OF BINARY GROUPING TO PRODUCE AFFECTING INPUTS

If all affecting inputs produced by a coder are of the same type and have consecutive identifying numbers not necessarily corresponding with the numbers that would have been shown at the outputs of the coder, use can be made of the binary grouping symbol. *k* external lines effectively generate 2^k internal inputs. The bracket is followed by the letter denoting the type of dependency followed by m1/m2. The m1 is to be replaced by the smallest identifying number and the m2 by the largest one, as shown in Figure 31.

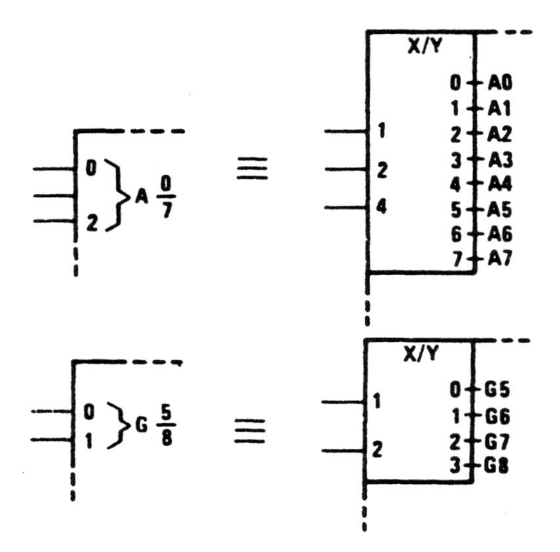

Figure 31. Use of the Binary Grouping Symbol

9.0 SEQUENCE OF INPUT LABELS

If an input having a single functional effect is affected by other inputs, the qualifying symbol (if there is any) for that functional effect is preceded by the labels corresponding to the affecting inputs. The left-to-right order of these preceding labels is the order in which the effects or modifications must be applied. The affected input has no functional effect on the element if the logic state of any one of the affecting inputs, considered separately, would cause the affected input to have no effect, regardless of the logic states of other affecting inputs.

If an input has several different functional effects or has several different sets of affecting inputs, depending on the mode of action, the input may be shown as often as required. However, there are cases in which this method of presentation is not advantageous. In those cases the input may be shown once with the different sets of labels separated by solidi (Figure 32). No meaning is attached to the order of these sets of labels. If one of the functional effects of an input is that of an unlabeled input to the element, a solidus will precede the first set of labels shown.

If all inputs of a combinational element are disabled (caused to have no effect on the function of the element), the internal logic states of the outputs of the element are not specified by the symbol. If all inputs of a sequential element are disabled, the content of this element is not changed and the outputs remain at their existing internal logic states.

Labels may be factored using algebraic techniques (Figure 33).

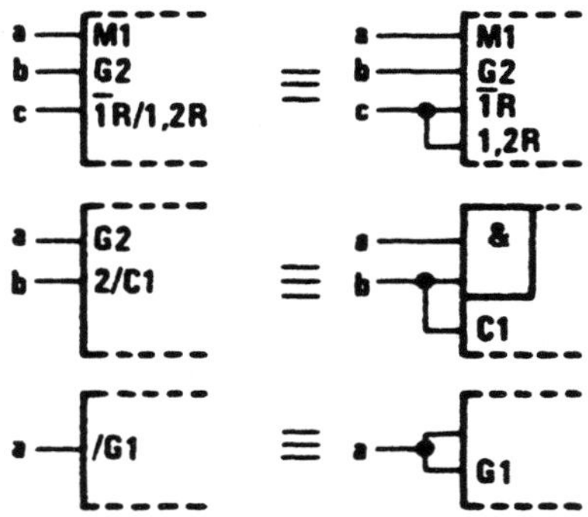

Figure 32. Input Labels

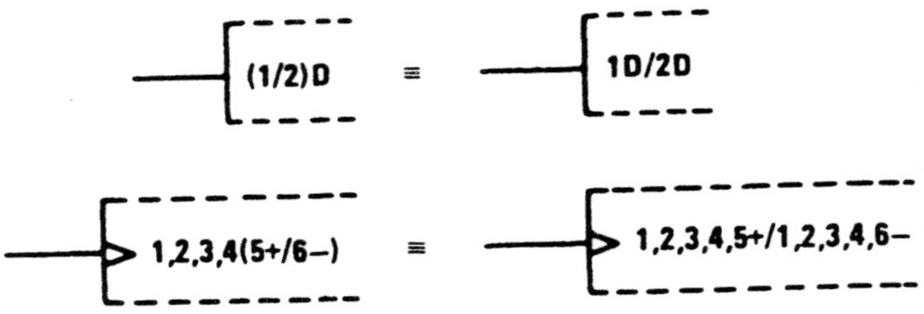

(1/2)D ≡ 1D/2D

1,2,3,4(5+/6–) ≡ 1,2,3,4,5+/1,2,3,4,6–

Figure 33. Factoring Input Labels

10.0 SEQUENCE OF OUTPUT LABELS

If an output has a number of different labels, regardless of whether they are identifying numbers of affecting inputs or outputs or not, these labels are shown in the following order:

1) If the postponed output symbol has to be shown, this comes first, if necessary preceded by the indications of the inputs to which it must be applied
2) Followed by the labels indicating modifications of the internal logic state of the output, such that the left-to-right order of these labels corresponds with the order in which their effects must be applied
3) Followed by the label indicating the effect of the output on inputs and other outputs of the element.

Symbols for open-circuit or three-state outputs, where applicable, are placed just inside the outside boundary of the symbol adjacent to the output line (Figure 34).

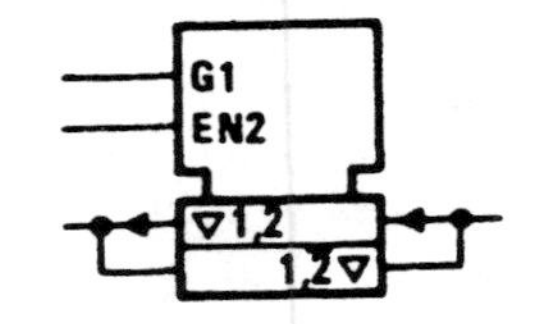

Figure 34. Placement of 3-State Symbols

If an output needs several different sets of labels that represent alternative functions (e.g., depending on the mode of action), these sets may be shown on different output lines that must be connected outside the outline. However, there are cases in which this method of presentation is not advantageous. In those cases the output may be shown once with the different sets of labels separated by solidi (Figure 35).

Two adjacent identifying numbers of affecting inputs in a set of labels that are not already separated by a nonnumeric character should be separated by a comma.

If a set of labels of an output not containing a solidus contains the identifying number of an affecting M*m* input standing at its internal 0 state, this set of labels has no effect on that output.

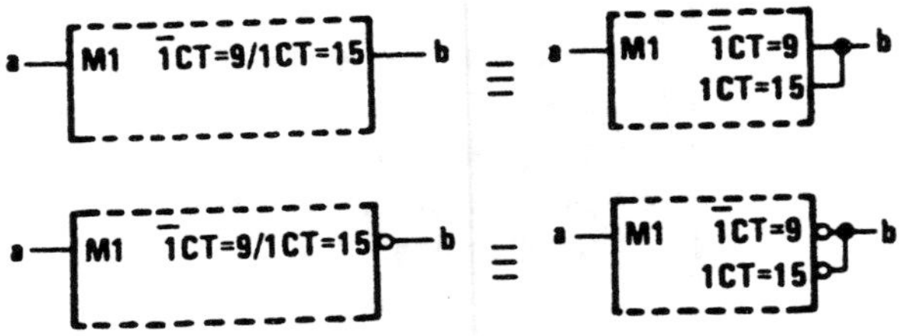

Labels may be factored using algebraic techniques (Figure 36).

a M1 $\overline{1}$CT=9/1CT=15 b ≡ a M1 $\overline{1}$CT=9 1CT=15 b

a M1 $\overline{1}$CT=9/1CT=15 b ≡ a M1 $\overline{1}$CT=9 1CT=15 b

Figure 35. Output Labels

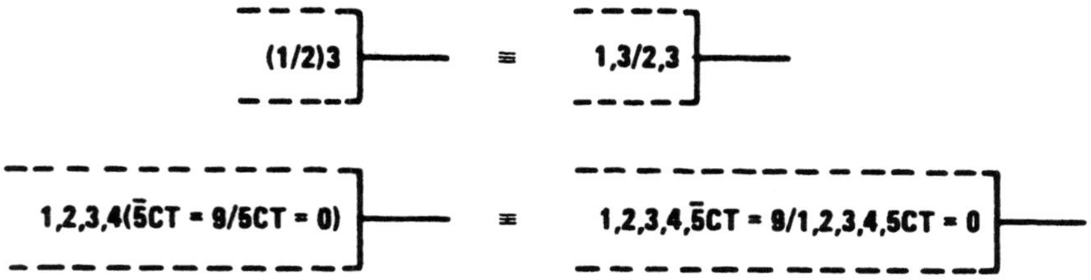

(1/2)3 ≡ 1,3/2,3

1,2,3,4($\bar{5}$CT = 9/5CT = 0) ≡ 1,2,3,4,$\bar{5}$CT = 9/1,2,3,4,5CT = 0

Figure 36. Factoring Output Labels

PART 6

Introduction to CUPL

Courtesy of Logical Devices, Inc.

Introduction To Programmable Logic
Chapter 1

1.1 WHAT IS PROGRAMMABLE LOGIC?

Programmable logic, as the name implies, is a family of components that contains arrays of logic elements (AND, OR, INVERT, LATCH, FLIP-FLOP) that may be configured into any logical function that the user desires and the component supports. There are several classes of programmable logic devices: ASICs, FPGAs, PLAs, PROMs, PALs, GALs, and complex PLDs.

ASICs

ASICs are *Application Specific Integrated Circuits* that are mentioned here because they are user definable devices. ASICs, unlike other devices, may contain analog, digital, and combinations of analog and digital functions. In general, they are mask programmable and not user programmable. This means that manufacturers will configure the device to the user specifications. They are used for combining a large amount of logic functions into one device. However, these devices have a high initial cost, therefore they are mainly used where high quantities are needed. Due to the nature of ASICs, CUPL and other programmable logic languages cannot fully support these devices.

Basic architecture of a user programmable device

First, a *user programmable device* is one that contains a pre-defined general architecture in which a user can program a design into the device using a set of development tools. The general architectures may vary but normally consists of one or more arrays of AND and OR terms for implementing logic functions. Many devices also contain combinations of flip-flops and latches which may be used as storage elements for inputs and outputs of a device. More complex devices contain *macrocells.* Macrocells allow the user to configure the type of inputs and outputs that are needed for a design.

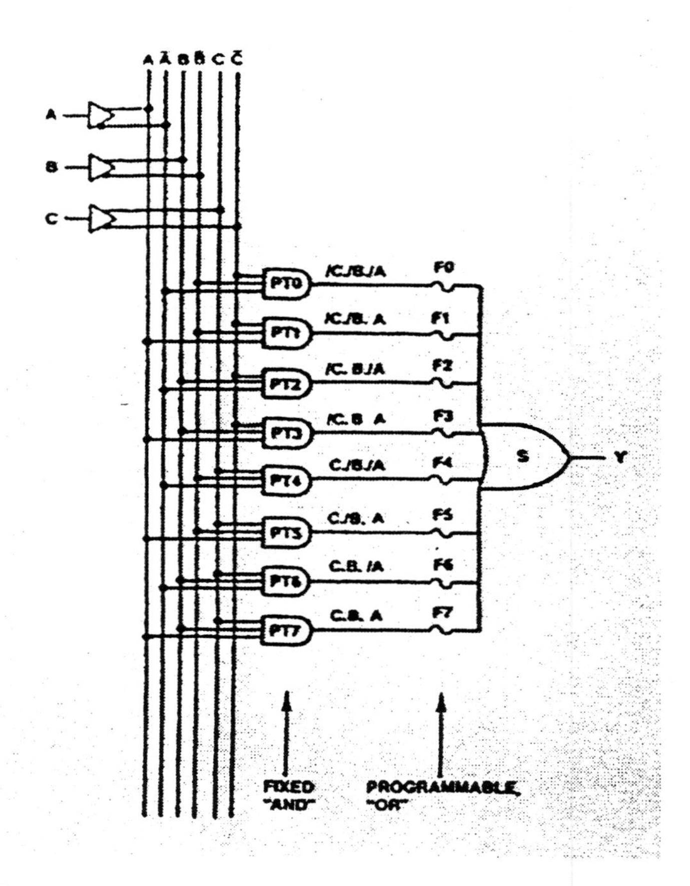

FIGURE 1-1 Elementary PROM architecture

PROMs

PROMs are *Programmable Read Only Memories.* Even though the name does not imply programmable logic, PROMs, are in fact logic. The architecture of most PROMs typically consists of a fixed number of AND array terms that feeds a programmable OR array. They are mainly used for decoding specific input combinations into output functions, such as memory mapping in microprocessor environments.

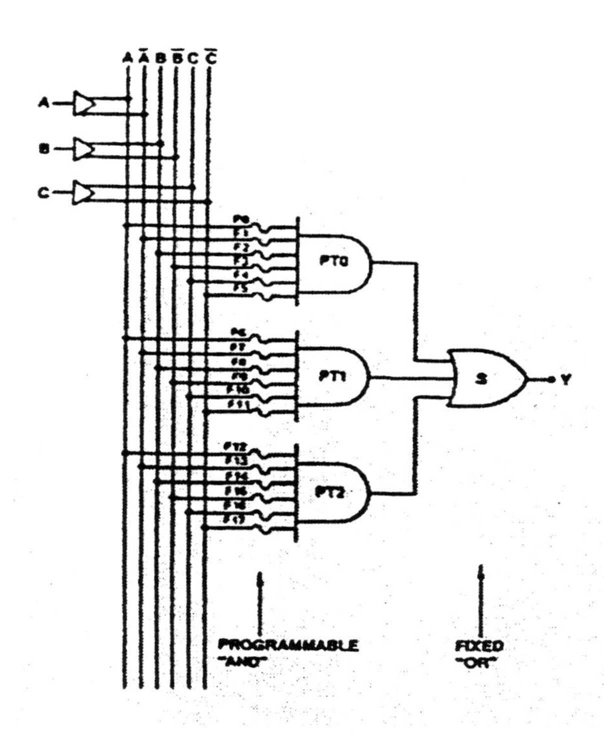

FIGURE 1-2. Elementary PAL architecture

PALs

PALs are *Programmable Array Logic* devices. The internal architecture consists of programmable AND terms feeding fixed OR terms. All inputs to the array can be ANDed together, but specific AND terms are dedicated to specific OR terms. PALs have a very popular architecture and are probably the most widely used type of user programmable device. If a device contains macrocells, it will usually have a PAL architecture. Typical macrocells may be programmed as inputs, outputs, or input/output (I/O) using a tri-state enable. They normally have output registers which may or may not be used in conjunction with the associated I/O pin. Other macrocells have more than one register, various type of feedback into the arrays, and occasionally feedback between macrocells. These devices are mainly used to replace multiple TTL logic functions commonly referred to as *glue logic*.

GALs

GALs are *Generic Array Logic* devices. They are designed to emulate many common PALs thought the use of macrocells. If a user has a design that is implemented using several common PALs, he may configure several of the same GALs to emulate each of the other devices. This will reduce the number of different devices in stock and increase the quantity purchased. Usually, a large quantity of the same device should lower the individual device cost. Also these devices are electrically erasable, which makes them very useful for design engineers.

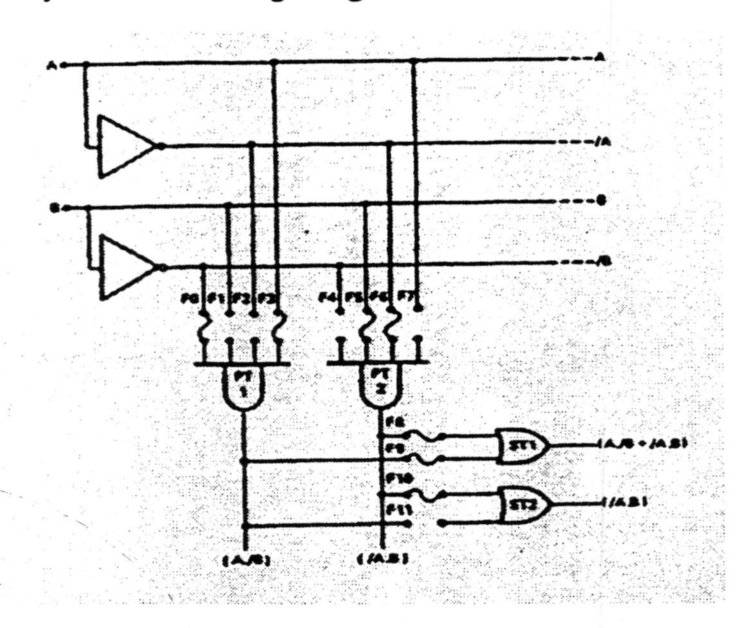

FIGURE 1-3. Elementary PLA architecture

PLAs

PLAs are *Programmable Logic Arrays.* These devices contain both programmable AND and OR terms which allow any AND term to feed any OR term. PLAs probably have the greatest flexibility of the other devices with regard to logic functionality. They typically have feedback from the OR array back into the AND array which may be used to implement asynchronous state machines. Most state machines, however, are implemented as synchronous machines. With this in mind, manufacturers created a type of PLA called a *Sequencer* which has registered feedback from the output of the OR array into the AND array.

Complex PLDs

Complex PLDs are what the name implies, *Complex Programmable Logic Devices.* They are considered very large PALs that have some characteristics of PLAs. The basic architecture is very much like a PAL with the capability to increase the amount of AND terms for any fixed OR term. This is accomplished by either stealing adjacent AND terms or using AND terms from an expander array. This allows for most any design to be implemented within these devices.

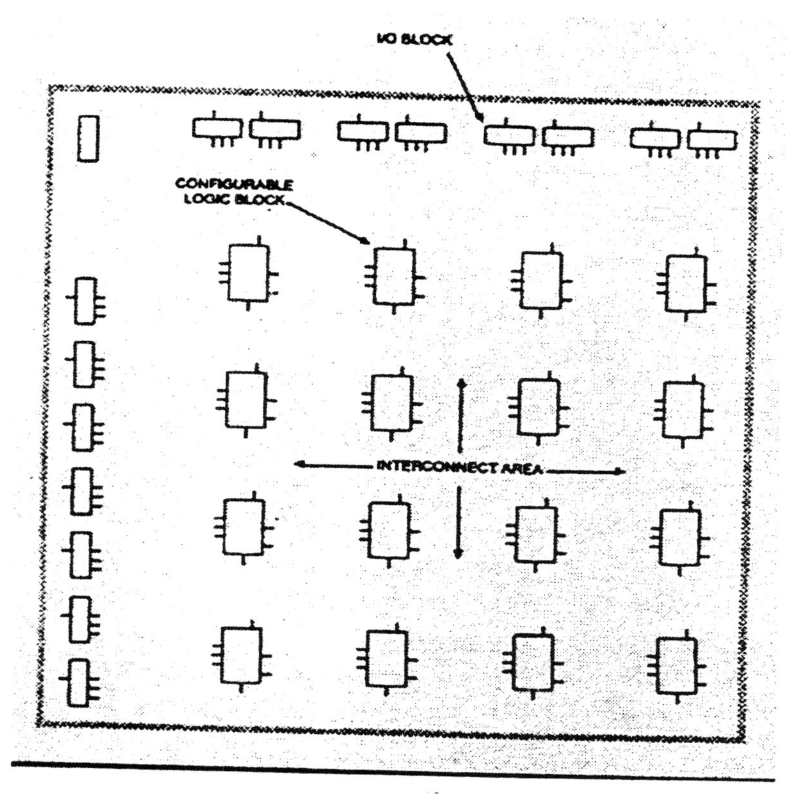

FIGURE 1-4. Elementary FPGA architecture

FPGAs

FPGAs are *Field Programmable Gate Arrays.* Simply put, they are electrically programmable gate array ICs that contain multiple levels of logic. FPGAs feature high gate densities, high performance, a large number of user-definable inputs and outputs, a flexible interconnect scheme, and a gate-array-like design environment. They are not constrained to the typical AND-OR array. Instead, they contain an interior matrix of configurable logic clocks (CLBs) and a surrounding ring of I/O blocks (IOBs). Each CLB contains programmable combinatorial logic and storage registers. The combinatorial logic section of the block is capable of implementing any Boolean function of its input variables. Each IOC can be programmed independently to be an input, and output with tri-state control or a bi-directional pin. It also contains flip-flops that can be used to buffer inputs and outputs. The interconnection resources are a network of lines that run horizontally and vertically in the rows and columns between the CLBs. Programmable switches connect the inputs and outputs of IOBs and CLBs to nearby lines. Long lines run the entire length or breadth of the device, bypassing interchanges to provide distribution of critical signals with minimum delay or skew. Designers using FPGAs can define logic functions of a circuit and revise these functions as necessary. Thus FPGAs can be designed and verified in a few days, as opposed to several weeks for custom gate arrays.

1.2 DEVICE TECHNOLOGIES AND PACKAGING

Device Technologies

Some of the technologies available are CMOS (Complimentary Metal Oxide Semiconductor), bipolar TTL, GaAs (Gallium Arsenide), and ECL (Emitter Coupled Logic) as well as combination fabrications like BiCMOS and ECL/bipolar. The two fastest semiconductor technologies are ECL and GaAs. However, these are also the most power hungry. Generally speed is proportional to power consumption.

Device Packaging

The packaging for devices fall into two categories: erasability and physical configuration. Certain devices have the capability of being erased and reprogrammed. These devices are erased by either applying UV light or a high voltage to re-fuse the cross-connection link. A UV erasable device will have a "window" in the middle of the device that allows the UV light to enter inside. An electrically erasable device usually need to have a high voltage applied to certain pins to erase the device. A device that cannot be erased is called One Time Programmable (OTP). As the name suggests, these devices can only be programmed once. Recent advances allow reprogramming without the use of high voltages

FIGURE 1-5. Picture of DIP and LCC devices

Programmable devices come in many shapes and sizes. Most devices come in the following physical configurations: DIP (Dual Inline Package), SKINNY-DIP, LCC (Leaded Chip Carrier), PLCC (Plastic Leaded Chip Carrier), QFP (Quad Flat Pack), BGA (Ball Grid Array), SOIC (Small Outline I.C.), TSOP (Thin Small Outline), and PGA (Pin Grid Array). These devices can be rectangular with pins on two sides, square with pins on all sides, or square with pins on the underside. It is important for the hardware and software development tools to fully support as many device types as possible to take full advantage of the myriad of devices on the market.

1.3 PROGRAMMING LOGIC DEVICES

Programmable logic devices are programmed by either shorting or opening connections within a device array, thus connecting or disconnecting inputs to a gate. Most hardware programmers receive a fuse information file from a software development package in ASCII format. The ASCII file could either be in JEDEC format for PLDs or HEX format for PROMs. This file contains the information necessary for the programmer to program the device. The JEDEC file contains fuse connections that are represented by an address followed by a series of 1's and 0's where a "1" indicates a disconnected state and a "0" indicates a connected state. The JEDEC file can also contain information that allows the hardware programmer the ability to perform a functional test on the device.

1.4 FUNCTIONALLY TESTING LOGIC DEVICES

A functional test may be performed after programming a device, provided that the hardware and software development package can support the generation and use of test vectors. Test vectors consist of a list of pins for the design, input values for each step of the functional test, and a list of expected outputs from the circuit. The programmer sequences through the input values, looks for the predicted outputs, and reports the results to the user. This allows design engineers and production crews the ability to verify that the programmed device works as designed.

Designing With The CUPL Language
Chapter 2

When creating any design, it is generally considered good practice to implement the design using a "Top-Down" approach. A Top-Down design is characterized by starting with a global definition of the design, then repeating the global definition process for each element of the main definition, etc., until the entire project has been defined. CUPL offers many features that accommodate this type of design. This chapter describes the instructions that CUPL offers for implementing a design.

2.1 DECLARATION OF LANGUAGE ELEMENTS

This section describes the elements that comprise the CUPL logic description language.

Pin/node Definition

Since the PIN definitions must be declared at the beginning of the source file, their definition is a natural starting point for a design. Pinnodes, used to define buried registers, should also be declared at the beginning of the source file. Pin assignment needs to be done if the designer already knows the device he wants to use. However, when creating a VIRTUAL design only the variable names that will later be assigned to pins need to be filled in. The area that normally contains the pin numbers will be left blank.

Defining Intermediate Variables

Intermediate variables are variables that are assigned an equation, but are not assigned to a PIN or PINNODE. These are used to define equations that are used by many variables or to provide an easier understanding of the design.

Using Indexed Variables

Variable names that end in a decimal number from 0 to 31 are referred to as indexed variables. They can be used to represent a group of address lines, data lines, or other sequentially numbered items. When indexed variables are used in bit field operations the variable with index number 0 is always the lowest order bit.

Number	Base	Decimal Value
'b'0	Binary	0
'B'1101	Binary	13
'O'663	Octal	435
'D'92	Decimal	92
'h'BA	Hexadecimal	186
'O'[300..477]	Octal (range)	192..314
'H'7FXX	Hexadecimal (range)	32512..32767

Table 2-1. Using Number Bases

Using Number Bases

All operations involving numbers in the CUPL compiler are done with 32-bit accuracy. Therefore, the numbers may have a value from 0 to $2^{32}-1$. A number may be represented in any one of the four common bases: binary, octal, decimal, or hexadecimal. The default base for all numbers used in the source file is hexadecimal, except for device pin numbers and indexed variables, which are always decimal. Binary, octal, and hexadecimal numbers can have don't care ("X") values intermixed with numerical values.

Using List Notation

A list is a shorthand method of defining groups of variables. It is commonly used in pin and node declarations, bit field declarations, logic equations, and set operations. Square brackets are used to delimit items in the list.

```
FIELD ADDRESS = [A7, A6, A5, A4, A3, A2, A1, A0];

FIELD DATA = [D7..D0];

FIELD Mode = [Up, Down, Hold];
```

Figure 2-2. Using The FIELD Statement

Using Bit Fields

A bit field declaration assigns a single variable name to a group of bits. After making a bit field assignment using the *FIELD* keyword, the name can be used in an expression; the operation specified in the expression is applied to each bit in the group. When a FIELD statement is used, the compiler generates a single 32-bit field internally. This is used to represent the variables in the bit field. Each bit represents one member of the bit field. The bit number which represents a member of a bit field is the same as the index number if indexed variables are used. This is mainly used for defining and manipulating address and data busses. This means that A0 will always occupy bit 0 in the bit field. Because of the mechanism, different indexed variables should not be included in the same bit field. A bit field containing A2 and B2 will assign both of these variables to the same bit position.

2.2 USAGE OF THE LANGUAGE SYNTAX

This section will discuss the logic and arithmetic operators and functions that are needed to create a Boolean equation design.

Using Logical Operators

Four standard logical operators are available for use: NOT, AND, OR, and XOR. The following table lists the operators and their order of precedence, from highest to lowest.

Operator	Examples	Description	Precedence
!	!A	NOT	1
&	A & B	AND	2
#	A # B	OR	3
$	A $ B	XOR	4

Table 2-3. Logical Operators

Using Arithmetic Operators And Functions

Six standard arithmetic operators are available for use in $repeat and $macro commands. The following table lists these operators and their order of precedence, from highest to lowest.

Operator	Examples	Description	Precedence
**	2**3	Exponentiation	1
*	2*I	Multiplication	2
/	4/2	Division	2
%	9%8	Modulus	2
+	2+4	Addition	3
-	4-I	Subtraction	3

One arithmetic function is available to use in arithmetic expressions being used in $repeat and $macro commands. The following table shows the arithmetic function and its bases

Function	Base
LOG2	Binary
LOG8	Octal
LOG16	Hexadecimal
LOG	Decimal

The LOG function returns an integer value. For example:

LOG2(32) = 5 <==> 2**5 = 32

LOG2(33) = ceil(5.0444) = 6 <==> 2**6 = 64

Ceil(x) returns the smallest integer not less than x.

Using Variable Extensions

Extensions can be added to variable names to indicate specific functions associated with the major nodes inside a programmable device, including such capabilities as flip-flop description and programmable tri-state enables. The compiler checks the usage of the extension to determine whether it is valid for the specified device and whether its usage conflicts with some other extension used. CUPL uses these extensions to configure the macrocells within a device. This way the designer does not have to know what fuses control what in the macrocells. To know what extensions are available for a particular device, use CBLD with the -e flag. A complete list of CUPL extensions can be found in the *CUPL PLD/FPGA Language Compiler* manual in the Extensions section of the CUPL Language chapter.

Extension	Side Used	Description
.AP	L	Asynchronous preset of flip-flop
.AR	L	Asynchronous reset of flip-flop
.APMUX	L	Asynchronous preset multiplexer selection
.ARMUX	L	Asynchronous reset multiplexer selection
.BYP	L	Programmable register bypass
.CA	L	Complement array
.CE	L	CE input of enabled D-CE type flip-flop
.CK	L	Programmable clock of flip-flop
.CKMUX	L	Clock multiplexer selection
.D	L	D nput of D-type flip-flop
.DFB	R	D registered feedback path selection
.DQ	R	Q output of D-type flip-flop
.IMUX	L	Input multiplexer selection of two pins
.INT	R	Internal feedback path for registered macrocell
.IO	R	Pin feedback path selection
.IOAR	L	Asynchronous reset for pin feedback register
.IOAP	L	Asynchronous preset for pin feedback register
.IOCK	L	Clock for pin feedback register

.IOD	R	Pin feedback path through D register
.IOL	R	Pin feedback path through latch
.IOSP	L	Synchronous preset for pin feedback register
.IOSR	L	Synchronous reset for pin feedback register
.J	L	J input of JK-type output flip-flop
.K	L	K input of JK-type output flip-flop
.L	L	D input of transparent latch
.LE	L	Programmable latch enable
.LEMUX	L	Latch enable multiplexer selection
.LFB	R	Latched feedback path selection
.LQ	R	Q output of transparent input latch
.OBS	L	Programmable observability of buried nodes
.OE	L	Programmable output enable
.OEMUX	L	Tri-state multiplexer selection
.PR	L	Programmable preload
.R	L	R input of SR-type output flip-flop
.S	L	S input of SR-type output flip-flop
.SP	L	Synchronous preset of flip-flop
.SR	L	Synchronous reset of flip-flop
.T	L	T input of toggle output flip-flop
.TEC	L	Technology-dependent fuse selection
.TFB	R	T registered feedback path selection
.T1	L	T1 input of 2-T flip-flop
.T2	L	T2 input of 2-T flip-flop

Table 2-6. Extensions

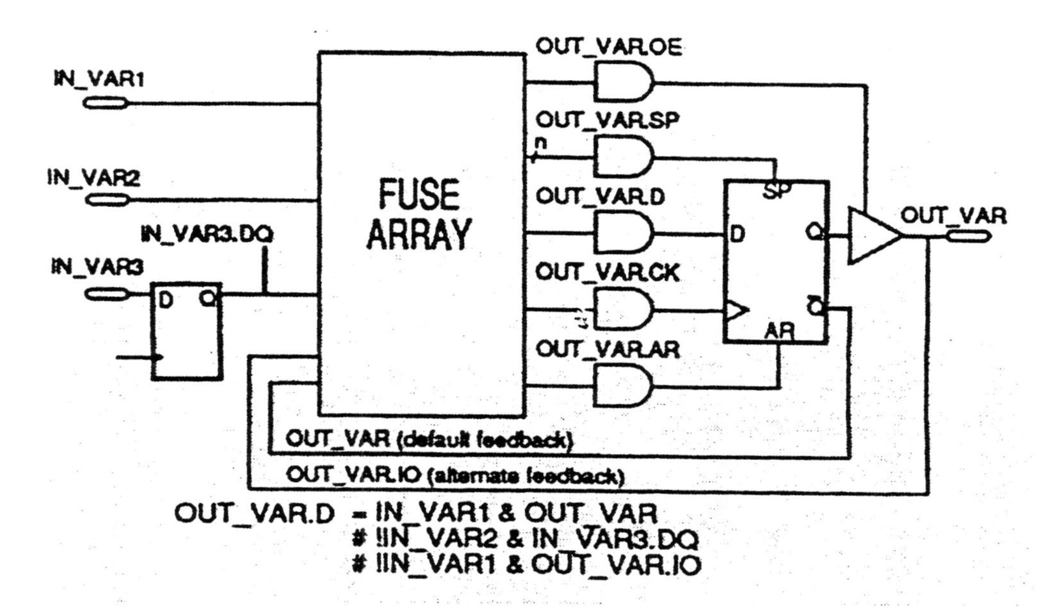

FIGURE 2-6. Circuit Illustrating Extensions

Figure 2-6 shows the use of extensions. Note that this figure does not represent an actual circuit, but shoes how to use extensions to write equations for different functions in a circuit.

Defining Logic Equations

Logic equations are the building clocks of the CUPL language. The form for logic equations is as follows:

[!] var [.ext] = exp;

where:

var is a single variable or a list of indexed or non-indexed variables defined according to the rules for list notation. When a variable list is used, the expression is assigned to each variable in the list

.ext is an option variable extension to assign a function to the major nodes inside programmable devices.

exp is an expression; that is, a combination of variables and operators.

= is the assignment operator; it assigns the value of an expression to a variable or set of variables.

In standard logic equations, normally only one expression is assigned to a variable. The *APPEND* statement enables multiple expressions to be assigned to a single variable. The APPENDed logic equation is logically ORed to the original equation for that variable. The format for using the APPEND statement is identical to defining a logic equation except the keyword APPEND appears before the logic equation begins.

Place logic equations in the "Logic Equation" section of the source file provided by the template file.

Using Set Operations

All operations that are performed on a single bit of information, for example, an input pin, a register, or an output pin, may be applied to multiple bits of information grouped into sets. Set operations can be performed between a set and a variable or expression, or between two sets.

The result of an operation between a set and a single variable (or expression) is a new set in which the operation is performed between each element of the set and the variable (or expression).

When an operation is performed on two sets, the sets must be the same size (that is, contain the same number of elements). The result of an operation between two sets is a new set in which the operation is performed between elements of each set.

When numbers are used in set operations, they are treated as sets of binary digits. A single octal number represents a set of three binary digits, and a single decimal or hexadecimal number represents a set of four binary digits.

Using Equality Operations

Unlike other set operations, the equality operation evaluates to a single Boolean expression. It checks for bit equality between a set of variables and a constant. The bit positions of the constant number are checked against the corresponding positions in the set. Where the bit position is a binary 1, The set element is unchanged. Where the bit position is a binary 0, the set element is negated. Where the bit position is a binary X, the set element is removed. The resulting elements are then ANDed together to create a single expression.

The equality operator can also be used with a set of variables that are to be operated upon identically. For example, the following three expressions:

[A3, A2, A1, A0]:&

[B3..B0]:#

[C3, C2, C1, C0]:$

are equivalent respectively to:

A3 & A2 & A1 & A0

B3 # B2 # B1 # B0

C3 $ C2 $ C1 $ C0

Using Range Operations

The range operation is similar to the equality operation except that the constant field is a range of values instead of a single value. The check for bit equality is made for each constant value in the range.

First, define the address bus, as follows:

FIELD address = [A3..A0];

Then write the *RANGE* equation:

select = address:[C..F];

This is equivalent to the following equation:

select = address:C # address:D # address:E # address:F;

2.3 ADVANCED LANGUAGE SYNTAX

This section describes the advanced CUPL language syntax. It explains how to use truth tables, state machines, condition statements, and user-defined functions to create a PLD design.

Defining Truth Tables

Sometimes the clearest way to express logic descriptions is in tables of information. CUPL provides the *TABLE* keyword to create tables of information. First, define relevant input and output variable lists , and then specify one-to-one assignments between decoded values of the input and output variable lists. Don't-care values are supported for the input decode value, but not for the output decoded value.

A list of input values can be specified to make multiple assignments in a single statement. The following block describes a simple hex-to-BCD code converter:

```
FIELD input = [in3..0];

FIELD output = [out3..0];

TABLE input => output {

        0=> 00; 1=>01; 2=>02; 3=>03;

        4=>04; 5=>05; 6=>06; 7=>07;

        8=>08; 9=>09; A=>10; B=>11;

        C=>12; D=>13; E=>14; F=>15;

}
```

Defining State Machines

A state machine, according to AMD/MMI, is "a digital device which traverses through a predetermined sequence of states in an orderly fashion". A synchronous state machine is a logic circuit with flip-flops. Because its output can be fed back to its own or some other flip-flop's input, a flip-flop's input value may depend on both its own output and that of other flip-flops; consequently, its final output value depends on its own previous values, as well as those of other flip-flops.

The CUPL state-machine model, as shown in Figure 4-7, uses six components: inputs, combinatorial logic, storage registers, state bits, registered outputs, and non-registered outputs.

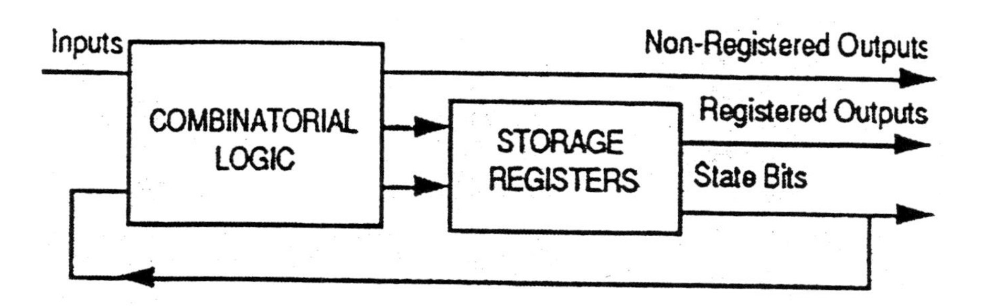

FIGURE 2-7. State Machine Model

Inputs - are signals entering the device that originate in some other device.

Combinatorial Logic - is any combination of logic gates (usually AND-OR) that produces an output signal that is valid Tpd (propagation delay time) nsec after any of the signals that drive these gates changes. Tpd is the delay between the initiation of an input or feedback event and the occurrence of a non-registered output.

State Bits - are storage register outputs that are fed back to drive the combinatorial logic. They contain the present-state information.

Storage Registers - are any flip-flop elements that receive their inputs from the state machine's combinatorial logic. Some registers are used for state bits: other are used for registered outputs. The registered output is valid Tco (clock to out time) nsec after the clock pulse occurs. Tco is the time delay between the initiation of a clock signal and the occurrence of a valid flip-flop output.

To implement a state machine, CUPL supplies a syntax that allows the describing of any function in the state machine. The *SEQUENCE* keyword identifies the outputs of a state machine and is followed by statements that define the function of the state machine. The SEQUENCE keyword causes the storage registers and registered output types generated to be the default type for the target device. Along with the SEQUENCE keyword are the *SEQUENCED, SEQUENCEJK, SEQUENCERS,* and *SEQUENCET keywords.* Respectively, they force the state registers and registered outputs to be generated as D, J-K, S-R, and T-type flip-flops. The format for the SEQUENCE syntax is as follows:

```
SEQUENCE state_var_list {

        PRESENT state_n0

                IF (condition1)  NEXT  state_n1;

                IF (condition2)  NEXT  state_n2  OUT  out_n0;

                DEFAULT          NEXT  state_n0;

        PRESENT state_n1

                NEXT  state_n2;

        .

        .

        .

        PRESENT state_nn statements;

}
```

where

state_var_list is a list of the state bit variables used in the state machine block. The variable list can be represented by a field variable.

state_n is the state number and is a decode value of the **state_variable_list** and must be unique for each present statement.

statements are any of the conditional, next, or output statements described in the following subsection.

Defining Multiple State Machines

The CUPL syntax allows for more than one state machine to be defined within the same PLD design. When multiple state machines are defined, occasionally the designer would like to have the state machines communicate with each other. That is, when one state machine reaches a certain state another state machine may begin. There are two methods of accomplishing state machine communication: using set operations on the state bits or defining a "global" register that can be accessed by both state machines.

In one state machine a conditional statement can contain another state machine's name followed by a state number or range of state numbers. The conditional statement will become TRUE when the other state machine reaches that particular state or states. The same case is true when using a register that is accessed by multiple state machines. However, this method requires the use one of the devices output or buried registers. Depending on the situation, the global register could also be combinatorial which may make a difference as to when the state machine receives the information from another state machine.

Using Condition Statement

The *CONDITION* syntax provides a higher-level approach to specifying logic functions than does writing standard Boolean logic equations for combinatorial logic. The format is as follows:

```
CONDITION {
        IF      expr0   OUT   var;
        .
        .
        IF      exprn   OUT   var;
        DEFAULT         OUT   var;
        }
```

The CONDITION syntax is equivalent to the asynchronous conditional output statements of the state machine syntax, except that there is no reference to any particular state. The variable is logically asserted whenever the expression or DEFAULT condition is met.

Defining A Function

The *FUNCTION* keyword permits the creating of personal keywords by encapsulating some logic as a function and giving it a name. This name can then be used in a logic equation to represent the function. The format for user-defined functions is a follows:

FUNCTION name ([Parameter$_0$,...,Parameter$_n$])

{ **body** }

The statements in the body may assign an expression to the function, or may be unrelated equations.

When using optional parameters, the number of parameters in the function definition and in the reference must be identical. The parameters defined in the body of the function are substituted for the parameters referenced in the logic equation. The function invocation variable is assigned an expression according to the body of the function. If no assignment is made in the body statements, the function invocation variable is assigned the value of 'h'0.

MIN Declaration Statements

The **MIN** declaration permits specifying different levels for different outputs in the same design, such as no reduction for outputs requiring redundant or contains product terms (to avoid asynchronous hazard conditions), and maximum reduction for a state machine application.

The **MIN** declaration statement overrides, for specified variables, the minimization level specified on the command line when running CUPL. The format is as follows:

MIN var [.ext] = level ;

MIN is a keyword to override the command line minimization level.

var is a single variable declared in the file or a list of variables grouped using the list notation; that is,

MIN [var, var, ... var] = level

.ext is an optional extension that identifies the function of the variable

level is an integer between 0 and 4.

; is a semicolon to mark the end of the statement.

The levels 0 to 4 correspond to the minimization levels available: None, Quick, Quine McClusky, Presto, Espresso.

The following are examples of valid **MIN** declarations.

```
MIN async_out      = 0;   /* no reduction */
MIN [outa, outb]   = 1;   /* Quine McClusky reduction */
MIN count.d        = 4;   /* Espresso reduction */
```

Note that the last declaration in the example above uses the .d extension to specify that the registered output variable is the one to be reduced.

2-4 USING THE SIMULATOR

The CUPL simulator, also known as **CSIM**, is a program that allows the user to create test vectors for the programmable logic device under design. Test vectors specify the expected functional operation of a PLD by defining the outputs as a function of the inputs. Test vectors are used both for simulation of the device logic before programming and for functional testing of the device once it has been programmed. **CSIM** can generate JEDEC-compatible downloadable test vectors.

A test specification source file (*filename.SI*) is the input to the CUPL simulator. It contains a functional description of the requirements of the device in the circuit. A template simulation file (***tmpl.**si*) is included for quick generation of test vectors.

```
Name            XXXXX;
Partno          XXXXX;
Revision        XX;
Date            XX/XX/XX;
Designer        XXXXX;
Company         XXXXX;
Location        XXXXX;
Assembly        XXXXX;
Device          XXXXX;

/*************************************************************/
/*                                                           */
/*                                                           */
/*************************************************************/


/*
 * Order:  define order, polarity, and output
 * spacing of stimulus and response values
 */


Order:


/*
 * Vectors:  define stimulus and response values, with header
 *           and intermediate messages for the simulator listing.
 *
 * Note: Don't Care state (X) on inputs is reflected in outputs
 *       where appropriate.
 */


Vectors:
```

FIGURE 2-9. Simulation template file.

Header information which is entered must be identical to the information in the corresponding CUPL logic description file. If any header information is different, a warning message will appear stating that the status of the logic equation could be inconsistent with the currect test vectos in the test specification file.

Use the **ORDER** keyord to list the variables to be used in the simulation table, and to define how they are displayed. Typically, the variable names are the same as those in the corresponding **CUPL** logic description file. When entering the variable you should retain the polarity status found in the **CUPL** logic description file.

ORDER: **InputA, InputB, outputC ;**
ORDER: **A0, A1, A2, A3, SELECT, !OUT0, !OUT1 ;**
ORDER: **A0..3, SELECT, !OUT0..1;**

Formatting characters spaces may be inserted into the simulation output file by using the % symbol and a decimal value between 1 and 80. Text can be inserted into the output file by putting a character string, enclosed by double quotes (" ",) into the **ORDER** statement. (Do not place text in the **ORDER** statement if waveform output will be used.)

ORDER: **"clock is ", clock, "and input is", A7..4, %2, A3..0, "output goes", D3..0,%2, !CS_SEL;**

produces the following results in the output file:

0001: Clock is C and input is 0110 1100 output goes HLHL H
0002: Clock is C and input is 0101 1001 output goes HLLH H

Use the **VECTORS** keyword to prefix the test vector table. Following the keyword, include test vectors made up of a single test value or quoted test values. Each vector must be containd on a single line. No semicolons follow test vectors. Table 2-5 list allowable test vector values.

Test Value	Description
0	Drive input LO (0 volts) (negate active-HI input)
1	Drive input HI (+5 volts) (assert active-HI input)
C	Drive (clock) input LO. HI, LO
K	Drive (clock) input HI. LO, HI
L	Test output LO (0 volts) (active-HI output negated)
H	Test output HI (+5 volts) (active-HI output asserted)
Z	Test output for high impedance.
X	Input HI or LO, output HI or LO NOTE: not all device programmers treat X on inputs the same; some put it to 0, some allow input to be pulled to 1, and some leave it at the previous value.
N	Output not tested.
P	Preload internal registers (value is applied to !Q output)
*	Outputs only -simulator determines test value and substitutes
' '	Enclose input values to be expanded to a specified BASE (octal, decimal, or hex). Valid values are 0-F and X.
" "	Enclose input values to be expanded to a specified BASE (octal, decimal, or hex). Valid values are 0-F, H, L, Z and X.

Figure 2-10. Simulation Test Values

The Following is an example of a test vector table:

VECTORS:
0 0 1 1 'F' Z "H" **/* Test outputs HI */**
0 1 0 0 '0' Z "L" **/* Test outputs LO */**

Use the **BASE** keyword to specfy how each quoted number is expanded. The format for the **BASE** statement is:

BASE: name ;

where:

name is either octal, decimal, or hex.

2-5 VIEWING WAVEFORMS WITH WCSIM

Compiling or simulationg a CUPL logic description file with the Display Waveform Option check will generate the waveform output graphically on the screen. The view of the waveform can be changed by:

Key	Action
→	Scroll right
←	Scroll left
↓	Scroll down
↑	Scroll up
PgUp	Shift screen up
PgDn	Shift screen down
F1	Decrease scale horizontally
F2	Enlarge scale horizontally
F3	Change current signal layer
F4	Exit to DOS
F5	Shift screen left
F6	Shift screen right
F7	Change signal orders
F8	Group signals into bus
F9	Create Waveform Hardcopy
F10	Waveform Legend
HOME	Show/hide fixed markers
INS	Show/hide moving marker
CTL/->	Move marker to the right
CTL<-	Move marker to the left

Any printable character key: Waveform Labels

Table 2-11. WCSIM Controls

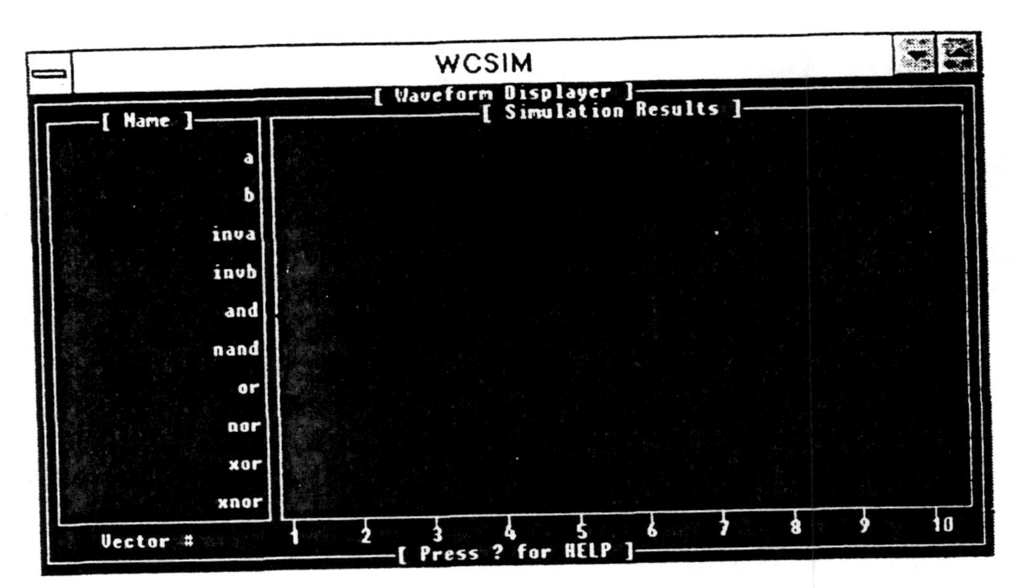

Figure 2-12. Waveform Display

The waveform display allows signal orders to be changed, grouped into busses, and colors modified. It is also possible to get a graphical output of the file for use with an extended ASCII printer. More information on the waveform display program can be located in the *CUPL PLD/FPGA Language Compiler Manual.*

CUPL Software Package
Chapter 3

This chapter briefly describes CUPL source file operations and the types of output that CUPL creates.

3.1 ABOUT THE COMPILER

The CUPL compiler is a program that takes a text file consisting of high level directives and commands and creates files consisting of more primitive information. This information is either used by a device programmer to program a logic function into a programmable logic device or by a simulator to simulate the design.

To start CUPL, double click on the CUPL icon in the program manager for Window 3.X operating system, or click on Start, <Programs>, <CUPL>, CUPL in Windows 95.

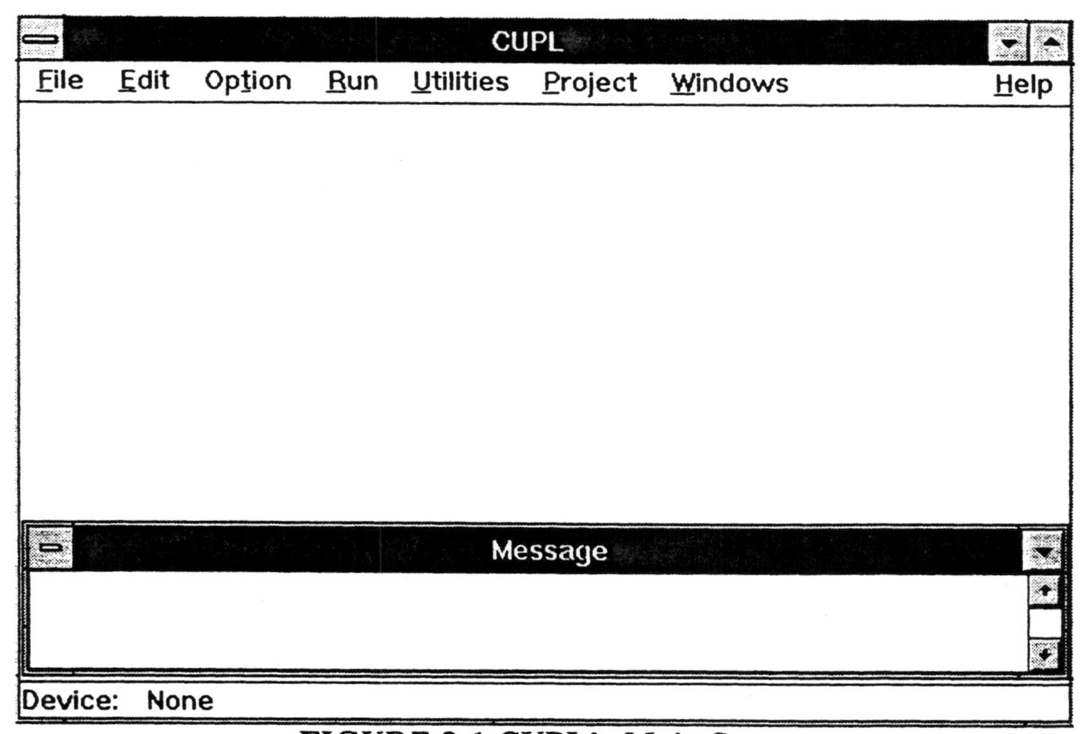

FIGURE 3-1 CUPL's Main Screen

CUPL's MENU

File Menu - Controls and features relating to general program manipulation.

New - Opens a template PLD file for a new design.

Open - Opens an existing file for modification.

Save - Saves the current file being modified.

Save As - Save the current file as a new file with a different name.

Print - Print the currently selected document.

Exit - Exit the program.

Edit Menu - Controls and utilities for editing files.

Cut - Moves the selected text to the clipboard.

Copy - Copies the selected text to the clipboard.

Paste - Paste text from the clipboard to the current cursor location.

Delete - Delete the selected text.

Copy Message - Copy the contents of the message window to the clipboard.

Search - Search for a text string in the body of text.

Line To - Advance to the line number selected.

Option Menu - Menu for selecting options related to CUPL's performance and compilation.

Compiler Options - Options directly affecting CUPL's compiler in minimization, optimization, and selecting output file formats.

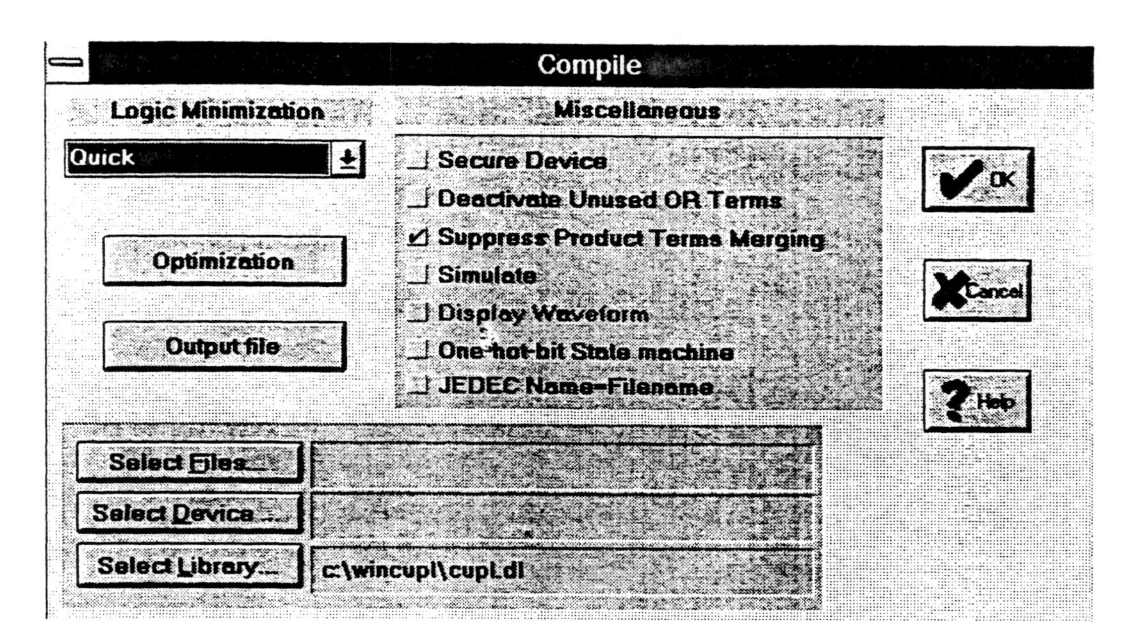

Figure 3-2. CUPL's Compiler Options

Logic Minimization - Select the level of minimization desired on the entire design. Please note that pin by pin minimization is available. See page 22 for more information on CUPL's minimization techniques.

Optimization - Select the optimizations desired on the entire design. Please note that pin by pin optimization is also available. See page 22 for more information on CUPL's optimization techniques.

Output file - Select the output files needed for the design.

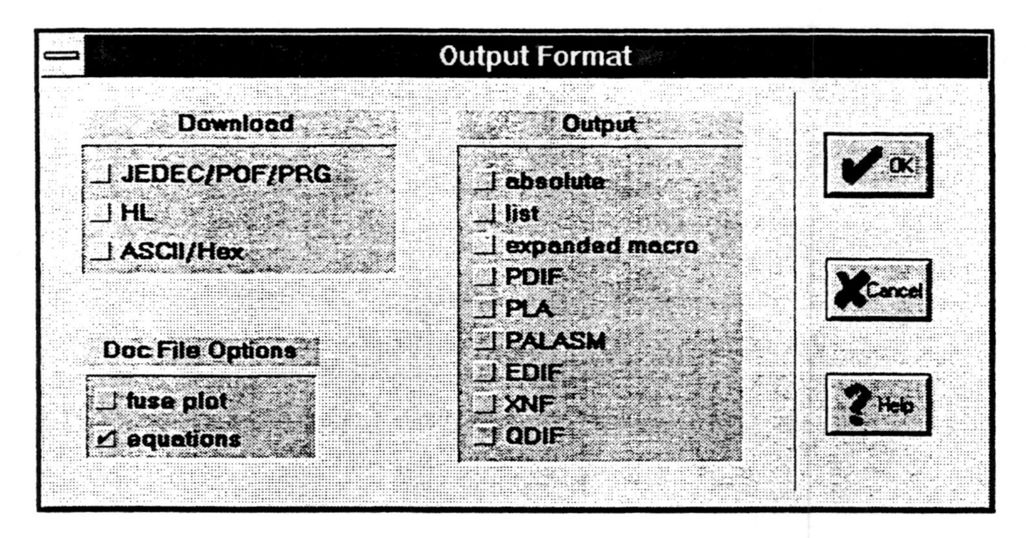

Figure 3-3. Output Format Files.

Download - Select the file type to download to the programmer.

DOC File Options - Select the options for the .DOC file.

Output - Several output formats are available from the compilation. *See page 37 for more information on output file formats.*

Simulator Options - Options related to simulation of the .PLD file.

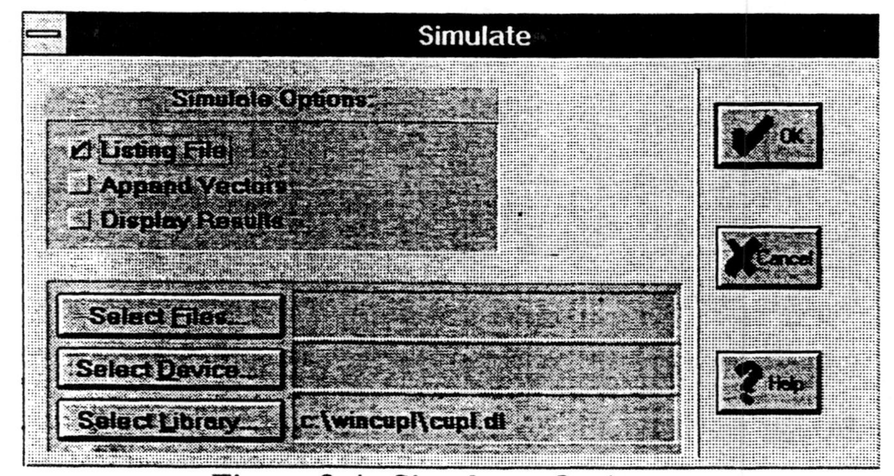

Figure 3-4. Simulator Options.

Listing File - Create a simulation output file (.SO).

Append Vectors - Add test vectors to a JEDEC file.

Display Results - Display the waveform outputs graphically.

VHDL Options - Options for using the VHDL compiler available separately from Logical Devices Inc. *This product available as a separate option.*

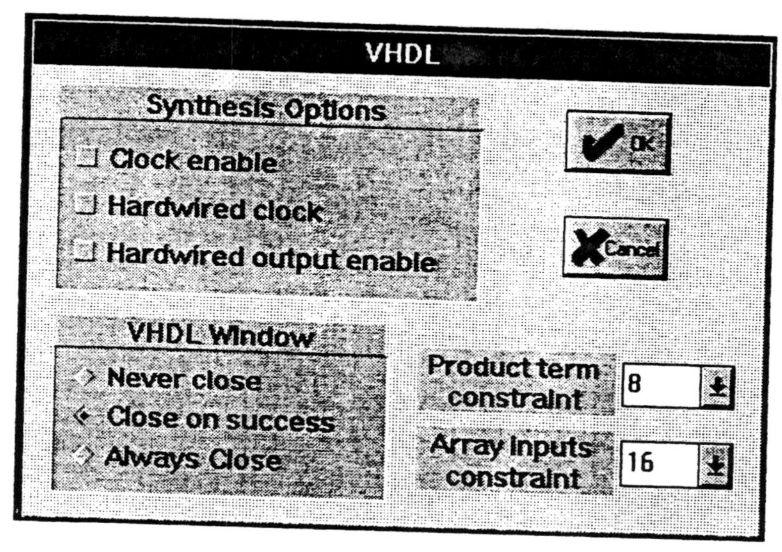

Figure 3-5. VHDL Options

Clock Enable - Enable the clock.

Hardwired Clock - Clock is set on a dedicated clock pin.

Hardwired output enable - Output enable is a dedicated pin.

VHDL Window - Specify when to close the VHDL window. Useful in debugging designs.

Product term constraint - Maximum number of product terms generated from the VHDL compiler.

Array inputs constraint - Maximum number of signals in an equation.

Select Device - Allows the user to select a device to target. *Using this option is not necessary and will override the device selection in the .PLD file.*

To select the device, click on the general type of PLD it is. Next select DIP or PLCC nmeumonic and specific type of device. Note that if the device type is only available in PLCC it only appears in the DIP section.

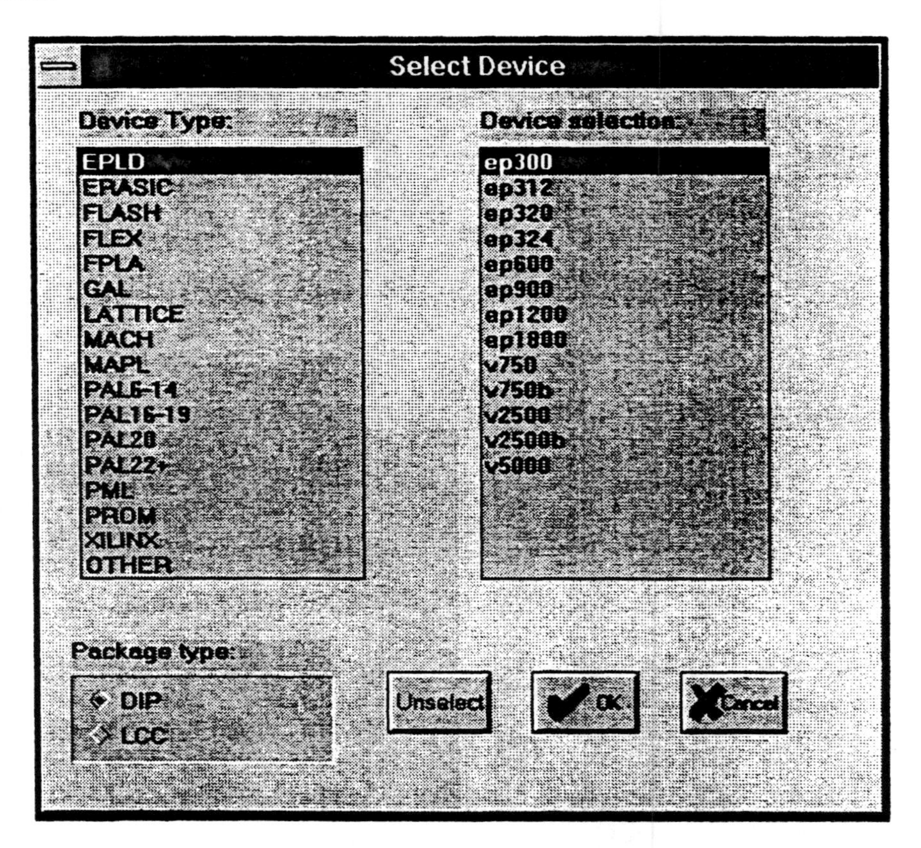

Figure 3-6. Device Selection Dialog Box.

Select Library - Allows the user to choose a user supplied library.

Select Files - Allows the user to specify which file should be compiler.

Preferences - User defined preferences affecting environment

Run Menu - Compile, simulate, and analysis.

Device Specific Compile - Compile the currently selected design for the specific device selected.

Device Specific Simulate - Simulate the currently selected design for the specific device selected.

Device Independent Compile - Compile the currently selected design for a virtual device.

Device Independent Simulate - Simulate the currently selected design as a virtual device.

Compile VHDL - Compile a VHDL design to JEDEC, PLA or selected output. *VHDL option not available with all packages.*

VHDL Analysis - Analysis of a VHDL design syntax validity.

Utilities Menu - Additional useful utilities.

DEVICE LIBRARY - utility for manipulating CUPL's device library.

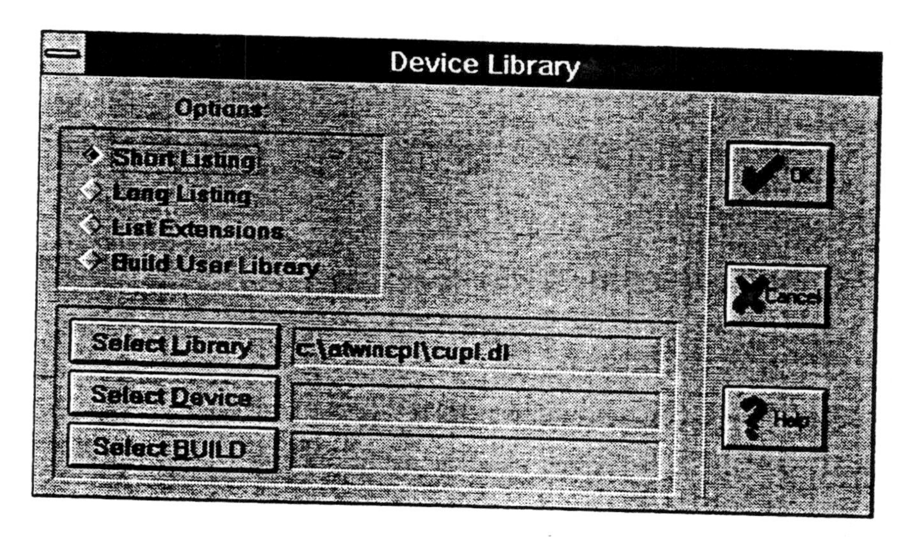

Figure 3-7. Device Library

Options - Allows the user to specify what action is to be taken.

Short Listing - Outputs to the message box a wide list of all the devices nmeumonics contained in the selected device library.

Long Listing - Outputs to the message box a list of all the device mnemonics in the selected library along with revision information, number of pins, number of fuses, and total number of available product terms in the device.

List extensions - Lists the extensions of the selected device, or if no device is selected it lists extensions of all devices, from the library selected to the message window.

Build User Library - Build a user library from an existing library.

Select Library - Allows the user to select a device library.

Select Device - Allows the user to select a device *(see figure 3-6)*.

Select Build - Select a user build file.

PALASM to CUPL - Derives CUPL syntax from a currently selected PALASM file.

Netlist to CUPL - Derives CUPL syntax from a currently selected EDIF netlist file.

Calculator - Calls Windows calculator.

File Manager - Calls Windows File Manager.

DOS Prompt - Calls Windows DOS Prompt.

Project - CUPL's project option.

Load - Loads a project file for a .PLD file.

Save - Saves a project file which includes compiler and simulator settings.

Windows - Manipulation of multiple document interface windows.

Cascade - Cascade open windows.

Tile - Tile all open windows.

Arrange Icons - Arrange the icons of minimized windows in the CUPL window.

Help - On-line help files and general information about CUPL.

Index - Open the help file for CUPL for Windows.

Using Help - Information on how to use the help menu.

About - Opens the about CUPL dialog box. Contains version information.

3.2 Output file format descriptions.

A JEDEC-compatible ASCII download file (***filename*.JED**) for input to a device programmer.

An ASCII Hex download file (***filename*.HEX**) available for PROMs.

An HL download file (***filename*.HL**) available for Signetics IFL devices.

An absolute file (***filename*.ABS**) for use by **CSIM**, the CUPL logic simulation program.

An error listing file (***filename*.LST**) that lists errors in the original source file.

A documentation file (***filename*.DOC**) that contains expanded logic equations, a variable symbol table, product term utilization, and fusemap information.

P-CAD PDIF file (*filename*.**PDF**) that can be translated by PDIFIN into a PC-CAPS symbol representing the pinouts of the programmable logic device..

A Berkeley PLA file (*filename*.**PLA**) for use by the Berkeley PLA tools.

A Open PLA file (*filename*.**PLA**) for use by various back end fitters.

A Xiline XNF file (*filename*.**XNF**) for use by Xilinx XACT software.

A EDIF file (*filename*.**EDF**) for use by ACTel software and Orcad Schematic.

CUPL Tutorial
Chapter 4

This section covers an example of how to use CUPL for Windows to compile a simple program and basically show the general flow of a design using CUPL.

4-1 TUTORIAL FOR GATES.

Start CUPL by double clicking on the CUPL icon in Window 3.X, or in Windows 95 click on Start, <programs>, <CUPL>, CUPL.

Under the file menu click on open.

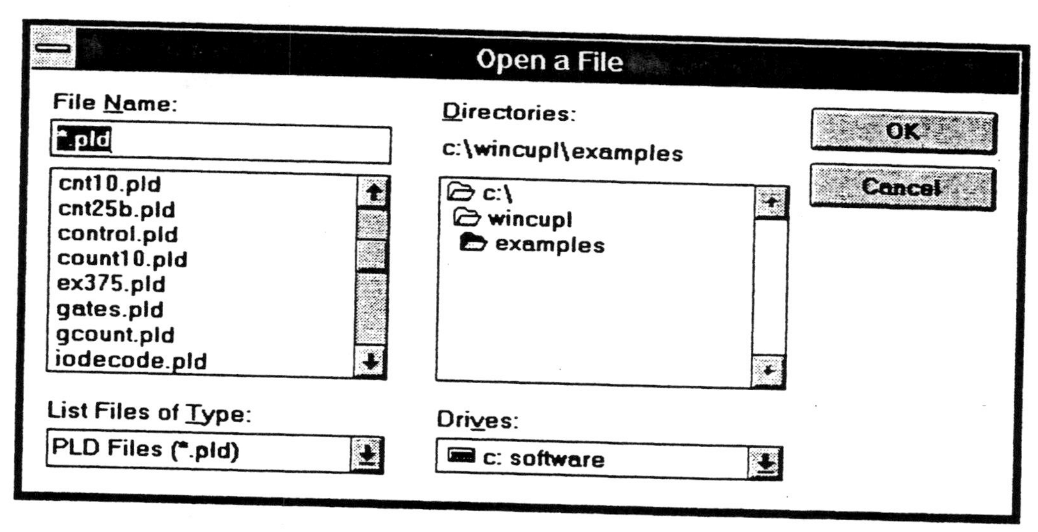

Figure 4-1. Open Dialog Box

Select gates.pld and click OK. The file gates.pld should appear in the CUPL windows. Take a second to look over the file. This file illustrates the use of CUPL's basic combinatorial logic.

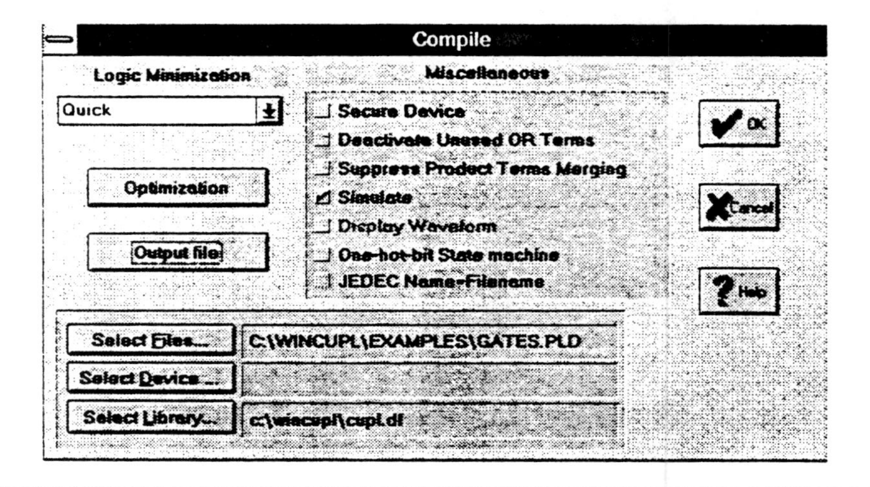

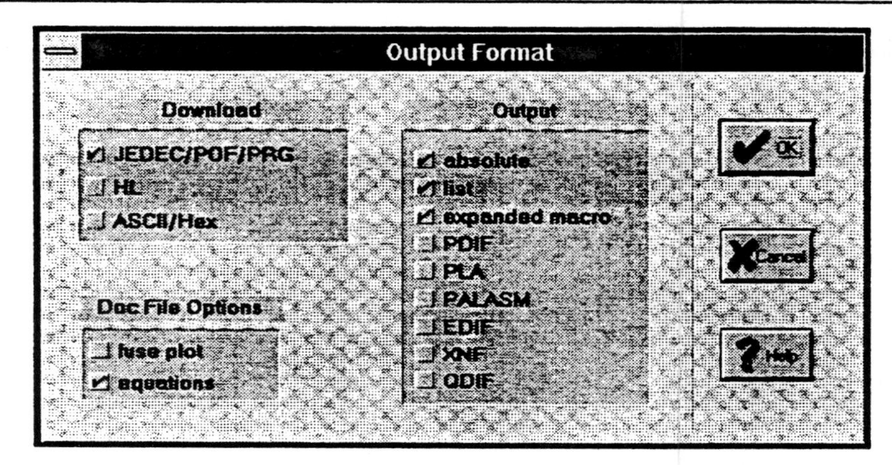

Figure 4-2. GATES.PLD Compiler Options

Click on <Options> and <Compiler Options> to bring up the compiler options dialog box. Click the simulate box and then click on <Output File>. Select JEDEC and absolute to produce a JEDEC file with test vectors. It is also useful to select Expanded Macro and listing files to get information on the compilation.

Once the compiler options are set, the file is ready to be compiled. Under the Compile menu, select Device Specific Compile, or press F9. If the file has been modified, it will need to be saved before a compile is performed. This function will compile the file, create the JEDEC file, simulate, and add the test vectors to the program.

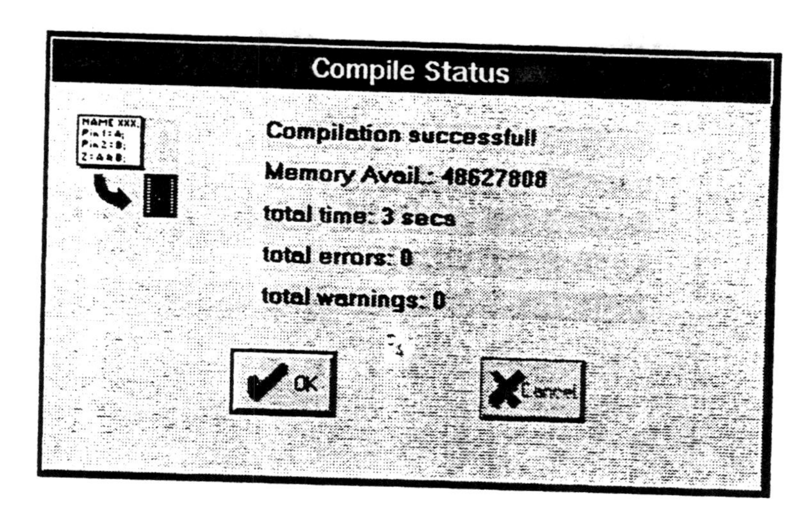

After the compilation is completed, there will be several files created. The file gates.jed is used to download to a programmer in order to program the devices and do the functional testing. The file gates.so is the simulation output file, and displays the logic simulation CUPL created. The Expanded Macro file, gates.doc, is the logic, after minimization, the CUPL implemented into the device. The last file created is the Listing file, gates.lst, and is used to display errors in the source file.

4-2 TUTORIAL FOR COUNT10

This tutorial covers advance CUPL syntax of **State Machines** and **Conditional Statements**.

Open up the file COUNT10.PLD and look it over. The file uses the CUPL state machine syntax and conditional statements to demonstrate a four bit up/down decade counter with synchronous clear capability. The up, down, and clear statements control the direction and reset of the counter. An asynchronous ripple carry is **OUT**ed when the counter loops.

Click on <Options> and <Compiler Options> to bring up the compiler options dialog box. Click on <Output File> and select JEDEC, absolute, simulation, Expanded Macro and listing files. Click on <Compile> and <Device Specific Compile> now to compile the file and generate the information files. Then open up the COUNT10.DOC file.

```
*******************************************************************************
                                   Count10
*******************************************************************************

CUPL(WM)        4.7a Serial# MW-67999999
Device          g16v8ms  Library DLIB-h-36-11
Created         Mon May 06 10:19:28 1996
Name            Count10
Partno          CA0018
Revision        02
Date            12/19/89
Designer        Kahl
Company         Logical Devices, Inc.
Assembly        None
Location        None

===============================================================================
                           Expanded Product Terms
===============================================================================

Q0.d  =>
    !Q0 & !Q1 & !Q2 & Q3 & !clr
  # !Q0 & !Q3 & !clr

Q1.d  =>
    !Q0 & !Q1 & !Q2 & Q3 & !clr & dir
  # Q0 & !Q1 & !Q3 & !clr & !dir
  # !Q0 & Q1 & !Q3 & !clr & !dir
  # Q0 & Q1 & !Q3 & !clr & dir
  # !Q0 & !Q1 & Q2 & !Q3 & !clr & dir

Q2.d  =>
    !Q0 & !Q1 & !Q2 & Q3 & !clr & dir
  # Q0 & Q1 & !Q2 & !Q3 & !clr & !dir
  # !Q1 & Q2 & !Q3 & !clr & !dir
  # Q0 & Q2 & !Q3 & !clr & dir
  # !Q0 & Q1 & Q2 & !Q3 & !clr

Q3.d  =>
    Q0 & !Q1 & !Q2 & Q3 & !clr & dir
  # !Q0 & !Q1 & !Q2 & !Q3 & !clr & dir
  # Q0 & Q1 & Q2 & !Q3 & !clr & !dir
  # !Q0 & !Q1 & !Q2 & Q3 & !clr & !dir

carry =>
    !Q0 & !Q1 & !Q2 & !Q3 & !clr & dir
  # Q0 & !Q1 & !Q2 & Q3 & !clr & !dir

clear =>
    clr
```

```
count =>
    Q3 , Q2 , Q1 , Q0

down =>
    !clr & dir

mode =>
    clr , dir

up =>
    !clr & !dir

carry.oe  =>
    1
```

Figure 4-4. COUNT10.DOC

This is an example of how CUPL translates a state machine into simple Boolean logic. The CUPL state machine syntax is a very useful tool in designing counters, processes, or any sequence of events.

4-3 TUTORIAL FOR SQUARE.PLD

To start this example, click on <File> and <New>. This brings up a template file for modification. Start by filling out all of the header information. It is generally good practice to use have the Name field the same as the file name.

After the header information is supplied the pin declarations need to be made. For this design we will need 4 inputs and 8 outputs. A 16V8 in simple mode will accommodate this. Declare 4 input pins as the input bus and all of the I/O pins available as the output bus. Please note that you cannot name a signal **OUT** because it is a CUPL reserved word.

The next step is to define the **Field** statements for the signals. To do this look at the listing of the PLD file on the next page. Having the fields set up we can now define the **Table**. A general direct match is used to do this with the equating symbol (=>). CUPL also supports a repeat state that allows the user to quickly go through values without computing the value manually.

```
Name        SQUARE;
Partno      XX;
Date        05/01/96;
Revision    01;
Designer    Chip Willman;
Company     Logical Devices Inc.;
Assembly    None;
Location    U1;
Device      G16V8;

/*******************************************************************/
/* This Design Example is an example of a lookup table to produce */
/* the square of a number coming in.                               */
/*                                                                 */
/*******************************************************************/
/*  Allowable Target Device Types:                                 */
/*******************************************************************/

/**  Inputs  **/

Pin [2..5]    = [I0..2]  ;      /*    Input bus line 4 bits      */

/**  Outputs  **/

Pin [12..19] = [Ot0..7] ;       /*    Output bus line 8 bits     */

/** Declarations and Intermediate Variable Definitions **/
Field input = [I3..0];
Field output = [Ot7..0];

/**  Logic Equations  **/

Table input=>output {
     `d'00 => `d'000;
     `d'01 => `d'001;
     `d'02 => `d'004;
$REPEAT A = [3..15]
     `d'{A} => `d'{A*A};
$REPEND
}
```

Figure 4-6. SQUARE.PLD file.

Included with this software are several other examples with useful demonstrations of CUPL syntax. The file EXAMPLES.TXT gives a description of most of the examples included in the package. Due to the continuing advance of PLDs, some examples may not be listed in the file, while other examples may not be present. Consult the README.TXT file on the installation disk 1.

CUPL SOFTWARE FEATURES 5

This chapter briefly describes the different CUPL software packages.

5-1 CUPL- PALexpert:

PALexpert contains the features mentioned in this package and supports 75 popular PAL, GAL and PROM architectures (approximately 1500 devices). This low cost CAE tool allows you to discover the benefits of designing with PLDs.

5-2 CUPL - PLDmaster:

PLDmaster supports over 250 PAL, GAL, FPLA and PROM architectures which equates to over 3000 devices. With the more complex devices, it is possible for the designer to impliment larger designs into a single device

5-3 CUPL - Total Designer:

Total Designer is the complete programmable logic design system including support most all industry devices including Complex PLDs and FPGAs. These include Actel, Altera, Max, AMD MACH, Intel FLEXlogic, Lattice pLSI, Xilinx, and others. Manufacturer specific place and route software and device fitters may not be included. In addition, partition software is provided for creating multiple PLD designs. Also SchemaQuik and ONCUPL are provided so that schematic entry designs can be created and translated into CUPL source files. All the programmable logic design software an engineer needs is in this package.

5-4 CUPL - Total Designer VHLD: OPTION for Total Designer

CUPL Total Designer VHDL transforms a VHDL design description into a Boolean design description and CUPL source design file. During processing the CUPL Total Designer VHDL program performs analysis, translation, and minimization. As an IEEE standard, VHDL descriptions are portable to other synthesis and simulation tools. VHDL allows the user to describe a design in any of three levels of abstraction: Structural (netlist like), data flow (like a PLD programming language), and behavioral (like a programming language).

5-5 ONCUPL

ONCUPL is a software tool that allows PLD designs to be done with schematic capture. A designer first draws the design with a schematic capture program. The schematic design is the converted into a netlist using a netlist extractor provided with the schematic capture program. ONCUPL then translate this netlist into a PLD file. This PLD file can be compiled with CUPL to produce any of the output files that CUPL is capable of producing.

ONCUPL is shipped with a library of symbols which can be used to implement the various macrocell architectures found in PLDs. Any design that will be processed by ONCUPL must be done using only ONCUPL symbols since these are specially structured for PLDs/ This usually means that existing TTL devices connected at the board level. Most often the design will have to be modified to some extent accommodate for the difference.

5-6 LIAISON

Liaison - Logic Input Algorithm Interface for Symbolic Object Netlists - is a software system for converting schematic netlists to CUPL PLD files. As a more sophisticated translation tool than ONCUPL, LIAISON converts EDIF 2 0 0 netlists generated by schematic capture tools into its own internal format and the proceeds to translate this format into a PLD design file.

LIAISON automatically adjusts for differences in the target architecture by examining the design and the desired target to determine if there is a match. It then proceeds to implement the desired logic as closely as possible. LIAISON also adjusts for different register types, resorting to emulation where necessary.

LIAISON provides sophisticated and flexible symbols called COMPLEX symbols. These can be used to create customized symbols like a variety of TTL symbols.

5-7 PLPartition

PLPartition is a logic synthesis tool that works with CUPL for producing designs that span multiple PLDs. A design is created in CUPL using the device independent compilation feature. The .DOC output from this process is read into PLPartition where the designer then directs the software as to how to divide the logic and what devices to choose. The designer will set partitioning criteria such as how many solutions to produce, the maximum number of devices the solution can use, the percentage of product terms to use per output in each target device and several other optimization features. PLPartition then produces a list of solutions that match the users specified criteria. The designer can then choose one of these solutions and PLPartition will divide the logic in that manner.

PLPartiton has several interesting features. It can do automatic product term splitting. This means that if an equation cannot fit on a particular output Y then part of the equations placed on an unused output Y1 which is then fedback to the output Y where it is combined with the other part of the equation. This can allow more complex logic to fit into a device than may have been thought possible. The partitioning process can be optimized for minimum pin usage or minimum product term usage. This can be used to increase device usage efficiency by some simple analysis of the design.

PLPartition report file provides a mechanism for showing the designer how the logic was placed and the efficiency of the fit. This report file can be read back in by PLPartition in a future use of the same design so that pin placements can be retained if they were unchanged. This can potentially save board rework by retaining pinouts.

5-8 PLA2CUPL

PLA2CUPL is one of the several translational programs available with all CUPL products. It is the solution to out dated compiler languages and multiple file translations. Compiler programs, such as ABEL, and schematic capture programs, such as ORCAD, which output a UC Berkley PLA file can be directly imported into CUPL PLD files. This useful tool is also available free of charge on our BBS and World Wide Web site.